# Principles of Fermentation Technology

## Second Edition

**Other books of related interest**

BIRCH, G. G., CAMERON, A. G. & SPENCER, M.
Food Science, 3rd Edition

COULSON, J. M. & RICHARDSON, J. F.
Chemical Engineering

GAMAN, P. M. & SHERRINGTON, K. B.
The Science of Food, 4th Edition

# Principles of Fermentation Technology

**PETER F. STANBURY**
B.Sc., M.Sc., D.I.C.

*Division of Biosciences,*
*University of Hertfordshire,*
*Hatfield, U.K.*


**ALLAN WHITAKER**
M.Sc., Ph.D., A.R.C.S., D.I.C.

*Division of Biosciences,*
*University of Hertfordshire*


**STEPHEN J. HALL**
B.Sc., M.Sc., Ph.D.

*Division of Chemical Sciences,*
*University of Hertfordshire*

BUTTERWORTH
HEINEMANN

OXFORD   AMSTERDAM   BOSTON   LONDON   NEW YORK   PARIS
SAN DIEGO   SAN FRANCISCO   SINGAPORE   SYDNEY   TOKYO

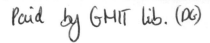
Butterworth-Heinemann
An imprint of Elsevier Science
Linacre House, Jordan Hill, Oxford OX2 8DP
200 Wheeler Road, Burlington, MA 01803

First published 1984
Reprinted 1986, 1987, 1989, 1993 (twice)
Second edition 1995
Reprinted 1999 (twice), 2000, 2003
Transferred to digital printing 2003

**British Library Cataloguing in Publication Data**
A catalogue record for this book is available from the British Library

**Library of Congress Cataloguing in Publication Data**
A catalogue record for this book is available from the Library of Congress

ISBN 0 7506 4501 6

For information on all Butterworth-Heinemann
publications  visit our website at www.bh.com

Printed and bound in Great Britain by Antony Rowe Ltd, Eastbourne

This book is dedicated to the memory of

David L. Cohen

Microbiologist, teacher, colleague and, above all, friend.

# Acknowledgements

We wish to thank the authors, publishers and manufacturing companies listed below for allowing us to reproduce either original or copyright material.

*Authors*

S. Abe (Fig. 3.13), A. W. Nienow (Figs 9.13 B–F and 7.10, 7.11 and 9.19 from *Trends in Biotechnology*, **8** (1990)), (Figs 5.2a, 5.2b, 5.2c, 5.3a, 5.3b, 5.5, 5.7, 7.18 and Table 5.2 from *Introduction to Industrial sterilization*, Academic Press, London (1968)), F. G. Shinskey (Fig. 8.11, R. M. Talcott (Figs 10.11, 10.12 and 10.13) and D. I. C. Wang (Table 12.7).

*Publishers and manufacturing companies*

Academic Press, London and new York: Fig. 1.2 from Turner, W. B. (1971) *Fungal Metabolites*; Fig. 6.7 from Norris, J. R. and Ribbons, D. W. (1972) *Methods in Microbiology*, **7b**, Fig. 7.9 from Solomons, G. L. (1969) *Materials and Methods in Fermentation*; Fig. 7.14 from *Journal of Applied Bacteriology* 21 (1958); Fig. 7.45 from Rose, A. H. (1978), Fig. 7.51 from *Economic Microbiology*, Vol. **4** (1979), *Economic Microbiology*, Vol. **2**; Figs 7.1 and 10.27 and Table 12.2 from Rose, A. H. (1979) *Economic Microbiology*, Vol. **3**.; Fig. 7.55 from Spiers, R. E. and Griffiths, J. B. (1988) *Animal Cell Biotechnology*, Vol. **3**; Figs 9.21 and 12.1 from Nisbet, L. J. and Winstanley, D. J. (1983) *Microbial Products 2. Development and Production*; Fig. 10.6 from *Advances in Applied Microbiology*, **12**; Table 4.5 from Cook, A. H. (1962) *Barley and Malt, Biochemistry and Technology*; Tables 8.3 from Aiba, S., Humphrey, A. E. (1973) *Biochemical Engineering* (2nd Edition).

Alfa Laval Engineering Ltd, Brentford: Figs 5.8, 5.9 and 5.11.

Alfa Laval Sharples Ltd, Camberley: Fig 10.16a, 10.16b, 10.17a, 10.17b and 10.20.

American Chemical Society: Fig. 7.43 reprinted with permission from *Industrial and Engineering Chemistry*, **43** (1951); Fig. 7.49 reprinted with permission from Ladisch, M. R. and Bose, A. (1992) *Harnessing Biotechnology for the 21st Century*. ACS Conference Proceedings Series.

American Society for Microbiology: Fig. 3.36, 5.11 and 9.17.

American Society for Testing and Materials: Fig. 6.11. Copyright ASTM, reprinted with permission.

Applikon dependable Instruments BV, Gloucester, UK: Fig. 7.16 and Table 7.5.

Blackwell Scientific Publications Ltd: Figs 1.1 and 2.8.

Bio/Technology: Table 3.6.

British Mycological Society: Fig. 7.48.

British Valve and Actuator Manufacturers Association (BVAMA): Fig. 7.28, 7.29, 7.30, 7.31, 7.32, 7.33, 7.34, 7.35, 7.37 and 7.38.

Butterworth-Heinemann: Fig. 6.10, 7.22 and 7.25 from Collins, C. H. and Beale, A. J. (1992) *Safety in Industrial Microbiology and Biotechnology*; Table 3.7 from Vanek, Z. and Hostelek, Z. (1986) *Overproduction of Microbial Metabolites. Strain Improvement and Process Strategies*.

Canadian Chemical News, Ottawa: Figs 10.33a and 10.33b.

Chapman and Hall: Fig. 7.46 from Hough, J. S. *et al.* (1971) *Malting and Brewing Science*.

Chilton Book Company Ltd, Radnor, Pennsylvania, USA: Figs 8.2, 8.3, 8.4, 8.5, 8.8 and 8.9 Reprinted from *Engineers Handbook*, Vol. 1 by B. Liptak. Copyright 1969 by the author. Reprinted with the permission of the publisher.

Marcel Dekker Inc.: Figs 6.4, 6.5 and 6.6. Reprinted with permission from Vandamme, E. J. (1984) *Biotechnology of Industrial Antibiotics*.

Elsevier Science Ltd, Kidlington: Fig. 2.12 reprinted from *Process Biochemistry*, **1** (1966); Figs 5.4, 5.5, 5.12, 5.20 reprinted from *Process Biochemistry*, **2** (1967); Fig. 10.4 reprinted from *Process Biochemistry*, **16** (1981); Table 6.2 reprinted from *Process Biochemistry*, **13** (1978); Table 2.3 reprinted from *Journal of Biotechnology*, **22** (1992); Fig. 7.47a from *Endeavour* (NS), **2** (1978), Fig. 8.12, 8.20, 8.22 and 8.24 reprinted from Cooney, C. L. and Humphrey, A. E. (1985) *Comprehensive Biotechnology*, Vol **2**; Figs 9.2 and 10.34 reprinted from Moo-Young, M. **et al.** (1980) *Advances in Biotechnology*, Vol. 1; Fig. 10.3 from Blanch, H. W. et al (1985) *Comprehensive Biotechnology* **3**, Figs 10.9a and 10.9b reprinted from Coulson, J. M. and Richardson, J. F. (1968) *Chemical Engineering* (2nd edition), Fig. 10.30 reprinted from *Journal of Chromatography*, **43** (1969).

Elsevier Trends Journals, Cambridge: Fig. 3.33, reprinted from *Trends in Biotechnology*, **10** (1992), 7.10, 7.11 and 9.19 reprinted from *Trends in Biotechnology*, **8** (1990).

Ellis Horwood: Fig. 9.16 and 10.5. Tables 3.5 and 9.3

Dominic Hunter, Birtley: Fig. 5.19.

Inceltech LH, Reading: Figs 7.4 and 7.17.

International Thomson Publishing Services: Figs 5.13 and 7.24 from Yu, P. L. (1990) *Fermentation Technologies; Industrial Applications*; Fig. 6.3 from Vandamme, E. J. (1989)

*Biotechnology of Vitamins, Pigments and Growth Factors* and Fig. 3.9 from Fogarty, W. M. and Kelly, K. T. (1990) *Microbial Enzymes and Biotechnology*, (2nd Edition).

Institute of Chemical Engineering: Fig. 11.6 from *Effluent Treatment in the Process Industries* (1983).

Institute of Water Pollution Control: Fig. 11.5.

IRL Press, Fig. 4.3. from Poole *et al*. *Microbial Growth Dynamics*, (1990), Fig. 6.1 from McNeil, B. and Harvey, L. M. *Fermentation—A Practical Approach* (1990), Fig. 8.26. from Bryant, T. N. and Wimpenny, J. W. T. *Computers in Microbiology: A Practical Approach* (1989).

Japan Society for Bioscience, Biotechnology and Agrochemistry: Fig. 3.23 from *Agricultural and Biological Chemistry*, **36** (1972).

Kluwer Academic Publishers: Fig. 7.52 reprinted with permission from Varder-Sukan, F. and Sukan, S. S. (1992) *Recent Advances in Biotechnology*.

Life Science Laboratories Ltd, Luton: Figs 7.6 and 7.7.

MacMillan: Table 1.1 from Prescott and Dunn's *Industrial Microbiology*, edited by Reed, G. (1982).

Marshall Biotechnology Ltd: Fig. 7.23.

Mcgraw Hill, New York: Fig. 7.27 reproduced with permission from *Chemical Engineering*, **94** (1987), also Fig. 7.36 from King, R. C. (1967) Piping Handbook (5th edition), also Figs 8.21 and 8.23 from Considine, D. M. (1974)*Process Instrumentation and Control Handbook* (2nd Edition) and also Fig. 10.10 from Perry, R. H. and Chilton, C. H. (1973) *Chemical Engineer's Handbook* (5th Edition).

Microbiology Research Foundation of Japan, Tokyo: Fig. 3.21 from *Journal of General and Applied Microbiology*, **19** (1973).

New Brunswick Ltd, Hatfield, Figs 7.5, 7.15, 7.26, 7.54.

New York Academy of Sciences: Figs 2.14, 3.3, 3.4 and 3.31.

Pall Process Filtration Ltd, Portsmouth: Figs 5.14, 5.15, 5.16, 5.17 and 5.18.

Royal Netherlands Chemical Society: Table 12.4

Royal Society of Chemistry: Fig. 6.9 and Table 3.8.

The Royal Society, London: Fig. 7.47b.

Science and Technology Letters, Northwood, UK: Figs 9.20a and 9.20b.

Society for General Microbiology: Figs 3.29, 3.34 and 3.35 and Tables 3.2 and 9.2.

Society for Industrial Microbiology, USA: Fig. 9.18.

Southern Cotton Oil Company, Memphis, USA: Table 4.8.

Spirax Sarco Ltd, Cheltenham, UK: Figs 7.39, 7.40, 7.41 and 7.42.

Springer Verlag GmbH and Co. KG: Table 7.4 reproduced from *Applied Microbiology and Biotechnology* **30** (1989), Fig. 8.7 reproduced from *Advances in Biochemical Engineering*, **13** (1979), Table 12.2 from *Advances in Chemical Engineering*, **37** (1988).

John Wiley and Sons Inc., New York: Fig. 3.2 from *Journal of Applied Chemistry Biotechnology* **22** (1972) Fig. 3.13 from Yamada, K. et al. (1972) *The Microbial Production of Amino Acids*.: Fig. 7.50 from *Biotechnology and Bioengineering*, **42** (1993), Fig. 7.53 from *Biotechnology and Bioengineering Symposium*, **4** (1974), Fig. 7.44 from *Biotechnology Bioengineering* **9** (1967), Fig. 9.4 from *Biotechnology and Bioengineering*, **12** (1970), Table 12.6 *from Biotechnology and Bioengineering*;, **15** (1973); Fig. 8.11 from Shinskey, F. G. (1973) *pH and pIon Control in Process and Waste Streams*.; Figs 10.11, 10.12 and 10.13 from *Kirk–Othmer Encyclopedia of Chemical Technology*, 3rd Edition (1980); Figs 10.21 and 10.22 from *Biotechnology and Bioengineering*, **16** (1974); Fig. 10.23 from *Biotechnology and Bioengineering*, **19** (1977); Table 12.7 from Wang, D. I. C. et al. (1979) *Fermentation and Enzyme Technology*.

We also wish to thank Mr Jim Campbell (Pall Process Filtration Ltd, Portsmouth), Mr Nelson Nazareth (Life Science Laboratories Ltd, Luton), Mr Peter Senior (Applikon Dependable Instruments BV, Tewkesbury) and Mr Nicholas Vosper (New Brunswick Ltd, Hatfield) for advice on fermentation equipment and Dr Geoffrey Leaver and Mr Ian Stewart (Warren Spring Laboratories, Stevenage) for advice on safety and containment and Mr Michael Whitaker for his comments on a student friendly book design.

Last but not least we wish to express our thanks to Lesley, John, David and Abigail Stanbury and Lorna, Michael and Ben Whitaker for their encouragement and patience during all stages in the preparation of this edition of the book.

December, 1994.

# Contents

# CHAPTER 1

# An Introduction to Fermentation Processes

THE TERM 'fermentation' is derived from the Latin verb *fervere*, to boil, thus describing the appearance of the action of yeast on extracts of fruit or malted grain. The boiling appearance is due to the production of carbon dioxide bubbles caused by the anaerobic catabolism of the sugars present in the extract. However, fermentation has come to have different meanings to biochemists and to industrial microbiologists. Its biochemical meaning relates to the generation of energy by the catabolism of organic compounds, whereas its meaning in industrial microbiology tends to be much broader.

The catabolism of sugars is an oxidative process which results in the production of reduced pyridine nucleotides which must be reoxidized for the process to continue. Under aerobic conditions, reoxidation of reduced pyridine nucleotide occurs by electron transfer, via the cytochrome system, with oxygen acting as the terminal electron acceptor. However, under anaerobic conditions, reduced pyridine nucleotide oxidation is coupled with the reduction of an organic compound, which is often a subsequent product of the catabolic pathway. In the case of the action of yeast on fruit or grain extracts, NADH is regenerated by the reduction of pyruvic acid to ethanol. Different microbial taxa are capable of reducing pyruvate to a wide range of end products, as illustrated in Fig. 1.1. Thus, the term fermentation has been used in a strict biochemical sense to mean an energy-generation process in which organic compounds act as both electron donors and terminal electron acceptors.

The production of alcohol by the action of yeast on malt or fruit extracts has been carried out on a large scale for very many years and was the first 'industrial' process for the production of a microbial metabolite. Thus, industrial microbiologists have extended the term fermentation to describe any process for the production of product by the mass culture of a micro-organism. Brewing and the production of organic solvents may be described as fermentations in both senses of the word but the description of an aerobic process as a fermentation is obviously using the term in the broader, microbiological, context and it is in this sense that the term is used in this book.

## THE RANGE OF FERMENTATION PROCESSES

There are five major groups of commercially important fermentations:

(i) Those that produce microbial cells (or biomass) as the product.
(ii) Those that produce microbial enzymes.
(iii) Those that produce microbial metabolites.
(iv) Those that produce recombinant products.
(v) Those that modify a compound which is added to the fermentation — the transformation process.

The historical development of these processes will be considered in a later section of this chapter, but it is first necessary to include a brief description of the five groups.

### Microbial biomass

The commercial production of microbial biomass may be divided into two major processes: the production of yeast to be used in the baking industry and the production of microbial cells to be used as human or animal food (single-cell protein). Bakers' yeast has been

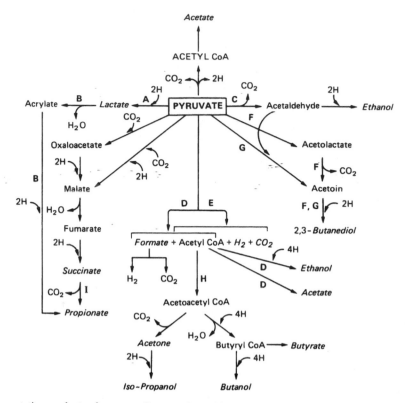

FIG. 1.1. Bacterial fermentation products of pyruvate. Pyruvate formed by the catabolism of glucose is further metabolized by pathways which are characteristic of particular organisms and which serve as a biochemical aid to identification. End products of fermentations are italicized (Dawes and Large, 1982).

A Lactic acid bacteria (*Streptococcus, Lactobacillus*)
B *Clostridium propionicum*
C Yeast, *Acetobacter, Zymomonas, Sarcina ventriculi, Erwinia amylovora*
E Clostridia

F *Klebsiella*
G Yeast
H Clostridia (butyric, butylic organisms)
I Propionic acid bacteria

produced on a large scale since the early 1900s and yeast was produced as human food in Germany during the First World War. However, it was not until the 1960s that the production of microbial biomass as a source of food protein was explored to any great depth. As a result of this work, reviewed briefly in Chapter 2, a few large-scale continuous processes for animal feed production were established in the 1970s. These processes were based on hydrocarbon feedstocks which could not compete against other high protein animal feeds, resulting in their closure in the late 1980s (Sharp, 1989). However, the demise of the animal feed biomass fermentations was balanced by ICI plc and Rank Hovis McDougal establishing a process for the production of fungal biomass for human food. This process was based on a more stable economic platform and appears to have a promising future.

### Microbial enzymes

Enzymes have been produced commercially from plant, animal and microbial sources. However, microbial enzymes have the enormous advantage of being able to be produced in large quantities by established fermentation techniques. Also, it is infinitely easier to improve the productivity of a microbial system compared with a plant or animal one. Furthermore, the advent of recombinant DNA technology has enabled enzymes of animal origin to be synthesized by micro-organisms (see Chapter 3). The uses to which microbial enzymes have been put are summarized in Table 1.1, from which it may be seen that the majority of applications are in the food and related industries. Enzyme production is closely controlled in micro-organisms and in order to improve productivity these controls may

TABLE 1.1. *Commercial applications of enzymes* (Modified from Boing, 1982)

| Industry | Application | Enzyme | Source |
|---|---|---|---|
| Baking and milling | Reduction of dough viscosity, acceleration of fermentation, increase in loaf volume, improvement of crumb softness and maintenance of freshness | Amylase | Fungal |
| | Improvement of dough texture, reduction of mixing time, increase in loaf volume | Protease | Fungal/bacterial |
| Brewing | Mashing | Amylase | Fungal/bacterial |
| | Chillproofing | Protease | Fungal/bacterial |
| | Improvement of fine filtration | β-Glucanase | Fungal/bacterial |
| Cereals | Precooked baby foods, breakfast foods | Amylase | Fungal |
| Chocolate and cocoa | Manufacture of syrups | Amylase | Fungal/bacterial |
| Coffee | Coffee bean fermentation | Pectinase | Fungal |
| | Preparation of coffee concentrates | Pectinase, hemicellulase | Fungal |
| Confectionery | Manufacture of soft centre candies | Invertase, pectinase | Fungal/bacterial |
| Corn syrup | Manufacture of high-maltose syrups | Amylase | Fungal |
| | Production of low D.E. syrups | Amylase | Bacterial |
| | Production of glucose from corn syrup | Amyloglycosidase | Fungal |
| | Manufacture of fructose syrups | Glucose isomerase | Bacterial |
| Dairy | Manufacture of protein hydrolysates | Protease | Fungal/bacterial |
| | Stabilization of evaporated milk | Protease | Fungal |
| | Production of whole milk concentrates, icecream and frozen desserts | Lactase | Yeast |
| | Curdling milk | Protease | Fungal/bacterial |
| Eggs, dried | Glucose removal | Glucose oxidase | Fungal |
| Fruit juices | Clarification | Pectinases | Fungal |
| | Oxygen removal | Glucose oxidase | Fungal |
| Laundry | Detergents | Protease, lipase | Bacterial |
| Leather | Dehairing, baiting | Protease | Fungal/bacterial |
| Meat | Tenderization | Protease | Fungal |
| Pharmaceutical | Digestive aids | Amylase, protease | Fungal |
| | Anti-blood clotting | Streptokinase | Bacterial |
| | Various clinical tests | Numerous | Fungal/bacterial |
| Photography | Recovery of silver from spent film | Protease | Bacterial |
| Protein hydrolysates | Manufacture | Proteases | Fungal/bacterial |
| Soft drinks | Stabilization | Glucose oxidase, catalase | Fungal |
| Textiles | Desizing of fabrics | Amylase | Bacterial |
| Vegetables | Preparation of purees and soups | Pectinase, amylase, cellulase | Fungal |

have to be exploited or modified. Such control systems as induction may be exploited by including inducers in the medium (see Chapter 4), whereas repression control may be removed by mutation and recombination techniques. Also, the number of gene copies coding for the enzyme may be increased by recombinant DNA techniques. Aspects of strain improvement are discussed in Chapter 3.

### Microbial metabolites

The growth of a microbial culture can be divided into a number of stages, as discussed in Chapter 2.

After the inoculation of a culture into a nutrient medium there is a period during which growth does not appear to occur; this period is referred to as the lag phase and may be considered as a time of adaptation. Following a period during which the growth rate of the cells gradually increases the cells grow at a constant, maximum rate and this period is known as the log, or exponential, phase. Eventually, growth ceases and the cells enter the so-called stationary phase. After a further period of time the viable cell number declines as the culture enters the death phase. As well as this kinetic description of growth, the behaviour of a culture may also be described according to the products which it produces during the various stages of the growth curve. During the log phase of growth the products produced are essential to the growth of the cells and include amino acids, nucleotides, proteins, nucleic acids, lipids, carbohydrates, etc. These products are referred to as the primary products of metabolism and the phase in which they are produced (equivalent to the log, or exponential phase) as the trophophase (Bu'Lock et al., 1965).

Many products of primary metabolism are of considerable economic importance and are being produced by fermentation, as illustrated in Table 1.2. The synthesis of primary metabolites by wild-type micro-organisms is such that their production is sufficient to meet the requirements of the organism. Thus, it is the task of the industrial microbiologist to modify the wild-type organism and to provide cultural conditions to improve the productivity of these compounds. This aspect is considered in Chapter 3.

During the deceleration and stationary phases some microbial cultures synthesize compounds which are not produced during the trophophase and which do not appear to have any obvious function in cell metabolism. These compounds are referred to as the secondary compounds of metabolism and the phase in which they are produced (equivalent to the stationary phase) as the idiophase (Bu'Lock et al., 1965). It is important to realize that secondary metabolism may occur in continuous cultures at low growth rates and is a property of slow-growing, as well as non-growing, cells. When it is appreciated that micro-organisms grow at relatively low growth rates in their natural environments, it is tempting to suggest that it is the idiophase state that prevails in nature rather than the trophophase, which may be more of a property of micro-organisms in culture. The inter-relationships between primary and secondary metabolism are illustrated in Fig. 1.2, from which it may be seen that secondary metabolites tend to be elaborated from the intermediates and products of primary metabolism. Although the primary biosynthetic routes illustrated in Fig. 1.2 are common to the vast majority of micro-organisms, each secondary product would be synthesized by only a very few different microbial species. Thus, Fig. 1.2 is a representation of the secondary metabolism exhibited by a very wide range of different micro-organisms. Also, not all micro-organisms undergo secondary metabolism — it is common amongst the filamentous bacteria and fungi and the sporing bacteria but it is not found, for example, in the Enterobacteriaceae. Thus, the taxonomic distribution of secondary metabolism is quite different from that of primary metabolism. It is important to appreciate that the classification of microbial products into primary and secondary metabolites is a convenient, but in some cases, artificial system. To quote Bushell (1988), the classification "should not be allowed to act as a conceptual straitjacket, forcing the reader to consider all products as either primary or secondary metabolites". It is sometimes difficult to categorize a product as primary or secondary and the kinetics of synthesis of certain compounds may change depending on the cultural conditions.

The physiological role of secondary metabolism in the producer cells has been the subject of considerable debate, but the importance of these metabolites to the fermentation industry is the effects they have on organisms other than those that produce them. Many secondary metabolites have antimicrobial activity, others are specific enzyme inhibitors, some are growth promoters and many have pharmacological properties. Thus, the products of secondary metabolism have formed the basis of a number of fermentation processes. As is the case for primary metabolites, wild-type micro-organisms tend to produce only low concentrations of secondary metabolites, their synthesis being controlled by induction, catabolite repression and feedback systems. The techniques which have been developed to improve secondary metabolite production are considered in Chapters 3 and 4.

TABLE 1.2. *Some primary products of microbial metabolism and their commercial significance*

| Primary metabolite | Commercial significance |
| --- | --- |
| Ethanol | 'Active ingredient' in alcoholic beverages |
| | Used as a motor-car fuel when blended with petroleum |
| Citric acid | Various uses in the food industry |
| Glutamic acid | Flavour enhancer |
| Lysine | Feed supplement |
| Nucleotides | Flavour enhancers |
| Phenylalanine | Precursor of aspartame, sweetener |
| Polysaccharides | Applications in the food industry |
| | Enhanced oil recovery |
| Vitamins | Feed supplements |

### Recombinant products

The advent of recombinant DNA technology has extended the range of potential fermentation products. Genes from higher organisms may be introduced into

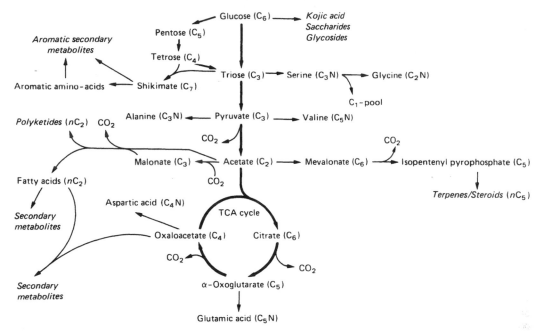

FIG. 1.2. The interrelationships between primary and secondary metabolism. Primary catabolic routes are shown in heavy lines and secondary products are italicized (Turner, 1971).

microbial cells such that the recipients are capable of synthesizing 'foreign' (or heterologous) proteins. A wide range of microbial cells have been used as hosts for such systems including *Escherichia coli*, *Saccharomyces cerevisiae* and filamentous fungi. Products produced by such genetically engineered organisms include interferon, insulin, human serum albumin, factors VIII and IX, epidermal growth factor, calf chymosin and bovine somatostatin. Important factors in the design of these processes include the secretion of the product, minimization of the degradation of the product and control of the onset of synthesis during the fermentation, as well as maximizing the expression of the foreign gene. These aspects are considered in more detail in Chapters 3 and 4.

### Transformation processes

Microbial cells may be used to convert a compound into a structurally related, financially more valuable, compound. Because micro-organisms can behave as chiral catalysts with high positional specificity and stereospecificity, microbial processes are more specific than purely chemical ones and enable the addition, removal or modification of functional groups at specific sites on a complex molecule without the use of chemical protection. The reactions which may be catalysed include dehydrogenation, oxidation, hydroxylation, dehydration and condensation, decarboxylation, amination, deamination and isomerization. Microbial processes have the additional advantage over chemical reagents of operating at relatively low temperatures and pressures without the requirement for potentially polluting heavy-metal catalysts. Although the production of vinegar is the most well-established microbial transformation process (conversion of ethanol to acetic acid) the majority of these processes involve the production of high-value compounds including steroids, antibiotics and prostaglandins.

The anomaly of the transformation fermentation process is that a large biomass has to be produced to catalyse a single reaction. Thus, many processes have been streamlined by immobilizing either the whole cells, or the isolated enzymes which catalyse the reactions, on an inert support. The immobilized cells or enzymes may then be considered as catalysts which may be reused many times.

### THE CHRONOLOGICAL DEVELOPMENT OF THE FERMENTATION INDUSTRY

The chronological development of the fermentation industry may be represented as five overlapping stages as illustrated in Table 1.3. The development of the industry prior to 1900 is represented by stage 1, where the products were confined to potable alcohol and vinegar. Although beer was first brewed by the ancient Egyptians, the first true large-scale breweries date from the early 1700s when wooden vats of 1500 barrels capacity were introduced (Corran, 1975). Even some process control was attempted in these early breweries, as indicated by the recorded use of thermometers in 1757 and the development of primitive heat exchangers

TABLE 1.3. *The stages in the chronological development of the fermentation industry*

| Stage | Main products | Vessels | Process control | Culture method | Quality control | Pilot plant facilities | Strain selection |
|---|---|---|---|---|---|---|---|
| 1 Pre-1900 | Alcohol | Wooden, up to 1500 barrels capacity Copper used in later breweries | Use of thermometer, hydrometer and heat exchangers | Batch | Virtually nil | Nil | Pure yeast cultures used at the Carlsberg brewery (1886) |
| | Vinegar | Barrels, shallow trays, trickle filters | | Batch | Virtually nil | Nil | Fermentations inoculated with 'good' vinegar |
| 2 1900–1940 | Bakers' yeast glycerol, citric acid, lactic acid and acetone/butanol | Steel vessels of up to 200 m$^3$ for acetone/butanol Air spargers used for bakers' yeast Mechanical stirring used in small vessels | pH electrodes with off-line control Temperature control | Batch and fed-batch systems | Virtually nil | Virtually nil | Pure cultures used |
| 3 1940–date | Penicillin, streptomycin, other antibiotics, gibberelin, amino acids, nucleotides, transformations, enzymes | Mechanically aerated vessels, operated aseptically—true fermenters | Sterilizable pH and oxygen electrodes. Use of control loops which were later computerized | Batch and fed-batch common Continuous culture introduced for brewing and some primary metabolites | Very important | Becomes common | Mutation and selection programmes essential |
| 4 1964–date | Single-cell protein using hydrocarbon and other feedstocks | Pressure cycle and pressure jet vessels developed to overcome gas and heat exchange problems | Use of computer linked control loops | Continuous culture with medium recycle | Very important | Very important | Genetic engineering of producer strains attempted |
| 5 1979–date | Production of heterologous proteins by microbial and animal cells Monoclonal antibodies produced by animal cells | Fermenters developed in stages 3 and 4. Animal cell reactors developed | Control and sensors developed in stages 3 and 4 | Batch, fed-batch or continuous Continuous perfusion developed for animal cell processes | Very important | Very important | Introduction of foreign genes into microbial and animal cell hosts. *In vitro* recombinant DNA techniques used in the improvement of stage 3 products |

in 1801. By the mid-1800s the role of yeasts in alcoholic fermentation had been demonstrated independently by Cagniard-Latour, Schwann and Kutzing but it was Pasteur who eventually convinced the scientific world of the obligatory role of these micro-organisms in the process. During the late 1800s Hansen started his pioneering work at the Carlsberg brewery and developed methods for isolating and propagating single yeast cells to produce pure cultures and established sophisticated techniques for the production of starter cultures. However, use of pure cultures did not spread to the British ale breweries and it is true to say that many of the small, traditional, ale-producing breweries still use mixed yeast cultures at the present time but, nevertheless, succeed in producing high quality products.

Vinegar was originally produced by leaving wine in shallow bowls or partially filled barrels where it was slowly oxidized to vinegar by the development of a natural flora. The appreciation of the importance of air in the process eventually led to the development of the 'generator' which consisted of a vessel packed with an inert material (such as coke, charcoal and various types of wood shavings) over which the wine or beer was allowed to trickle. The vinegar generator may be considered as the first 'aerobic' fermenter to be developed. By the late 1800s to early 1900s the initial medium was being pasteurized and inoculated with 10% good vinegar to make it acidic, and therefore resistant to contamination, as well as providing a good inoculum (Bioletti, 1921). Thus, by the beginning of the twentieth century the concepts of process control were well established in both the brewing and vinegar industries.

Between the years 1900 and 1940 the main new products were yeast biomass, glycerol, citric acid, lactic acid, acetone and butanol. Probably the most important advances during this period were the developments in the bakers' yeast and solvent fermentations. The production of bakers' yeast is an aerobic process and it was soon recognized that the rapid growth of yeast cells in a rich wort led to oxygen depletion in the medium which, in turn, resulted in ethanol production at the expense of biomass formation. The problem was minimized by restricting the initial wort concentration such that the growth of the cells was limited by the availability of the carbon source rather than oxygen. Subsequent growth of the culture was then controlled by adding further wort in small amounts. This technique is now called fed-batch culture and is widely used in the fermentation industry to avoid conditions of oxygen limitation. The aeration of these early yeast cultures was also improved by the introduction of air

through sparging tubes which could be steam cleaned (de Becze and Liebmann, 1944).

The development of the acetone–butanol fermentation during the First World War by the pioneering efforts of Weizmann led to the establishment of the first truly aseptic fermentation. All the processes discussed so far could be conducted with relatively little contamination provided that a good inoculum was used and reasonable standards of hygiene employed. However, the anaerobic butanol fermentation was susceptible to contamination in the early stages by aerobic bacteria, and by acid-producing anaerobic ones once anaerobic conditions had been established in the later stages of the process. The fermenters employed were vertical cylinders with hemispherical tops and bottoms constructed from mild steel. They could be steam sterilized under pressure and were constructed to minimize the possibility of contamination. Two-thousand-hectalitre fermenters were commissioned which presented the problems of inoculum development and the maintenance of aseptic conditions during the inoculation procedure. The techniques developed for the production of these organic solvents were a major advance in fermentation technology and paved the way for the successful introduction of aseptic aerobic processes in the 1940s.

The third stage of the development of the fermentation industry arose as a result of the wartime need to produce penicillin in submerged culture under aseptic conditions. The production of penicillin is an aerobic process which is very vulnerable to contamination. Thus, although the knowledge gained from the solvent fermentations was exceptionally valuable, the problems of sparging a culture with large volumes of sterile air and mixing a highly viscous broth had to be overcome. Also, unlike the solvent fermentations, penicillin was synthesized in very small quantities by the initial isolates and this resulted in the establishment of strain-improvement programmes which became a dominant feature of the industry in subsequent years. Process development was also aided by the introduction of pilot-plant facilities which enabled the testing of new techniques on a semi-production scale. The development of a large-scale extraction process for the recovery of penicillin was another major advance at this time. The technology established for penicillin fermentation provided the basis for the development of a wide range of new processes. This was probably the stage when the most significant changes in fermentation technology took place resulting in the establishment of many new processes over the period, including other antibiotics, vitamins, gibberellin, amino acids, enzymes and steroid transformations. From the 1960s onwards microbial products were screened for activities other than simply antimicrobial properties and screens became more and more sophisticated. These screens have evolved into those operating today utilizing miniaturized culture systems, robotic automation and elegant assays.

In the early 1960s the decisions of several multi-national companies to investigate the production of microbial biomass as a source of feed protein led to a number of developments which may be regarded as the fourth stage in the progress of the industry. The largest mechanically stirred fermentation vessels developed during stage 3 were in the range 80,000 to 150,000 dm$^3$. However, the relatively low selling price of microbial biomass necessitated its production in much larger quantities than other fermentation products in order for the process to be profitable. Also, hydrocarbons were considered as potential carbon sources which would result in increased oxygen demands and high heat outputs by these fermentations (see Chapters 4 and 9). These requirements led to the development of the pressure jet and pressure cycle fermenters which eliminated the need for mechanical stirring (see Chapter 7). Another feature of these potential processes was that they would have to be operated continuously if they were to be economic. At this time batch and fed-batch processes were common in the industry but the technique of growing an organism continuously by adding fresh medium to the vessel and removing culture fluid had been applied only to a very limited extent on a large scale. The brewers were also investigating the potential of continuous culture at this time, but its application in that industry was short-lived. Several companies persevered in the biomass field and a few processes came to fruition, of which the most long-lived was the ICI Pruteen animal feed process which utilized a continuous 3,000,000-dm$^3$ pressure cycle fermenter for the culture of *Methylophilus methylotrophus* with methanol as carbon source (Smith, 1981; Sharp, 1989). The operation of an extremely large continuous fermenter for time periods in excess of 100 days presented a considerable aseptic operation problem, far greater than that faced by the antibiotic industry in the 1940s. The aseptic operation of fermenters of this type was achieved as a result of the high standards of fermenter construction, the continuous sterilization of feed streams and the utilization of computer systems to control the sterilization and operation cycles, thus minimizing the possibility of human error. However, although the Pruteen process was a technological triumph it became an economic failure because the

product was out-priced by soybean and fishmeal. Eventually, in 1989, the plant was demolished, marking the end of a short, but very exciting, era in the fermentation industry.

Whilst biomass is a very low-value, high-volume product, the fifth stage in the progress of the industry resulted in the establishment of very high-value, low-volume products. The developments in *in vitro* genetic manipulation, commonly known as genetic engineering, enabled the expression of human and mammalian genes in micro-organisms, thereby enabling the large scale production of human proteins which could then be used therapeutically. According to Dykes (1993) it was the small, venture-capital biotechnology companies that pioneered the development of heterologous proteins for therapeutic use. The established pharmaceutical companies used the new genetic engineering techniques to help in the discovery of natural products and in the rational design of drugs; for example, mammalian receptor proteins have been cloned and used in *in vitro* detection systems.

Table 3.8 (Chapter 3) lists the recombinant proteins licensed for therapeutic use. According to Dykes (1993), insulin and human growth hormone have been the two most successful products but other products have far greater potential. Erythropoietin and the myeloid colony stimulating factors (CSFs) control the production of blood cells by stimulating the proliferation, differentiation and activation of specific cell types. Erythropoietin has been used to treat renal-failure anaemia and may have application in the treatment of the platelet deficiency associated with cancer chemotherapy; it is expected to become the top-selling therapeutic protein by the mid-1990s with annual sales of around $1200 million (Dykes, 1993). Granulocyte-colony stimulating factor (G-CSF), which is used during cancer chemotherapy, generated sales of over $230 million in 1991 and, due to other uses, its sales could reach $1000 million by 1996. A number of different growth factors are involved in wound healing and recombinant forms of these proteins would be expected to yield significant returns during the 1990s.

The commercial exploitation of recombinant proteins has necessitated the design of contained production facilities. Thus, these processes are drawing on the experience of vaccine fermentations where pathogenic organisms have been grown on relatively large scales. Also, recombinant proteins have been classified as biologicals, not as drugs, and thus come under the same regulatory authorities as do vaccines. The major difference between the approval of drugs and biologicals is that the process for the production of a biological must be precisely specified and carried out in a facility that has been inspected and licensed by the regulatory authority, which is not the case for the production of drugs (antibiotics, for example) (Bader, 1992). Thus, any changes which a manufacturer wishes to incorporate into a licensed process must receive regulatory approval. For drugs, only major changes require approval prior to implementation. The result of these containment and regulatory requirements is that the cost of developing a recombinant protein process is extremely high. Buckland (1992) illustrated this point in his claim that "It now costs as much to build a 3000 dm$^3$ scale facility for Biologics as for a 200,000 dm$^3$ scale facility for an antibiotic. Also, even though titres are now reasonable for a recombinant protein (1 g dm$^{-3}$), the cost of manufacture kg$^{-1}$ of bulk drug is about two orders of magnitude higher than that of an antibiotic at 10 g dm$^{-3}$ titre". Also, the development time for a recombinant protein is considerably longer than that for an antibiotic. For example, Bader (1992) claimed that it is feasible for an antibiotic plant to begin production four years after the initiation of the plant design whereas seven years would be required before a recombinant protein could be produced.

The exploitation of genetic engineering approximately coincided with another major development in biotechnology which influenced the progress of the fermentation industry — the production of monoclonal antibodies. The availability of monoclonal antibodies opened the door to sophisticated analytical techniques and raised hopes for their use as therapeutic agents. Although the promise of therapeutic agents has yet to be realized (only one monoclonal antibody has been licensed for clinical use, OKT3, used in the treatment of acute renal allograft rejection (Webb, 1993)), their use as tools in biological research has increased exponentially. Thus, animal cell culture processes were established to produce monoclonals on a commercial scale. Subsequently, animal cells were also used as hosts for the production of some human proteins, especially where post-translational modification was essential for protein activity. Although these animal cell processes were based on microbial fermentation technology a number of novel problems had to be solved — animal cells are extremely fragile compared with microbial cells, the achievable cell density is very much less than in a microbial process and the media are very complex. These aspects are considered in Chapters 4 and 7.

The outstanding developments in recombinant fermentations (stage 5) have tended to overshadow the progress which has been made in recent years in establishing new fermentations based on conventional microbial products (the continuing development of stage 4). However, the appreciation by the pharmaceutical industry that the activity of microbial metabolites extended well beyond antibacterials has resulted in a number of new microbial products reaching the marketplace in the late 1980s and early 1990s. Buckland (1992) listed four secondary metabolites which were launched in the 1980s: cyclosporin, an immunoregulant used to control rejection of transplanted organs; imipenem, a modified carbapenem which has the widest antimicrobial spectrum of any antibiotic; lovastatin, a drug used for reducing cholesterol levels and ivermectin, an anti-parasitic drug which has been used to prevent 'African River Blindness' as well as in veterinary practice. Buckland summarized these developments succinctly, "One of the best kept secrets (unintentionally kept as a secret) in the 1980s in Biochemical Engineering was that working on secondary metabolites was a fascinating, important and rewarding experience. Furthermore the four products listed added together have higher sales than all of the recombinant products added together". Thus, it is still relevant to heed Foster's warning (1949) "never underestimate the power of the microbe".

## THE COMPONENT PARTS OF A FERMENTATION PROCESS

Regardless of the type of fermentation (with the possible exception of some transformation processes) an established process may be divided into six basic component parts:

(i) The formulation of media to be used in culturing the process organism during the development of the inoculum and in the production fermenter.

(ii) The sterilization of the medium, fermenters and ancillary equipment.

(iii) The production of an active, pure culture in sufficient quantity to inoculate the production vessel.

(iv) The growth of the organism in the production fermenter under optimum conditions for product formation.

(v) The extraction of the product and its purification.

(vi) The disposal of effluents produced by the process.

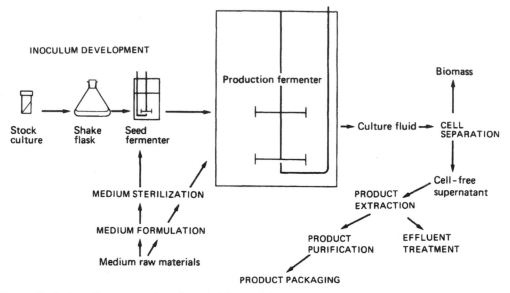

FIG. 1.3. A generalized schematic representation of a typical fermentation process.

The interrelationships between the six component parts are illustrated in Fig 1.3.

However, one must also visualize the research and development programme which is designed to gradually improve the overall efficiency of the fermentation. Before a fermentation process is established a producer organism has to be isolated, modified such that it produces the desired product in commercial quantities, its cultural requirements determined and the plant designed accordingly. Also, the extraction process has to be established. The development programme would involve the continual improvement of the process organism, the culture medium and the extraction process.

The subsequent chapters in this book consider the basic principles underlying the component parts of a fermentation. Chapter 2 considers growth, comprehension of which is crucial to understanding many aspects of the process, other than simply the growth of the organism in the production fermenter. The isolation and improvement of commercial strains is considered in Chapter 3 and the design of media in Chapter 4. The sterilization of the medium, fermenters and air is considered in Chapter 5 and the techniques for the development of inocula are discussed in Chapter 6. Chapters 7, 8 and 9 consider the fermenter as an environment for the culture of micro-organisms; Chapter 7 considers the design and construction of fermenters including contained systems and animal cell fermenters, Chapter 8 discusses the instrumentation involved in monitoring and maintaining a controlled environment in a fer-

menter, while the provision of oxygen to a culture is investigated in Chapter 9. The recovery of fermentation products is dealt with in Chapter 10 and the disposal of effluents of processes is covered in Chapter 11. Finally, the economics of fermentation processes are discussed in Chapter 12. Throughout the book examples are drawn from a very wide range of fermentations to illustrate the applications of the techniques being discussed but it has not been attempted to give detailed considerations of specific processes as this is well covered elsewhere, for example in the Biotechnology series edited by Rehm and Reed. We hope that the approach adopted in this book will give the reader an understanding of the basic principles underlying the techniques used for the large-scale production of microbial products.

## REFERENCES

BADER, F. G. (1992) Evolution in fermentation facility design from antibiotics to recombinant proteins. In *Harnessing Biotechnology for the 21st Century*, pp. 228–231 (Editors Ladisch, M. R. and Bose, A.). American Chemical Society, Washington, DC.

BIOLETTI, F. T. (1921) The manufacture of vinegar. In *Microbiology*, pp. 636–648 (Editor Marshall, C. E.). Churchill, London.

BOING, J. T. P. (1982) Enzyme production. In *Prescott and Dunn's Industrial Microbiology* (4th edition), pp. 634–708 (Editor Reed, G.). MacMillan, New York.

BUCKLAND, B. C. (1992) Reduction to practice. In *Harnessing Biotechnology for the 21st Century*, pp. 215–218 (Editors Ladisch, M. R. and Bose, A.). American Chemical Society, Washington, DC.

BU'LOCK, J. D., HAMILTON, D., HULME, M. A., POWELL, A. J., SHEPHERD, D., SMALLEY, H. M. and SMITH, G. N. (1965) Metabolic development and secondary biosynthesis in *Penicillium urticae*. *Can. J. Micro.* **11**, 765–778.

BUSHELL, M. E. (1988) Application of the principles of industrial microbiology to biotechnology. In *Principles of Biotechnology*, pp. 5–43 (Editor Wiseman, A.). Chapman and Hall, New York.

CORRAN, H. S. (1975) *A History of Brewing*. David and Charles, Newton Abbott.

DAWES, I. and LARGE, P. J. (1982) Class 1 reactions: Supply of carbon skeletons. In *Biochemistry of Bacterial Growth*, pp. 125–158 (Editors Mandelstam, J., McQuillen, K. and Dawes, I.). Blackwell, Oxford.

DE BECZE and LIEBMANN, A. J. (1944) Aeration in the production of compressed yeast. *Ind. Eng. Chem.* **36**, 882–890.

DYKES, C. W. (1993) Molecular biology in the pharmaceutical industry. In *Molecular Biology and Biotechnology*, pp. 155–176 (Editors Walker, J. M. and Gingold, E. B.). Royal Society of Chemistry, Cambridge.

FOSTER. J. W. (1949) *Chemical Activities of the Fungi*. Academic Press, New York.

REHM, H.-J. and REED, G. (1993) *Biotechnology* (2nd edition), Volumes 1–12. VCH, Weinheim.

SHARP, D. H. (1989) *Bioprotein Manufacture : A Critical Assessment*. Ellis Horwood, Chichester.

SMITH, S. R. L. (1981) Some aspects of ICI's single cell protein process. In *Microbial Growth on $C_1$ Compounds*, pp. 342–348 (Editor Dalton, H.). Heyden, London.

TURNER, W. B. (1971) *Fungal Metabolites*. Academic Press, London.

WEBB, M. (1993) Monoclonal antibodies. In *Molecular Biology and Biotechnology*, pp. 357–386 (Editors Walker, J. M. and Gingold, E. B.). Royal Society of Chemistry, Cambridge.

# CHAPTER 2

# Microbial Growth Kinetics

As OUTLINED in Chapter 1, fermentations may be carried out as batch, continuous and fed-batch processes. The mode of operation is, to a large extent, dictated by the type of product being produced. This chapter will consider the kinetics and applications of batch, continuous and fed-batch processes.

### BATCH CULTURE

Batch culture is a closed culture system which contains an initial, limited amount of nutrient. The inoculated cuture will pass through a number of phases, as illustrated in Fig. 2.1. After inoculation there is a period during which it appears that no growth takes place; this period is referred to as the lag phase and may be considered as a time of adaptation. In a commercial process the length of the lag phase should be reduced as much as possible and this may be achieved by using a suitable inoculum, as described in Chapter 6. Following a period during which the growth rate of the cells gradually increases, the cells grow at a constant, maximum, rate and this period is known as the log, or exponential, phase. The exponential phase may be described by the equation:

$$dx/dt = \mu x \qquad (2.1)$$

where $x$ is the concentration of microbial biomass,
  $t$ is time, in hours
and $\mu$ is the specific growth rate, in hours$^{-1}$.
On integration equation (2.1) gives:

$$x_t = x_0 e^{\mu t} \qquad (2.2)$$

where $x_0$ is the original biomass concentration,
  $x_t$ is the biomass concentration after the time interval, $t$ hours,
  e is the base of the natural logarithm.

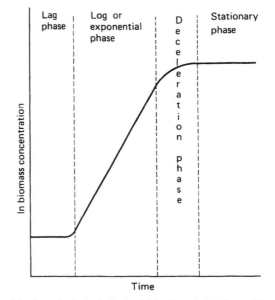

FIG. 2.1. Growth of a typical microbial culture in batch conditions.

On taking natural logarithms, equation (2.2) becomes:

$$\ln x_t = \ln x_0 + \mu t. \qquad (2.3)$$

Thus, a plot of the natural logarithm of biomass concentration against time should yield a straight line, the slope of which would equal $\mu$. During the exponential phase nutrients are in excess and the organism is growing at its maximum specific growth rate, $\mu_{max}$, for the prevailing conditions. Typical values of $\mu_{max}$ for a range of micro-organisms are given in Table 2.1.

It is easy to visualize the exponential growth of single celled organisms which replicate by binary fission. Indeed, animal and plant cells in suspension culture will behave very similarly to unicellular micro-organisms (Griffiths, 1986; Petersen and Alfermann, 1993). However, it is more difficult to appreciate that

13

TABLE 2.1. *Some representative values of $\mu_{max}$ (obtained under the conditions specified in the original reference) for a range of organisms*

| Organism | $\mu_{max}$ (h$^{-1}$) | Reference |
|---|---|---|
| *Vibrio natriegens* | 4.24 | Eagon (1961) |
| *Methylomonas methanolytica* | 0.53 | Dostalek *et al.* (1972) |
| *Aspergillus nidulans* | 0.36 | Trinci (1969) |
| *Penicillium chrysogenum* | 0.12 | Trinci (1969) |
| *Fusarium graminearum* Schwabe | 0.28 | Trinci (1992) |
| Plant cells in suspension culture | 0.01–0.046 | Petersen and Alfermann (1993) |
| Animal cells | 0.01–0.05 | Lavery (1990) |

mycelial organisms which show apical growth also grow exponentially. Plomley (1959) was the first to suggest that filamentous fungi have a 'growth unit' which is replicated at a constant rate and is composed of the apex of the hypha and a short length of supporting hypha. Trinci (1974) demonstrated that the total hyphal length of a mycelium and the number of tips increased exponentially at approximately the same rate. Thus, when the volume of the hyphal growth unit exceeds a critical volume a new branch, and hence, a new growing point, is initiated. This is equivalent to the division of a single cell when the cell reaches a critical volume. Hence, the rate of increase in hyphal mass, total length and number of tips is dictated by the specific growth rate and:

$$dx/dt = \mu x,$$

$$dH/dt = \mu H,$$

$$dA/dt = \mu A$$

where $H$ is total hyphal length and $A$ is the number of growing tips.

In submerged culture (shake flask or fermenter) a mycelial organism may grow as dispersed hyphal fragments or as pellets (see also Chapters 6 and 9). The growth of pellets will be exponential until the density of the pellet results in diffusion limitation. Under such limitation the central biomass of the pellet will not receive a supply of nutrients, nor will potentially toxic products diffuse out. Thus, the growth of the pellet proceeds from the outer shell of biomass which is the actively growing zone and was described by Pirt (1975) as:

$$M^{1/3} = kt + M_0^{1/3}$$

where $M_0$ and $M$ are the mycelium mass at time 0 and $t$, respectively. Thus, a plot of the cube root of mycelial mass against time will give a straight line, the slope of which equals $k$.

It is possible for new pellets to be generated by the fragmentation of old pellets and, thus, the behaviour of a pelleted culture may be intermediate between exponential and cube root growth.

Whether the organism is unicellular or mycelial the foregoing equations predict that growth will continue indefinitely. However, growth results in the consumption of nutrients and the excretion of microbial products; events which influence the growth of the organism. Thus, after a certain time the growth rate of the culture decreases until growth ceases. The cessation of growth may be due to the depletion of some essential nutrient in the medium (substrate limitation), the accumulation of some autotoxic product of the organism in the medium (toxin limitation) or a combination of the two.

The nature of the limitation of growth may be explored by growing the organism in the presence of a range of substrate concentrations and plotting the biomass concentration at stationary phase against the initial substrate concentration, as shown in Fig. 2.2. From Fig. 2.2 it may be seen that over the zone A to B an increase in initial substrate concentration gives a proportional increase in the biomass produced at stationary phase. The situation may be described by the equation:

$$x = Y(S_R - s) \tag{2.4}$$

where $x$  is the concentration of biomass produced,
$Y$  is the yield factor (g biomass produced g$^{-1}$ substrate consumed),
$S_R$  is the initial substrate concentration, and
$s$  is the residual substrate concentration.

Over the zone A to B in Fig. 2.2, $s$ equals zero at the point of cessation of growth. Thus, equation (2.4) may be used to predict the biomass which may be produced

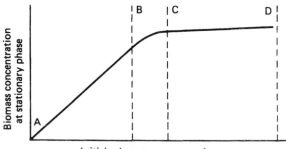

FIG. 2.2. The effect of initial substrate concentration on the biomass concentration at the onset of stationary phase, in batch culture.

from a certain amount of substrate. Over the zone C to D an increase in the initial substrate concentration does not give a proportional increase in biomass. This may be due to either the exhaustion of another substrate or the accumulation of toxic products. Over the zone B to C the utilization of the substrate is deleteriously affected by either the accumulating toxins or the availability of another substrate.

The yield factor ($Y$) is a measure of the efficiency of conversion of any one substrate into biomass and it can be used to predict the substrate concentration required to produce a certain biomass concentration. However, it is important to appreciate that $Y$ is not a constant — it will vary according to growth rate, pH, temperature, the limiting substrate and the concentration of the substrates in excess.

The decrease in growth rate and the cessation of growth, due to the depletion of substrate, may be described by the relationship between $\mu$ and the residual growth-limiting substrate, represented in equation (2.5) and in Fig. 2.3 (Monod, 1942):

$$\mu = \mu_{max} s / (K_s + s) \qquad (2.5)$$

where  $s$  is the residual substrate concentration,

$K_s$ is the substrate utilization constant, numerically equal to substrate concentration when $\mu$ is half $\mu_{max}$ and is a measure of the affinity of the organism for its substrate.

The zone A to B in Fig. 2.3 is equivalent to the exponential phase in batch culture where substrate concentration is in excess and growth is at $\mu_{max}$. The zone C to A in Fig. 2.3 is equivalent to the deceleration phase of batch culture where the growth of the organism has resulted in the depletion of substrate to a

growth-limiting concentration which will not support $\mu_{max}$. If the organism has a very high affinity for the limiting substrate (a low $K_s$ value) the growth rate will not be affected until the substrate concentration has declined to a very low level. Thus, the deceleration phase for such a culture would be short. However, if the organism has a low affinity for the substrate (a high $K_s$ value) the growth rate will be deleteriously affected at a relatively high substrate concentration. Thus, the deceleration phase for such a culture would be relatively long. Typical values of $K_s$ for a range of organisms and substrates are shown in Table 2.2, from which it may be seen that such values are usually very small and the affinity for substrate is high. It will be appreciated that the biomass concentration at the end of the exponential phase is at its highest and, thus, the decline in substrate concentration will be very rapid so that the time period during which the substrate concentration is close to $K_s$ is very short.

The stationary phase in batch culture is that point where the growth rate has declined to zero. However, as Bull (1974) pointed out, the stationary phase is a misnomer in terms of the physiology of the organism, as the population is still metabolically active during this phase and may produce products called secondary metabolites, which are not produced during the exponential phase. Bull suggested that this phase be termed the maximum population phase. The metabolic activity of the stationary phase has been recognized in the physiological descriptions of microbial growth presented by Borrow *et al.* (1961) and Bu'Lock *et al.* (1965). Borrow *et al.* investigated the biosynthesis of gibberellic acid by *Gibberella fujikuroi* and divided the growth of the organism into several phases:

(i) The balanced phase; equivalent to the early to middle exponential phase.
(ii) The storage phase; equivalent to the late exponential phase where the increase in mass is due to the accumulation of lipid and carbohydrate.

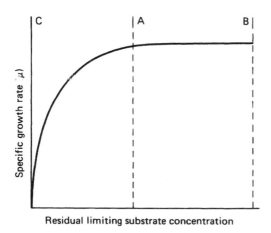

Residual limiting substrate concentration

FIG. 2.3. The effect of residual limiting substrate concentration on the specific growth rate of a hypothetical bacterium.

TABLE 2.2. *Some representative values of* $K_s$ *for a range of micro-organisms and substrates*

| Organism | Substrate | $K_s$ (mg dm$^{-3}$) | References |
|---|---|---|---|
| *Escherichia coli* | Glucose | $6.8 \times 10^{-2}$ | Shehata and Marr (1971) |
| *Saccharomyces cerevisiae* | Glucose | 25.0 | Pirt and Kurowski (1970) |
| *Pseudomonas* sp. | Methanol | 0.7 | Harrison (1973) |

(iii) The maintenance phase; equivalent to the stationary phase.

Gibberellic acid (a secondary metabolite) was synthesized only towards the end of the storage phase and during the maintenance phase. As discussed in Chapter 1, Bu'Lock *et al.* (1965) coined the terms trophophase, to refer to the exponential phase, and idiophase to refer to the stationary phase where secondary metabolites are produced. The idiophase was depicted as the period subsequent to the exponential phase in which secondary metabolites were synthesized. However, it is now obvious that the culture conditions may be manipulated to induce secondary metabolism during logarithmic growth, for example by the use of carbon sources which support a reduced maximum growth rate (see Chapter 4).

Pirt (1975) has discussed the kinetics of product formation by microbial cultures in terms of growth-linked products and non-growth-linked products. Growth-linked may be considered equivalent to primary metabolites which are synthesized by growing cells and non-growth-linked may be considered equivalent to secondary metabolites. The formation of a growth-linked product may be described by the equation:

$$dp/dt = q_p x \qquad (2.6)$$

where $p$ is the concentration of product
and $q_p$ is the specific rate of product formation (mg product $g^{-1}$ biomass $h^{-1}$)
Also, product formation is related to biomass production by the equation:

$$dp/dx = Y_{p/x} \qquad (2.7)$$

where $Y_{p/x}$ is the yield of product in terms of biomass (g product $g^{-1}$ biomass)
Multiply equation (2.7) by $dx/dt$, then:

$$dx/dt \cdot dp/dx = Y_{p/x} \cdot dx/dt$$

and $\qquad dp/dt = Y_{p/x} \cdot dx/dt.$

But $dx/dt = \mu x$ and therefore:

$$dp/dt = Y_{p/x} \cdot \mu x$$

and $\qquad dp/dt = q_p \cdot x$

and therefore:

$$q_p \cdot x = Y_{p/x} \cdot \mu x,$$

$$q_p = Y_{p/x} \cdot \mu. \qquad (2.8)$$

From equation (2.8) it may be seen that when product formation is growth associated the specific rate of product formation increases with specific growth rate. Thus, productivity in batch culture will be greatest at $\mu_{max}$ and improved product output will be achieved by increasing both $\mu$ and biomass concentration. Non-growth linked product formation is related to biomass concentration and, thus, increased productivity in batch culture should be associated with an increase in biomass. However, it should be remembered that non-growth related secondary metabolites are produced only under certain physiological conditions — primarily under limitation of a particular substrate so that the biomass must be in the correct 'physiological state' before production can be achieved. The elucidation of the environmental conditions which create the correct 'physiological state' is extremely difficult in batch culture and this aspect is developed in a later section.

Thus, batch fermentation may be used to produce biomass, primary metabolites and secondary metabolites. For biomass production, cultural conditions supporting the fastest growth rate and maximum cell population would be used; for primary metabolite production conditions to extend the exponential phase accompanied by product excretion and for secondary metabolite production, conditions giving a short exponential phase and an extended production phase, or conditions giving a decreased growth rate in the log phase resulting in earlier secondary metabolite formation.

## CONTINUOUS CULTURE

Exponential growth in batch culture may be prolonged by the addition of fresh medium to the vessel. Provided that the medium has been designed such that growth is substrate limited (i.e. by some component of the medium), and not toxin limited, exponential growth will proceed until the additional substrate is exhausted. This exercise may be repeated until the vessel is full. However, if an overflow device were fitted to the fermenter such that the added medium displaced an equal volume of culture from the vessel then continuous production of cells could be achieved. If medium is fed continuously to such a culture at a suitable rate, a steady state is achieved eventually, that is, formation of new biomass by the culture is balanced by the loss of cells from the vessel. The flow of medium into the vessel is related to the volume of the vessel by the term dilution rate, $D$, defined as:

$$D = F/V \qquad (2.9)$$

where $F$ is the flow rate (dm$^3$ h$^{-1}$)

and     $V$ is the volume ($dm^3$).

Thus, $D$ is expressed in the units $h^{-1}$.

The net change in cell concentration over a time period may be expressed as:

$$dx/dt = \text{growth} - \text{output}$$

or

$$dx/dt = \mu x - Dx. \qquad (2.10)$$

Under steady-state conditions the cell concentration remains constant, thus $dx/dt = 0$ and:

$$\mu x = Dx \qquad (2.11)$$

and

$$\mu = D. \qquad (2.12)$$

Thus, under steady-state conditions the specific growth rate is controlled by the dilution rate, which is an experimental variable. It will be recalled that under batch culture conditions an organism will grow at its maximum specific growth rate and, therefore, it is obvious that a continuous culture may be operated only at dilution rates below the maximum specific growth rate. Thus, within certain limits, the dilution rate may be used to control the growth rate of the culture.

The growth of the cells in a continuous culture of this type is controlled by the availability of the growth limiting chemical component of the medium and, thus, the system is described as a chemostat. The mechanism underlying the controlling effect of the dilution rate is essentially the relationship expressed in equation (2.5), demonstrated by Monod in 1942:

$$\mu = \mu_{max}s/(K_s + s).$$

At steady state, $\mu = D$, and, therefore,

$$D = \mu_{max}\bar{s}/(K_s + \bar{s})$$

where $\bar{s}$ is the steady-state concentration of substrate in the chemostat, and

$$\bar{s} = K_sD/(\mu_{max} - D). \qquad (2.13)$$

Equation (2.13) predicts that the substrate concentration is determined by the dilution rate. In effect, this occurs by growth of the cells depleting the substrate to a concentration that supports the growth rate equal to the dilution rate. If substrate is depleted below the level that supports the growth rate dictated by the dilution rate, the following sequence of events takes place:

(i)   The growth rate of the cells will be less than the dilution rate and they will be washed out of the vessel at a rate greater than they are being produced, resulting in a decrease in biomass concentration.

(ii)   The substrate concentration in the vessel will rise because fewer cells are left in the vessel to consume it.

(iii)   The increased substrate concentration in the vessel will result in the cells growing at a rate greater than the dilution rate and biomass concentration will increase.

(iv)   The steady state will be re-established.

Thus, a chemostat is a nutrient-limited self-balancing culture system which may be maintained in a steady state over a wide range of sub-maximum specific growth rates.

The concentration of cells in the chemostat at steady state is described by the equation:

$$\bar{x} = Y(S_R - \bar{s}) \qquad (2.14)$$

where $\bar{x}$ is the steady-state cell concentration in the chemostat.

By combining equations (2.13) and (2.14), then:

$$\bar{x} = Y[S_R - \{K_sD/(\mu_{max} - D)\}]. \qquad (2.15)$$

Thus, the biomass concentration at steady state is determined by the operational variables, $S_R$ and $D$. If $S_R$ is increased, $\bar{x}$ will increase but $\bar{s}$, the residual substrate concentration in the chemostat, will remain the same. If $D$ is increased, $\mu$ will increase ($\mu = D$) and the residual substrate at the new steady state would have increased to support the elevated growth rate; thus, less substrate will be available to be converted into biomass, resulting in a lower steady state value.

An alternative type of continuous culture to the chemostat is the turbidostat, where the concentration of cells in the culture is kept constant by controlling the flow of medium such that the turbidity of the culture is kept within certain, narrow limits. This may be achieved by monitoring the biomass with a photoelectric cell and feeding the signal to a pump supplying medium to the culture such that the pump is switched on if the biomass exceeds the set point and is switched off if the biomass falls below the set point. Systems other than turbidity may be used to monitor the biomass concentration, such as $CO_2$ concentration or pH in which case it would be more correct to term the culture a biostat. The chemostat is the more commonly used system because it has the advantage over the biostat of not requiring complex control systems to maintain a steady state. However, the biostat may be advantageous in continuous enrichment culture in avoiding the total washout of the culture in its early stages and this aspect is discussed in Chapter 3.

The kinetic characteristics of an organism (and,

therefore, its behaviour in a chemostat) are described by the numerical values of the constants $Y$, $\mu_{max}$ and $K_s$. The value of $Y$ affects the steady-state biomass concentration; the value of $\mu_{max}$ affects the maximum dilution rate that may be employed and the value of $K_s$ affects the residual substrate concentration (and, hence, the biomass concentration) and also the maximum dilution rate that may be used. Figure 2.4 illustrates the continuous culture behaviour of a hypothetical bacterium with a low $K_s$ value for the limiting substrate, compared with the initial limiting substrate concentration. With increasing dilution rate, the residual substrate concentration increases only slightly until $D$ approaches $\mu_{max}$ when $s$ increases significantly. The dilution rate at which $x$ equals zero (that is, the cells have been washed out of the system) is termed the critical dilution rate ($D_{crit}$) and is given by the equation:

$$D_{crit} = \mu_{max} S_R / (K_s + S_R). \quad (2.16)$$

Thus, $D_{crit}$ is affected by the constants, $\mu_{max}$ and $K_s$, and the variable, $S_R$; the larger $S_R$ the closer is $D_{crit}$ to $\mu_{max}$. However, $\mu_{max}$ cannot be achieved in a simple steady state chemostat because substrate limited conditions must always prevail.

Figure 2.5 illustrates the continuous culture behaviour of a hypothetical bacterium with a high $K_s$ for the limiting substrate compared with the initial limiting substrate concentration. With increasing dilution rate, the residual substrate concentration increases significantly to support the increased growth rate. Thus, there is a gradual increase in $s$ and a decrease in $x$ as $D$ approaches $D_{crit}$. Figure 2.6 illustrates the effect of increasing the initial limiting substrate concentration

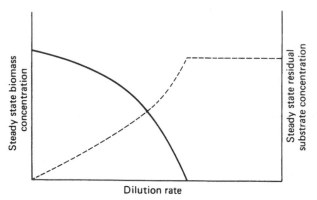

FIG. 2.5. The effect of dilution rate on the steady-state biomass and residual substrate concentrations in a chemostat of a micro-organism with a high $K_s$ value for the limiting substrate, compared with the initial substrate concentration.
_____ Steady-state biomass concentration.
− − − Steady-state residual substrate concentration.

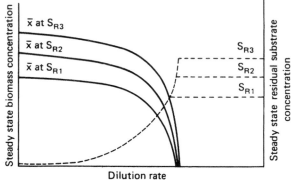

FIG. 2.6. The effect of the increased initial substrate concentration on the steady-state biomass and residual substrate concentrations in a chemostat.
_____ Steady-state biomass concentrations.
− − − Steady-state residual substrate concentrations.
$S_{R1}$, $S_{R2}$ and $S_{R3}$ represent increasing concentrations of the limiting substrate in the feed medium.

on $\bar{x}$ and $\bar{s}$. As $S_R$ is increased, so $\bar{x}$ increases, but the residual substrate concentration is unaffected. Also, $D_{crit}$ increases slightly with an increase in $S_R$.

The results of chemostat experiments may differ from those predicted by the foregoing theory. The reasons for these deviations may be anomalies associated with the equipment or the theory not predicting the behaviour of the organism under certain circumstances. Practical anomalies include imperfect mixing and wall growth. Imperfect mixing would cause an

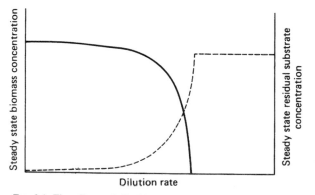

FIG. 2.4. The effects of dilution rate on the steady-state biomass and residual substrate concentrations in a chemostat culture of a micro-organism with a low $K_s$ value for the limiting substrate, compared with the initial substrate concentration.
_____ Steady-state biomass concentration.
− − − Steady-state residual substrate concentration.

increase in the degree of heterogeneity in the fermenter with some organisms being subject to nutrient excess whilst others are under severe limitation. This phenomenon is particularly relevant to very low dilution rate systems when the flow of medium is likely to be very intermittent. This problem may be overcome by the use of feed-back systems, as discussed later in this Chapter. Wall growth is another commonly encountered practical difficulty in which the organism adheres to the inner surfaces of the reactor resulting, again, in an increase in heterogeneity. The immobilized cells are not subject to removal from the vessel but will consume substrate resulting in the suspended biomass concentration being lower than predicted. Wall growth may be limited by coating the inner surfaces of the vessel with Teflon.

A frequent observation in carbon and energy limited chemostats is that the biomass concentration at low dilution rates is lower than predicted. This is attributed to the phenomenon of micro-organisms utilizing a greater proportion of substrate for maintenance at low dilution rates. Effectively, the yield factor decreases at low dilution rates. Bull (1974) has reviewed the major causes of deviations from basic chemostat theory.

The basic chemostat may be modified in a number of ways, but the most common modifications are the addition of extra stages (vessels) and the feedback of biomass into the vessel.

### Multistage systems

A multistage system is illustrated in Fig. 2.7. The advantage of a multistage chemostat is that different conditions prevail in the separate stages. This may be

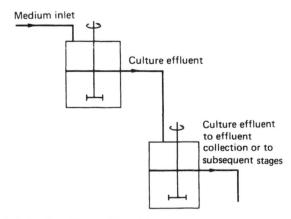

FIG. 2.7. A multistage chemostat.

advantageous in the utilization of multiple carbon sources and in the production of secondary metabolites. Harte and Webb (1967) demonstrated that when *Klebsiella aerogenes* was grown on a mixture of glucose and maltose only the glucose was utilized in the first stage and maltose in the second. Secondary metabolism may occur in the second stage of a dual system in which the second stage acts as a holding tank where the growth rate is much smaller than that in the first stage. The adoption of multistage systems in research and industry has been extremely limited, due to the complexity of the systems. One example of the industrial application of the technique is in continuous brewing which is described in a later section.

### Feedback systems

A chemostat incorporating biomass feedback has been modified such that the biomass in the vessel reaches a concentration above that possible in a simple chemostat, that is, greater than $Y(S_R - s)$. Biomass concentration may be achieved by:

(i) Internal feedback. Limiting the exit of biomass from the chemostat such that the biomass in the effluent stream is less concentrated than in the vessel.

(ii) External feedback. Subjecting the effluent stream to a biomass separation process, such as sedimentation or centrifugation, and returning a portion of the concentrated biomass to the growth vessel.

Pirt (1975) gave a full kinetic description of these feedback systems and this account summarizes his analysis.

INTERNAL FEEDBACK

A diagrammatic representation of an internal feedback system is shown in Fig. 2.8a. Effluent is removed from the vessel in two streams, one filtered, resulting in a dilute effluent stream (and, thus, a concentration of biomass in the reactor) and one unfiltered. The proportion of the outflow leaving via the filter and the effectiveness of the filter then determines the degree of feedback. The flow rate of incoming medium is designated $F$ (dm$^3$ h$^{-1}$) and the fraction of the outflow which is not filtered is designated $c$; thus the outflow rate of the unfiltered stream is $cF$ and that of the

**(a)**

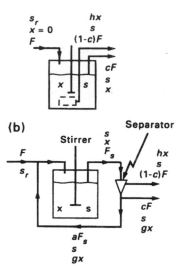

FIG. 2.8. Diagrammatic representations of chemostats with feedback (Pirt, 1975).
(a) Internal feedback.
$F$ = flow rate of incoming medium (dm³ h⁻¹)
$c$ = fraction of the outflow which is not filtered
$x$ = biomass concentration in the vessel and in the unfiltered stream
$hx$ = biomass concentration in the filtered stream
(b) External feedback.
$F$ = flow rate from the medium reservoir (dm³ h⁻¹)
$F_s$ = flow rate of the effluent upstream of the separator
$x$ = biomass concentration in the vessel and upstream of the separator
$hx$ = biomass concentration in the dilute stream from the separator
$g$ = factor by which the separator concentrates the biomass
$a$ = proportion of the flow which is fed back to the fermenter
$s$ = substrate concentration in the vessel and effluent lines
$S_R$ = substrate concentration in the medium reservoir

filtered stream is $(1 - c)F$. The concentration of the biomass in the fermenter and in the unfiltered stream is $x$ and the concentration of biomass in the filtered stream is $hx$. The biomass balance of the system is:

Change in biomass = Growth − Output in unfiltered stream − Output in filtered stream,

which may be expressed as:

$$dx/dt = \mu x - cDx - (1 - c)Dhx \quad (2.17)$$

or:

$$dx/dt = \mu x - D\{c(1 - h) + h\}x.$$

At steady state $dx/dt = 0$, thus:

$$\mu \bar{x} = D\{c(1 - h) + h\}\bar{x}$$

and

$$\mu = D\{c(1 - h) + h\}.$$

If the term '$c(1 - h) + h$' is represented by '$A$', then:

$$\mu = AD. \quad (2.18)$$

When the filtered effluent stream is cell free then h = 0 and $A = c$. However, if the filter removes very little biomass then $h$ will approach 1, and when $h = 1$ there is no feedback and $A = h$. Thus, the range of values of $A$ is $c$ to $h$ and when feedback occurs $A$ is less than 1, which means that $\mu$ is less than $D$.

The concentration of the growth limiting substrate in the vessel at steady state is then given by:

$$\bar{s} = K_s AD/(\mu_{max} - AD) \quad (2.19)$$

and the biomass concentration at steady state is given by:

$$\bar{x} = Y/A(S_R - \bar{s}). \quad (2.20)$$

EXTERNAL FEEDBACK

A diagrammatic representation of an external feedback system is shown in Fig. 2.8b. The effluent from the fermenter is fed through a separator, such as a continuous centrifuge or filter, which produces two effluent streams — a concentrated biomass stream and a dilute one. A fraction of the concentrated stream is then returned to the vessel. The flow rate from the medium reservoir is $F$ (dm³ h⁻¹); the flow rate of the effluent upstream of the separator is $F_s$ (dm³ h⁻¹) and the concentration of biomass in the stream (and in the fermenter) is $x$; $a$ is the proportion of the flow which is fed back to the fermenter and $g$ is the factor by which the separator concentrates the biomass. Biomass balance in the system will be:

Change = growth − output + feedback

or

$$dx/dt = \mu x - F_s x/V + aF_s gx/V. \quad (2.21)$$

The culture outflow (before separation), $F_s$, from the chemostat is:

$$F_s = F + aF_s$$

or

$$F_s = F/(1 - a),$$

substituting $F/(1 - a)$ for $F_s$ in equation (2.21) and remembering that $D = F/V$:

$$dx/dt = \mu x - Dx/(1 - a) + agDx/(1 - a). \quad (2.22)$$

If all the cells are returned to the fermenter then biomass will continue to accumulate in the vessel. However, if the feedback is partial then a steady state may be achieved, $dx/dt = 0$ and:

$$\mu = BD \quad (2.23)$$

where $B = (1 - ag)/(1 - a)$.

The steady-state substrate and biomass concentrations in a fermenter with feedback are then given by the following equations:

$$\bar{s} = BDK_s/(\mu_{max} - D), \qquad (2.24)$$

$$\bar{x} = Y/B(S_R - s). \qquad (2.25)$$

From the equations describing $\mu$, $\bar{s}$ and $\bar{x}$ (2.18, 2.19, 2.20, 2.23, 2.24, 2.25) in a fermenter with either external or internal feedback it can be appreciated that:

(i) Dilution rate is greater than growth rate.

(ii) Biomass concentration in the vessel is increased.

(iii) The increased biomass concentration results in a decrease in the residual substrate compared with a simple chemostat.

(iv) The maximum output of biomass and products is increased.

(v) Because $D$ is less than $\mu$, the critical dilution rate (the dilution rate at which washout occurs) is increased.

Biomass feedback is applied widely in effluent treatment systems where the advantages of feedback contribute significantly to the process efficiency. The outlet substrate concentration is considerably less and the feedback of biomass may improve stability in effluent treatment systems where mixed substrates of varying concentration are used. The system will also result in increased productivity of microbial products as illustrated by Major and Bull (1989) who reported very high lactic acid productivities in laboratory biomass recycle fermentations. Anaerobic processes are particularly suited to feedback continuous culture because the elevated biomass is not susceptible to oxygen limitation.

Feedback systems seem particularly attractive for animal cell culture where low growth rates and low cell densities limit productivity. A number of internal feedback systems have been developed based on immobilized cells on either hollow fibres or microcarriers (see also Chapter 7) and it is claimed that with the rapid developments in centrifuge design, centrifugal separation and feedback in suspension cultures may be scaled-up (Griffiths, 1992). The potential for continuous animal-cell processes is considered in the next section of this Chapter.

### Comparison of batch and continuous culture in industrial processes

BIOMASS PRODUCTIVITY

The productivity of a culture system may be de-

scribed as the output of biomass per unit time of the fermentation. Thus, the productivity of a batch culture may be represented as:

$$R_{batch} = (x_{max} - x_0)/(t_i - t_{ii}) \qquad (2.26)$$

where $R$ is the output of the culture (g biomass $dm^{-3}$ $h^{-1}$),

$x_{max}$ is the maximum cell concentration achieved at stationary phase,

$x_0$ is the initial cell concentration at inoculation,

$t_i$ is the time during which the organism grows at $\mu_{max}$,

and $t_{ii}$ is the time during which the organism is not growing at $\mu_{max}$ and includes the lag phase, the deceleration phase and the periods of batching, sterilizing and harvesting.

The productivity of a continuous culture may be represented as:

$$R_{cont} = D\bar{x}(1 - t_{iii}/T) \qquad (2.27)$$

where $R_{cont}$ is the output of the culture (g biomass $dm^{-3}$ $h^{-1}$),

$t_{iii}$ is the time period prior to the establishment of a steady state and includes vessel preparation, sterilization and operation in batch culture prior to continuous operation,

and $T$ is the time period during which steady-state conditions prevail.

The term $D\bar{x}$ increases with increasing dilution rate until it reaches a maximum value, after which any further increase in $D$ results in a decrease in $D\bar{x}$, as illustrated in Fig. 2.9. Thus, maximum productivity of biomass may be achieved by the use of the dilution rate giving the highest value of $D\bar{x}$.

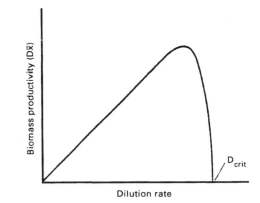

FIG. 2.9. The effect of dilution rate on biomass productivity in steady-state continuous culture.

The output of a batch fermentation described by equation (2.26) is an average over the period of the fermentation and, because the rate of biomass production is dependent on initial biomass ($dx/dt = \mu x$), the vast proportion of the biomass is produced towards the end of the fermentation. Thus, productivity in batch culture is at its maximum only towards the end of the process. For a continuous culture operating at the optimum dilution rate, under steady-state conditions, the productivity will be constant and always maximum. Thus, the productivity of the continuous system must be greater than the batch. A continuous system may be operated for a very long time period (several weeks or months) so that the negative contribution of the unproductive time, $t_{iii}$, to productivity would be minimal. However, a batch culture may be operated for only a limited time period so that the negative contribution of the time, $t_{ii}$, would be very significant, especially when it is remembered that the batch culture would have to be re-established many times during the time-course of a continuous run. Thus, the superior productivity of biomass by a continuous culture, compared with a batch culture, is due to the maintenance of maximum output conditions throughout the fermentation and the insignificance of the non-productive period associated with a long-running continuous process.

The steady state achievable in a continuous process also adds to the advantage of improved biomass productivity, as discussed by Hospodka (1966). Cell concentration, substrate concentration, product concentration and toxin concentration should remain constant throughout the fermentation. Thus, once the culture is established the demands of the fermentation, in terms of process control, should be constant. In a batch fermentation, the demands of the culture vary during the fermentation — at the beginning, the oxygen demand is low but towards the end the demand is high, due to the high biomass and the increased viscosity of the broth. Also, the amount of cooling required will increase during the process, as will the degree of pH control. In a continuous process oxygen demand, cooling requirements and pH control should remain constant. Thus, the use of continuous culture should allow for the easier introduction of process automation.

A batch process requires periods of intensive labour during medium preparation, sterilization, batching and harvesting but relatively little during the fermentation itself. However, a continuous process results in a more constant labour demand in that medium is supplied continuously sterilized (see Chapter 5), the product is continuously extracted and the relative time spent on equipment preparation and sterilization is very small.

The argument against continuous biomass processes is that the duration of a continuous fermentation is very much longer than a batch one so that there is a greater probability of a contaminating organism entering the continuous process and a greater probability of equipment failure. However, problems of contamination and equipment reliability are related to equipment design, construction and operation and, provided sufficiently rigorous standards are applied, these problems can be overcome. In fact, the fermentation industry has recognized the superiority of continuous culture for the production of biomass and several large-scale processes have been established. This aspect is considered in more detail in a later section of this Chapter.

METABOLITE PRODUCTIVITY

Theoretically, a fermentation to produce a metabolite should also be more productive in continuous culture than in batch because a continuous culture may be operated at the dilution rate which maintains product output at its maximum, whereas in batch culture product formation may be a transient phenomenon during the fermentation. The kinetics of product formation in continuous culture have been reviewed by Pirt (1975) and Trilli (1990). Product formation in a chemostat may be described as:

Change in product concentration = production − output:

or:

$$dp/dt = q_p x - Dp \qquad (2.28)$$

where $p$ is the concentration of product
and $q_p$ is the specific rate of product formation (mg product g$^{-1}$ biomass h$^{-1}$).
At steady state, $dp/dt = 0$, and thus:

$$\bar{p} = q_p \cdot x/D \qquad (2.29)$$

where $\bar{p}$ is the steady-state product concentration.
If $q_p$ is strictly related to $\mu$, then as $D$ increases so will $q_p$; thus, the steady-state product concentration ($\bar{p}$) and product output ($Dp$) will behave in the same way as $x$ and $Dx$, as shown in Fig 2.10a. If $q_p$ is independent of $\mu$ then it will be unaffected by $D$ and thus concentration will decline with increasing $D$ but output will remain constant, as shown in Fig. 2.10b. If product formation occurs only within a certain range of growth rates (dilution rates) then a more complex relationship is produced.

Thus, from this consideration a chemostat process for the production of a product can be designed to optimize either output (g dm$^{-3}$ h$^{-1}$) or product concentration. However, as Heijnen et al. (1992) explained,

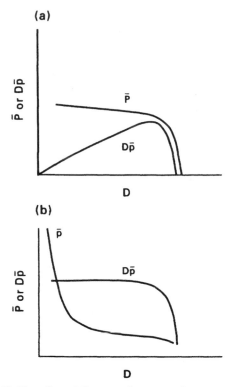

FIG. 2.10. The effect of $D$ on steady-state product concentration ($\bar{p}$) and product output ($D_{\bar{p}}$) when:
(a) $q_p$ is growth related.
(b) $q_p$ is independent of growth rate.

when $q_p$ is growth-related the advantage of high productivity obtained at high dilution rates must be balanced against the disadvantage of low product concentration resulting in increased downstream processing costs. The other arguments presented for the superiority of continuous culture for biomass production also hold true for product synthesis — ease of automation and the advantages of steady state conditions. The question that then arises is 'Why has the fermentation industry not adopted continuous culture for the manufacture of microbial products?' It can be appreciated that the arguments cited previously against continuous culture (contamination and equipment reliability) are not valid as these difficulties have been overcome in the large-scale continuous biomass processes. The answer to the question lies in the highly selective nature of continuous culture. We have already seen that $\mu$ is determined by $D$ in a steady state chemostat and that $\mu$ and $D$ are related to substrate concentration according to the equation:

$$D = \mu = \mu_{max}\bar{s}/(K_s + \bar{s})$$

The effect of substrate concentration on specific growth rate for two organisms, A and B, is shown in Fig. 2.11. A is capable of growing at a higher specific growth rate at any substrate concentration. The self-balancing properties of the chemostat mean that the organism reduces the substrate concentration to the value where $\mu = D$. Thus, at dilution rate $X$, organism A would reduce the substrate concentration to $Z$. However, at this substrate concentration, organism B could grow only at a $\mu$ of $Y$. Therefore, if organisms A and B were introduced into a chemostat operating at dilution rate $X$, A would reduce the substrate concentration to $Z$ at which B could not maintain a $\mu$ of $X$ and would be washed out at a rate of $(X - Y)$ and a monoculture of A would be established eventually. The same situation would occur if A and B were mutant strains arising from the same organism. Commercial organisms have been selected and mutated to produce metabolites at very high concentrations (see Chapter 3) and, as a result, tend to grow inefficiently with low $\mu_{max}$ values and, possibly, high $K_s$ values. Back mutants of production strains produce much lower concentrations of product and, thus, grow more efficiently. If such back mutants arise in a chemostat industrial process, then the production strain will be displaced from the fermentation as described in the foregoing scenario.

Calcott (1981) described this phenomenon as "contamination from within" and this type of 'contamination' cannot be solved by the design of more 'secure' fermenters. Thus, it is the problem of strain degeneration which has limited the application of large scale continuous culture to biomass and, to a lesser extent, potable and industrial alcohol. The production of alcohol by continuous culture is feasible because it is a byproduct of energy generation and, thus, is not a drain on the resources of the organism. However, it is possi-

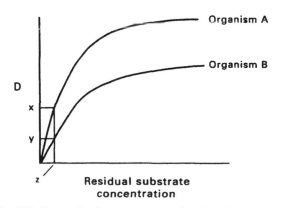

FIG. 2.11. Competition between two organisms in a chemostat.

ble that the technique could be exploited for other products provided strain degeneration is controlled; this may be possible in certain genetically engineered strains. It has been reported that the development of a continuous process for the manufacture of polyhydroxybutyrate, a biopolymer. Other processes have been developed using chemostat culture.

The adoption of continuous culture for animal cell products is even more complex than for microbial systems. Griffiths (1992) compared the following process options for producing an animal cell product:

(i) Batch culture.
(ii) Semi-continuous culture where a portion of the culture is harvested at regular intervals and replaced by an equal volume of medium.
(iii) Fed-batch culture where medium is fed to the culture resulting in an increase in volume (see later section)
(iv) Continuous perfusion where an immobilized cell population is perfused with fresh medium and is equivalent to an internal feedback continuous system.
(v) Continuous culture.

The characteristics of all five modes of operation are shown in Table 2.3, from which it may be seen that the perfusion continuous system appears extremely attractive. However, the practicalities of running a large scale continuous perfusion system present considerable difficulties. The process has to be reliable and able to operate aseptically for the long periods necessary to exploit the advantage of a continuous process. Also, the licensing of a continuous process may present some difficulties where a consignment of product must be traceable to a batch of raw materials. In a long-term continuous process several different batches of media would have to be used which presents the problem of associating product with raw material. Furthermore, it is difficult to monitor the genetic stability of cells which are immobilized in a large reactor system. Thus, large scale (kg quantities) animal cell products are still produced by batch methods. However, where very high value products are required and production can be satisfied on a small scale, the continuous perfusion system is a very attractive proposition (Griffiths, 1992).

Continuous brewing and biomass production which are the major industrial applications of continuous microbial culture will now be considered in more detail.

## CONTINUOUS BREWING

The brewing industry in the United Kingdom has had a relatively brief 'courtship' with continuous culture. Two types of continuous brewing have been used:

(i) The cascade or multistage system.
(ii) The tower system.

Hough *et al.* (1976) described the cascade system utilized at Watneys' Mortlake brewery in London. The process utilized three vessels, the first two for fermentation and the third for separation of the yeast biomass. The specific gravity of the wort was reduced from 1040 to 1019 in the first vessel and from 1019 to 1011 in the second vessel. The residence time for the system was 15 to 20 hours, using worts in the specific gravity range of 1035 to 1040, and it could be run continuously for 3 months. However, it is believed that the system was abandoned due to problems of excessive biomass production. This process appears to have been used widely in New Zealand with greater success (Kirsop, 1982), but apparently newer installations are of the batch type.

A typical tower fermenter for brewing is illustrated in Fig. 2.12. The system is partially closed in that

TABLE 2.3. *Comparison of the performance of different operational modes of an animal cell fermentation (After Griffiths, 1992)*

| Operational mode | Cell No. ($\times 10^{-6}$ cm$^{-3}$) | Product yield (mg day$^{-1}$) | Product yield (mg month$^{-1}$) | Length of run (days) |
|---|---|---|---|---|
| Batch | 3 | 100 | 200 | 7 |
| Semi-continuous | 3 | 200 | 600 | 21 |
| Fed-batch | 6 | 200 | 500 | 14 |
| Perfusion | 30 + | 3000 | 12000 | > 30 |
| Continuous culture | 2 | 300 | 1200 | > 100 |

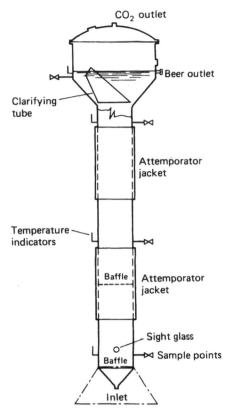

FIG. 2.12. A schematic representation of a tower fermenter for the brewing of beer (Royston, 1966).

relatively little yeast leaves the fermenter due to the highly flocculent nature of the strains employed. Thus, the system is a type of internal feed-back. Wort is introduced into the base of the tower and passes through a porous plug of yeast. As the wort rises through the vessel it is progressively fermented and leaves the fermenter via a yeast-separation zone, which is twice the diameter of the rest of the tower. Hough *et al.* (1976) described the protocol employed in the establishment of a yeast plug in the tower and the subsequent operating conditions. Prior to the fermentation, the tower is thoroughly cleaned and steam sterilized, aseptic operation being more important for the continuous process than the batch one. The vessel is filled partially with sterile wort and inoculated with a laboratory culture. The initial stages are designed to encourage high biomass production by the periodical addition of wort over about a 9-day period. A porous plug of yeast develops at the base of the tower. The flow rate of the wort is then gradually increased over a further 9 to 12 days, by which time an approximate steady state

may be achieved. Following the establishment of a high biomass in the fermenter the system is operated such that the wort is converted to beer with the formation of approximately the same amount of yeast as would be produced in the batch process. The beer produced during the 3-week start-up period is usually below specification and would have to be blended with high-quality beer. Thus, more than 3 months' continuous operation is necessary to compensate for the initial losses of the process.

The major advantage of the continuous tower process was that the wort residence time could be reduced from about 1 week to 4 to 8 hours as compared with the batch system. However, the development of the cylindro-conical vessel (initially described by Nathan in the 1930s, but not introduced until the 1970s, see also Chapter 7) led to the shortening of the batch fermentation time to approximately 48 hours. Although this is still considerably longer than the residence time in a tower fermenter, it should be remembered that beer conditioning and packaging takes considerably longer than the fermentation stage, so that the difference in the overall processing time between the tower and the cylindro-conical batch process is not sufficient to justify the disadvantages of the tower. The major disadvantages of the tower system are the long start-up time, the technical complexity of the plant, the requirement for more highly skilled personnel than for a batch plant, the inflexibility of the system in that a long time delay would ensue between changing from one beer fermentation to another and, finally, the difficulty in matching the flavour of the continuously produced product with that of the traditional batch product. Thus, the continuous-tower system has fallen from use, with virtual universal adoption of the cylindro-conical batch process.

## CONTINUOUS CULTURE AND BIOMASS PRODUCTION

Microbial biomass which is produced for human or animal consumption is referred to as single cell protein (SCP). Although yeast was produced as food on a large scale in Germany during the First World War (Laskin, 1977) the concept of utilizing microbial biomass as food was not thoroughly investigated until the 1960s. Since the 1960s, a large number of industrial companies have explored the potential of producing SCP from a wide range of carbon sources. Almost without exception, these investigations have been based on the use of continuous culture as the growth technique.

As previously discussed, continuous culture is the

ideal method for the production of microbial biomass. The superior productivity of the technique, compared with that of batch culture, may be exploited fully and the problem of strain degeneration is not as significant as in the production of microbial metabolites. The selective pressure in the chemostat would tend to work to the advantage of the industrialist producing SCP, in that the most efficient strain of the organism would be selected, although this is not necessarily the case for mycelial processes. The development of SCP processes generated considerable research into large-scale chemostat design and the behaviour of the production organism in these very large vessels. Many 'novel' fermenters have been designed for SCP processes and these are considered in more detail in Chapter 7.

A very wide range of carbon sources have been investigated for the production of SCP. Whey has been used as a carbon source for biomass production since the 1940s and such fermentations have been shown to be economic in that they provide a high-grade feed product, whilst removing an, otherwise, troublesome waste product of the cheese industry (Meyrath and Bayer, 1979). Cellulose has been investigated extensively as a potential carbon source for SCP production and this work has been reviewed by Callihan et al. (1979) and Woodward (1987). The major difficulty associated with the use of cellulose as a substrate is its recalcitrant nature.

An enormous amount of research has been conducted into the use of hydrocarbons as sources of carbon for biomass processes; the hydrocarbons investigated being methane, methanol and n-alkanes. A large number of commercial firms were involved in this research field but very few created viable, commercial processes based on SCP production from hydrocarbons because of the economic difficulties involved (Sharp, 1989). At the start of this research, hydrocarbons were relatively cheap but, following the 1973 Middle East War, oil prices escalated and transformed the economic basis of biomass production from petroleum sources. ICI were successful in developing a commercial process for production of bacterial biomass (Pruteen) from methanol at an annual rate of 54,000 to 70,000 tonnes. The process utilized a novel air-lift, pressure cycle fermenter, of 3000 m³ capacity, and was the first commercial process to produce SCP from methanol (King, 1982). The fermentation was run successfully for periods in excess of 100 days without contamination (Howells, 1982). Regrettably, the economics of the process were such that when the price of soya and fishmeal declined Pruteen could not compete as an animal feed. Selling in bulk ceased in 1985 (Sharp, 1989) and the 3000 m³ vessel was eventually demolished.

The expertise developed by ICI during the Pruteen project and RHM's research into the use of a fungus, *Fusarium graminearum*, for the production of human food formed the basis of a joint venture between the two companies. The ICI pressure cycle pilot-plant was used to produce the fungal biomass (Myco-protein, marketed as Quorn) in continuous culture. The advantage of fungal biomass is that it may be processed to give a textured protein which is acceptable for human consumption. The low shear properties of the air-lift vessel conserve the desirable morphology of the fungus. The process is operated at a dilution rate of between 0.17 and 0.20 h$^{-1}$ ($\mu_{max}$ is 0.28 h$^{-1}$). The phenomenon of mutation and intense selection in the chemostat has proved to be problematical in Myco-protein fermentation, because highly branched mutants have arisen in the vessel resulting in the loss of the desirable morphology. However, the process may still be operated in chemostat culture for 1000 hours on the full scale (Trinci, 1992).

### Comparison of batch and continuous culture as investigative tools

Although the use of continuous culture on an industrial scale is very limited it is an invaluable investigative technique. The principle characteristic of batch culture is change. Even during the log phase cultural conditions are not constant and it is only the constant maximum specific growth rate which gives the semblance of stability — biomass concentration, substrate concentration and microbial products all change exponentially. During the deceleration phase the onset of nutrient limitation causes the growth rate to decline from its maximum to zero in a very short time, so it is virtually impossible to study the physiological effects of nutrient limitation in batch culture. As Trilli (1990) pointed out, adaptation of an organism to change is not instantaneous, so that the activity of a batch culture is not in equilibrium with the composition of its environment. Physiological events in a batch culture may have been initiated by a change in the environment which took place some significant time before the change was observed. Thus, it is very difficult to relate 'cause and effect'. The major feature of continuous culture, on the other hand, is 'the steady state' — biomass, substrate and product concentration should remain constant over very long periods of time. Specific growth rate is controlled by dilution rate and growth is nutrient limited. However, it is important not to exaggerate the significance of the steady state because a constant biomass level does not necessarily indicate that the culture is physiologically stable (Malek *et al.*, 1988). It is possible to separate the effects of growth rate and other environmental conditions, for example temperature, pH and dissolved oxygen concentration. Furthermore, because any of a wide range of substrates may be used to limit growth in the chemostat the effects of $\mu$ and substrate concentration may be distinguished.

Continuous cultures may generate valuable physiological information on an industrial strain which may be used in the optimization of the commercial process. An excellent example is the effect of growth rate and limiting substrate on metabolite formation. The interpretation of secondary metabolites as compounds produced in the idiophase of batch culture may lead one to suppose that the specific production rates ($q_p$) of such compounds are inversely linked to specific growth rate ($\mu$). The testing of this supposition may be achieved in chemostat culture and it has been shown to be correct for cephamycin and thienamycin synthesis by *Streptomyces cattleya* (Lilley *et al.*, 1981) and gibberellin by *Gibberella fujikuroi* (Bu'Lock *et al.*, 1974). However, different relationships have been demonstrated in other systems. Pirt and Righelato (1967) and Ryu and Hospodka (1980) showed that the $q_p$ of penicillin is positively correlated with $\mu$ up to a specific growth rate of 0.013 h$^{-1}$, after which it is independent of $\mu$. Pirt (1990) suggests that the apparent negative correlation may be related to penicillin degradation. These observations suggest that the growth rate in a commercial penicillin process should not decline below 0.013 h$^{-1}$. Positive correlations between $\mu$ and $q_p$ have been obtained for chlortetracycline production by *Streptomyces aureofaciens* (Sikyta *et al.*, 1961), oxytetracycline by *Streptomyces rimosus* (Rhodes, 1984) and erythromycin A by *Streptomyces erythraeus* (Trilli *et al.*, 1987).

The fiercely selective nature of the chemostat, which is its major disadvantage for industrial production, makes it an excellent tool for the isolation and improvement of micro-organisms. The use of continuous culture in this context is considered in Chapter 3, from which it may be seen that continuous enrichment culture offers considerable advantages over batch enrichment techniques and that continuous culture may be used very successfully to select strains producing higher yields of certain microbial enzymes.

## FED-BATCH CULTURE

Yoshida *et al.* (1973) introduced the term fed-batch culture to describe batch cultures which are fed continuously, or sequentially, with medium, without the removal of culture fluid. A fed-batch culture is established initially in batch mode and is then fed according to one of the following feed strategies:

(i) The same medium used to establish the batch culture is added, resulting in an increase in volume.

(ii) A solution of the limiting substrate at the same concentration as that in the initial medium is added, resulting in an increase in volume.

(iii) A concentrated solution of the limiting substrate is added at a rate less than in (i) and (ii), resulting in an increase in volume.

(iv) A very concentrated solution of the limiting substrate is added at a rate less than in (i), (ii) and (iii), resulting in an insignificant increase in volume.

Fed-batch systems employing strategies (i) and (ii) are described as variable volume, whereas a system employing strategy (iv) is described as fixed volume. The use of strategy (iii) gives a culture intermediate between the two extremes of variable and fixed volume.

The kinetics of the two basic types of fed-batch culture, variable volume and fixed volume, will now be described.

### Variable volume fed-batch culture

The kinetics of variable volume fed-batch culture have been developed by Dunn and Mor (1975) and Pirt (1974, 1975, 1979). The following account is based on that of Pirt (1975). Consider a batch culture in which growth is limited by the concentration of one substrate; the biomass at any point in time will be described by the equation:

$$x_t = x_0 + Y(S_R - s) \qquad (2.30)$$

where $x_t$ is the biomass concentration after time, $t$ hours,

and $x_0$ is the inoculum concentration.

The final biomass concentration produced when $s = 0$ may be described as $x_{max}$ and, provided that $x_0$ is small compared with $x_{max}$:

$$x_{max} \simeq Y \cdot S_R \qquad (2.31)$$

If, at the time when $x = x_{max}$, a medium feed is started such that the dilution rate is less than $\mu_{max}$, virtually all the substrate will be consumed as fast as it enters the culture, thus:

$$FS_R \simeq \mu(X/Y) \qquad (2.32)$$

where $F$ is the flow rate of the medium feed,
and $\quad X$ is the total biomass in the culture, described by $X = xV$, where $V$ is the volume of the culture medium in the vessel at time $t$.

From equation (2.32) it may be concluded that input of substrate is equalled by consumption of substrate by the cells. Thus, $(ds/dt) \simeq 0$. Although the total biomass in the culture $(X)$ increases with time, cell concentration $(x)$ remains virtually constant, that is $(dx/dt) \simeq 0$ and therefore $\mu \simeq D$. This situation is termed a quasi steady state. As time progresses the dilution rate will decrease as the volume increases and $D$ will be given the expression:

$$D = F/(V_0 + Ft) \qquad (2.33)$$

where $V_0$ is the original volume. Thus, according to Monod kinetics, residual substrate should decrease as $D$ decreases resulting in an increase in the cell concentration. However, over most of the range of $\mu$ which will operate in fed-batch culture, $S_R$ will be much larger than $K_s$ so that, for all practical purposes, the change in residual substrate concentration would be extremely small and may be considered as zero. Thus, provided that $D$ is less than $\mu_{max}$ and $K_s$ is much smaller than $S_R$, a quasi steady state may be achieved. The quasi steady state is illustrated in Fig. 2.13a. The major difference between the steady state of a chemostat and the quasi steady state of a fed-batch culture is that $\mu$ is constant in the chemostat but decreases in the fed-batch.

Pirt (1979) has expressed the change in product concentration in variable volume fed-batch culture in the same way as for continuous culture (see equation 2.28):

$$dp/dt = q_p x - Dp.$$

Thus, product concentration changes according to the balance between production rate and dilution by the feed. However, in the genuine steady state of a chemostat, dilution rate and growth rate are constant whereas in a fed-batch quasi steady state they change over the time of the fermentation. Product concentration in the chemostat will reach a steady state, but in a fed-batch system the profile of the product concentration over the time of the fermentation will be dependent on the relationship between $q_p$ and $\mu$ (hence $D$). If $q_p$ is

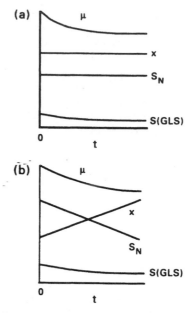

FIG. 2.13. Time profiles of fed-batch cultures.
$\mu$ = specific growth rate
$x$ = biomass concentration
$S(GLS)$ = growth limiting substrate
$S_N$ = any other substrate than $S(GLS)$
(a) Variable volume fed-batch culture.
(b) Fixed volume fed-batch culture.
(Pirt, 1979).

strictly growth related then it will change as $\mu$ changes with $D$ and, thus, the product concentration will remain constant. However, if $q_p$ is constant and independent of $\mu$, then product concentration will decrease at the start of the cycle when $Dp$ is greater than $q_p x$ but will rise with time as $D$ decreases and $q_p x$ becomes greater than $Dp$. These relationships are shown in Fig. 2.14a. If $q_p$ is related to $\mu$ in a complex manner, then the product concentration will vary according to that relationship. Thus, the feed strategy of a fed-batch system would be optimized according to the relationship between $q_p$ and $\mu$.

### Fixed volume fed-batch culture

Pirt (1979) described the kinetics of fixed volume fed-batch culture as follows. Consider a batch culture in which the growth of the process organism has depleted the limiting substrate to a limiting level. If the limiting substrate is then added in a concentrated feed such that the broth volume remains almost constant, then:

$$dx/dt = GY \qquad (2.34)$$

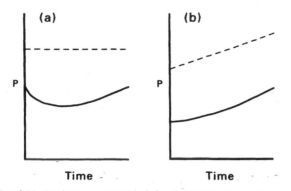

FIG. 2.14. Product concentration ($p$) in fed-batch culture when $q_p$ is growth related ($-----$) or non-growth related, i.e. $q_p$ constant ($———$)
(a) Variable volume fed-batch culture.
(b) Fixed volume fed-batch culture.
(Modified from Pirt, 1979.)

where $G$ is the substrate feed rate (g dm$^{-3}$ h$^{-1}$)
and $Y$ is the yield factor.
But $dx/dt = \mu x$, thus substituting for $dx/dt$ in equation (2.34) gives:

$$\mu x = GY, \text{ and thus:}$$

$$\mu = GY/x \qquad (2.35)$$

Provided that $GY/x$ does not exceed $\mu_{max}$ then the limiting substrate will be consumed as soon as it enters the fermenter and $ds/dt \simeq 0$. However, $dx/dt$ may not be equated to zero, as in the case of variable volume fed-batch, because the biomass concentration, as well as the total amount of biomass in the fermenter, will increase with time. Biomass concentration is given by the equation:

$$x_t = x_{\dot{a}} + GYt \qquad (2.36)$$

where $x_t$ is the biomass after operating in fed batch for $t$ hours
and $x_{\dot{a}}$ is the biomass concentration at the onset of fed-batch culture.

As biomass increases then the specific growth rate will decline according to equation (2.35). The behaviour of a fixed volume fed-batch culture is illustrated in Fig. 2.13b from which it may be seen that $\mu$ declines (according to equation (2.35)), the limiting substrate concentration remains virtually constant, biomass increases and the concentration of the non-limiting nutrients declines.

Pirt (1979) described the product balance in a fixed volume fed-batch system as:

$$dp/dt = q_p x,$$

but substituting for $x$ from equation (2.36) gives:

$$dp/dt = q_p(x_{\dot{a}} + GYt).$$

If $q_p$ is strictly growth-rate related then product concentration will rise linearly as for biomass. However, if $q_p$ is constant then the rate of increase in product concentration will rise as growth rate declines, i.e. as time progresses and $x$ increases. These relationships are shown in Fig. 2.14b. If $q_p$ is related to $\mu$ in a complex manner then the product concentration will vary according to that relationship. As in the case of variable volume fed-batch the feed profile would be optimized according to the relationship between $q_p$ and $\mu$.

### Cyclic fed-batch culture

The life of a variable volume fed-batch fermentation may be extended beyond the time it takes to fill the fermenter by withdrawing a portion of the culture and using the residual culture as the starting point for a further fed-batch process. The decrease in volume results in a significant increase in the dilution rate (assuming that the flow rate remains constant) and thus, eventually, in an increase in the specific growth rate. The increase in $\mu$ is then followed by its gradual decrease as the quasi steady state is re-established. Such a cycle may be repeated several times resulting in a series of fed-batch fermentations. Thus, the organism would experience a periodic shift-up in growth rate followed by a gradual shift-down. This periodicity in growth rate may be achieved in fixed volume fed-batch systems by diluting the culture when the biomass reaches a concentration which cannot be maintained under aerobic conditions. Dilution results in a decline in $x$ and, thus, according to equation (2.35) an increase in $\mu$. Subsequently, as feeding continues, the growth rate will decline gradually as biomass increases and approaches the maximum sustainable in the vessel once more, at which point the culture may be diluted again. Dilution would be achieved by withdrawing culture and refilling to the original level with sterile water or medium not containing the feed substrate.

### Application of fed-batch culture

The use of fed-batch culture by the fermentation industry takes advantage of the fact that the concentration of the limiting substrate may be maintained at a very low level, thus avoiding the repressive effects of high substrate concentration. Furthermore, the fed-batch system also gives some control over the organisms' growth rate, which is also related to the specific oxygen uptake rate giving some control over the oxygen demand of the fermentation (see Chapter 9). Both variable and fixed volume systems result in low limiting substrate concentrations, but the quasi steady-state of the variable volume system has the advantage of maintaining the concentrations of both the biomass and the non-limiting nutrients constant. Pirt (1979) cites this as an important feature because the concentrations of substrates other than those which limit growth can have a significant effect on biomass composition and product formation.

The obvious advantage of cyclic fed-batch culture is that the productive phase of a process may be extended under controlled conditions. However, a further advantage lies in the controlled periodic shifts in growth rate which may provide an opportunity to optimize product synthesis. Dunn and Mor (1975) pointed out that changes in the rates of chemical processes can give rise to increases in intermediate concentrations and similar effects may be possible in microbial systems. This observation is particularly relevant to secondary metabolite production which is maximal in batch culture during the deceleration phase. Bushell (1989) suggested that optimum conditions for secondary metabolite synthesis may occur during the transition phase after the withdrawal of a volume of broth from the vessel and before the re-establishment of the steady-state following the resumption of the nutrient feed. During this period the dilution rate will be greater than the growth rate but, according to Bushell, the rate of uptake of the growth-limiting substrate should respond immediately to the increased substrate concentration. Thus, an imbalance results between the substrate uptake rate and the specific growth rate — this imbalance then contributing to a diversion of intermediates into secondary metabolism.

The advantages of cyclic fed-batch culture must be weighed against difficulties inherent in the system. Care has to be taken in the design of cyclic fed-batch processes to ensure that toxins do not accumulate to inhibitory levels and that nutrients other than those incorporated into the feed medium become limiting (Queener and Swartz, 1979). Also, a prolonged series of fed-batch cycles may result in the accumulation of non-producing or low-producing variants.

The early fed-batch systems that were developed did not incorporate any form of feedback control and relied on the inherent quasi steady state to maintain process stability. However, the use of concentrated feeds resulting in very high biomass concentrations and the development of more sophisticated feeding programmes has necessitated the introduction of feedback control techniques. In such feedback controlled fermentations a process parameter directly related to the organism's physiological state is monitored continuously by an on-line sensor. The signal generated by the sensor is then used in a control loop (see Chapter 8) to control the medium feed rate. Parameters which have been utilized in this way include dissolved oxygen concentration, pH, effluent gas composition and limiting substrate concentration. Examples of these control systems are included in the next section and in Chapter 9.

### Examples of the use of fed-batch culture

Fed-batch culture was used in the production of bakers' yeast as early as 1915. It was recognized that an excess of malt in the medium would lead to too high a growth rate resulting in an oxygen demand in excess of that which could be met by the equipment. This resulted in the development of anaerobic conditions and the formation of ethanol at the expense of biomass production (Reed and Nagodawithana, 1991). Thus, the organism was grown in an initially weak medium to which additional medium was added at a rate less than the maximum rate at which the organism could use it. Thus, this process fulfils the criteria stipulated in Pirt's (1975) kinetic description for the establishment of a quasi steady state, that is, a substrate limited culture and the use of a feed rate equivalent to a dilution rate less than $\mu_{max}$. It is now recognized that bakers' yeast is very sensitive to free glucose and respiratory activity may be repressed at a concentration of about 5 $\mu$g dm$^{-3}$ (Crabtree, 1929). Thus, a high glucose concentration represses respiratory activity as well as giving rise to a high growth rate, the oxygen demand of which cannot be met. In modern fed-batch processes for yeast production the feed of molasses is under strict control based on the automatic measurement of traces of ethanol in the exhaust gas of the fermenter. Although such systems may result in low growth rates, the biomass yield is near the theoretical maximum obtainable (Fiechter, 1982).

It is interesting to note that the production of recombinant proteins from yeast may be achieved using fed-batch culture techniques very similar to that developed for the bakers' yeast fermentation. Gu et al. (1991) reported the production of hepatitis B surface antigen (HBsAg) in a 0.9 dm$^3$ fed-batch reactor where the feed rate was increased exponentially to maintain a relatively high growth rate. HBsAg production was growth associated in this strain and good productivity was achieved by maintaining a high growth rate whilst keeping the glucose concentration below that which would repress respiratory activity. Ibba *et al.* (1993) reported a cyclic fed-batch process for recombinant hirudin from *S. cerevisiae*, under the control of a constitutive promoter. The cyclic fed-batch process gave three times the hirudin activity of a continuous fermentation, the superior productivity being due to increased transcription although a genetic explanation of the results could not be offered.

Penicillin fermentation provides an excellent example of the use of feed systems in the production of a secondary metabolite. The fermentation may be divided into two phases — the 'rapid-growth' phase, during which the culture grows at $\mu_{max}$, and the 'slow-growth' or 'production' phase. Glucose feeds may be used to control the metabolism of the organism during both phases. During the rapid-growth phase an excess of glucose causes an accumulation of acid and a biomass oxygen demand greater than the aeration capacity of the fermenter, whereas glucose starvation may result in the organic nitrogen in the medium being used as a carbon source, resulting in a high pH and inadequate biomass formation (Queener and Swartz, 1979). The accumulation of hexose may be prevented by the use of a slowly hydrolysed carbohydrate such as lactose in simple batch culture (Matelova, 1976). However, according to Queener and Swartz, considerable increases in productivity have been achieved by the use of computer controlled feeding of glucose during the rapid growth phase, such that the dissolved oxygen or pH is maintained within certain limits. Both control parameters essentially measure the same activity in that both oxygen concentration and pH will fall when glucose is in excess, due to an increased respiration rate and the accumulation of organic acids when the respiration rate exceeds the aeration capacity of the fermenter. Both systems appear to work well in controlling feed rates during the rapid-growth phase.

During the production phase of the penicillin fermentation the feed rates utilized should limit the growth rate and oxygen consumption such that a high rate of penicillin synthesis is achieved, and sufficient dissolved oxygen is available in the medium. The control factor in this phase is normally dissolved oxygen because pH is less responsive to the effect of dissolved oxygen on penicillin synthesis than on growth. As the fed-batch process proceeds then the total biomass, viscosity and oxygen demand increase until, eventually, the fermentation is oxygen limited. However, limitation may be delayed by reducing the feed rate as the fermentation progresses and this may be achieved by the use of computer controlled systems.

Suzuki *et al.* (1987) developed a pH feedback fed-batch system for the production of thiostrepton from *Streptomyces laurentii*. When glucose was exhausted in the fermentation the pH rose immediately and this event was used as the signal for the addition of more feed which consisted of a concentrated glucose, corn steep liquor, soy bean meal and mineral mixture. This process maintained a biomass level of 157 g dm$^{-3}$ and a thiostrepton concentration of 10.5 g dm$^{-3}$ with a productivity nine times that of a conventional batch culture.

Many enzymes are subject to catabolite repression, where enzyme synthesis is prevented by the presence of rapidly utilized carbon sources (Aunstrup *et al.*, 1979). It is obvious that this phenomenon must be avoided in enzyme fermentations and fed-batch culture is the major technique used to achieve this. Concentrated medium is fed to the culture such that the carbon source does not reach the threshold for catabolite repression. For example, Waki *et al.* (1982) controlled the production of cellulase by *Trichoderma reesei* in fed-batch culture utilizing $CO_2$ production as the control factor and Suzuki *et al.* (1988) achieved high lipase production from *Pseudomonas fluorescens* also using $CO_2$ production to control the addition of an oil feed.

Shioya (1990) developed a method for the optimization of fed-batch systems based on the relationship between $\mu$ and $q_p$, the product specific production rate. Once the relationship between the two parameters was established a computer control system was used to maintain the fed-batch at the optimum specific growth rate (feed rate). The system was tested for a number of fermentations including histidine (*Brevibacterium flavum*), acid phosphatase and glutathione (*S. cerevisiae*), and lysine (*Corynebacterium glutamicum*). Specific growth rate was maintained constant using a feed-forward control profile which was updated throughout the fermentation as data were collected. Very promising results were obtained but difficulties were experienced in generating sufficiently accurate data to up-date the control system and maintain the specific growth rate constant.

## REFERENCES

AUNSTRUP, K., ANDRESEN, O., FALCH, E. A. and NIELSEN, T. K. (1979) Production of microbial enzymes. In *Microbial Technology*, Vol. 1 (2nd edition), pp. 282–309 (Editors

Peppler, H. J. and Perlman, D.). Academic Press, New York.

BORROW, A., JEFFERYS, E. G., KESSEL, R. H. J., LLOYD, E. C., LLOYD, P. B. and NIXON, I. S. (1961) Metabolism of *Gibberella fujikuroi* in stirred culture. *Can. J. Micro.* **7**, 227–276.

BULL, A. T. (1974) Microbial growth. In *Companion to Biochemistry, Selected Topics for Further Study*, pp. 415–442 (Editors Bull, A. T., Lagnado, J. R., Thomas, J. O. and Tipton, K. F.). Longman, London.

BU'LOCK, J. D., HAMILTON, D., HULME, M. A., POWELL, A. J., SHEPHERD, D., SMALLEY, H. M. and SMITH, G. N. (1965) Metabolic development and secondary biosynthesis in *Penicillium urticae. Can. J. Micro.* **11**, 765–778.

BU'LOCK, J. D., DETROY, R. W., HOSTALEK, Z. and MUNIM-AL-SHAKARCHI, A. (1974) Regulation of biosynthesis in *Gibberella fujikuroi. Trans. Br. Mycol. Soc.* **62**, 377–389.

BUSHELL, M. E. (1989) The process physiology of secondary metabolite production. In *Microbial Products: New Approaches. Soc. Gen. Microbiol. Symp.* 44, pp. 95–120 (Editors Baumberg, S., Hunter, I. S. and Rhodes, P. M.). Cambridge University Press, Cambridge.

CALCOTT, P. H. (1981) The construction and operation of continuous cultures. In *Continuous Culture of Cells,* Vol. 1, pp.13–26 (Editor Calcott, P. H.). CRC Press, Boca Raton.

CALLIHAN, C. D. and CLEMMER, J. E. (1979) Biomass from cellulosic materials. In *Economic Microbiology*, Vol. 4, *Microbial Biomass*, pp. 208–270 (Editor Rose, A. H.) . Academic Press, London.

CRABTREE, H. G. (1929) Observations on the carbohydrate metabolism of tumours. *Biochem. J.* **23**, 536–545.

DOSTALEK, M., HAGGSTROM, C. and MOLIN, N. (1972) Optimisation of biomass production from methanol. *Fermentation Technology Today*, Proc. IV Int. Fermentation Symp., pp. 497–511 (Editor Terui, G.).

DUNN, I. J. and MOR, J.-R. (1975) Variable-volume continuous cultivation. *Biotechnol. Bioeng.* **17**, 1805–1822.

EAGON, R. G. (1961) *Pseudomonas natriegens*, a marine bacterium with a generation time of less than ten minutes. *J. Bacteriol.* **83**, 736–737.

FIECHTER, A. (1982) Regulatory aspects in yeast metabolism. In *Advances in Biotechnology*, 1. *Scientific and Engineering Principles*, pp. 261–267 (Editors Moo-Young, M., Robinson, C. W. and Vezina, C.). Pergamon Press, Toronto.

GU, M.B., PARK, M. H. and KIM, D.-I. (1991) Growth rate control in fed-batch cultures of recombinant *Saccharomyces cerevisiae* producing hepatitis B surface antigen (HBsAg). *Appl. Microbiol. Biotechnol.* **35**, 46–50.

GRIFFITHS, J. B. (1986) Scaling-up of animal cell cultures. In *Animal Cell Culture, a Practical Approach*, pp. 33–70 (Editor Freshney, R. I.). IRL Press, Oxford.

GRIFFITHS, J. B. (1992) Animal cell processes — batch or continuous? *J. Biotechnol.* **22**, 21–30.

HARRISON, D. E. F. (1973) Studies on the affinity of methanol and methane utilising bacteria for their carbon substrates. *J. Appl. Bacteriol.* **36**, 309–314.

HARTE, M. J. and WEBB, F. C. (1967) Utilisation of mixed sugars in continuous fermentations. *Biotech. Bioeng.* **9**, 205–221.

HEIJNEN, J. J., TERWISSCHA VAN SCHELTINGA, A. H. and STRAATHOF, A. J. (1992) Fundamental bottlenecks in the application of continuous bioprocesses. *J. Biotechnol.* **22**, 3–20.

HOSPODKA, J. (1966) Industrial applications of continuous fermentation. In *Theoretical and Methodological Basis of Continuous Culture of Microorganisms*, pp. 493–645 (Editors Malek, I and Fencl, Z.). Academic Press, New York.

HOUGH, J. S., KEEVIL, C. W., MARIC, V., PHILLISKIRK, G. and YOUNG, T. W. (1976) Continuous culture in brewing. In *Continuous Culture*, 6, *Applications and New Fields*, pp. 226–238. (Editors Dean, A. C. R., Ellwood, D. C., Evans, C. G. T. and Melling, J.). Ellis Horwood, Chichester.

HOWELLS, E. R. (1982) Single-cell protein and related technology. *Chem. Indust.*, **7**, 508–511.

IBBA, M., BONARIUS, D., KUHLA, J., SMITH, A. and KUENZI, M. (1993) Mode of cultivation is critical for the optimal expression of recombinant hirudin by *Saccharomyces cerevisiae. Biotech. Letts.* **15** (7), 667–672.

KIRSOP, B. H. (1982) Developments in beer fermentation. *Topics in Enzyme and Fermentation Biotechnology*, **6**, 79–131 (Editor Wiseman, A.). Ellis Horwood, Chichester.

KING, P. P. (1982) Biotechnology: an industrial view. *J. Chem. Tech. Biotechnol.* **32**, 2–8.

LASKIN, A. I. (1977) Single cell protein. *Ann. Reports on Fermentation Processes*, **1**, 151–180 (Editor Perlman, D.). Academic Press, New York.

LAVERY, M. (1990) Animal cell fermentation. In *Fermentation: A Practical Approach*, pp. 205–220. (Editors McNeil, B. and Harvey, L. M.). IRL Press, Oxford.

LILLEY, G., CLARK, A. E. and LAWRENCE, G. C. (1981) Control of the production of cephamycin C and thienamycin by *Streptomyces cattleya* NRRL 8057. *J. Chem. Tech. Biotechnol.* **31**, 127–134.

MAJOR, N. C. and BULL, A. T. (1989) The physiology of lactate production by *Lactobacillus delbreukii* in a chemostat with cell recycle. *Biotech. Bioeng.* **34**, 592–599.

MALEK, I., VOTRUBA, J. 'and RICICA, J. (1988) The continuality principle and the role of continuous cultivation of microorganisms in basic research. In *Continuous Culture*, pp. 95–104. (Editors Kyslik, P., Dawes, E. A., Krumphanzl, V. and Novak, M.). Academic Press, London.

MATELOVA, V. (1976) Utilization of carbon sources during penicillin biosynthesis. *Folia Microbiologica*, **21**, 208–209.

MEYRATH, J. and BAYER, K. (1979) Biomass from whey. In *Economic Microbiology, 4. Microbial Biomass*, pp. 208–270. (Editor Rose, A. H.). Academic Press, New York.

MONOD, J. (1942) *Recherches sur les Croissances des Cultures Bacteriennes* (2nd edition). Hermann and Cie, Paris.

NATHAN, L. (1930) Improvements in the fermentation and maturation of beers. *J. Inst. Brewing*, **36**, 538–550.

PETERSEN, M. and ALFERMANN, A. W. (1993) Plant cell cultures. In *Biotechnology*, Vol. 1, *Biological Fundamentals*, (2nd edition), pp. 577–614 (Editors Rhem, H.-J. and Reed, G.). VCH, Weinheim.

PIRT, S. J. (1974) The theory of fed batch culture with reference to the penicillin fermentation. *J. Appl. Chem. Biotechnol.* **24**, 415–424.

PIRT, S. J. (1975) *Principles of Microbe and Cell Cultivation*. Blackwell, Oxford.

PIRT, S. J. (1979) Fed-batch culture of microbes. *Ann. N.Y. Acad. Sci*, **326**, 119–125.

PIRT, S. J. (1990) The dynamics of microbial processes: a personal view. In *Microbial Growth Dynamics*, pp. 1–16. (Editors Poole, R. K., Bazin, M. J. and Keevil, C. W.). IRL Press, Oxford.

PIRT, S. J. and KUROWSKI, W. M. (1970) An extension of the theory of the chemostat with feedback of organisms. Its experimental realisation with a yeast culture. *J. Gen. Micro.* **63**, 357–366.

PIRT, S. J. and RIGHELATO, R. C. (1967) Effect of growth rate on the synthesis of penicillin by *Penicillium chrysogenum* in batch and continuous culture. *Appl. Micro.*, **15**, 1284–1290.

PLOMLEY, N. J. B. (1959). Formation of the colony in the fungus *Chaetomium*. *Aust. J. Biol. Sci.*, **12**, 53–64.

QUEENER, S. and SWARTZ, R. (1979) Penicillins; Biosynthetic and semisynthetic. In *Economic Microbiology*, Vol. 3. *Secondary Products of Metabolism*, pp. 35–123 (Editor Rose, A. H.). Academic Press, London.

REED, G. and NAGODAWITHANA, T.W. (1991) *Yeast Technology*, (2nd edition), pp. 413–437. Van Nostrand Reinhold, New York.

RHODES, P. M. (1984) The production of oxytetracycline in chemostat culture. *Biotech. Bioeng.* **26**, 382–385.

ROYSTON, M. G. (1966) Tower fermentation of beer. *Process Biochem.* **1**(4), 215–221.

RYU, D. D. and HOSPODKA, J. (1980) Quantitative physiology of *Penicillium chrysogenum* in penicillin fermentation. *Biotechnol Bioeng.*, **22**, 289–298.

SHARP, D. H. (1989) *Bioprotein Manufacture. A Critical Assessment*. Ellis Horwood, Chichester.

SHEHATA, T. E. and MARR, A. G. (1971) *J. Bacteriol.* **107**, 215–221.

SHIOYA, S. (1990) Optimization and control in fed-batch reactors. *Adv. Biochem. Eng.*

SIKYTA, B., SLEZAK, J. and HEROLD, H. (1961) Growth of *Streptomyces aureofaciens* in continuous culture. *Appl. Micro.* **9**, 233–238.

SUZUKI, T., YAMANE, T. and SHIMIZU, S. (1987) Mass production of thiostrepton by fed-batch culture of *Streptomyces laurentii* with pH-stat modal feeding of multi-substrate. *Appl. Microbiol. Biotechnol.* **25**, 526–531.

SUZUKI, T., MUSHIGA, Y., YAMANE, T. and SHIMIZU, S. (1988). Mass production of lipase by fed-batch culture of *Pseudomonas fluorescens*. *Appl. Microbiol. Biotechnol.* **27**, 417–422.

TRILLI, A. (1990) Kinetics of secondary metabolite production. In *Microbial Growth Dynamics*, pp. 103–126 (Editors, Poole, R. K., Bazin, M. J. and Keevil, C. W.). IRL Press, Oxford.

TRILLI, A., CROSSLEY, M. V. and KONTAKOU, M. (1987) Relation between growth and erythromycin production in *Streptomyces erythraeus*. *Biotechnol. Lett.* **9**, 765–770

TRINCI, A. P. J. (1969) A kinetic study of growth of *Aspergillus nidulans* and other fungi. *J. Gen. Micro.* **57**, 11–24.

TRINCI, A. P. J. (1974) A study of the kinetics of hyphal extension and branch initiation of fungal mycelia. *J. Gen. Micro.* **81**, 225–236.

TRINCI, A. P. J. (1992) Mycoprotein: A twenty year overnight success story. *Mycol. Res.* **96**(1), 1–13.

WAKI, T., LUGA, K. and ICHIKAWA, K. (1982) Production of cellulase in fed-batch culture. In *Advances in Biotechnology*, Vol. 1. *Scientific and Engineering Principles*, pp 359–364 (Editors Moo-Young, M., Robinson, C. W. and Vezina, C.). Pergamon Press, Toronto.

WOODWARD, J. (1987) Utilisation of cellulose as a fermentation substrate: problems and potential. In *Carbon Substrates in Biotechnology*, pp.45–66 (Editors Stowell, J. D., Beardsmore, A. J., Keevil, C. W. and Woodward, J. R.). IRL Press, Oxford.

YOSHIDA, F., YAMANE, T. and NAKAMOTO, K. (1973) Fed-batch hydrocarbon fermentations with colloidal emulsion feed. *Biotech. Bioeng.* **15**, 257–270.

# CHAPTER 3

# The Isolation, Preservation and Improvement of Industrially Important Micro-organisms

## THE ISOLATION OF INDUSTRIALLY IMPORTANT MICRO-ORGANISMS

THE MOST publicized advances in biotechnology over the last ten years have been those in recombinant DNA technology. Whilst these advances have resulted in the development of extremely valuable new commercial processes and have improved many others, it is worthwhile reiterating Buckland's (1992) comment that the combined sales of four new microbial secondary metabolites introduced in the 1980s was greater than the sales of all the recombinant products added together. Thus, the diversity of micro-organisms may be exploited still by searching for strains from the natural environment able to produce products of commercial value. The first stage in the screening for micro-organisms of potential industrial application is their isolation. Isolation involves obtaining either pure or mixed cultures followed by their assessment to determine which carry out the desired reaction or produce the desired product. In some cases it is possible to design the isolation procedure in such a way that the growth of producers is encouraged or that they may be recognized at the isolation stage, whereas in other cases organisms must be isolated and producers recognized at a subsequent stage. However, it should be remembered that the isolate must eventually carry out the process economically and therefore the selection of the culture to be used is a compromise between the productivity of the organism and the economic con-straints of the process. Bull *et al.* (1979) cited a number of criteria as being important in the choice of organism:

1. The nutritional characteristics of the organism. It is frequently required that a process be carried out using a very cheap medium or a pre-determined one, e.g. the use of methanol as an energy source. These requirements may be met by the suitable design of the isolation medium.
2. The optimum temperature of the organism. The use of an organism having an optimum temperature above 40° considerably reduces the cooling costs of a large-scale fermentation and, therefore, the use of such a temperature in the isolation procedure may be beneficial.
3. The reaction of the organism with the equipment to be employed and the suitability of the organism to the type of process to be used.
4. The stability of the organism and its amenability to genetic manipulation.
5. The productivity of the organism, measured in its ability to convert substrate into product and to give a high yield of product per unit time.
6. The ease of product recovery from the culture.

Points 3, 4, and 6 would have to be assessed in detailed tests subsequent to isolation and the organism most well suited to an economic process chosen on the basis of these results. However, before the process may be

TABLE 3.1. *Major culture collections*

| Culture collection | Address |
| --- | --- |
| National Collection of Type Cultures (NCTC) | PHLS Central Public Health Laboratory, 61 Colindale Avenue, London NW9 5HT, UK |
| National Collections of Industrial and Marine Bacteria Ltd. (NCIB, NCMB) | 23 St Machar Drive, Aberdeen AB2 1RY, UK |
| National Collection of Yeast Cultures (NCYC) | AFRC Institute of Food Research, Norwich Laboratory, Colney Lane, Norwich NR4 7UA, UK |
| Collection of International Mycological Institute (IMI) | Culture Collection and Industrial Service Division, Ferry Lane, Kew, Surrey TW9 3AF, UK |
| American Type Culture Collection (ATCC) | 12301 Parklawn Drive, Rockville, MD 20852, USA |
| Deutsche Sammlung von Mikroorganismen und Zelkulturen (DSM) | Mascheroder Weg 1 b, D-3300 Braunschweig, Germany |
| Centraalbureau voor Schimmelcultures (CBS) | P.O. Box 273, Oosterstraat 1, NL-3740 AG Baarn, Netherlands |
| Czechoslovak Collection of Microorganisms (CCM) | Masaryk University, Jostova 10, 662 43 Brno, Czech Republic |
| Collection Nationale de Cultures de Microorganisms (CNCM) | Institut Pasteur, 25, rue du Docteur Roux, F-75724 Paris Cedex 15, France |
| Japan Collection of Microorganisms (JCM) | Riken, Wako-shi, Saitama, 351-01 Japan |
| Culture Collection of the Institute for Fermentation (IFO) | Institute for Fermentation, 17-85 Jugo-Honchmachi, 2-chome, Yodogawa-ku, Osaka, Japan |

put into commercial operation the toxicity of the product and the organism has to be assessed.

The above account implies that cultures must be isolated, in some way, from natural environments. However, the industrial microbiologist may also 'isolate' micro-organisms from culture collections. Kirsop and Doyle (1991) have provided a comprehensive list of collections and Table 3.1 cites some examples. Such collections may provide organisms of known characteristics but may not contain those possessing the most desirable features, whereas the environment contains a myriad of organisms, very few of which may be satisfactory. It is certainly cheaper to buy a culture than to isolate from nature, but it is also true that a superior organism may be found after an exhaustive search of a range of natural environments. The economic considerations are discussed in more detail in Chapter 12. However, it is always worthwhile to purchase cultures demonstrating the desired characteristics, however weakly, as they may be used as model systems to develop culture and assay techniques which may then be applied to the assessment of natural isolates.

The ideal isolation procedure commences with an environmental source (frequently soil) which is highly probable to be rich in the desired types, is so designed as to favour the growth of those organisms possessing the industrially important characteristic (i.e. the industrially useful characteristic is used as a selective factor) and incorporates a simple test to distinguish the most desirable types. Selective pressure may be used in the isolation of organisms which will grow on particular substrates, in the presence of certain compounds or under cultural conditions adverse to other types. However, if it is not possible to apply selective pressure for the desired character it may be possible to design a procedure to select for a microbial taxon which is known to show the characteristic at a relatively high frequency, e.g. the production of antibiotics by streptomycetes. Alternatively, the isolation procedure may be designed to exclude certain microbial 'weeds' and to encourage the growth of more novel types. Indeed, as pointed out by Bull (1992) for screening programmes to continue to generate new products it is becoming increasingly more important to concentrate on lesser known microbial taxa or to utilize very specific screening tests to identify the desired activity. During the 1980s significant advances have been made in the establishment of taxonomic databases describing the properties of microbial groups and these databases have been used to predict the cultural conditions which would select for the growth of particular taxa. Thus, the advances in the taxonomic description of taxa have allowed the rational design of procedures for the isolation of strains which may have a high probability of being productive or are representatives of unusual groups. The advances in pharmacology and molecular biology have also enabled the design of more effective screening tests to identify productive strains amongst the isolated organisms.

## Isolation methods utilizing selection of the desired characteristic

In this section we consider methods which take direct advantage of the industrially relevant property exhibited by an organism to isolate that organism from an environmental source. Isolation methods depending on the use of desirable characteristics as selective factors are essentially types of enrichment culture. Enrichment culture is a technique resulting in an increase in the number of a given organism relative to the numbers of other types in the original inoculum. The process involves taking a mixed population and providing conditions either suitable for the growth of the desired type, or unsuitable for the growth of the other organisms, e.g. by the provision of particular substrates or the inclusion of certain inhibitors. Prior to the culture stage it is often advantageous to subject the environmental source (normally soil) to conditions which favour the survival of the organisms in question. For example, air-drying the soil will favour the survival of actinomycetes.

ENRICHMENT LIQUID CULTURE

Enrichment liquid culture is frequently carried out in shake flasks. However, the growth of the desired type from a mixed inoculum will result in the modification of the medium and therefore changes the selective force which may allow the growth of other organisms, still viable from the initial inoculum, resulting in a succession. The selective force may be re-established by inoculating the enriched culture into identical fresh medium. Such sub-culturing may be repeated several times before the dominant organism is isolated by spreading a small inoculum of the enriched culture onto solid medium. The time of sub-culture in an enrichment process is critical and should correspond to the point at which the desired organism is dominant.

The prevalence of an organism in a batch enrichment culture will depend on its maximum specific growth rate compared with the maximum specific growth rates of the other organisms capable of growth in the inoculum. Thus, provided that the enrichment broth is sub-cultured at the correct times, the dominant organism will be the fastest growing of those capable of

growth. However, it is not necessarily true that the organism with the highest specific growth rate is the most useful, for it may be desirable to isolate the organism with the highest affinity for the limiting substrate.

The problems of time of transfer and selection on the basis of maximum specific growth rate may be overcome by the use of a continuous process where fresh medium is added to the culture at a constant rate. Under such conditions the selective force is maintained at a constant level and the dominant organism will be selected on the basis of its affinity for the limiting substrate rather than its maximum growth rate.

The basic principles of continuous culture are considered in Chapter 2 from which it may be seen that the growth rate in continuous culture is controlled by the dilution rate and is related to the limiting substrate concentration by the equation:

$$\mu = \mu_{max} s / (K_s + s). \qquad (3.1)$$

Equation (3.1) is represented graphically in Fig. 3.1. A model of the competition between two organisms capable of growth in a continuous enrichment culture is represented in Fig. 3.2. Consider the behaviour of the two organisms, A and B, in Fig. 3.2. In continuous culture the specific growth rate is determined by the substrate concentration and is equal to the dilution rate, so that at dilution rates below point Y in Fig. 3.2 strain B would be able to maintain a higher growth rate than strain A, whereas at dilution rates above Y strain A would be able to maintain a higher growth rate. Thus, if A and B were present in a continuous enrichment culture, limited by the substrate depicted in Fig. 3.2, strain A would be selected at dilution rates above Y and strain B would be selected at dilution rates below Y. Thus, the organisms which are isolated by continuous enrichment culture will depend on the dilution rate employed which may result in the isolation of

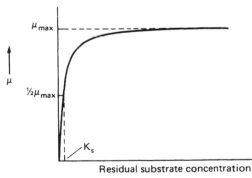

FIG. 3.1. The effect of substrate concentration on the specific growth rate of a micro-organism.

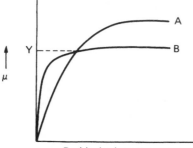

F<small>IG</small>. 3.2. The effect of substrate concentration on the specific growth rates of two micro-organisms A and B.

organisms not so readily recovered by batch techniques.

Continuous enrichment techniques are especially valuable in isolating organisms to be used in a continuous-flow commercial process. Organisms isolated by batch enrichment and purification on solid media frequently perform poorly in continuous culture (Harrison *et al.*, 1976), whereas continuous enrichment provides an organism, or mixture of organisms, adapted to continuous culture. The enrichment procedure should be designed such that the predicted isolate meets as many of the criteria of the proposed process as possible and both Johnson (1972) and Harrison (1978) have discussed such procedures for the isolation of organisms to be used for biomass production. Johnson emphasized the importance of using the carbon source to be employed in the subsequent commercial process as the sole source of organic carbon in the enrichment medium, and that the medium should be carbon limited. The inclusion of other organic carbon sources, such as vitamins or yeast extract, may result in the isolation of strains adapted to using these, rather than the principal carbon source, as energy sources. The isolation of an organism capable of growth on a simple medium should also form the basis of a cheaper commercial process and should be more resistant to contamination — a major consideration in the design of a commercial continuous process. The use of as high as possible an isolation temperature should also result in the isolation of a strain presenting minimal cooling problems in the subsequent process.

The main difficulty in using a continuous-enrichment process is the washout of the inoculum before an adapted culture is established. Johnson (1972) suggested that the isolation process should be started in batch culture using a 20% inoculum and as soon as growth is observed, the culture should be transferred to fresh medium and the subsequent purification and stabilization of the enrichment performed in continuous culture. The continuous system should be periodically inoculated with soil or sewage which may not only be a source of potential isolates but should also ensure that the dominant flora is extremely resistant to contamination.

Harrison (1978) proposed two solutions to the problem of early washout in continuous isolation processes: The first uses a turbidostat and the second uses a two-stage chemostat (see Chapter 2). A turbidostat is a continuous-flow system provided with a photoelectric cell to determine the turbidity of the culture and maintain the turbidity between set points by initiating or terminating the addition of medium. Thus, washout is avoided as the medium supply will be switched off if the biomass falls below the lower fixed point. The use of a turbidostat will result in selection on the basis of maximum specific growth rate as it operates at high levels of limiting substrate. Thus, although the use of the turbidostat removes the danger of washout it is not as flexible a system as the chemostat which may be used at a range of dilution rates. The two-stage chemostat described by Harrison (1978) is very similar to Johnson's (1972) procedure. The first stage of the system was used as a continuous inoculum for the second stage and consisted of a large bottle containing a basic medium inoculated with a soil infusion. Continuous inoculation was employed until an increasing absorbance was observed in the second stage. Bull (1992) advocated the use of feed-back continuous systems for the isolation of strains with particularly high affinity for substrate and this approach would also guard against premature washout.

The use of continuous enrichment culture has frequently resulted in the selection of stable, mixed cultures presumably based on some form of symbiotic relationship. It is extremely unlikely that such mixed, stable systems could be isolated by batch techniques so that the adoption of continuous enrichment may result in the development of novel, mixed culture fermentations. Harrison *et al.* (1972) isolated a mixed culture using methane as the carbon source in a continuous enrichment and demonstrated that the mixture contained one methylotroph and a number of non-methylotrophic symbionts. The performance of the methylotroph in pure culture was invariably poorer than the mixture in terms of growth rate, yield and culture stability.

Continuous enrichment has also been used for the isolation of organisms to be used in systems other than biomass production; Rowley and Bull (1977) used the technique to isolate an *Arthrobacter* sp. producing a yeast Iysing enzyme complex. The technique has been used widely for the

## ENRICHMENT CULTURES USING SOLIDIFIED MEDIA

Solidified media have been used for the isolation of certain enzyme producers and these techniques usually involve the use of a selective medium incorporating the substrate of the enzyme which encourages the growth of the producing types. Aunstrup *et al.* (1972) isolated species of *Bacillus* producing alkaline proteases. Soils of various pHs were used as the initial inoculum and, to a certain extent, the number of producers isolated correlated with the alkalinity of the soil sample. The soil samples were pasteurized to eliminate non-sporulating organisms and then spread onto the surface of agar media at pH 9–10, containing a dispersion of an insoluble protein. Colonies which produced a clear zone due to the digestion of the insoluble protein were taken to be alkaline protease producers. The size of the clearing zone could not be used quantitatively to select high producers as there was not an absolute correlation between the size of the clearing zone and the production of alkaline protease in submerged culture. However, this example demonstrates the importance of choice of starting material, the use of a selective force in the isolation and the incorporation of a preliminary diagnostic test, albeit of limited use.

### Isolation methods not utilizing selection of the desired characteristic

The synthesis of some products does not give the producing organism any selective advantage which may be exploited directly in an isolation procedure. Examples include the production of antibiotics and growth promoters. Therefore, a pool of organisms has to be isolated and subsequently tested for the desired characteristic. The major problem faced by industrial microbiologists in this situation is the reisolation of strains which have already been screened many times before. However, this 'reinvention of the wheel' syndrome may be minimized in two major ways:

(i) Developing procedures to favour the isolation of unusual taxa which are less likely to have been screened previously.

(ii) Identifying selectable features correlated with the unselectable industrial trait thus enabling an enrichment process to be developed.

Much use has been made of numerical taxonomic data bases to design media selective for particular taxa. For example, Williams' group at Liverpool University (U.K.) used such databases to design isolation media to either encourage the growth of uncommon streptomycetes or to discourage the growth of *Streptomyces albidoflavus* (Vickers *et al.*, 1984; Williams and Vickers, 1988). Several groups of workers have taken advantage of the antibiotic sensitivity information stored in taxonomic databases to design media selective for particular taxa, as shown in Table 3.2. The incorporation of particular antibiotics into isolation media may result in the selection of the resistant taxa. Such techniques are reviewed by Goodfellow and O'Donnell (1989) and Bull (1992). Bull emphasized that the taxonomic approach may be optimized and developed according to Fig. 3.1. The isolates from an isolation procedure would be screened for activity and then the growth requirements of the positive cultures determined. This knowledge can then be used to optimize the isolation medium and the cycle begins again.

Whilst taxonomic databases are a convenient source of information, it is important to appreciate Huck *et al.*'s (1991) observation that these systems were designed to provide information for taxonomic differentiation within a group. Thus, some of the diagnostic data may not be applicable to isolation systems. More significantly, the reactions of organisms outside the taxon in question would not be listed. Of course, an environmental sample contains a vast variety of organisms and the design of isolation media based on knowledge of only one taxon may inadvertently result in the preferential isolation of undesirable types. Huck *et al.* (1991) used the statistical stepwise discrimination analysis (SDA) technique to design media for the positive selection of antibiotic producing soil isolates. This was achieved by characterizing a collection of eubacterial and actinomycete soil isolates according to 43 physiological and nutritional tests. Certain features were identified which, when used as selective factors, enhanced the probability of either isolating actinomycetes or antibiotic producing actinomycetes. These features are shown in Table 3.3. Using this approach several media were developed which enhanced the isolation of antibiotic producers.

The most desirable isolation medium would be one

TABLE 3.2 *Antibacterial compounds used in selective media for the isolation of actinomycetes* (Goodfellow and O'Donnell, 1989)

| Selective agent | Target organism | Reference |
|---|---|---|
| Bruneomycin | *Micromonospora* | Preobrazhenskayai *et al.* (1975) |
| Dihydroxymethylfuratriazone | *Microtetraspora* | Tomita *et al.* (1980) |
| Gentamycine | *Micromonospora* | Bibikova *et al.* (1981) |
| Kanamycin | *Actinomadura* | Chormonova (1978) |
| Nitrofurazone | *Streptomyces* | Yoshioka (1952) |
| Novobiocin | *Micromonospora* | Sveshnikova *et al.* (1976) |
| Tellurite | *Actinoplanes* | Willoughby (1971) |
| Tunicamycin | *Micromonospora* | Wakkisaka *et al.* (1982) |

which selects for the desired types and also allows maximum genetic expression. Cultures grown on such media could then be used directly in a screen. However, it is more common that, once isolated, the organisms are grown on a range of media designed to enhance productivity. Nisbet (1982) put forward some guidelines for the design of such media and these are summarized in Table 3.4.

### Screening methods

Early screening strategies tended to be empirical, labour intensive and had relatively low success rates. As the number of commercially important compounds isolated increased, the success rates of such screens decreased further. Thus, new screening methods have been developed which are more precisely targeted to identify the desired activity. The evolution of antibiotic

screens serves as an excellent illustration of the development of more precise, targeted systems.

Antibiotics were initially detected by growing the potential producer on an agar plate in the presence of an organism (or organisms) against which antimicrobial action was required. Production of the antibiotic was detected by inhibition of the test organism(s). Alternatively, the microbial isolate could be grown in liquid culture and the cell-free broth tested for activity. This approach was extended by using a range of organisms to detect antibiotics with a defined antibacterial spectrum. For example, Zahner (1978) discussed the use of test organisms to detect the production of antibiotics with confined action spectra. The use of *Bacillus subtilis* and *Streptomyces viridochromogenes* or *Clostridium pasteurianum* allows the identification of antibiotics with a low activity against *B. subtilis* and a high activity against the other test organisms. Such antibiotics may be new because they would not be isolated by the more common tests using *B. subtilis* alone. The kirromycin group of antibiotics were discovered using methods of this type.

In the 15 years prior to 1971, no novel naturally occurring β-lactams were discovered (Nisbet and Porter, 1989). However, the advances made in the

TABLE 3.3 *Selective substrates for the isolation of actinomycetes and antibiotic-producing actinomycetes*

| Substrates selective for Actinomycetes | Substrates selective for antibiotic-producing actinomycetes |
|---|---|
| Proline | Proline |
| Glucose (1.0%) | Glucose (1.0%) |
| Glycerol | Glycerol |
| Starch | Starch |
| Humic acid (0.1%) | Humic acid (0.1%) |
| Propionate (0.1%) | Zinc |
| Methanol | Alanine |
| Nitrate | Potassium |
| Calcium | Vitamins |
| | Cobalt (0.05%) |
| | Phenol (0.01%) |
| | Asparagine |

TABLE 3.4 *Guidelines for 'overproduction media'* (Nisbet, 1982)

1. Prepare a range of media in which different types of nutrients become growth-limiting e.g. C, N, P, O
2. For each type of nutrient depletion use different forms of the growth-sufficient nutrient
3. Use a polymeric or complexed form of the growth-limiting nutrient
4. Avoid the use of readily assimilated forms of carbon (glucose) or nitrogen ($NH_4^+$) that may cause catabolite repression
5. Ensure that known cofactors are present ($Co^{3+}$, $Mg^{2+}$, $Mn^{2+}$, $Fe^{2+}$).
6. Buffer to minimize pH changes.

understanding of cell-wall biosynthesis and the mode of action of antibiotics allowed the development of mode of action screens in the 1970s which resulted in a very significant increase in the discoveries of new $\beta$-lactam antibiotics. Nagarajan *et al*.'s (1971) discovery of the cephamycins was based on the detection of compounds which induced morphological changes in susceptible bacteria. The appreciation that the mode of action of penicillins was the inhibition of the transpeptidase enzyme crosslinking mucopeptide molecules led to the development of enzyme inhibitor assays, which were particularly attractive because they could be automated. Fleming *et al*. (1982) described the devlopment of an automated screen for the detection of carboxypeptidase inhibitors which has led to the detection of several novel cephamycin and carbapenem compounds.

The increasing frequency of penicillin and cephalosporin resistance amongst clinical bacteria led to the development of mechanism based screens for the isolation of more effective antibacterials. The logic of the Beechams Pharmaceuticals group of Brown *et al*. (1976) was to search for a compound which would inhibit $\beta$-lactamase and could be incorporated with ampicillin as a combination therapeutic agent. Samples were tested for their ability to increase the inhibitory effect of ampicillin on a $\beta$-lactamase producing *Klebsiella aerogenes* and this strategy resulted in the discovery of clavulanic acid. Further examples of targeted antibiotic screens are given in the excellent review by Nisbet and Porter (1989).

The concept of using the inhibition of enzymes as a screening mechanism was pioneered by Umezawa (1972) in his search for microbial products inhibiting key enzymes of human metabolism. His approach was based on the logic that, if a compound inhibits a key human enzyme *in vitro*, it may have a pharmacological action *in vivo*. Such screens have been applied to a wide range of pharmacological targets and have resulted in the isolation of several important drugs. From *Aspergillus*, Alberts *et al*. (1980) isolated mevinolin which is an inhibitor of hydroxy-methyl-glutaryl reductase, the rate limiting step in the biosynthesis of sterols. It is now marketed as an agent to lower high cholesterol levels. Bull (1992) list the following clinical situations in which microbial products have been shown to inhibit key enzymes: hypercholesterolaemia, hypertension, gastric inflammation, muscular dystrophy, benign prostate hyperplasia and systemic lupus erythemosus. A further specific example is provided by the work of Hashimoto *et al*. (1990). The activity of carbapenem antibiotics is lost in therapy due to renal dehydropeptidase activity. These workers isolated microbial products capable of inhibiting the enzyme which could then be administered along with the antibiotic to maintain its clinical activity.

The detection of pharmacological agents by receptor–ligand binding assays has been developed rapidly by pharmaceutical companies (Bull, 1992). These are extensions of the enzyme inhibitor approach but agents which block receptor sites are likely to be more effective at very low concentrations. The gastrointestinal hormone cholecystokinin (CCK) controls a range of digestive activities such as pancreatic secretion and gall bladder contraction. Receptor screening identified a fungal metabolite, aperlicin (from *Aspergillus alliaceus*) that had a very high affinity for CCK receptors. Although the fungal product did not prove to be a suitable drug it was used as a model for the design of analogues which were receptor binders and pharmacologically acceptable.

The progress in molecular biology, genetics and immunology has also contributed extensively to the development of innovative screens, by enabling the construction of specific detector strains, increasing the availability of enzymes and receptors and constructing extremely sensitive assays. Bull (1992) summarized the major contributions as follows:

(i) The provision of test organisms that have increased sensitivities, or resistances, to known agents. For example, the use of super-sensitive strains for the detection of $\beta$-lactam antibiotics.

(ii) The cloning of genes coding for enzymes or receptors that may be used in inhibitor or binding screens makes such materials more accessible and available in much larger amounts.

(iii) The development of reporter gene assays. A reporter gene is one which codes for an easily assayable product so that it can be used to detect the activation of a control sequence to which it is fused. Such systems have been used in the search for metabolites that disrupt viral replication.

(iv) Molecular probes for particular gene sequences may enable the detection of organisms capable of producing certain product groups. This information may be used to focus the search on these organisms in an attempt to find novel representatives of an already known commercially attractive chemical family.

(v) The development of immunologically based assays such as ELISA.

It is important to appreciate at this stage that the advances in the biological sciences which have enabled the design of sophisticated screening tests have been paralleled by the development of robotic automation systems. This means that the engineering now exists to automate such tests, resulting in an enormous throughput. Indeed, Nisbet and Porter (1989) claimed that "the modern microbial discovery programme must achieve rates of $10^5$ tests per year, for as many as 20 different assay systems, to compete in the race for novel therapeutics."

## THE PRESERVATION OF INDUSTRIALLY IMPORTANT MICRO-ORGANISMS

The isolation of a suitable organism for a commercial process may be a long and very expensive procedure and it is therefore essential that it retains the desirable characteristics that led to its selection. Also, the culture used to initiate an industrial fermentation must be viable and free from contamination. Thus, industrial cultures must be stored in such way as to eliminate genetic change, protect against contamination and retain viability. An organism may be kept viable by repeated sub-culture into fresh medium, but, at each cell division, there is a small probability of mutations occurring and because repeated sub-culture involves very many such divisions, there is a high probability that strain degeneration would occur. Also, repeated sub-culture carries with it the risk of contamination. Thus, preservation techniques have been developed to maintain cultures in a state of 'suspended animation' by storing either at reduced temperature or in a dehydrated form. Full details of the techniques are given by Kirsop and Doyle (1991).

### Storage at reduced temperature

STORAGE ON AGAR SLOPES

Cultures grown on agar slopes may be stored in a refrigerator (5°) or a freezer (−20°) and sub-cultured at approximately 6-monthly intervals. The time of sub-culture may be extended to I year if the slopes are covered with sterile medicinal grade mineral oil.

STORAGE UNDER LIQUID NITROGEN

The metabolic activities of micro-organisms may be reduced considerably by storage at the very low temperatures (−150° to −196°) which may be achieved using a liquid nitrogen refrigerator. Snell (1991) claimed that this aproach is the most universally applicable of all preservation methods. Fungi, bacteriophage, viruses, algae, yeasts, animal and plant cells and tissue cultures have all been successfully preserved. The technique involves growing a culture to the maximum stationary phase, resuspending the cells in a cryoprotective agent (such as 10% glycerol) and freezing the suspension in sealed ampoules before storage under liquid nitrogen. Some loss of viability is suffered during the freezing and thawing stages but there is virtually no loss during the storage period. Thus, viability may be predictable even after a period of many years. Snell (1991) suggested that liquid nitrogen is the method of choice for the preservation of valuable stock cultures and may be the only suitable method for the long term preservation of cells that do not survive freeze-drying. Although the equipment is expensive the process is economical on labour. However, the method has the major disadvantage that liquid nitrogen evaporates and must be replenished regularly. If this is not done, or the apparatus fails, then the consequences are the loss of the collection.

### Storage in a dehydrated form

DRIED CULTURES

Dried soil cultures have been used widely for culture preservation, particularly for sporulating mycelial organisms. Moist, sterile soil may be inoculated with a culture and incubated for several days for some growth to occur and then allowed to dry at room temperature for approximately 2 weeks. The dry soil may be stored in a dry atmosphere or, preferably, in a refrigerator. The technique has been used extensively for the storage of fungi and actinomycetes and Pridham et al. (1973) observed that of 1800 actinomycetes dried on soil about 50% were viable after 20-years storage.

Malik (1991) described methods which extend the approach using substrates other than soil. Silica gel and porcelain beads are suggested alternatives and detailed methods are given for these simple, inexpensive techniques in Malik's discussion.

LYOPHILIZATION

Lyophilization, or freeze-drying, involves the freezing of a culture followed by its drying under vacuum, which results in the sublimation of the cell water. The technique involves growing the culture to the maximum stationary phase and resuspending the cells in a protective medium such as milk, serum or sodium glutamate. A few drops of the suspension are transferred to an ampoule, which is then frozen and subjected to a high vacuum until sublimation is complete, after which the ampoule is sealed. The ampoules may be stored in a refrigerator and the cells may remain viable for 10 years or more (Perlman and Kikuchi, 1977).

Lyophilization is very convenient for service culture collections (Snell, 1991) because, once dried, the cultures need no further attention and the storage equipment (a refrigerator) is cheap and reliable. Also, the freeze dried ampoules may be dispatched as such, still in a state of 'suspended animation' whereas liquid nitrogen stored cultures begin to deteriorate. However, freeze-dried cultures are tedious to open and revitalize and several sub-cultures may be needed before the cells regain their typical characteristics. Overall, the technique appears to be second only to liquid nitrogen storage and even when liquid nitrogen is used makes an excellent insurance against the possibility of the breakdown of the nitrogen freezer. The technique is considered in detail by Rudge (1991).

### Quality control of preserved stock cultures

Whichever technique is used for the preservation of an industrial culture it is essential to be certain of the quality of the stocks. Each batch of newly preserved cultures should be routinely checked to ensure their quality and such a procedure has been outlined by Lincoln (1960): a single colony of the culture to be preserved is inoculated into a shake flask and the growth of the culture observed to ensure a typical growth pattern; after a further shake-flask sub-culture the broth is used to prepare a large number of storage ampoules. At least 3% of the ampoules are reconstituted and the cultures assessed for purity, viability and productivity. If the samples fail any one of these tests the entire batch should be destroyed. Thus, by the use of such a quality-control system stock cultures may be retained, and used, with confidence.

### THE IMPROVEMENT OF INDUSTRIAL MICRO-ORGANISMS

Natural isolates usually produce commercially important products in very low concentrations and there-fore every attempt is made to increase the productivity of the chosen organism. Increased yields may be achieved by optimizing the culture medium and growth conditions, but this approach will be limited by the organism's maximum ability to synthesize the product. The potential productivity of the organism is controlled by its genome and, therefore, the genome must be modified to increase the potential yield. The cultural requirements of the modified organism would then be examined to provide conditions that would fully exploit the increased potential of the culture, while further attempts are made to beneficially change the genome of the already improved strain. Thus, the process of strain improvement involves the continual genetic modification of the culture, followed by reappraisals of its cultural requirements.

Genetic modification may be achieved by selecting natural variants, by selecting induced mutants and by selecting recombinants. There is a small probability of a genetic change occurring each time a cell divides and when it is considered that a microbial culture will undergo a vast number of such divisions it is not surprising that the culture will become more heterogeneous. The heterogeneity of some cultures can present serious problems of yield degeneration because the variants are usually inferior producers compared with the original culture. However, variants have been isolated which are superior producers and this has been observed frequently in the early stages in the development of a natural product from a newly isolated organism. An explanation of this phenomenon for mycelial organisms may be that most new isolates are probably heterokaryons (contain more than one type of nucleus) and the selection of the progeny of uninucleate spores results in the production of homokaryons (contain only one type of nucleus) which may be superior producers. However, the phenomenon is also observed with unicellular isolates which are certainly not heterokaryons. Therefore, it is worthwhile to periodically plate out the producing culture and screen a proportion of the progeny for productivity; this practice has the added advantage that the operator tends to become familiar with morphological characteristics associated with high productivity and, by selecting 'typical' colonies, a strain subject to yield degeneration may still be used with consistent results.

Therefore, selection of natural variants may result in increased yields but it is not possible to rely on such improvements, and techniques must be employed to increase the chances of improving the culture. These techniques are the isolation of induced mutants and recombination. The most dramatic examples of strain improvement come from the applications of recombinant DNA technology which has resulted in organisms producing compounds which they were not able to produce previously. Furthermore, the advances in these techniques have resulted in very significant improvements in the production of conventional fermentation products. However, it should be remembered that these methods have not replaced mutant isolation but have made an invaluable addition to an impressive repertoire. The techniques of mutant isolation have contributed enormously to the development of present-day industrial strains and these techniques will be considered first. The applications of recombination systems will then be discussed along with the interactions between mutant isolation and recombination in strain improvement advances.

Dulaney and Dulaney (1967) compared the spread in productivity of chlortetracycline of natural variants of *Streptomyces viridifaciens* with the spread in productivity of the survivors of an ultraviolet treatment. The results of their comparison are shown in Figs 3.3 and 3.4, from which it may be seen that although there are more inferior producers amongst the survivors of the ultraviolet treatment there are also strains producing more than twice the parental level, far greater than the best of the natural variants. The use of ultraviolet light is only one of a large number of physical or chemical agents which increase the mutation-rate — such agents are termed mutagens. The reader is referred to Baltz (1986) and Birge (1988) for accounts of the modes of action of mutagens. The vast majority of induced muta-

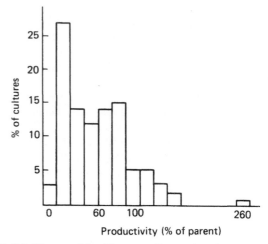

FIG. 3.4. The spread in chlortetracycline productivity of the survivors of a UV-treated population of *Streptomyces viridifaciens* (Dulaney and Dulaney, 1967).

tions are deleterious to the yield of the desired product but, as shown in Fig. 3.3, a minority are more productive than the parent. The problem of obtaining the high-yielding mutants may be approached from two theoretical standpoints; the number of desirable mutants may be increased by 'directed mutation', i.e. the use of a technique which will preferentially produce particular mutants at a high rate; or techniques may be developed to improve the separation of the few desirable types from the large number of mediocre producers.

Inherent in the concept of directed mutation is the assumption that a mutation programme can be optimized to produce mutants of a particular kind. The choice of mutagen was demonstrated to affect the success of mutation programmes early in the history of strain improvement schemes. For example, Hostalek (1964) claimed that ultraviolet radiation was the most effective mutagen for increasing the yield of tetracycline by strains of *Streptomyces aureofaciens*. DeWitt *et al.* (1989) emphasized that as well as certain mutagens being more beneficial, the dose will affect the generation of the desired types. Despite these observations it is frequently the case that it is difficult to predict what type of mutation is required at the molecular level to improve a strain, and therefore it is extremely unlikely that the concept of directed mutation can be applied in these circumstances. Thus, it is the second approach specified above that is likely to provide the solution to this type of problem, i.e. the development of selection techniques.

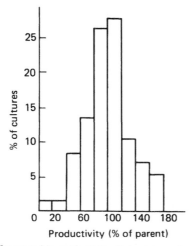

FIG. 3.3. The spread in productivity of chlortetracycline of natural variants of *Streptomyces viridifaciens* (Dulaney and Dulaney, 1967).

However, the concept of directed mutation is of considerable relevance when the genes to be modified are known and the organism is genetically well documented (Shortle *et al.*, 1981). In these systems a cloned DNA molecule may be subjected to *in vitro* enzymatic cleavage and manipulation such that the mutation is present at a particular site. Thus, reintroduction of the modified DNA into the cell should result in the production of a cell mutated at a particular site. The incorporation of a selectable marker into the cloned DNA would assist in the recovery of cells incorporating the modified gene(s). Thus, such systems still depend on the development of methods to recover cells displaying the desirable feature.

The separation of desirable mutants from the other survivors of a mutation treatment is similar to the isolation of desirable organisms from nature. Where possible, the mutant isolation procedure should use the improved characteristic of the desired mutant as a selective factor. Presumably, superior productivity is a result of a diversion of precursors into the product and/or a modification of the control mechanisms limiting the level of production. Thus, a knowledge of the biosynthetic route and the mechanisms of control of the biosynthesis of the product should enable the prediction of a 'blueprint' of the desirable mutant. Such a 'blueprint' might then enable the design of isolation techniques which would give the desired mutants a selective advantage over the other types present. Knowledge of biosynthetic routes and control mechanisms are more detailed for primary metabolites and, therefore, the use of selective pressure in mutant isolation is more common in the fields of amino acid, nucleotide and enzyme production than in secondary metabolite production. However, considerable progress has also been made in the design of such procedures for the isolation of secondary metabolite over-producers.

### The selection of induced mutants synthesizing improved levels of primary metabolites

Before considering the methods used for the selection of mutants producing improved levels of primary metabolites it is necessary to study the mechanisms of control of their biosynthesis such that the 'blueprints', referred to above, may be drawn accurately. The levels of primary metabolites in micro-organisms are regulated by feedback control systems. The major systems involved are feedback inhibition and feedback repression. Feedback inhibition is the situation where the end product of a biochemical pathway inhibits the activity of an enzyme catalysing one of the reactions (normally the first reaction) of the pathway. Inhibition acts by the end product binding to the enzyme at an allosteric site which results in interference with the attachment of the enzyme to its substrate. Feedback repression is the situation where the end product (or a derivative of the end product) of a biochemical pathway prevents the synthesis of an enzyme (or enzymes) catalysing a reaction (or reactions) of the pathway. Repression occurs at the gene level by a derivative of the end product combining with the genome in such a way as to prevent the transcription of the gene into messenger RNA, thus resulting in the prevention of enzyme synthesis.

Feedback inhibition and repression frequently act in concert in the control of biosynthetic pathways, where inhibition may be visualized as a rapid control which switches off the biosynthesis of an end product and repression as a mechanism to then switch off the synthesis of temporarily redundant enzymes. The control of pathways giving rise to only one product (i.e. unbranched pathways) is normally achieved by the first enzyme in the sequence being susceptible to inhibition by the end product and the synthesis of all the enzymes being susceptible to repression by the end product, as shown in Fig. 3.5.

The control of biosynthetic pathways giving rise to a number of end products (branched pathways) is more complex than the control of simple, unbranched sequences. The end products of the same, branched biosynthetic pathway are rarely required by the micro-organism to the same extent, so that if an end product exerts control over a part of the pathway common to two, or more, end products then the organism may suffer deprivation of the products not participating in the control. Thus, mechanisms have evolved which enable the level of end products of branched pathways to be controlled without depriving the cell of essential intermediates. The following descriptions of these mechanisms are based on the effect of the control, which may be arrived at by inhibition, repression or a combination of both systems.

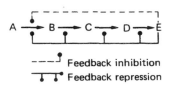

FIG. 3.5. The control of a biosynthetic pathway converting precursor A to end product E via the intermediates B, C and D.

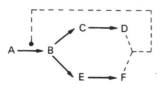

--- → Feedback control

FIG. 3.6. The control of a biosynthetic pathway by the concerted effects of products D and F on the first enzyme of the pathway.

*Concerted or multivalent feedback control.* This control system involves the control of the pathway by more than one end product — the first enzyme of the pathway is inhibited or repressed only when all end products are in excess, as shown in Fig. 3.6.

*Co-operative feedback control.* The system is similar to concerted control except that weak control may be effected by each end product independently. Thus, the presence of all end products in excess results in a synergistic repression or inhibition. The system is illustrated in Fig. 3.7 and it may be seen that for efficient control to occur when one product is in excess there should be a further control operational immediately after the branch point to the excess product. Thus, the reduced flow of intermediates will be diverted to the product which is still required.

*Cumulative feedback control.* Each of the end products of the pathway inhibits the first enzyme by a certain percentage independently of the other end products. In Fig. 3.8 both D and F independently reduce the activity of the first enzyme by 50%, resulting in total inhibition when both products are in excess. As in the case of co-operative control, each end product must exert control immediately after the branch point so that the common intermediate, B, is diverted away from the pathway of the product in excess.

*Sequential feedback control.* Each end product of the pathway controls the enzyme immediately after the

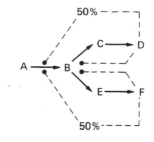

--- 50% --→ Inhibition of 50% of the activity of the enzyme
----- → Total inhibition of enzyme activity

FIG. 3.8. The control of a biosynthetic pathway by the cumulative control of products D and F.

branch point to the product. The intermediates which then build up as a result of this control earlier enzymes in the pathway. Thus, in Fig. 3.9, D inhibits the conversion of B to C, and F inhibits the conversion of B to E. The inhibitory action of D, F, or both, would result in an accumulation of B which, in turn, would inhibit the conversion of A to B.

*Isoenzyme control.* Isoenzymes are enzymes which catalyse the same reaction but differ in their control characteristics. Thus, if a critical control reaction of a pathway is catalysed by more than one isoenzyme, then the different isoenzymes may be controlled by the different end products. Such a control system should be very efficient, provided that control exists immediately after the branch point so that the reduced flow of intermediates is diverted away from the product in excess. An example of the system is shown in Fig. 3.10.

Thus, the levels of microbial metabolites may be controlled by a variety of mechanisms, such that end products are synthesized in amounts not greater than those required for growth. However, the ideal industrial micro-organism should produce amounts far greater than those required for growth and, as sug-

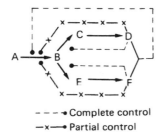

--- → Complete control
—x— → Partial control

FIG. 3.7. The control of a biosynthetic pathway by the co-operative control by end products D and F.

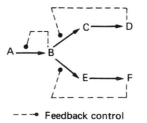

--- → Feedback control

FIG. 3.9. The control of a biosynthetic pathway by sequential feedback control.

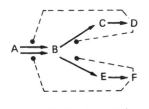

- - -•  Feedback control

FIG. 3.10. The control of two isoenzymes (catalysing the conversion of A to B) by end products D and F.

gested earlier, an understanding of the control of production of a metabolite may enable the construction of a 'blueprint' of the most useful industrial mutant, i.e. one where the production of the metabolite is not restricted by the organism's control system. Such postulated mutants may be modified in three ways:

1. The organism may be modified such that the end products which control the key enzymes of the pathway are lost from the cell due to some abnormality in the permeability of the cell membrane.
2. The organism may be modified such that it does not produce the end products which control the key enzymes of the pathway.
3. The organism may be modified such that it does not recognize the presence of inhibiting or repressing levels of the normal control metabolites.

## MODIFICATION OF THE PERMEABILITY

The best example of the modification of the permeability of a micro-organism is provided by the glutamic acid fermentation. Kinoshita *et al.* (1957a) isolated a biotin-requiring, glutamate-producing organism, subsequently named *Corynebacterium glutamicum*, the permeability of which could be modified by the level of biotin. Provided that the level of biotin in the production medium was below 5 $\mu$g dm$^{-3}$ then the organism would excrete glutamate, but at concentrations of biotin optimum for growth the organism produced lactate (Kinoshita and Nakayama, 1978). Thus, the permeability of *C. glutamicum* may be controlled by the composition of the culture medium and, as such, is not an example of strain improvement in the sense that the term has been used in this chapter. However, the isolation of *C. glutamicum* was a major advance in the microbial production of amino acids and the demonstration of the role of permeability in the production of

glutamate suggested the possibility of the genetic modification of permeability to achieve high levels of productivity. Thus, it is relevant at this stage to consider the physiology of glutamate production by biotin-limited culture.

Kinoshita's isolate was not only deficient in its ability to produce biotin but also in the level of $\alpha$-ketoglutarate dehydrogenase, which normally converts $\alpha$-ketoglutarate to succinate in the tricarboxylic acid cycle. Thus, a metabolic block results in the accumulation of a large concentration of $\alpha$-ketoglutarate which the organism converts to glutamic acid. Oxaloacetate is regenerated in glutamate producers by the activity of the glyoxylate pathway as shown in Fig. 3.11.

When *C. glutamicum* was grown in a medium containing a high concentration of biotin, the organism synthesized glutamate at a level of 25–36 $\mu$g mg$^{-1}$ dry weight of cells, further production being assumed to be prevented by some form of feedback control by glutamate of its own synthesis (Demain and Birnbaum,

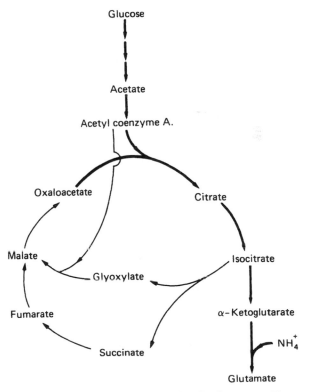

FIG. 3.11. Biosynthesis of glutamate by *C. glutamicum*. Heavy lines indicate the route to glutamate and the light lines indicate the route in the regeneration of oxaloacetate via the glyoxylate cycle.

1968). Under conditions of biotin limitation, glutamate was released from the cells and accumulated to a level of up to 50 g dm$^{-3}$. The increased permeability of the biotin-limited cells to glutamate corresponds with a change in the fatty acid and phospholipid content of the cell envelopes. The crucial factor in the control of permeability appears to be the synthesis of membranes deficient in phospholipids (Nakao *et al.*, 1973).

The role of biotin in the glutamate fermentation obviously necessitates the inclusion of limiting levels of biotin in the medium. Thus, the use of crude carbon sources such as cane molasses, which are rich in biotin, presented difficulties which have been overcome by the modification of the permeability by means other than biotin limitation. The inclusion of penicillin or fatty acid derivatives such as polyoxyethylene sorbitan mono-oleate (Tween 80) during logarithmic growth in biotin-rich medium caused an aberration of the cell envelope permeability resulting in the release of glutamate (Udagawa *et al.*, 1962).

The production of glutamate from hydrocarbons (as carbon sources) has also presented difficulties in the control of permeability, in that the assimilation of hydrocarbons results in the production of fatty acids which effectively bypasses the site of biotin control (Nakao *et al.*, 1972). The permeability of the producing organism may be controlled by the addition of penicillin (Wang *et al.*, 1979) but a genetic solution to this problem has also been found. Nakao *et al.* (1970) isolated a glycerol requiring auxotrophic mutant of *Corynebacterium alkanolyticum* in which phospholipid synthesis was controlled by the supply of glycerol. The mutant produced about 40 g dm$^{-3}$ glutamate when grown on n-paraffins in the presence of 0.01% glycerol (Nakao *et al.*, 1972). Thus, an understanding of the mode of action of biotin limitation in the glutamate fermentation has led to the use of genetic modification of permeability as a method to overcome the normal control mechanisms of the producing organism.

## THE ISOLATION OF MUTANTS WHICH DO NOT PRODUCE FEEDBACK INHIBITORS OR REPRESSORS

Mutants which do not produce certain feedback inhibitors or repressors may be useful for the production of intermediates of unbranched pathways and intermediates and end products of branched pathways. Demain (1972) presented several 'blue-prints' of hypothetical mutants producing intermediates and end products of biosynthetic pathways and these are illustrated in Fig. 3.12. The mutants illustrated in Fig. 3.12 do not produce some of the inhibitors or repressors of the pathways considered and, thus, the control of the pathway is lifted, but, because the control factors are also essential for growth, they must be incorporated into the medium at concentrations which will allow growth to proceed but will not evoke the normal control reactions.

In the case of Fig. 3.12(1) the unbranched pathway is normally controlled by feedback inhibition or repression of the first enzyme of the pathway by the end product, E. However, the organism represented in Fig. 3.12(1) is auxotrophic for E due to the inability to convert C to D so that control of the pathway is lifted and C will be accumulated provided that E is included in the medium at a level sufficient to maintain growth but insufficient to cause inhibition or repression.

Figure 3.12(2) is a branched pathway controlled by the concerted inhibition of the first enzyme in the pathway by the combined effects of E and G. The mutant illustrated is auxotrophic for E due to an inability to convert C to D, resulting in the removal of the concerted control of the first enzyme. Provided that E is included in the medium at a level sufficient to allow growth but insufficient to cause inhibition then C will be accumulated due to the control of the end product G on the conversion of C to F. The example shown in Fig. 3.12(3) is similar to that in Fig. 3.12(2) except that it is a double auxotroph and requires the feeding of both E and G. Figure 3.12(4) is, again, the same pathway and illustrates another double mutant with the deletion for the production of G occurring between F and G, resulting in the accumulation of F.

Figure 3.12(5) illustrates the accumulation of an end product of a branched pathway which is normally controlled by the feedback inhibition of the first enzyme in the pathway by the concerted effects of E and I. The mutant illustrated is auxotrophic for I and G due to an inability to convert C to F and, thus, provided G and I are supplied in quantities which will satisfy growth requirements without causing inhibition, the end product, E, will be accumulated.

All the hypothetical examples discussed above are auxotrophic mutants and, under certain circumstances, may accumulate relatively high concentrations of intermediates or end products. Therefore, the isolation of auxotrophic mutants may result in the isolation of high-producing strains, provided that the mutation for auxotrophy occurs at the correct site, e.g. between C and D in Figs 3.12(1) and (2). The recovery of auxotrophs is a simpler process than is the recovery of high producers, as such, so that the best approach is to

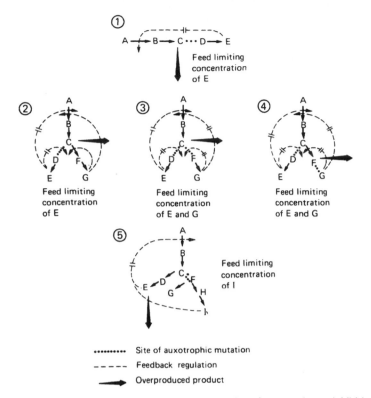

········· Site of auxotrophic mutation

– – – – Feedback regulation

➞ Overproduced product

FIG. 3.12. Overproduction of primarymetabolites by decreasing the concentration of a repressing or inhibiting end product. ···· Site of auxotrophicmutations; ————,feedback regulation; → overproduced product (Demain, 1972).

design a procedure to select relevant auxotrophs from the survivors of a mutation and subsequently screen the selected auxotrophs for productivity. Productive strains amongst the auxotrophs may be detected by over-layering colonies of the mutants with agar suspensions of bacteria auxotrophic for the required product. The high-producing mutants may be identified by the growth of the overlay around the producer. The most commonly used methods for the recovery of auxotrophic mutants are the use of some form of enrichment culture or the use of a technique to visually identify the mutants.

The enrichment processes employed are based on the provision of conditions which adversely affect the prototrophic cells but do not damage the auxotrophs. Such conditions may be achieved by exposing the population, in minimal medium, to an antimicrobial agent which only affects dividing cells which should result in the death of the growing prototrophs but the survival of the non-growing auxotrophs. Several techniques have been developed using different antimicrobials suitable for use with a range of micro-organisms.

Davis (1949) developed an enrichment technique utilizing penicillin as the inhibitory agent. The survivors of a mutation treatment were first cultured in complete medium, harvested by centrifugation, washed and resuspended in minimal medium plus penicillin. Only the growing prototrophic cells were susceptible to the penicillin and the non-growing auxotrophs survived. The cells were harvested by centrifugation, washed (to remove the penicillin and products released from lysed cells) and resuspended in complete medium to allow the growth of the auxotrophs, which could then be purified on solid medium. The nature of the auxotrophs isolated may be determined by the design of the so-called complete medium; if only one addition is made to the minimal medium then mutants auxotrophic for the additive should be isolated.

Abe (1972) described the use of Davis' technique to isolate auxotrophic mutants of the glutamic acid producing organism *C. glutamicum*. The procedure is outlined in Fig. 3.13.

Advantage has also been taken of the fact that the ungerminated spores of some organisms are more resistant to certain compounds than are the germinated spores. Thus, by culturing mutated spores in minimal

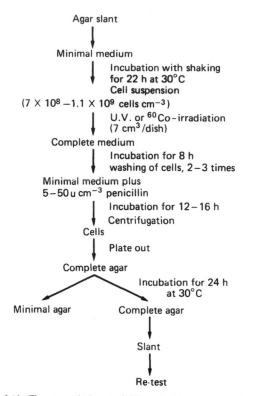

Agar slant

↓

Minimal medium

↓ Incubation with shaking
for 22 h at 30°C
Cell suspension

$(7 \times 10^8 - 1.1 \times 10^9 \text{ cells cm}^{-3})$

↓ U.V. or $^{60}$Co-irradiation
$(7 \text{ cm}^3/\text{dish})$

Complete medium

↓ Incubation for 8 h
washing of cells, 2–3 times

Minimal medium plus
$5-50 \text{ u cm}^{-3}$ penicillin

↓ Incubation for 12–16 h
Centrifugation

Cells

↓ Plate out

Complete agar

↙ ↘ Incubation for 24 h
at 30°C

Minimal agar    Complete agar

↓

Slant

↓

Re-test

FIG. 3.13. The use of the penicillin selection method for the isolation of auxotrophic mutants of *C. glutamicum* (Abe, 1972).

medium only the prototrophs will germinate and subsequent treatment of the spore suspension with a suitable compound would kill the germinated prototrophic spores but leave the ungerminated auxotrophic spores unharmed. The auxotrophic spores may then be isolated by washing, to remove the inhibitor, and cultured on supplemented medium. Ganju and Iyengar (1968) developed a technique of this type using sodium pentachlorophenate against the spores of *Penicillium chrysogenum*, *Streptomyces aureofaciens*, *S. olivaceus* and *Bacillus subtilis*.

The mechanical separation of auxotrophic and prototrophic spores of filamentous organisms has been achieved by the 'filtration enrichment method' (Catcheside, 1954). Liquid minimal medium is inoculated with mutated spores and shaken for a few hours, during which time the prototrophs will germinate but the auxotrophs will not. The suspension may then be filtered through a suitable medium, such as sintered glass, which will tend to retain the germinated spores resulting in a concentration of auxotrophic spores in the filtrate.

The visual identification of auxotrophs is based on the alternating exposure of suspected colonies to supplemented and minimal media. Colonies which grow on supplemented media, but not on minimal, are auxotrophic. The alternating exposure of colonies to supplemented and minimal medium has been achieved by replica plating (Lederberg and Lederberg, 1952). The technique consists of allowing the survivors of a mutation treatment to develop colonies on petri dishes of supplemented medium and then transferring a portion of each colony to minimal medium. The transfer process may be 'mechanized' by using some form of replicator. For bacteria the replicator is a sterile velvet pad attached to a circular support and replication is achieved by inverting the petri dish on to the pad, thus leaving an imprint of the colonies on the pad which may be used to inoculate new plates by pressing the plates on to the pad. It may be possible to replicate fungal and streptomycete cultures using a velvet pad, but, if unsatisfactory results are obtained, a steel pin replicator may be more appropriate.

Visual identification of auxotrophs has also been achieved by the so-called 'sandwich technique'. The survivors of a mutation treatment are seeded in a layer of minimal agar in a petri dish and covered with a layer of sterile minimal agar. The plate is incubated for 1 or 2 days and the colonies developed are marked on the base of the plate, after which a layer of supplemented agar is poured over the surface. The colonies which then appear after a further incubation period are auxotrophic, as they were unable to grow on the minimal medium.

## EXAMPLES OF THE USE OF AUXOTROPHS FOR THE PRODUCTION OF PRIMARY METABOLITES

Many auxotrophic mutants have been produced from *C. glutamicum* for the synthesis of both amino acids and nucleotide related compounds. *C. glutamicum* is a biotin-requiring organism which will produce glutamic acid under biotin-limited conditions but it is important to remember that mutants of this organism, employed for the production of other amino acids, must be supplied with levels of biotin optimum for growth. Biotin-limited conditions will result in these mutants producing glutamate and not the desired amino acid.

Auxotrophic mutants of *C. glutamicum* have been used for the production of lysine. The control of the production of the aspartate family of amino acids in *C. glutamicum* is shown in Fig. 3.14. Aspartokinase, the

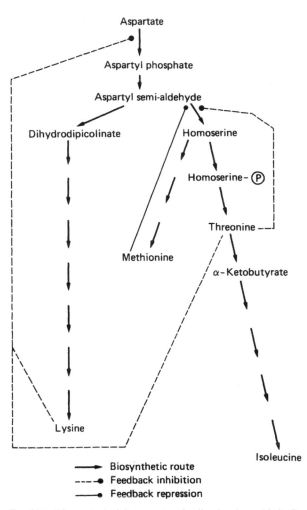

FIG. 3.14. The control of the aspartate family of amino acids in *C. glutamicum*.

concerted feedback inhibition of aspartokinase. Kinoshita and Nakayama (1978) quoted the homoserine auxotroph, *C. glutamicum 901*, as producing 44 g dm$^{-3}$ lysine.

The control of the production of arginine in *C. glutamicum* is shown in Fig. 3.15. The major control of the pathway is the feedback inhibition of the second enzyme in the sequence, acetylglutamic acid phosphorylating enzyme, although the first enzyme may also be subject to regulation.

Kinoshita *et al.* (1957b) isolated a citrulline requiring auxotroph of *C. glutamicum* which would accumulate ornithine at a molar yield of 36% from glucose, in the presence of limiting arginine and excess biotin. The mutant lacked the enzyme converting ornithine to citrulline which resulted in the cessation of arginine synthesis and, therefore, the removal of the control of the pathway.

Inosine monophosphate has been shown to demonstrate flavour-enhancing qualities and is produced commercially by the chemical phosphorylation of ino-

first enzyme in the pathway, is controlled by the concerted feedback inhibition of lysine and threonine. Homoserine dehydrogenase is subject to feedback inhibition by threonine and repression by methionine. The first enzyme in the route from aspartate semialdehyde to lysine is not subject to feedback control. Thus, the control system found in *C. glutamicum* is a relatively simple one. Nakayama *et al.* (1961) selected a homoserine auxotroph of *C. glutamicum*, by the penicillin selection and replica plating method, which produced lysine in a medium containing a low level of homoserine, or threonine plus methionine. The mutant lacked homoserine dehydrogenase which allowed aspartic semi-aldehyde to be converted solely to lysine and the resulting lack of threonine removed the

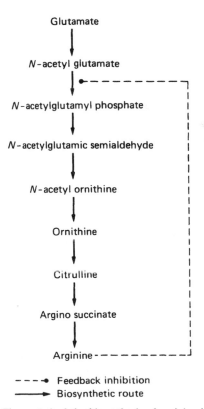

FIG. 3.15. The control of the biosynthesis of arginine in *C. glutamicum*.

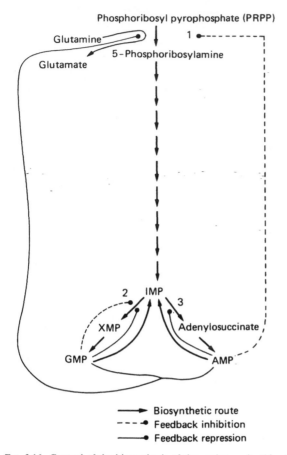

FIG. 3.16. Control of the biosynthesis of the purine nucleotides in *Bacillus subtilis*. 1. Reaction catalysed by PRPP amidinotransferase. 2. Reaction catalysed by IMP dehydrogenase. 3. Reaction catalysed by adenylosuccinate synthase. AMP = Adenosine monophosphate. IMP = Inosine monophosphate. XMP = Xanthine monophosphate. GMP = Guanosine monophosphate.

sine (Shibai *et al.* 1978) which is produced from auxotrophic strains of *Bacillus subtilis* The control of the production of purine nucleotides is shown in Fig. 3.16. The main sites of control shown in Fig. 3.16 are:

(i) Phosphoribosyl pyrophosphate (PRPP) amidinotransferase (the first enzyme in the sequence) is feedback inhibited by AMP but only very slightly by GMP.

(ii) The synthesis of PRPP amidinotransferase is repressed by the co-operative action of AMP and GMP, as are the syntheses of the other enzymes indicated in Fig. 3.16.

(iii) IMP dehydrogenase is feedback inhibited and repressed by GMP.

(iv) Adenylosuccinate synthase is repressed by AMP but is not significantly inhibited.

Mutants which are auxotrophic for AMP or doubly auxotrophic for AMP and GMP have been isolated which will excrete inosine at levels of up to 15 g dm$^{-3}$ (Demain, 1978). AMP auxotrophs, lacking adenylosuccinate synthase activity, require the addition of small quantities of adenosine but will accumulate inosine due to the removal of the inhibition and co-operative repression of PRPP amidinotransferase, as shown in Fig. 3.17. AMP and GMP double auxotrophs will produce inosine due to the removal of the controls normally imposed by the two end products, as illustrated in Fig. 3.18. Such double auxotrophs require the addition of both adenosine and guanosine, in small concentrations.

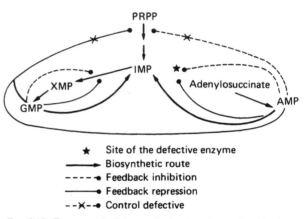

FIG. 3.17. The control of the synthesis of purine nucleotides in a mutant with a defective adenylosuccinate synthase.

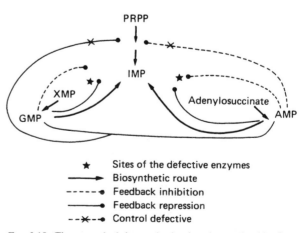

FIG. 3.18. The control of the synthesis of purine nucleotides in a mutant defective in adenylosuccinate synthase and IMP dehydrogenase activities.

## THE ISOLATION OF MUTANTS THAT DO NOT RECOGNIZE THE PRESENCE OF INHIBITORS AND REPRESSORS

The use of auxotrophic mutants has resulted in the production of many microbial products in large concentrations, but, obviously, such mutants are not suitable for the synthesis of products which control their own synthesis independently. A hypothetical example is shown in Fig. 3.19 where the end product P controls its own biosynthesis by feedback inhibition of the first enzyme in the pathway. If it is required to produce the intermediate F in large concentrations then this may be achieved by the isolation of a mutant auxotrophic for P, blocked between F and P. However, if P is required to be synthesized in large concentrations it is quite useless to produce an auxotrophic mutant. The solution to this problem is to modify the organism such that the first enzyme in the pathway no longer recognizes the presence of inhibiting levels of P. The isolation of mutants altered in the recognition of control factors has been achieved principally by the use of two techniques:

(i) The isolationOB of analogue resistant mutants.
(ii) The isolation of revertants.

An analogue is a compound which is very similar in structure to another compound. Analogues of amino acids and nucleotides are frequently growth inhibitory, and their inhibitory properties may be due to a number of possible mechanisms. For example, the analogue may be used in the biosynthesis of macromolecules resulting in the production of defective cellular components. In some circumstances the analogue is not incorporated in place of the natural product but interferes with its biosynthesis by mimicing its control properties. For example, consider the pathway illustrated in Fig. 3.19 where the end product, P, feedback inhibits the first enzyme in the pathway. If P* were an analogue of P (which could not substitute for P in biosynthesis) and were to inhibit the first enzyme in a similar way to P, then the biosynthesis of P may be prevented by P* which could result in the inhibition of the growth of the organism.

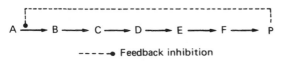

A $\longrightarrow$ B $\longrightarrow$ C $\longrightarrow$ D $\longrightarrow$ E $\longrightarrow$ F $\longrightarrow$ P

- - - - • Feedback inhibition

FIG. 3.19. The control of the production of an end product P.

Mutants may be isolated which are resistant to the inhibitory effects of the analogue and, if the site of toxicity of the analogue is the mimicing of the control properties of the natural product, such mutants may overproduce the compound to which the analogue is analogous. To return to the example of the biosynthesis of P where P* is inhibitory due to its mimicing the control properties of P; a mutant may be isolated which may be capable of growing in the presence of P* due to the fact that the first enzyme in the pathway is no longer susceptible to inhibition by the analogue. The modified enzyme of the resistant mutant may not only be resistant to inhibition by the analogue but may also be resistant to the control effects of the natural end product, P, resulting in the uninhibited production of P. If the control system were the repression of enzyme synthesis, then the resistant mutant may be modified such that the enzyme synthesis machinery does not recognize the presence of the analogue. However, the site of resistance of the mutant may not be due to a modification of the control system; for example, the mutant may be capable of degrading the analogue, in which case the mutant would not be expected to overproduce the end product. Thus, analogue resistant mutants may be expected to overproduce the end product to which the analogue is analogous provided that:

(i) The toxicity of the analogue is due to its mimicing the control properties of the natural product.
(ii) The site of resistance of the resistant mutant is the site of control by the end product.

Resistant mutants may be isolated by exposing the survivors of a mutation treatment to a suitable concentration of the analogue in growth medium and purifying any colonies which develop. Sermonti (1969) described a method to determine the suitable concentration. The organism was exposed to a range of concentrations of the toxic analogue by inoculating each of a number of agar plates containing increasing levels of the analogue with $10^6$ to $10^9$ cells. The plates were incubated for several days and examined to determine the lowest concentration of analogue which allowed only a very few isolated colonies to grow, or completely inhibited growth. The survivors of a mutation treatment may then be challenged with the pre-determined concentration of the analogue on solid medium. Colonies which develop in the presence of the analogue may be resistant mutants.

Szybalski (1952) constructed a method of exposing

the survivors of a mutation to a range of analogue concentrations on a single plate. Known as the gradient plate technique, it consists of pouring 20 cm³ of molten agar medium, containing the analogue, into a slightly slanted petri dish and allowing the agar to set at an angle. After the agar has set, a layer of medium not containing the analogue is added and allowed to set with the plate level. The analogue will diffuse into the upper layer giving a concentration gradient across the plate and the survivors of a mutation treatment may be spread over the surface of the plate and incubated. Resistant mutants should be detected as isolated colonies appearing beyond a zone of confluent growth, as indicated in Fig. 3.20. Whichever method is used for the isolation of analogue-resistant mutants, great care should be taken to ensure that the isolates are genuinely resistant to the analogue by streaking them, together with analogue-sensitive controls, on both analogue-supplemented and analogue-free media. The resistant isolates should then be screened for the production of the desired compound by overlayering them with a bacterial strain requiring the compound; producers may then be recognized by a halo of growth of the indicator strain.

Sano and Shiio (1970) investigated the use of lysine analogue-resistant mutants of *Brevibacterium flavum* for the production of lysine. The control of the biosynthesis of the aspartate family of amino acids in *B. flavum* is as illustrated for *C. glutamicum* in Fig. 3.14. The main control of lysine synthesis is the concerted feedback inhibition of aspartokinase by lysine and threonine. Sano and Shiio demonstrated that S-(2 aminoethyl) cysteine (AEC) completely inhibited the growth of *B. flavum* in the presence of threonine, but only partially in its absence. Also, the inhibition by AEC and threonine could be reversed by the addition of lysine. This evidence suggested that the inhibitory effect of AEC was due to its mimicing lysine in the concerted inhibition of aspartokinase. AEC-resistant mutants were isolated by plating the survivors of a mutation treatment on minimal agar containing 1 mg cm⁻³ of both AEC and threonine. A relatively large number of the resistant isolates accumulated lysine, the best producers synthesizing more than 30 g dm⁻³. Investigation of the lysine producers indicated that their aspartokinases had been desensitized to the concerted inhibition by lysine and threonine.

The development of an arginine-producing strain of *B. flavum* by Kubota *et al.* (1973) provides an excellent example of the selection of a series of mutants resistant to increasing levels of an analogue. The control of the

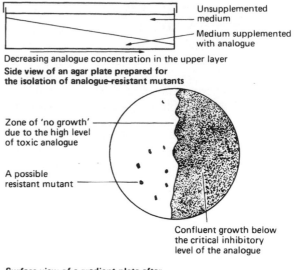

Decreasing analogue concentration in the upper layer

**Side view of an agar plate prepared for the isolation of analogue-resistant mutants**

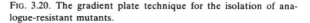

**Surface view of a gradient plate after inoculation and incubation.**

FIG. 3.20. The gradient plate technique for the isolation of analogue-resistant mutants.

biosynthesis of arginine in *B. flavum* is similar to that shown for *C. glutamicum* in Fig. 3.15. Kubota *et al.* selected mutants resistant to the arginine analogue, 2-thiazolealanine, and the genealogy of the mutants is shown in Fig. 3.21. Strain number 352 produced 25.3 g dm⁻³ arginine. Presumably, the mutants were altered in the susceptibility of the second enzyme in the pathway to inhibition by arginine.

A classic example of the rationale of analogue resistant mutant isolation is seen in the development of biotin overproducing strains. Currently, biotin is produced commercially by chemical synthesis, but considerable effort has been (and is being) expended to develop a competitive biotechnological process. The screening of natural isolates for their ability to accumulate biotin-vitamers led to the isolation of *Bacillus sphaericus* (Ogata *et al.*, 1965). Repression by biotin was shown to be an important control mechanism and, thus, attempts were made to isolate mutants resistant to biotin analogues. Mutants of *B. sphaericus* resistant to acidomycin (ACM) and/or 5-(2-thienyl)-*n*-valeric acid (TVA) were capable of synthesizing up to 11 times (0.4 mg cm⁻³) the biotin level of the wild-type (Tanaka *et al.*, 1988). More recently, Sakurai *et al.* (1993) isolated dual ACM/TVA resistant strains of *Serratia marcescens* capable of synthesizing 20 mg cm⁻³ biotin. The use of genetic engineering techniques to further develop these

B. flavum ATCC 14067

↓   X–ray irradiation

No. 33038 (guanine⁻)

↓   NG treatment, selection with TA at 5 mg dm⁻³

No. 112 (guanine⁻, TA resistant)

↓   NG treatment, selection with TA at 10 mg dm⁻³

No. 179 (guanine⁻, TA resistant) L–arginine producer at 14.3 g dm⁻³

↓   Diethyl sulphate treatment

No. 352 (guanine⁻, TA resistant) L–arginine producer at 25.3 g dm⁻³

FIG. 3.21. The genealogy of L-arginine-producing mutants of *B. flavum*. TA, thiazolealanine; NG, *N*-methyl-*N'*-nitro-*N*-nitroso-guanidine (Kubota *et al.*, 1973).

strains is considerd in a later section of this chapter.

The second technique used for the isolation of mutants altered in the recognition of control factors is the isolation of revertant mutants. Auxotrophic mutants may revert to the phenotype of the mutant 'parent'. Consider the hypothetical pathway illustrated in Fig. 3.19 where P controls its own production by feedback inhibiting the first enzyme (*a*) of the pathway. A mutant does not produce the enzyme, *a*, and is, therefore, auxotrophic for P. However, a revertant of the mutant produces large concentrations of P. The explanation of the behaviour of the revertant is that, with two mutations having occurred at loci concerned with the production of enzyme *a*, the enzyme of the revertant is different from the enzyme of the original prototrophic strain and is not susceptible to the control by P. Revertants may occur spontaneously or mutagenic agents may be used to increase the frequency of occurrence, but the recognition of the revertants would be achieved by plating millions of cells on medium which would allow the growth of only the revertants, i.e. in the above example, on medium lacking P.

Shiio and Sano (1969) investigated the use of prototrophic revertants of *B. flavum* for the production of lysine. These workers isolated prototrophic revertants from a homoserine dehydrogenase-defective mutant. The revertants were obtained as small-colony forming strains and produced up to 23 g dm⁻³ lysine. The overproduction of lysine was shown to be due to the very low level of homoserine dehydrogenase in the revertants which, presumably, resulted in the synthesis of threonine and methionine in quantities sufficient for some growth, but insufficient to cause inhibition or repression.

Mutant isolation programmes for the improvement of strains producing primary metabolites have not relied on the use of only one selection technique. Most projects employed a number of methods including the selection of natural variants and the selection of induced mutants by a variety of means. The selection of bacteria overproducing threonine provides a good example of the use of a variety of selection techniques. Attempts to isolate auxotrophic mutants of *C. glutamicum* producing threonine were unsuccessful despite the fact that productive auxotrophic strains of *Escherichia coli* had been isolated. Huang (1961) demonstrated threonine production at a level of 2–4 g dm⁻³ by a diaminopimelate and methionine double auxotroph of *E. coli*. Kase *et al.* (1971) isolated a triple auxotrophic mutant of *E. coli* which required diaminopimelate, methionine and isoleucine and produced between 15 and 20 g dm⁻³ threonine. The control of the production of the aspartate family of amino acids in *E. coli* is shown in Fig. 3.22 and that in *C. glutamicum* in Fig. 3.14. The mechanism of control in *E. coli* involves a system of isoenzymes, three isoenzymic forms of aspartokinase and two of homoserine dehydrogenase, under the influence of different end products. However, in *C. glutamicum* control is effected by the concerted inhibition of a single aspartokinase by threonine and lysine; by the inhibition of homoserine dehydrogenase by threonine and the repression of homoserine dehydrogenase by methionine. Thus, the control of homoserine dehydrogenase may not be removed by auxotrophy without the loss of threonine production. However, in *E. coli* methionine auxotrophy would remove control of the methionine-sensitive homoserine dehydrogenase and aspartokinase which would allow threonine production, despite the control of the threonine-sensitive isoenzymes by threonine. *E. coli* mutants

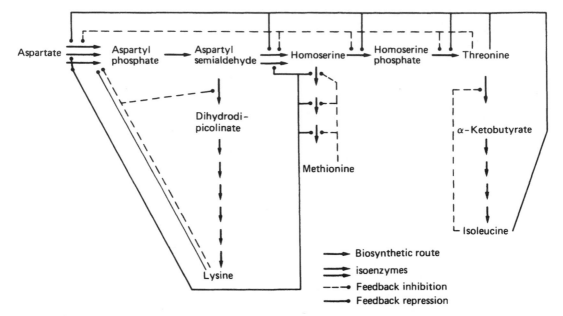

FIG. 3.22. Control of the aspartate family of amino acids in *Escherichia coli*.

also lacking lysine and isoleucine would be relieved of the control of the lysine-sensitive aspartokinase and the degradation of threonine to isoleucine.

The production of threonine by *C. glutamicum* has been achieved by the use of combined auxotrophic and analogue resistant mutants. A good example of the approach is given by Kase and Nakayama (1972) who obtained stepwise improvements in productivity by the imposition of resistance to α-amino-β-hydroxy valeric acid (a threonine analogue) and *S*-(β-aminoethyl)-L-

cysteine (a lysine analogue) on a methionine auxotroph of *C. glutamicum*. The genealogy of the mutants is shown in Fig. 3.23. The analogue-resistant strains were shown to be altered in the susceptibility of aspartoki-nase and homoserine dehydrogenase to control, and the lack of methionine removed the repression control of homoserine dehydrogenase. The use of recombinant DNA technology has resulted in the construction of far more effective threonine producers and these strains are considered in a later section of this chapter.

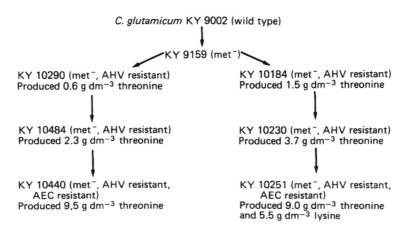

FIG. 3.23. The genealogy of mutants of *C. glutamicum* producing L-threonine or L-threonine plus L-lysine. AHV, α-amino-β-hydroxy valeric acid: AEC, *S*-(β-aminoethyl)-L-cysteine (Kase and Nakayama, 1972).

The development of strains producing guanosine in high levels provides a further example of the use of auxotrophs and analogue-resistant mutants in the same programme. Guanosine production has been achieved using auxotrophic and analogue resistant mutants of *B. subtilis*. The control of the production of purine nucleotides in *B. subtilis* is shown in Fig. 3.16 (Demain, 1978). The major points of control to be modified to achieve guanosine overproduction are:

(i) Removal of AMP inhibition of phosphoribosyl pyrophosphate amidino transferase by adenine auxotrophy.
(ii) Elimination of GMP reductase.
(iii) Bypassing the inhibition of IMP dehydrogenase by GMP.

This approach is demonstrated in the work of Shiio (cited by Demain, 1978) who obtained analogue-resistant mutants from an adenine auxotroph of *B. subtilis*. Mutants were isolated in two stages — first, for resistance to low levels of 8-azaguanine (an analogue of GMP) and then for resistance to high levels of the analogue. The resulting mutant produced 9 g dm$^{-3}$ guanosine due to the removal of control of phosphoribosyl pyrophosphate amidinotransferase and IMP dehydrogenase.

### The isolation of induced mutants producing improved yields of secondary metabolites where directed selection is difficult to apply

The discussion so far has considered the isolation of mutants producing products whose biosynthesis and control have been sufficiently understood to prepare 'blueprints' of the desirable mutants which have enabled the construction of suitable selection procedures. In contrast, important secondary metabolites were being produced long before their biosynthetic pathways, and certainly the control of those pathways, had been elucidated. Thus, strain improvement programmes had to be developed without this fundamental knowledge which meant that they depended on the random selection of the survivors of mutagen exposure. Elander and Vournakis (1987) described these techniques as "hit or miss methods that require brute force, persistence and skill in the art of microbiology". However, despite the limited knowledge underlying these approaches they were extremely effective in increasing the yields of antibiotics, as illustrated in Table 3.5 (Riviere, 1977).

TABLE 3.5. *Improvement of antibiotic yields during the first 20 years of antibiotic development (Riviere, 1977)*

| Antibiotic | Initial yield at time of discovery (units cm$^{-3}$) | Improved yield in France, 1972 (units cm$^{-3}$) |
|---|---|---|
| Penicillin | 20 (1943) | 12,000–15,000 |
| Steptomycin | 50 (1945) | 12,000–15,000 |
| Chlortetracycline | 200 (1948) | 12,000–15,000 |
| Erythromycin | 100 (1955) | 3000 |

More rational approaches have been developed which reduce the empirical nature of strain-improvement programmes. These developments include the streamlining of the empirical techniques and the use of more directed selection methods. Before discussing the attempts at directed selection the empirical approach will be considered along with the attempts made to improve this approach, including miniaturized programmes.

The empirical approach to strain improvement involves subjecting a population of the micro-organism to a mutation treatment and then screening a proportion of the survivors of the treatment for improved productivity. The assessment of the chosen survivors was usually carried out in shake flasks, resulting in the procedure being costly, both in terms of time and personnel. According to Fantini (1966) the two questions which arise in the design of such programmes are:

(i) How many colonies from the survivors of a mutation treatment should be isolated for testing?
(ii) Which colonies should be isolated?

In attempting to answer the first question Fantini dismissed statistical approaches as impractical and claimed that the number of colonies isolated for testing is determined by the practical limiting factors of personnel, incubator and shaker space and time. However, Davies (1964) demonstrated that a statistical approach could give valuable guidelines for the efficient utilization of physical resources in strain-development programmes. Davies based a computer simulation of a mutation and screening programme on the availability of 200 shaker spaces and practical results of error variance and the distribution of yield likely to occur amongst the mutants. He assumed that the majority of the progeny of a mutation would give a small, rather than a large, increase in productivity and it would, therefore, be more feasible to screen a small number in the hope of obtaining a small increase rather than screen a large number in the hope of obtaining a large increase. The major difficulty inherent in this approach is the error involved in determining small increases which implies the replication of screening tests and, therefore, the use of more facilities.

Davies used the computer simulation to investigate the merits of replication in screening programmes and finally proposed the use of a two-stage scheme where mutants were tested singly in the first stage and then the better producers were tested in quadruplicate in the second stage. Davies concluded that such two-stage schemes were adequate over a wide range of conditions, although a one-stage screen could be used if the testing error were small and the frequency of occur-

rence of favourable types high, and a three-stage screen if the testing error were high and the frequency of occurrence of favourable types low. A screening programme based on Davies' proposals is shown in Fig. 3.24.

The answer to Fantini's second question (which colonies should be isolated?) is extremely difficult in the field of secondary metabolites and in Davies' scheme the colonies were chosen at random. The selection of colonies on the basis of changed morphology has been considered by a number of workers and it appears that this is an undesirable technique. Elander (1966) demonstrated that it was preferable to isolate normal morphological types as, although a morphologi-

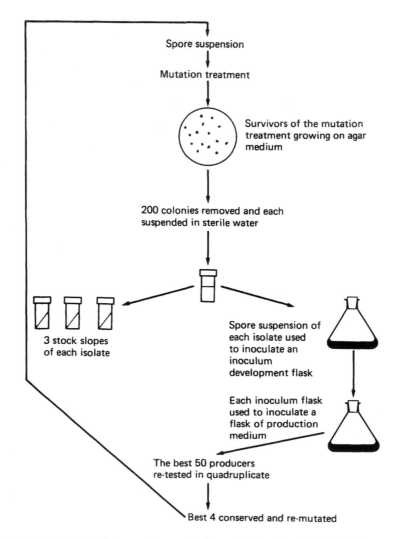

Fig. 3.24. A strain-improvement programme for a secondary metabolite producing culture (Davies, 1964).

cal variant may be a superior producer, it might require considerable fermentation development to materialize the increased production. Also, Alikhanian (1962) claimed that the vast majority of morphological mutants of the antibiotic producing actinomycetes tested were inferior producers. The most common type of shake flask programme quoted is similar to that of Davies where the choice of colonies is random. However, there are now many reports of screens which miniaturize the procedure for the improvement of secondary metabolite producing strains. Such miniaturized systems are designed to enable the productivity of all (or a significant proportion) of the survivors to be assessed which should eliminate (or reduce) the problem of choice of colonies for assessment and increase the throughput of cultures.

The basis of the miniaturized techniques is to grow the survivors of the mutation treatment either in a very low volume of liquid medium or on solidified (agar) medium. If the product is an antibiotic, the agar-grown colonies may be overlayed with an indicator organism sensitive to the antibiotic produced, allowing assay to be done in situ. The level of antibiotic is assessed by the degree of inhibition of the overlayed indicator. The system is simple to apply to strains producing low levels of antibiotic but must be modified to allow the screening of high producers where very large zones of inhibition would be obtained. Also a system should be used to free the superior producers from contaminating indicator organisms. Dulaney and Dulaney (1967) used overlay techniques in the isolation of mutants producing chlortetracycline. Mutated spores were cultured on an agar medium in petri dishes for 6 days and then covered with pieces of sterile cellophane. An overlay of agar containing the indicator organism was then added and the plates incubated overnight. The mutant colonies were kept free from contamination by the cellophane and the size of the inhibition zone could be controlled by the depth of the base layer, the age of the colonies when over-layed, the depth of the overlay and the temperature of incubation. The system was calibrated initially such that a single colony of a non-mutated strain would not produce an inhibition zone but that two such adjacent colonies would. In practice, the size of the inhibition zone was controlled by the depth of the overlay. Dulaney and Dulaney obtained a far greater enrichment in the number of desired phenotypes by the overlay technique than by random selection and testing in liquid medium.

Ichikawa et al. (1971) screened for the increased production of the antibiotic kasugamycin by a Streptomyces sp., using a miniaturized technique termed the agar piece method. In order to prevent interference between colonies, mutated spores were grown on plugs of agar which were then placed on assay plates containing agar seeded with the indicator organism, levels of the antibiotic being determined by the size of inhibition zones. By combining this technique with a medium improvement programme the authors obtained a tenfold increase in productivity. Ichikawa's method was modified by Ditchburn et al. (1974) for the isolation of Aspergillus nidulans mutants synthesizing improved levels of penicillin. These workers obtained promising results using the technique and claimed that its potential for the recovery of higher-yielding mutants, for a given expenditure of labour time, was greater than the shake-flask method. The level of production of penicillin by A. nidulans is very small compared with P. chrysogenum and the technique would have to be modified considerably for commercial use.

Ball and McConagle (1978) developed a miniaturized technique suitable for the assay of penicillin production by 'high-yielding' strains of P. chrysogenum (producing up to 6000 units $cm^{-3}$). These workers highlighted the design of the solidified medium as a critical factor in the optimization of the method. They claimed that the growth of a colony on agar-solidified medium would be unlikely to modify the medium to the same extent as the growth of the organism on the same medium in submerged liquid culture. Thus, the nutrient-limiting conditions that favour the onset of antibiotic production might not be achieved by the growth of the culture on solid medium which would not allow the full production potential of the culture to be detected. However, nutrient-limiting conditions were achieved in the solidified medium by omitting the main carbon source and reducing the corn steep-liquor content. Mutated spores were plated over the surface of petri dishes of the nutrient limited medium such that ten to twenty colonies developed per plate. The time of incubation of the plates was not quoted but "when the colonies were of a size suitable for accurate measurement" a pasteurized spore suspension of B. subtilis containing 0.16 units $cm^{-3}$ of penicillinase (to limit the size of the inhibition zones) was dispersed over the surface of the dish. The plates were then incubated for 18 to 24 hours and the inhibition zones examined. Suitable colonies were freed of contaminating B. subtilis by culturing on nutrient agar containing sodium penicillin and streptomycin sulphate. The use of this technique enabled three operators to scan 15,000 survivors from ultraviolet irradiated spore populations in 3 months.

The major disadvantages of the miniaturized solidified medium technique approach is that productivity expressed on the solid medium may not be expressed in subsequent liquid culture and conversely, colonies not showing activity on solid media may be highly productive in liquid medium. Despite these limitations the above workers have demonstrated that the approach has considerable merit and Ball (1978) claimed that the increase in throughput may be as much as 20 times that of a conventional shake-flask programme. Some recent work by Bushell's group (Pickup et al., 1993) provides an interesting insight into the problem of certain streptomycete isolates producing a secondary metabolite on agar but not in liquid culture. It is important to appreciate that the strains were natural isolates and not the survivors of mutation treatments but the conclusions do have relevance to mutant development. Working on the premise that non-production in liquid medium may be due to a more fragmented morphology these strains were subjected to a filtration enrichment procedure. Liquid cultures were filtered through a linen filter and the filtrate and retentate mycelium re-cultured in fresh medium. The procedure was then repeated sequentially. The enriched retentate mycelium of several isolates which were previously unable to produce antibiotic in liquid culture gave rise to stable filamentous types synthesizing in liquid medium. Although this approach could not be used routinely in a high throughput mutant screen it may be applied to a few high-producing strains which have not fulfilled their promise (detected in an agar screen) in liquid media.

An excellent example of a miniaturized screening programme is given by the work of Dunn-Coleman et al. (1991) of Genencor International. These workers developed a low-volume liquid medium system to isolate mutants of Aspergillus niger var. awamori capable of improved secretion of bovine chymosin. The Aspergillus strain had been constructed using recombinant DNA technology and this construction will be considered in a later section of this Chapter. Mutated spore suspensions were diluted and inoculated into 96-well microtitre plates using robots. The dilutions were such that each well should have contained one viable spore. The plates were incubated in a static incubator and then harvested using a robotic pipetting station and assayed for product. Typically, 50–60,000 mutated viable spores were assessed in each screen. The most promising 10–50 strains from the miniaturized screen were then tested in shake flask culture. The best producer from the shake-flask screen was then used as the starter for a further round of mutagenesis. The results of seven rounds of mutation and selection are shown in Table 3.6. Investigation of the improved strains showed

TABLE 3.6. *Chymosin production by NTG mutated strains of Aspergillus niger var awamori (Dunn-Coleman et al., 1991)*

| Rounds of NTG mutagenesis and recurrent selection | Chymosin concentration ($\mu g\ cm^{-3}$) | |
|---|---|---|
| | Microtitre plate | Shake flask |
| Parent | 42 | 286 |
| First | 50 | 273 |
| Second | 64 | 280 |
| Third | 136 | 377 |
| Fourth | 170 | 475 |
| Fifth | 200 | 510 |
| Sixth | 256 | 646 |

that they were capable of secreting a range of enzymes and, thus, the strains were presumed to be modified in their secretion mechanisms. It is interesting to note that an improved secreter of a heterologous protein has been isolated using a traditional mutation and selection approach which has been miniaturized and automated.

In several cases the examination of randomly selected high-yielding mutants has indicated that their superiority may have been due to modifications of their control systems. Goulden and Chataway (1969) demonstrated that a mutant of P. chrysogenum, producing high levels of penicillin, was less sensitive to the control of acetohydroxyacid synthase by valine. Valine is one of the precursors of penicillin and its synthesis is controlled by its feedback inhibition of acetohydroxyacid synthase. Thus, removal of the control of valine synthesis may result in the production of higher valine levels and, hence, greater production of penicillin. Pruess and Johnson (1967) demonstrated that higher-yielding strains of P. chrysogenum also contained higher levels of acyltransferase, the enzyme which catalyses the addition of phenylacetic acid to 6-aminopenicillanic acid. Dulaney (1954) reported that the best producer of streptomycin amongst Streptomyces griseus mutants was auxotrophic for vitamin $B_{12}$ and Demain (1973) claimed that this auxotrophic mutation was still a characteristic of production strains being used in 1969. Thus, it appears that some strains isolated by random selection, and overproducing secondary metabolites, are altered in ways similar to those strains isolated by directed selection techniques and overproducing primary metabolites.

There are many examples in the literature where a more directed selection approach has been adopted for the improvement of secondary metabolite producers.

The techniques used include the isolation of auxotrophs, revertants and analogue-resistant mutants.

## THE ISOLATION OF AUXOTROPHIC MUTANTS

Although mutation of secondary metabolite producers to auxotrophy has resulted frequently in their producing lower yields, cases of improved productivity have also been demonstrated. For example, Alikhanian *et al.* (1959) investigated the tetracycline producing abilities of fifty-three auxotrophic mutants, all of which produced significantly less tetracycline than the parent strain in normal production medium. Supplementation of the medium with the growth requirement resulted in one mutant expressing productivity superior to that of the parent.

It is sometimes difficult to explain the precise reason for the effect of mutation to auxotrophy on the production of secondary metabolites, but in the majority of cases it has been demonstrated to be an effect on the secondary metabolic system rather than, simply, an effect on the growth of the organism. The simplest explanation for the deleterious effect on secondary metabolite yield is that the auxotroph is blocked in the biosynthesis of a precursor of the end product, for example, Polsinelli *et al.* (1965) demonstrated that auxotrophs of *Streptomyces antibioticus* which required any of the precursors of actinomycin (isoleucine, valine or threonine) were poor producers of the antibiotic.

Many secondary metabolites may be considered as end products of branched pathways which also give rise to primary metabolites. Thus, a mutation to auxotrophy for the primary end product may also influence the production of the secondary product. In *P. chrysogenum*, lysine and penicillin share the same common biosynthetic route to α-aminoadipic acid, as shown in Fig. 3.25. This biosynthetic route may explain Bonner's (1947) observation that 25% of the lysine auxotrophs of *P. chrysogenum* he isolated could not produce penicillin. The role of lysine in the penicillin fermentation has also led to the investigation of lysine auxotrophs as potential superior penicillin producers. Demain (1957) demonstrated that lysine was inhibitory to the biosynthesis of pencillin. The explanation of this phenomenon is considered to be the inhibition of homocitrate synthase by lysine resulting in the depletion of α-aminoadipic acid required for penicillin synthesis (Demain and Masurekar, 1974). It may be postulated that lysine auxotrophs blocked immediately after α-aminoadipic acid would produce higher levels of penicillin due to the diversion of the intermediate towards

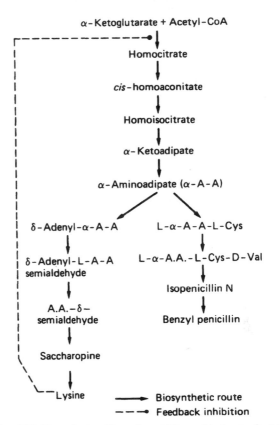

FIG. 3.25. Biosynthesis of benzyl penicillin and lysine in *Penicillium chrysogenum*.

penicillin synthesis and the removal of any control of homocitrate synthase by endogenous lysine. O'Sullivan and Pirt (1973) investigated the production of penicillin by lysine auxotrophs of *P. chrysogenum* in continuous culture but were unable to demonstrate improved productivity in a range of lysine-supplemented media. Luengo *et al.* (1979) examined the lysine regulation of penicillin biosynthesis in low-producing and industrial strains of *P. chrysogenum*. The industrial strain was capable of producing up to 12,000 units cm$^{-3}$ penicillin. These workers demonstrated that although the onset of penicillin synthesis in the industrial strain was less sensitive to lysine than in the low-producing strain, the empirical industrial selection procedures had not completely removed the mechanisms of lysine regulation of penicillin biosynthesis. Luengo *et al.* expressed the possibility that overproducers of penicillin may be obtained by the selection of mutants to lysine regulation. However, it should be noted that the industrial strain employed was a representative of only one series of penicillin producers and that other series may have already been modified with respect to the regulatory effects of lysine. Nevertheless, this study does indicate the possibility that selection of strains resistant to lysine control may be overproducers of penicillin.

It is far more difficult to explain the effect of auxotrophy for factors not associated with the biosynthesis of the secondary metabolite. Polsinelli *et al.* (1965) demonstrated that seven out of twenty-seven auxotrophs of *S. griseus* produced more actinomycin than did the prototrophic parent. Dulaney and Dulaney (1967) demonstrated increased chlortetracycline yields in eight out of eleven auxotrophs of *S. viridifaciens* when grown in supplemented media. In neither of these cases were the auxotrophic requirements directly involved in the biosynthesis of the secondary metabolites. Demain (1973) put forward two possible explanations to attempt to account for the behaviour of such auxotrophs. The first explanation is that the auxotrophic factors were involved in 'cross-pathway' regulation with the secondary metabolite or its precursors. Demain quoted several examples of cross-pathway regulation in primary metabolism where the activity of one pathway is affected by the product of an apparently unrelated sequence. The alternative explanation is that the effect on secondary metabolism is not due to the auxotrophy but to a second mutation accompanying the auxotrophy, i.e. a double mutation. Demain cited two attempts to determine whether the effects of auxotrophy on secondary metabolism were due to double mutations or to the auxotrophy. MacDonald *et al.* (1963) reverted a low-producing thiosulphate-requiring mutant of *P. chrysogenum* to thiosulphate independence and examined penicillin productivity by the revertants. Approximately half of the revertants re-acquired their 'grandparents' production level, whereas the other half retained their poor productivity. Polsinelli *et al.* (1965) reverted five isoleucine–valine auxotrophs of *S. antibioticus* (which also produced low levels of actinomycin compared with the parent strain) to prototrophy and discovered that some were returned to normal production and others to higher production levels than the 'grandparent'. Thus, in the case of Polsinelli *et al.*'s mutants it is unlikely that the effect on secondary metabolism was due to a double mutation, but it is possible that this was the case for some of MacDonald *et al.*'s strains. However, it should be remembered that both these groups of auxotrophs were poor secondary metabolite producers blocked in routes directly involved with the secondary biosynthetic pathway. It may be more relevant to examine the nature of auxotrophic strains blocked in apparently unrelated pathways and

produce improved levels of the secondary metabolite. It is not possible to say whether any of the auxotrophic mutants previously discussed produced superior levels of the secondary metabolite as a result of double mutations, but it may be possible to exploit this possibility in the future.

The treatment of bacterial cells with nitrosoguanidine (NTG) has been demonstrated to result in clusters of mutations around the replicating fork of the chromosome (Guerola *et al.*, 1971). Thus, if one of the mutations were selectable (for example, auxotrophy) it may be possible to isolate a strain containing the selectable mutation along with other non-selectable mutations which map close by. The efficient application of this technique would depend on the accurate mapping of the gene involved in producing the secondary metabolite so that neighbouring mutations may be selected. This technique may be valuable for the selection of mutants of the bacilli and streptomycetes producing high levels of antibiotics where mutations affecting synthesis may be mapped for each biosynthetic step. Co-mutation by NTG may then be followed by selection for changes to genes adjacent to those loci known to be involved in production of the particular secondary metabolite.

THE ISOLATION OF RESISTANT MUTANTS

The isolation of analogue-resistant mutants has already been discussed in the field of primary metabolism, where the rationale was that a mutant resistant to the inhibitory effects of an analogue, which mimics the control characteristics of the natural metabolite, may overproduce the natural metabolite. This approach has been adopted, or may be adopted, in a number of guises in the field of secondary metabolism.

(i) Mutants may be isolated which are resistant to the analogues of primary metabolic precursors of the secondary metabolite, thus increasing the availability of the precursor.

(ii) Mutants may be isolated which are resistant to the feedback effects of the secondary metabolite.

(iii) Mutants may be selected which are resistant to the toxic effects of the secondary metabolite when added to the trophophase of the producing organism.

(iv) Mutants may be isolated which are resistant to the toxic effects of a compound due to the production of the secondary metabolite.

#### (i) *Mutants resistant to the analogues of primary metabolic precursors of the secondary metabolite*

This approach has been adopted by Elander *et al.* (1971) in the isolation of mutants of *Pseudomonas aureofaciens* overproducing the antibiotic pyrrolnitrin. Tryptophan is a precursor of pyrrolnitrin and although it is stimulatory to production it is uneconomic to use as an additive in an industrial process. Thus, Elander *et al.* (1971) isolated mutants resistant to tryptophan analogues using the gradient plate technique described previously. A strain was eventually isolated which produced two to three times more antibiotic than the parent and was resistant to feedback inhibition by tryptophan. Addition of tryptophan to the improved strain would not result in higher pyrrolnitrin synthesis, indicating that tryptophan supply was no longer the limiting factor and the organism was producing sufficient endogenous tryptophan for antibiotic synthesis.

The role of tryptophan in the control of the candicidin fermentation has been investigated by Martin and co-workers (Martin, 1978). Candicidin is a polyene macrolide antibiotic containing an aromatic *p*-aminoacetophenone moiety derived from chorismic acid, a precursor of the aromatic amino acids, so that tryptophan and candicidin may be considered to be end products of a branched biosynthetic pathway. Tryptophan has been demonstrated to inhibit the biosynthesis of the antibiotic which led Martin *et al.* (1979) to isolate mutants resistant to tryptophan analogues. Such mutants produced more candicidin than the parent strain, apparently due to the removal of tryptophan control

Godfrey (1973) isolated mutants of *Streptomyces lipmanii* resistant to the valine analogue, trifluoroleucine. These mutants produced higher levels of cephamycin than the parent strain, and appeared to be deregulated for the isoleucine, leucine, valine biosynthetic pathway, indicating that valine may have been a rate-limiting step in cephamycin synthesis.

Methionine has been demonstrated to stimulate the biosynthesis of cephalosporin by *Acremonium chrysogenum* and superior producers have been isolated in the form of methionine analogue-resistant mutants (Chang and Elander, 1979). Lysine analogue resistant mutants have yielded a greater frequency of superior β-lactam antibiotic producers compared with random selection (Elander, 1989).

#### (ii) *Mutants resistant to the feedback effects of the secondary metabolite*

There are many cases in the literature of a secondary metabolite preventing its own synthesis. Martin (1978) cited the following examples: chloramphenicol, aurodox, cycloheximide, staphylomycin, ristomycin, puromycin, fungicidin, candihexin, mycophenolic acid and penicillin. The precise mechanisms of these controls is not clear but they appear to be specific against the synthesis of the secondary metabolite and not against the general metabolism of the cells. The mechanism of the control of its own synthesis by chloramphenicol appears to be the repression of arylamine synthetase (the first enzyme in the pathway from chorismic acid to chloramphenicol) by chloramphenicol. Jones and Vining (1976) demonstrated that arylamine synthetase was fully repressed by 100 mg dm$^{-3}$ chloramphenicol, a level of antibiotic which neither affected cell growth nor the activities of the other enzymes of the chloramphenicol pathway.

Martin (1978) quoted several examples of correlations between the level of secondary metabolite accumulation and the level which causes 'feedback' control, which may imply that the factor limiting the yields of some secondary metabolites is the feedback inhibition by the end product. The selection of mutants resistant to feedback inhibition by a secondary metabolite is a far more difficult task than the isolation of strains resistant to primary metabolic control. It is extremely unlikely that a toxic analogue of the secondary metabolite could be found where the toxicity lay in the mimicking of the feedback control of the secondary metabolite, a compound which would not be necessary for growth. However, the detection of mutants resistant to feedback inhibition by antibiotics may be achieved by the use of solidified media screening techniques, similar to the miniaturized screening techniques previously described. The technique would involve culturing the survivors of a mutation treatment on solidified medium containing hitherto repressing levels of the antibiotic and detecting improved producers by overlaying the colonies with an indicator organism. The difficulty inherent in this technique is that the incorporated antibiotic, itself, will inhibit the development of the indicator organism. This problem may be overcome by adjusting the depth of the overlay or the concentration of the indicator such that an inhibition zone could be produced only by a level of antibiotic greater than that incorporated in the original medium. However, it is unlikely that this would be a satisfactory solution for the selection of a high-producing commercial strain. Another approach would be to utilize an analogue of the antibiotic which mimicked the feedback control by the natural product but which did not have antimicrobial properties. Inhibition of antibiotic synthesis by analogues has been demonstrated in the cases of aurodox (Liu *et al.*, 1972) and penicillin (Gordee and Day, 1972). An alternative may be to use a mutant indicator bacterium for the overlay which is only sensitive to levels of the antibiotic in excess of that originally incorporated into the medium. In cases where it has been demonstrated that feedback control by the end product plays an important role in limiting productivity it would probably be worthwhile to design such procedures.

### (iii) *The isolation of mutants resistant to the toxic effects of the secondary metabolite in the trophophase*

It has been demonstrated for many secondary metabolite producing organisms that the secondary metabolite is toxic to the producing cell when it is present in the trophophase (growth phase) (Demain, 1974). Thus, it appears that a 'switch' in the metabolism of the organism in the idiophase enables it to produce an otherwise 'autotoxic' product. Furthermore, it has been demonstrated that, in some cases, the higher the resistance to the secondary metabolite in the growth phase the higher is productivity in the production phase. Dolezilova *et al.* (1965) demonstrated that the level of production of nystatin by various strains of *S. noursei* was related to the resistance of the strain to the antibiotic in the growth phase; a non-producing mutant was inhibited by 20 units $cm^{-3}$, the parent strain produced 6000 units $cm^{-3}$ and was inhibited by 2000 units $cm^{-3}$ in the growth phase and a mutant producing 15,000 units $cm^{-3}$ was found to be resistant to 20,000 units $cm^{-3}$.

The possible relationship between antibiotic resistance and productivity may be used to advantage in the selection of high-producing mutants by culturing the survivors of a mutation treatment in the presence of a high level of the end product. Those strains capable of growth in the presence of a high level of the antibiotic may also be capable of high productivity in the idiophase. This approach has been used successfully for antifungal agesterols (Bu'Lock, 1980), streptomycin (Woodruff, 1966) and ristomycin (Trenina and Trutneva, 1966) but without success for novobiocin (Hoeksema and Smith, 1961).

A similar rationale was used by McGuire *et al.* (1980) for the improvement of daunorubicin production, an anthracycline antitumour agent synthesized by the red pigmented streptomycete *Streptomyces peuceticus*. The red pigment is presumably anthracycline which shares its oligoketide origin with daunorubicin. The directed selection was based on the ability of the antibiotic cerulenin to suppress oligoketide synthesis. This was indicated in *S. peuceticus* by the lack of the

red pigment on cerulenin agar. Thus, mutants which were still capable of oligoketide synthesis in the presence of the inhibitor would remain red. Approximately a third of the resistant mutants were superior daunorubicin producers.

### (iv) *The isolation of mutants in which secondary metabolite synthesis gives resistance to toxic compounds*

A potentially toxic compound may be rendered harmless by an organism converting it to a secondary metabolite or a secondary metabolite complexing with it. Ions of heavy metals such as $Hg^{2+}$, $Cu^{2+}$ and related organometallic ions are known to complex with $\beta$-lactam antibiotics (Elander and Vournakis, 1986) and such agents have been used to select resistant mutants. The logic of this selection process is that overproduction of a $\beta$-lactam may be the mechanism for increased resistance to the heavy metal. Elander and Vournakis (1986) reported that the frequency of superior cephalosporin C producers was greater amongst mutants resistant to mercuric chloride than amongst random samples of the survivors of ultraviolet treatment.

The conversion of the penicillin precursor, phenylacetic acid, to penicillin is thought to be an example of detoxification by conversion to a secondary metabolite. It appears that strains capable of withstanding higher concentrations of the precursor may be able to synthesize higher levels of the end product. Polya and Nyiri (1966) applied this hypothesis in selecting phenylacetic acid-resistant mutants of *P. chrysogenum* and demonstrated that 7% of the isolates showed enhanced penicillin production. Barrios-Gonzalez *et al.* (1993) investigated the same phenomenon and developed methods for the enrichment of both spores and early idiophase mycelium resistant to phenyl acetic acid. Of the resistant spore population, 16.7% were superior penicillin producers as compared with 50% of the resistant idiophase population. This suggests that the selective force may be more 'directed' by using a population already committed to penicillin synthesis. Although the best mutants contained elevated levels of acyltransferase (the enzyme that directly detoxifies PAA) they also showed higher levels of isopenicillin N synthetase (cyclase) which the authors claimed to be the limiting step. Thus, it was claimed that the screening method was not specific to selecting for acyltransferase elevation but could select for strains having a faster carbon flow through the pathway enabling them to use PAA faster.

Ball (1978) stressed that the major difficulty in the selection of toxic precursor resistant mutants is that the site of resistance of a mutant may not result in the organism overproducing the end product — for example, the resistance may be due to an alteration in the permeability of the mutant or due to the mutants' ability to degrade the precursor to a harmless metabolite unrelated to the desired end product. Barrios-Gonzalez's approach of using mycelial fragments already committed to penicillin synthesis decreases the likelihood of such 'false selection' occurring.

## THE ISOLATION OF REVERTANT MUTANTS

As discussed in the section on primary metabolites, a mutant may revert to the phenotype of its 'parent', but the genotype of the revertant may not, necessarily, be the same as the original 'parent'. Some revertant auxotrophs have been demonstrated to accumulate primary metabolites (Shiio and Sano, 1969) and attempts have been made to apply the technique to the isolation of mutants overproducing secondary metabolites. Two approaches have been used in the field of secondary metabolites with respect to the isolation of revertants:

(i)   The isolation of revertants of mutants auxotrophic for primary metabolites which may influence the production of a secondary metabolite.

(ii)  The reversion of mutants which have lost the ability to produce the secondary metabolite

### (i) The isolation of revertants of mutants auxotrophic for primary metabolites which may influence the production of a secondary metabolite

As previously discussed, it is difficult to account for some of the effects of auxotrophy on secondary metabolism and it appears that some may be due to as yet unresolved cross-pathway phenomena and others due to the expression of other mutations associated with the auxotrophy. Similarly, revertant mutants may affect secondary metabolism in a number of ways — direct effects on the pathway, cross-pathway effects and effects due to mutations other than the detected reversion.

Dulaney and Dulaney (1967) investigated the tetracycline productivities of a population of prototroph revertants of *S. viridifaciens* derived from each of five auxotrophs. These workers predicted that some revertants may be productive due to direct influence of the mutations on tetracycline biosynthesis but that others may be so because they contained other lesions. Superior producers were obtained in all the prototroph-re-

vertant populations apart from those derived from a homocysteine auxotroph. However, the frequency of the occurrence of the superior producers was similar to that obtained by the random selection of the survivors of a mutation treatment in all but one of the populations. The exceptional population was the revertants of a methionine auxotroph, 98% of which produced between 1.2 and 3.2 times as much tetracycline as the original prototrophic culture. A possible explanation of the very favourable titres of the population may be the role of methionine as the methyl donor in tetracycline biosynthesis and that methionine availability limited the production of the secondary metabolite in the original prototroph. However, addition of exogenous methionine to the prototroph did not result in superior productivity.

Polsinelli *et al.* (1965) also demonstrated that reversion of five mutants of an actinomycin producing strain of *S. antibioticus* blocked in the isoleucine–valine pathway resulted in the isolation of some superior mutants. Godfrey (1973) reported that the reversion of a cysteine auxotrophic mutant of *S. lipmanii* resulted in improved production of cephamycin.

These studies provide promising evidence that the selection of revertants of auxotrophs of primary metabolites involved in secondary metabolism may yield a high number of productive mutants.

### (ii) The isolation of revertants of mutants which have lost the ability to produce the secondary metabolite

The reversion of non-producing strains may result in the detection of a high-producing mutant as that mutant would have undergone at least two mutations associated with the production of the secondary metabolite. Dulaney and Dulaney (1967) plated the progeny of a mutation of a non-producing strain of *S. viridifaciens* onto solidified production medium and screened for superior tetracycline producers by an overlay technique. A mutant was isolated which produced nine times the tetracycline yield of the original 'parent'. The major difficulty inherent in this technique is that non-producing mutants of high-yielding strains may be incapable of being reverted due to extreme deficiencies in their metabolism. Indeed, Rowlands (1992) suggested that strains which are non (or low) producers and illustrate other effects such as poor growth and sporulation are best discarded, but that revertant mutants not showing pleiotropic effects are perhaps more likely to possess genuine increases in biosynthetic activity than those produced by any other technique, having been mutated twice in genes directly affecting product formation.

### The use of recombination systems for the improvement of industrial micro-organisms

Hopwood (1979) defined recombination, in its broadest sense, as "any process which helps to generate new combinations of genes that were originally present in different individuals". The use of recombination mechanisms for the improvement of industrial strains has increased significantly due to the developments in recombinant DNA technology and the necessity to develop new methods of strain improvement as the returns generated from mutation and selection programmes decreased. However, it should be appreciated that mutation and selection techniques are frequently used in association with recombination systems in a strain improvement programme. The parasexual cycle in the filamentous fungi has been applied to strain development as have protoplast fusion techniques in a wide range of micro-organisms.

### THE APPLICATION OF THE PARASEXUAL CYCLE

Many industrially important fungi do not possess a sexual stage and therefore it would appear difficult to achieve recombination in these organisms. However, Pontecorvo *et al.* (1953) demonstrated that nuclear fusion and gene segregation could take place outside, or in the absence of, the sexual organs. The process was termed the parasexual cycle and has been demonstrated in the imperfect fungi, *A. niger* and *P. chrysogenum*, as well as the sexual fungus *A. nidulans*. In order for parasexual recombination to take place in an imperfect fungus, nuclear fusion must occur between unlike nuclei in the vegetative hyphae of the organism. Thus, recombination may be achieved only in an organism in which at least two different types of nuclei coexist, i.e. a heterokaryon. The heterozygous diploid nucleus resulting from the fusion of the two different haploid nuclei may give rise to a diploid clone and, in rare cases, a diploid nucleus in the clone may undergo an abnormal mitosis resulting in mitotic segregation and the development of recombinant clones which may be either diploid or haploid.

Recombinant clones may be detected by their display of recessive characteristics not expressed in the heterokaryon. Analysis of the recombinants normally demonstrates them to be segregant for only one, or a few linked, markers and culture of the segregants results in the development of clones displaying more recessive characters than the initial segregant. The process of recombination during the growth of the heterozygous diploid may occur in two ways: mitotic crossing over, which results in diploid recombinants, and haploidization, which results in haploid recombinants.

Mitotic crossing over is the result of an abnormal mitosis. The normal mitosis of a heterozygous diploid cell is shown in Fig. 3.26. During mitosis, each pair of homologous chromosomes replicate to produce two pairs of chromatids and a chromatid of one pair migrates to a pole of the cell with a chromatid of the other pair. Division of the cell at the equator results in the production of two cells, both of which are heterozygous for all the genes on the chromosome. Mitotic crossing over involves the exchange of distal segments between chromatids of homologous chromosomes as shown in Fig. 3.27. This process may result in the production of daughter nuclei homozygous for a portion of one pair of chromosomes and in the expression of any recessive alleles contained in that portion. Thus, the clone arising from the partial homozygote will be recombinant and further mitotic crossing over in the recombinant will result in the expression of more recessive alleles.

Haploidization is a process which results in the unequal distribution of chromatids between the progeny of a mitosis. Thus, of the four chromatids of an homologous chromosome pair, three may migrate to one pole and one to another resulting in the formation of two nuclei, one containing $2n + 1$ chromosomes and the

Diploid cell where 2n = 2    Chromatid replication    Spindle formation    Separation of chromatids

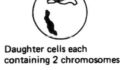

Daughter cells each containing 2 chromosomes

FIG. 3.26. Diagrammatic representation of the mitotic division of a eukaryotic cell containing two chromosomes. The nuclear membrane has not been portrayed in the figure.

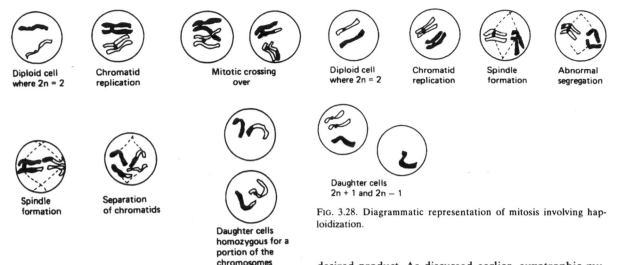

FIG. 3.27. Diagrammatic representation of mitosis including mitotic crossing over.

FIG. 3.28. Diagrammatic representation of mitosis involving haploidization.

other containing $2n - 1$ chromosomes. The $2n - 1$ nuclei tend towards the haploid state by the progressive random loss of further chromosomes. Thus, the resulting haploid nucleus will contain a random assortment of the homologues of the chromosomes of the organism. A representation of the process is shown in Fig. 3.28.

Therefore, the major components of the parasexual cycle are the establishment of a heterokaryon, vegetative nuclear fusion and mitotic crossing over or haploidization resulting in the formation of a recombinant. In practice, the occurrence and detection of these stages may be enhanced by the use of auxotrophic markers. The parents of the cross are made auxotrophic for different requirements and cultured together on minimal medium. The auxotrophs will grow very slightly due to the carry over of their growth requirements from the previous media, but if a heterokaryon is produced, by anastomoses forming between the two parents, then it will grow rapidly. The frequency of vegetative nuclear fusion in the heterokaryon may be enhanced by the use of agents such as camphor vapour or ultraviolet light; mitotic segregation may be enhanced by the use of agents such as X-rays, nitrogen mustard, p-fluorophenylalanine and ultraviolet light (Sermonti, 1969).

The application of the parasexual cycle to industrially important fungi has been hindered by a number of problems (Elander, 1980). A major difficulty is the influence of the auxotrophic markers (used for the selection of the heterokaryon) on the synthesis of the desired product. As discussed earlier, auxotrophic mutations have been observed to have quite unpredictable results on the production of some secondary metabolites. Even conidial colour markers have been shown to have deleterious effects on product synthesis (Elander, 1980). Even when suitable markers are available, the induction of heterokaryons in some industrial fungi has been demonstrated to be a difficult process and specialized techniques had to be employed to increase the probabilities of heterokaryon formation (see MacDonald and Holt, 1976). However, the development of protoplast fusion methods in the late 1970s enabled efficient heterokaryon formation to be achieved and removed the major barrier to the application of the parasexual cycle to strain improvement. These developments are considered in more detail in the next section on protoplast fusion.

Despite the early difficulties of inducing the parasexual cycle in some industrial fungi it was used in two ways to study these organisms. The cycle was used to investigate the genetics of the producing strains as well as to develop recombinant superior producers. Information obtained on the basic genetics of the industrial fungi using the parasexual cycle included the number of chromosomes (or linkage groups), the allocation of genes, important in product synthesis, to particular chromosomes and the mapping of important genes on a chromosome. Sermonti (1969) and MacDonald and Holt (1976) have described the techniques which were used to achieve these objectives.

Initial studies on haploid strains of *P. chrysogenum* derived from parasexual crosses demonstrated that most of the progeny exhibited the genotype of one of the parents, i.e. no recombination had occurred (MacDonald, 1968). MacDonald suggested that one of the reasons for this lack of success may have been the differences in gross chromosomal morphology between the parents caused by certain mutagens used in the development of the strains. Thus, one precaution to be adopted in the development of strains which may be used in a parasexual cross is the avoidance of mutagens which may cause gross changes in chromosomal morphology. Subsequently, Ball (1978) suggested that the careful choice of markers and the use of strains giving similar titres and being 'not too divergent' may result in achieving more recombinants.

Although diploids produced by the parasexual cycle are frequently unstable, stable diploids have been reported to have been used for the industrial production of penicillin. Elander (1967) isolated a diploid from a sister cross of *P. chrysogenum* which was shown to be a better penicillin producer than the parents and was morphologically more stable (a sister cross is one where the two strains differ only in the markers used to induce the heterokaryon). One explanation of the superior performance of the diploid may have been its resistance to strain degeneration caused by deleterious recessive mutations. Such mutations would only have been expressed in the diploid if both alleles had been modified. Calam *et al.* (1976) demonstrated that a diploid strain of *P. chrysogenum* was more stable than haploid mutants and mutation and selection of the diploid gave rise to a diploid strain producing higher levels of penicillin than the parents. Ball (1978) claimed that the usefulness of diploids may only be short term, presumably implying problems of degeneration, and may best be used as stepping stones to a recombinant haploid.

The advantages to be gained from the industrial use of parasexual recombinants are not confined to the amalgamation of different yield improving mutations. Equally advantageous would be the introduction of characteristics which would make the process more economic, for example low viscosity, sporulation and the elimination of unwanted products. The development of these issues will be considered in the next section on protoplast fusion.

## THE APPLICATION OF PROTOPLAST FUSION TECHNIQUES

Protoplasts are cells devoid of their cell walls and may be prepared by subjecting cells to the action of wall degrading enzymes in isotonic solutions. Protoplasts may regenerate their cell walls and are then capable of growth as normal cells. Cell fusion, followed by nuclear fusion, may occur between protoplasts of strains which would otherwise not fuse and the resulting fused protoplast may regenerate a cell wall and grow as a normal cell. Thus, protoplasts may be used to overcome some recombination barriers. Protoplast fusion has been demonstrated in a large number of industrially important organisms including *Streptomyces* spp. (Hopwood *et al.*, 1977), *Bacillus* spp. (Fodor and Alfoldi, 1976), corynebacteria (Karasawa *et al.*, 1986), filamentous fungi (Ferenczy *et al.*, 1974) and yeasts (Sipiczki and Ferenczy, 1977).

Fusion of fungal protoplasts appears to be an excellent technique to obtain heterokaryons between strains where conventional techniques have failed, or, indeed, as the method of choice. Thus, this approach has allowed the use of the parasexual cycle for breeding purposes in situations where it had not been previously possible. This situation is illustrated by the work of Peberdy *et al.* (1977) who succeeded in obtaining heterokaryons between *P. chrysogenum* and *P. cyaneofulvum* and demonstrated the formation of diploids which gave rise to recombinants after treatment with *p*-fluorophenylalanine or benomyl. Although it has been claimed that *P. chrysogenum* and *P. cyaneofulvum* are not different species of *Penicillium* (Samson *et al.*, 1977), Peberdy *et al.* still demonstrated that protoplast fusion could be successful where conventional techniques had failed.

A demonstration of the use of protoplast fusion for an industrial fungus is provided by the work of Hamlyn and Ball (1979) on the cephalosporin producer, *C. acremonium*. These workers compared the effectiveness of conventional techniques of obtaining nuclear fusion between strains of *C. acremonium* with the protoplast fusion technique. The results from conventional techniques suggested that nuclear fusion was difficult to achieve. Electron microscopic examination of fused protoplasts indicated that up to 1% underwent immediate nuclear fusion. Recombinants were obtained in both sister and divergent crosses. A cross between an asporulating, slow-growing strain with a sporulating fast-growing strain which only produced one-third of the cephalosporin level of the first strain eventually resulted in the isolation of a recombinant which combined the desirable properties of both strains, i.e. a strain which demonstrated good sporulation, a high growth rate and produced 40% more antibiotic than the higher-yielding parent. Chang *et al.* (1982) utilized protoplast fusion to combine the desirable qualities of two strains of *Penicillium chrysogenum*. Protoplasts from two strains, differing in colony morphology and the abilities to produce penicillin V (the desired product) and OH-V penicillin (an undesirable product), were fused, followed by plating on a non-selective medium. Out of 100 stable colonies which were scored, two possessed the desirable morphology, high penicillin V and low OH-V penicillin productivities.

Lein (1986) reported the penicillin strain improvement programme adopted by Panlabs, Inc. This programme included random and directed selection as well as protoplast fusion. Table 3.7 illustrates the properties of the two strains used in a protoplast fusion and one of the recombinants selected. To avoid any adverse effects no selective genetic markers were used and the regenerated colonies were screened on the basis of colony morphology and spore colour. A total of 238 colonies judged to be recombinants were screened for penicillin V production and the culture with the best combination of properties is shown in the table. Thus, the desirable characteristics of each strain were combined in the recombinant.

Protoplasts are also useful in the filamentous fungi for manipulations other than cell fusion. Rowlands (1992) suggested that they may be used in mutagenesis of non-sporulating fungi. Spores are the cells of choice for the mutagenesis of filamentous fungi but this is obviously impossible for non-sporulating strains. Mycelial fragments may be used but these will be multinucleate and very high mutagen doses are required. Although some protoplasts will be non-nucleate or multi-nucleate at least some will be uninucleate which will express any modified genes after mutation. Also, protoplasts will take up DNA in *in vitro* genetic manipulation experiments and this aspect will be discussed in a later section of this chapter.

Recombination can take place between actinomycetes by conjugation (Hopwood, 1976) and phage transduction (Studdard, 1979). However, both these mechanisms involve the transfer of only small regions of the bacterial chromosome. Furthermore, the low frequencies at which recombination occurs necessitate the use of selectable genetic markers such as auxotro-

phy or antibiotic resistance. The introduction of such markers is time-consuming but also they can detrimentally affect the synthetic capacity of the organism. Protoplast fusion has particular advantages over conjugation in that the technique involves the participation of the entire genome in recombination. Also, Hopwood (1979) has developed techniques which have resulted in the recovery of a very high proportion of recombinants from the fusion products of *Streptomyces coelicolor* protoplasts. By subjecting protoplasts to an exposure of ultraviolet light, sufficient to kill 99% of them prior to fusion, Hopwood has claimed a tenfold increase in recombinant detection for strains normally giving a low yield of recombinants (1%) and a doubling of the recombination frequency for a cross normally yielding 20% recombinants. Such yields of recombinants means that they would be detectable by simply screening a random proportion of the progeny of a protoplast fusion and the use of selectable markers to 'force' out the recombinants would not be necessary. Examples of the application of the technique to actinomycete strain improvement include cephamycin C yield enhancement in *Nocardia lactamdurans* (Wesseling, 1982) and the improvement of lignin degradation in *Streptomyces viridosporus* (Petty and Crawford, 1984).

Protoplast fusion has also been applied to the improvement of amino acid producing strains. Karasawa *et al.* (1986) used the technique to improve the fermentation rates of lysine producers developed using repeated mutation and directed selection. Such strains were good lysine producers but showed low glucose consumption and growth rates, undesirable features which had been inadvertently introduced during the selection programme. A protoplast fusion was performed between the lysine producer and a fast growing strain and a fusant was isolated displaying the desirable characteristics of high lysine production and high glucose consumption rate resulting in a much faster fermentation. The same authors used protoplast fusion to produce a superior threonine producing *Brevibacterium*

TABLE 3.7. *The use of protoplast fusion for the improvement of a pencillin V producer* (Lein, 1986)

| Characteristic | Parent A | Parent B | Best recombinant |
|---|---|---|---|
| Spores per slant ($\times 10^8$) | 2.2 | 2.5 | 7.5 |
| Germination frequency (%) | 99 | 40 | 49 |
| Colour of sporulating colonies | Green | Pale green | Deep green |
| Seed growth | Good | Poor | Good |
| Penicillin V yield (mg cm$^{-3}$) | 11.7 | 18.5 | 18.0 |
| Phenylacetic oxidation | Yes | No | No |

*lactofermentum* strain. Lysine auxotrophy was introduced into a threonine and lysine overproducer by fusing it with a lysine auxotroph — the recombinant produced higher levels of threonine due to its lysine auxotrophy.

## THE APPLICATION OF RECOMBINANT DNA TECHNIQUES

The transfer of DNA between different species of bacteria has been achieved experimentally using both *in vivo* and *in vitro* techniques (Atherton *et al.*, 1979). Thus, genetic material derived from one species may be incorporated into another where it may be expressed. *In vivo* techniques make use of phage particles which will pick up genetic information from the chromosome of one bacterial species, infect another bacterial species and in so doing introduce the genetic information from the first host. The information from the first host may then be expressed in the second host. Whereas, the *in vivo* techniques depend on vectors collecting information from one cell and incorporating it into another, the *in vitro* techniques involve the insertion of the information into the vector by *in vitro* manipulation followed by the insertion of the carrier and its associated 'extra' DNA into the recipient cell. Because the DNA is incorporated into the vector by *in vitro* methods the source of the DNA is not limited to that of the host organism of the vector. Thus, DNA from human or animal cells may be introduced into the recipient cell. Atherton *et al.* (1979) listed the basic requirements for the *in vitro* transfer and expression of foreign DNA in a host micro-organism as follows:

(i) A 'vector' DNA molecule (plasmid or phage) capable of entering the host cell and replicating within it. Ideally the vector should be small, easily prepared and must contain at least one site where integration of foreign DNA will not destroy an essential function.

(ii) A method of splicing foreign genetic information into the vector.

(iii) A method of introducing the vector/foreign DNA recombinants into the host cell and selecting for their presence. Commonly used simple characteristics include drug resistance, immunity, plaque formation, or an inserted gene recognizable by its ability to complement a known auxotroph.

(iv) A method of assaying for the 'foreign' gene product of choice from the population of recombinants created.

The initial work focused on *E. coli* but subsequently techniques have been developed for the insertion of foreign DNA into a range of bacteria, yeasts, filamentous fungi and animal cells. The range of vectors has been discussed by Gingold (1993) for bacteria, Curran and Bugeja (1993) for yeasts, Hopwood *et al.* (1985a) for streptomycetes, Elander (1989) for filamentous fungi and Murray (1993) for animal cells. The insertion of information into the vector molecule is achieved by the action of restriction endonucleases and DNA ligase. Site-specific endonucleases produce specific DNA fragments which may be joined to another similarly treated DNA molecule using DNA ligase. The modified vector is then normally introduced into the recipient cell by transformation. Because the transformation process is an inefficient one, selectable genes must be incorporated into the vector DNA so that the transformed cells may be cultured preferentially from the mixture of transformed and parental cells. This is normally accomplished by the use of drug-resistant markers so that those cells containing the vector will be capable of growth in the presence of a certain antimicrobial agent. The process is shown diagrammatically in Fig. 3.29.

Once the desired gene has been introduced into the recipient cell the problem of expression of the gene arises. This is particularly difficult when a eukaryotic gene is introduced into a bacterium. In the late 1970s a large number of mammalian genes were successfully introduced into bacterial cells, but there was little evidence of any gene expression (Atherton *et al.*, 1979). The problem of eukaryotic gene expression in prokaryotic cells is due to the different structure of eukaryote genes which contain non-coding segments of DNA. Thus, the production of a eukaryotic product by a prokaryotic cell necessitates the incorporation of the genes coding for the product in a form that may be translated by the recipient cell. Two approaches have been adopted to construct eukaryotic genes in a prokaryotic form: the first is to synthesize DNA corresponding to the primary structure of the protein product of the gene, although this method is suitable only for the construction of genes coding for small peptides of known structure. Itakura *et al.* (1977) synthesized the gene coding for the human hormone, somatostatin, and succeeded in incorporating it in *E. coli* where it was expressed. The alternative technique is to synthesize DNA from the messenger RNA, corresponding to the gene, using the enzyme, reverse transcriptase. Eukaryote messenger RNA is similar to bacterial RNA in that it does not contain non-coding segments so that DNA synthesized from an eukaryotic messenger RNA template should be in a form which is transcribable by a

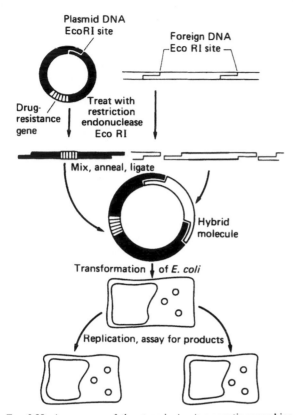

FIG. 3.29. A summary of the steps in *in vitro* genetic recombination. Both plasmid vector and foreign DNA are cut by the restriction endonuclease, EcoRI, producing linear double-stranded DNA fragments with single-stranded cohesive projections. EcoRI recognizes the oligonucleotide sequence $_{\text{GTTAAG}}^{\text{GAATTC}}$ and will cut any double-stranded DNA molecule to yield fragments with the same cohesive ends $_{\text{C}}^{\text{GAATT C}}$, $_{\text{TTAAG}}$. On mixing vector and foreign DNA, hybrids form into circular molecules which can be covalently joined using DNA ligase. Transformation of *E. coli* results in the low-frequency uptake of hybrid molecules whose presence can be detected by the ability of the plasmid to confer drug resistance on the host (Atherton *et al.*, 1979).

prokaryotic cell. Nagata *et al.* (1980) used the reverse transcriptase method to produce the genes coding for human interferon. Complementary single-stranded DNA was prepared from a mixture of messenger RNA extracted from virus-induced human lymphocytes. The DNA was then introduced into *E. coli* at random using a plasmid vector and the recipient cells screened for the presence of the interferon gene. Those colonies that contained the gene were then examined for interferon production. Using this method, Nagata's group succeeded in demonstrating the expression of the interferon gene in *E. coli*. Regrettably, a detailed explanation of the techniques involved in gene cloning is

outside the scope of this book, but excellent accounts of the various methods are given by Gingold (1993), Slater (1993) and Curran and Bugeja (1993), all in the same text (Walker and Gingold, 1993).

The most publicized application of recombinant DNA technology in the context of fermentation is the construction of strains capable of synthesizing foreign proteins. Although this chapter is concerned with strain improvement it is the obvious place to consider such chimeric strains. The use of the technique for the improvement of microbial product synthesis has also been very successful and this will be discussed in a later section.

### (i) *The production of heterologous proteins*

The rationale for the commercial production of foreign proteins in micro-organisms depends on the protein under consideration. The first commercial heterologous protein to be produced was human growth hormone (hGH) which is used to treat hypopituitary dwarfism and, prior to its manufacture by fermentation, was extracted from the brains of human cadavers. Naturally, this source was not readily available and carried the additional disadvantage of the risk of contamination with human pathogens. The successful production of recombinant hGH from *E. coli* both satisfied the demand for the compound and eliminated the risks associated with the human source (Dykes, 1993). Factor VIII is a blood clotting agent used in the treatment of haemophilia. Prior to its production as a recombinant protein it was extracted from human blood with the associated risk of contamination with HIV, so that the logic behind its production by fermentation is very similar to that of hGH — availability and safety.

The logic for the development of recombinant human insulin is not quite as clear because diabetes had been treated successfully for many years with animal insulin. However, it was assumed that the recombinant human product would cause fewer immunological difficulties and it would be pure and not contaminated with such pancreatic peptides as proinsulin, glucagon, somatostatin, pancreatic polypeptides and vaso-active intestinal peptides. Furthermore, the incidence of diabetes was also expected to increase due to changes in diet, the improved care of pregnant diabetic women (resulting in an increase of diabetes in the gene pool) and the increased life expectancy of diabetics (Dykes, 1993). Recombinant human insulin was first marketed in the U.K. in 1982 and by 1989 had become the most common form in use (Dykes, 1993).

Many other human proteins are synthesized at very low levels and the only practical way to produce them in sufficient quantities for use as therapeutic agents is as recombinant proteins. Examples of such products include the interferons and erythropoietin. Table 3.8 lists the recombinant proteins which have been licensed for therapeutic use. According to Dykes (1993) although insulin and hGH have been two of the most successful products (in terms of sales) some newer products have greater potential. For example, erythropoietin stimulates the production of erythrocytes from immature erythroid progenitor cells and has been used for the successful treatment of renal failure-induced anaemia. Predicted annual sales of the protein in the mid-1990s are in the region of $1200 million. The potential markets of some other recombinant human proteins are considered in Chapter 1.

Hepatitis B virus causes an infection of the liver giving rise to chronic viral hepatitis which may lead to progressive liver disease. There is no effective cure of the disease and, thus, vaccination is critical. The first hepatitis B vaccine became available in 1982, prepared from hepatitis B surface antigen (HBsAg) purified from the plasma of human carriers of the disease. Although the vaccine was successful its origin obviously presented the same difficulties of availability and contamination as discussed for hGH and Factor VIII. These difficulties provided the impetus to develop a recombinant vaccine which was achieved by expressing the HBsAg gene in *Saccharomyces cerevisiae*.

The vast majority of the early work on recombinant DNA technology concerned the transfer of genetic material into *E. coli* and this led to the adoption of the organism as the host for the production of several heterologous proteins. The bacterium had several advantages in its favour, primarily:

The availability of genetic knowledge.

The very wide range of vectors available.
A simple fermentation process using cheap media.
Promising protein yields in the range of 2–5 g $dm^{-3}$.

Despite these advantages *E. coli* also presents several problems for the production of heterologous proteins:

Proteins are formed as insoluble aggregates.
The proteins are not secreted.
Lack of post-translational modification.

By the mid-1980s the use of *E. coli* had declined considerably and yeasts, filamentous fungi and animal cells in culture were being investigated as alternative hosts. However, by 1992 *E. coli* was once again in favour. The protein yields obtained from animal cells were disappointing and the understanding of secretion and protein folding had progressed such that soluble proteins could be secreted by engineered *E. coli* (Hockney, 1994). However, mammalian cells are still the preferred host when the protein activity depends upon post-translational modification.

Whichever organism is used for the synthesis of an heterologous protein it is important that the strain is constructed such that expression of the gene may be controlled during the fermentation process. This may be achieved by inserting the gene into an inducible system, such as $\beta$-galactosidase, so that expression of the gene may be initiated by the addition of the inducer. This aspect is discussed in Chapter 4.

As an example of the production of an heterologous protein the development of the Genencor process for the production of recombinant chymosin will be con-

TABLE 3.8. *Recombinant proteins licensed for therapeutic use* (Dykes, 1993)

| Protein | Clinical use |
| --- | --- |
| Insulin | Diabetes |
| Growth hormone | Hypopituitary dwarfism |
| Tissue plasminogen activator | Clot lysis |
| Erythropoietin | Anaemia |
| G-CSF | Cancer chemotherapy |
| GM-CSF | Bone marrow transplantation |
| Factor VIII | Haemophilia |
| Interferon $\alpha$ | Cancers, hepatitis B, leukaemia |
| Interferon $\beta$ | Cancers, amyotrophic lateral sclerosis, genital warts |
| Interferon | Cancers, AIDS-related complex, osteopetrosis |
| Hepatitis B surface antigen | Hepatitis vaccine |

sidered (Ward *et al.*, 1990; Dunn-Coleman *et al.*, 1991). Bovine chymosin is an aspartyl protease extracted from the abomasum of unweaned calves and is used as a milk-clotting agent in the manufacture of cheese. Its natural source means that the enzyme is available only in limited quantities which made the possibility of producing it as an heterologous protein an attractive proposition. This process utilizes the filamentous fungus *Aspergillus niger* var. *awamori* which was claimed to have the following advantages as a potential host organism:

It is capable of secreting large amounts of protein.
The organism is regarded as safe.
Transformation systems were available.

Secretion from a eukaryotic cell is achieved via the endoplasmic reticulum and the Golgi apparatus. Entry into the secretory pathway is determined by the presence of a short, hydrophobic 'signal' sequence on the *N*-terminal end of secreted proteins. The signal sequence directs the protein to sites on the endoplasmic reticulum (ER) from where transport into the lumen of the ER occurs. The signal sequence is cleaved by a luminal signal peptidase and the protein passes through the ER and Golgi body, before being packaged into secretory vesicles and exported beyond the cell membrane (Curran and Bugeja, 1993).

Several expression vectors were constructed incorporating the organism's glucoamylase gene and prochymosin cDNA. The inclusion of the prochymosin cDNA between the glucoamylase promoter and terminator resulted in a strain which accumulated only 15 mg dm$^{-3}$ chymosin in the medium despite high intracellular mRNA and chymosin levels. This implied that the enzyme was not being secreted. To facilitate secretion, the prochymosin cDNA was incorporated between the glucoamylase coding region and the glucoamylase terminator. In this case, the organism should synthesize a 'fusion protein' consisting of chymosin attached to the carboxyl terminus of glucoamylase. The fusion protein was secreted and cleaved autocatalytically liberating the chymosin which was produced at 150 mg dm$^{-3}$.

It is fascinating to appreciate that further improvement of this strain has been achieved by mutant induction and selection. Mutagenesis followed by random selection resulted in the isolation of a superior producer which was subsequently shown to produce very low amounts of the native aspartyl protease. Armed with this information the Genencor group cloned the aspartyl protease and then deleted the gene from the strain. Thus, the insight gained from the random screen

allowed the directed *in vitro* approach to be used. The improved strain produced 250 mg dm$^{-3}$ chymosin. This strain was then used in further mutation and selection programmes. An automated microtitre plate screen was developed which has already been described in an earlier section of this chapter and the results are shown in Table 3.6.

The strains developed using the automated screen were then mutated and subjected to a directed screen. Spores which survived nitrosoguanidine exposure were plated onto glycerol media containing 1% deoxyglucose. Deoxyglucose is a toxic compound which Allen *et al.* (1989) used to isolate resistant mutants of *Neurospora crassa* capable of producing higher levels of extracellular enzymes normally repressed by glucose. The best resistant strains (termed dgr) produced in excess of 1200 mg dm$^{-3}$. Parasexual crosses between improved strains showed that two different loci were involved in resistance to deoxyglucose.

The chymosin story is an excellent example of the integration of recombinant DNA technology, mutant induction and automated random selection, directed selection and parasexual genetics in the development of a commercial strain.

(ii) *The use of recombinant DNA technology for the improvement of native microbial products*

Recombinant DNA technology has been used widely for the improvement of native microbial products. Frequently, this has involved 'self cloning' work where a chromosomal gene is inserted into a plasmid and the plasmid incorporated into the original strain and maintained at a high copy number. Thus, this is not an example of recombination because the engineered strain is altered only in the number of copies of the gene and does not contain genes which were present originally in a different organism. However, the techniques employed in the construction of these strains are the same as those used in the construction of chimeric strains, so it is logical to consider this aspect here.

The first application of gene amplification to industrial strains was for the improvement of enzyme production. Indeed, some regulatory mutants isolated by conventional means owed their productivity to their containing multiple copies of the relevant gene as well as the regulatory lesion. For example, the *E. coli* β-galactosidase constitutive mutants isolated by Horiuchi *et al.* (1963) in chemostat culture also contained up to four copies of the *lacZ* gene. According to Demain (1990), during the 1960s and early 1970s the number of gene copies were increased by using plasmids or transducing phage in the same species. The production of β-galactosidase, penicillinase, chloramphenicol transacetylase and aspartate transcarbamylase were all increased by transferring plasmids containing the structural gene into recipient cultures, especially when the plasmid replicated faster than the host chromosome. The advent of recombinant DNA technology increased the applicability of this approach by allowing the construction of vectors containing the desired gene and enabling the transfer of DNA to other species. Table 3.9 includes examples of gene amplification giving rise to improved enzyme yields.

The development of the process for the production of a lipase for use in washing powders by Novo Industri is an excellent example of the application of recombinant DNA technology to enzyme fermentations, although the technical details of the exercise are unavailable (Upshall, 1992). A lipase was isolated from a fungus which was unsuitable for commercial development. A cDNA clone coding for the lipase was prepared and transformed into an industrial *Aspergillus* strain used for enzyme production. The recombinant *Aspergillus* produced the lipase at high levels. Only 8 months elapsed between cloning the gene and the first commercial fermentation. Thus, recombinant DNA technology allowed the exploitation of a valuable microbial enzyme by facilitating its production in a well-established commercially acceptable strain.

The first successful application of genetic engineering techniques to the production of amino acids was obtained in threonine production with *E. coli*. Debabov (1982) investigated the production of threonine by a threonine analogue resistant mutant of *E. coli* Kl2. The entire threonine operon was introduced into a plasmid which was then incorporated into the organism by transformation. The plasmid copy number in the cell was approximately twenty and the activity of the threonine operon enzymes (measured as homoserine dehy-

drogenase activity) was increased 40–50 times. The manipulated organism produced 30 g dm$^{-3}$ threonine, compared with 2–3 g dm$^{-3}$ by the non-manipulated strain. Miwa *et al.* (1983) utilized similar techniques in constructing an *E. coli* strain capable of synthesizing 65 g dm$^{-3}$ threonine. It is important to appreciate that the genes which were amplified in these production strains were already resistant to feedback repression so that the multi-copies present in the modified organism were expressed and not subject to control. Thus, the recombinant DNA techniques built on the achievements made with directed mutant isolation.

The application of genetic engineering to the industrially important corynebacteria was hampered for some years by the lack of suitable vectors. However, vectors have been constructed from corynebacterial plasmids and transformation and selective systems developed. The first patents for suitable vectors were registered by the two Japanese companies Ajinomoto (1983) and Kyowa Hakko Kogyo (1983) and now a range of vectors is available with kanamycin, chloramphenicol and hygromycin as common selectable resistance markers (Martin, 1989). Transformation has been achieved using protoplasts and, more recently, electroporation has been used successfully to introduce the required DNA (Dunican and Shivnan, 1989). These systems have enabled not only the use of recombinant DNA technology for strain improvement but have also facilitated the detailed investigation of the molecular biology of these important amino acid and nucleotide producers; for example the molecular organization of the pathway to the aspartate family of amino acids in *C. glutamicum* has been elucidated.

It is not surprising that the improvement of threonine productio n was the first reported use of recombinant DNA technology with amino acid producing corynebacteria (Shiio and Nakamori, 1989). It may be recalled from the discussion of the development of the

TABLE 3.9. *Examples of the enhancement of enzyme production by gene amplification* (After Demain, 1990)

| Enzyme | Gene donor | Gene recipient | Increase (fold) | Reference |
|---|---|---|---|---|
| α-Amylase | *Bacillus amyloliquefaciens* | *Bacillus subtilis* | 10 | Sibakov *et al.* (1983) |
| α-Amylase | *Bacillus stearothermophilus* | *Bacillus stearothermophilus* | 5 | Aiba *et al.* (1983) |
| α-Amylase | *Bacillus stearothermophilus* | *Bacillus brevis* | 100 | Tsukagoshi *et al.* (1985) |
| *Eco*RI restriction endonuclease | *E. coli* | *E. coli* | 50–100 | Cheng *et al.* (1984) |
| DNA polymerase | *E. coli* | *E. coli* | 100 | *Kelley* et al. (1987) |

early threonine producers that *C. glutamicum* was not particularly amenable for threonine over-production using auxotrophs and analogue resistant mutants. Over-production was achieved by incorporating a DNA fragment coding for homoserine dehydrogenase from a *Brevibacterium lactofermentum* threonine producer into a plasmid and introducing the modified plasmid back into the *Brevibacterium*. A similar approach was used for homoserine kinase and a strain was developed with remarkably increased homoserine kinase and dehydrogenase activities which produced 33 g dm$^{-3}$ threonine (Morinaga *et al.*, 1987).

The application of recombinant DNA technology to the development of processes for the production of phenylalanine is an excellent illustration of the inter-relationship between mutant development and genetic engineering. Phenylalanine is a precursor of the sweetener, aspartame, and is thus an exceptionally important fermentation product. Backman *et al.* (1990) described the rationale used in the construction of an *E. coli* strain capable of synthesizing commercial levels of phenylalanine. *E. coli* was chosen as the producer because of its rapid growth, the availability of recombinant DNA techniques and the extensive genetic database.

The control of the biosynthesis of the aromatic family of amino acids in *E. coli* is shown in Fig. 3.30. The first step in the pathway is catalysed by three isoenzymes of dihydroxyacetone phosphate (DAHP) synthase, each being susceptible to one of the three end products of the aromatic pathway, phenylalanine, tyrosine or tryptophan. Control is achieved by both repression of enzyme synthesis and inhibition of enzyme activity. Within the common pathway to chorismic acid the production of shikimate kinase is also susceptible to repression. The conversion of chorismic acid to prephenic acid is catalysed by two isoenzymes of chorismate mutase, each being susceptible to feedback inhibition and repression by one of either tyrosine or phenylalanine. Each isoenzyme also carries an additional activity associated with either the phenylalanine or tyrosine branch. The tyrosine sensitive isoenzyme carries prephenate dehydrogenase activity (the next enzyme in the route to tyrosine) whilst the phenylalanine sensitive enzyme carries prephenate dehydratase (the next enzyme in the route to phenylalanine).

The regulation of gene expression was modified by:

(i)   Both tyrosine sensitive and phenylalanine sensitive DAHP synthase and shikimic kinase are regulated by the repressor protein coded by the *tyr*R gene. The *tyr*R gene had been cloned and

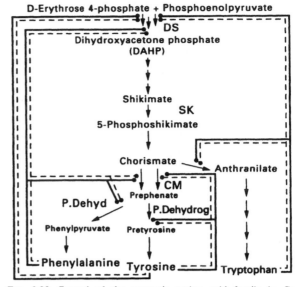

FIG. 3.30. Control of the aromatic amino acid family in *Escherichia coli*

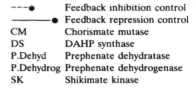

|         |                         |
|---------|-------------------------|
| ---●    | Feedback inhibition control |
| ———●    | Feedback repression control |
| CM      | Chorismate mutase       |
| DS      | DAHP synthase           |
| P.Dehyd | Prephenate dehydratase  |
| P.Dehydrog | Prephenate dehydrogenase |
| SK      | Shikimate kinase        |

was subjected to *in vitro* mutagenesis and the wild type gene replaced. This one mutation then resulted in the derepression of both tyrosine and phenylalanine DAHP synthase as well as shikimic kinase.

(ii)   The chorismic mutase/prephenate dehydratase protein is under both repression and attenuation control by phenylalanine. These controls were eliminated by replacing the normal promoter with one not containing the regulatory sequences, thus giving a ten times increase in gene expression.

The regulation of enzyme inhibition was modified by:

(i)   Chorismic mutase/prephenate dehydratase is subject to feedback inhibition by phenylalanine. It was shown that a certain tryptophan residue was particularly important in the manifestation of feedback inhibition. *In vitro* mutagenesis was used to delete the tryptophan codon and the modified gene introduced along with the substituted promoter referred to earlier. The enzyme produced by the modified gene was no longer susceptible to phenylalanine.

(ii) The rationale was to limitOA the availability of tyrosine such that the tyrosine-sensitive DAHP synthase isoenzyme would not be inhibited. Although the other two isoenzymes would still be susceptible to phenylalanine and tryptophan inhibition, sufficient DAHP would be synthesized by the third isoenzyme (in the absence of tyrosine) to facilitate overproduction of phenylalanine. Traditionally, this objective would have been achieved by using a tyrosine auxotroph fed with limiting tyrosine. However, these workers developed an excision vector system. An excision vector is a genetic element that can carry a cloned gene and can both integrate into, and be excised from, the bacterial chromosome. The vector is based on the bacteriophage lambda and the technology is explained in Fig. 3.31. The excision of the vector can be induced by a temperature shock. Thus, when the vector is excised the progeny of the cell will lose the inserted DNA. The tyrA gene was deleted from the production strain, inserted into the excision vector and transformed back into the organism where the vector became integrated into the chromosome. The fermentation could then be conducted using a cheap medium (the organism was not auxotrophic at inoculation) and allowing growth to an acceptable density. A heat shock may then be used to initiate vector excision, tyrosine auxtrophy and, hence, phenylalanine synthesis.

These efforts should have generated a high-producing strain. However, the tyrosine-sensitive DAHP synthase was susceptible to inhibition by high concentrations of phenylalanine. It will be recalled that the flow of DAHP was intended to come from the deregulated tyrosine sensitive isoenzyme. Thus, the final step in the development of the strain was to render this isoenzyme resistant to phenylalanine inhibition. This was achieved by the selection of mutants resistant to phenylalanine analogues. Thus, the strain was improved using a combination of gene cloning, *in vitro* mutagenesis and analogue resistance, indicating the importance of the contribution of a range of techniques to strain development. The final strain was capable of producing 50 g dm$^{-3}$ phenylalanine at a yield of 0.23 g g$^{-1}$ glucose and 2 g g$^{-1}$ biomass. The organism produced very low

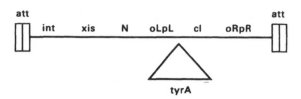

FIG. 3.31. Excision vector technology. An excision vector is represented as a line between two boxes. Above the line and boxes are indicated those genes from bacteriophage lambda that are carried on the vector. oRpR is required for the expression of a repressor determined by cl. This repressor binds to oRpR and oLpL to prevent the expression of other lambda genes. The other genes collectively act to form a recombination activity that allows the vector to integrate into or excise from the bacterial chromosome at the sites (indicated by the boxes) named *att*. Other genes, such as *tyr*A (indicated below the excision vector), can be cloned into the excision vector prior to its introduction into a target cell and can thereby be present or absent in the cell in coordination with the vector. Upon entering a cell, the genes for recombination (N, xis, and int) are expressed because there is not yet any repressor. As repressor accumulates, it shuts off the expression of those genes. In a fraction of the recipient cells, the recombination enzymes cause the vector to integrate into the cell chromosome before those enzymes decay away. That cell and all of its progeny inherit the vector and any gene it might carry (such as *tyr*A). If the repressor is inactivated, such as by high temperature, new recombination enzymes are formed that excise the vector from the chromosome. In such a cell and all of its progeny, the vector and the gene(s) it might carry are lost (Backman et al., 1990).

amounts of the other products and intermediates of the pathway which the workers claimed was due to the very precise manipulation of the strain which avoided the concomitant adverse characteristics associated with many highly mutated organisms.

Ikeda and Katsumata (1992) redesigned a tryptophan producing C. glutamicum strain such that it overproduced either phenylalanine or tyrosine. The regulation of the aromatic pathway in C. glutamicum is shown in Fig. 3.32. Phenylalanine producers which were resistant to phenylalanine analogues were used as sources of genes coding for enzymes resistant to control. Thus, a plasmid was constructed containing genes coding for the deregulated forms of DAHP synthase, chorismate mutase and prephenate dehydratase. The vector was introduced into a tryptophan overproducer having the following features:

(i) Chorismate mutase deficient and, therefore, auxotrophic for both tyrosine and phenylalanine.
(ii) Wild type DAHP synthase.
(iii) Anthranilate synthase partially desensitized to inhibition by tryptophan.

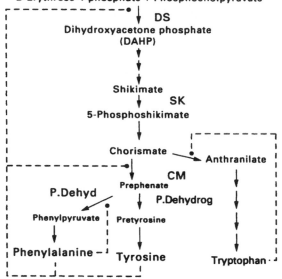

D-Erythrose 4-phosphate + Phosphoenolpyruvate

DS

Dihydroxyacetone phosphate
(DAHP)

Shikimate

SK

5-Phosphoshikimate

Chorismate → Anthranilate

CM

Prephenate

P.Dehyd          P.Dehydrog

Phenylpyruvate    Pretyrosine

Phenylalanine — Tyrosine          Tryptophan

FIG. 3.32. Control of the aromatic amino acid family in *Corynebacterium glutamicum*.

- – – ●      Feedback inhibition control
- CM            Chorismate mutase
- DS             DAHP synthase
- P.Dehyd      Prephenate dehydratase
- P.Dehydrog   Prephenate dehydrogenase
- SK             Skikimate kinase

The transformed strain was capable of producing 28 g $dm^{-3}$ phenylalanine and the levels of all three enzymes coded for by the vector were amplified approximately seven fold. If the original tryptophan producing strain was transformed with a plasmid containing only DAHP synthase and chorismate mutase, then tyrosine was overproduced (26 g $dm^{-3}$). Thus, genetic engineering techniques allowed a tryptophan producer to be redesigned into either a phenylalanine or tyrosine producer.

The development of *B. sphaericus* strains overproducing biotin was discussed in our consideration of the use of analogue resistant mutants. Although *B. sphaericus* is an intrinsically good biotin producer one of the limitations to its commercial exploitation is its complex nutritional requirements (Brown *et al.*, 1991). It uses carbon sources such as glucose inefficiently and requires complex organic nitrogen sources. Thus, workers have cloned the biotin genes from *B. sphaericus* into *E. coli*. Again, it is important to appreciate that it was the analogue resistant genes that were cloned allowing deregulated production in a more commercially amenable host (Osawa *et al.*, 1989; Brown *et al.*, 1991). The two gene cassettes which were cloned were under the control of inducible promoters so that production

could be initiated after the growth phase and therefore maintain plasmid stability. Yields of up to 45 mg $dm^{-3}$ were achieved.

A different application of recombinant DNA technology is seen in the modification of the ICI plc Pruteen organism, *Methylomonas methylotrophus*. The efficiency of the organisms' ammonia utilization was improved by the incorporation of a plasmid containing the glutamate dehydrogenase gene from *E. coli* (Windon *et al.*, 1980). The wild type *M. methylotrophus* contained only the glutamine synthetase/glutamate synthase system which, although having a lower $K_m$ value than glutamate dehydrogenase, consumes a mole of ATP for every mole of $NH_3$ incorporated. Glutamate dehydrogenase, on the other hand, has a lower affinity for ammonia but does not consume ATP. In the commercial process ammonia was in excess because methanol was the limiting substrate so the expenditure of ATP in the utilization of ammonia was wasteful of energy. The manipulated organism was capable of more efficient $NH_3$ metabolism, which resulted in a 5% yield improvement in carbon conversion. However, the strain was not used in the industrial process due to problems of scale-up.

The application of *in vitro* recombinant DNA technology to the improvement of secondary metabolite production may not be as advanced as it is for primary metabolites, but it has made a very significant contribution. Techniques have been developed for the genetic manipulation of streptomycetes (Hopwood *et al.*, 1985a) and the filamentous fungi (Elander, 1989) and a number of different strategies have been devised for cloning secondary metabolism genes (Hunter and Baumberg, 1989). In all the streptomycete systems so far studied the genes for the biosynthesis of a secondary metabolite are clustered. Furthermore, these clusters also contain the genes for regulation, resistance, export and extracellular processing. Work of this type has not only increased the basic understanding of the molecular genetics of secondary metabolism, but it has also facilitated strain improvement.

An excellent example of the application of recombinant DNA technology to the improvement of secondary metabolite production is provided by the work of the Lilly Research Laboratories group (Skatrud, 1992). These workers attempted to increase cephalosporin C synthesis by *Cephalosporium acremonium* by increasing the gene dosage at limiting steps in the pathway. Four critical steps were involved:

  (i)    Identifying the biochemical rate limiting step in the cephalosporin C industrial fermentation.

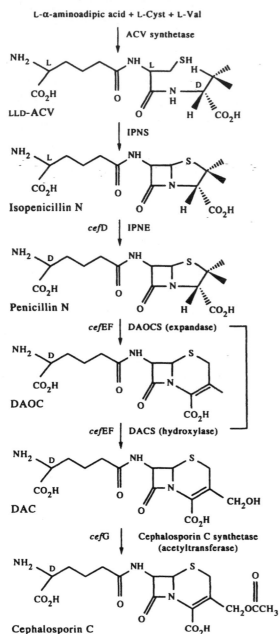

FIG. 3.33. The biosynthetic route to cephalosporin C in *Cephalosporium acremonium* (Skatrud, 1992).

Abbreviations:

| | |
|---|---|
| *cef* | Genes coding for cephalosporin synthesis |
| DAC | Deacetylcephalosporin C |
| DACS | DAC synthase (commonly known as hydroxylase) |
| DAOC | Deacetoxycephalosporin C |
| DAOCS | DAOC synthase (commonly known as expandase) |
| IPNS | Isopenicillin-N-synthase |

(ii) Cloning the gene coding for the enzyme catalysing the rate limiting step.

(iii) Constructing a vector containing the cloned gene and introducing it into the production strain.

(iv) Screening the transformants for increased productivity.

The biosynthetic route to cephalosporin C is shown in Fig. 3.33. Analysis of fermentation broths showed an accumulation of penicillin N which indicated that there was a bottleneck at the next reaction, i.e. the ring expansion step where the 5-membered penicillin N ring is expanded to the 6-membered ring of deacetoxycephalosporin C. This step is catalysed by deacetoxycephalosporin C synthase (DAOCS, commonly called expandase) coded for by the gene *cef* EF (already previously cloned by Samson *et al.*, 1987). *cef* EF also codes for the next enzyme in the route, deacetylcephalosporin C synthase (DACS, commonly called hydroxylase). Thus, if the expandase step were rate limiting, introduction of extra copies of *cef* EF should relieve the limitation and eliminate the accumulation of penicillin N.

The production strain of *C. acremonium* was transformed with a plasmid containing an exact copy of *cef* EF and the transformants screened for increased productivity. Approximately one in four transformants were superior producers and one produced almost 50% more antibiotic in a laboratory scale fermentation. Analysis of the transformant showed that a single copy of the transforming DNA had integrated into chromosome III, whereas native *cef* EF resides in chromosome II. In pilot scale fed-batch fermentations the transformant showed a 15% increase in yield which is still a very significant increase for an industrial strain. As predicted, the transformant did not accumulate penicillin N and the bottleneck appeared to have been relieved. The superior strain was among the first eight transformants examined, whereas 10,000 survivors of a mutagen exposure rendered no improved types. This work illustrates the enormous potential of recombinant DNA technology for secondary metabolite yield improvement, but it must be appreciated that such a programme involves a very considerable financial investment and the product must be sufficiently valuable to enable a good return to be realized on that investment.

IPNE    Isopenicillin-N-empimeraseThe brackets indicate that DAOCS and DACS enzyme activities are carried on one polypeptide coded for by the *cef* EF gene.

The major difficulty to be overcome in the use of genetically manipulated cultures is their potential instability in large-scale culture, especially continuous processes. The genetically manipulated strain of *Methylomonas methylotrophus* appears to be stable in continuous culture, but this is not surprising since the modification renders the cell more efficient and, thus, selection force in the chemostat would tend to work to its advantage. However, strains which have been manipulated to produce products which give the cell no selective advantage would be selected against in long-term culture, especially in chemostat systems. Thus, the development of techniques to stabilize manipulated cultures and the incorporation of selectable marker genes consistent with the use of cheap fermentation media are critical to the successful fulfilment of much of the promise of genetic-manipulation techniques. The improved cephalosporin C producer described earlier was transformed with an integrating vector which means that the transformant is stable even in the absence of selective pressure.

## The improvement of industrial strains by modifying properties other than the yield of product

The previous sections of this chapter have considered ways of increasing the yields of metabolites produced by industrial micro-organisms. However, the design and economics of a commercial process are influenced by properties of the organism other than its productivity. For example, although a strain may produce a very high level of a metabolite it would be unsuitable for a commercial process if its productivity were extremely unstable, or if the organism's oxygen demand were such that it could not be satisfied in the industrial fermenter available for the process. Therefore, characteristics of the producing organism which affect the process may be critical to its commercial success. Thus, it may be desirable to modify such characteristics of the producing organism which may be achieved by selecting natural and induced variants and recombinants. Naturally, it is crucial that the modified strain retains its desirable productivity so that the screening involved in these procedures should include assay of the yield as well as the characteristics being selected. Some examples of the characteristics which may be important in this context are: strain stability, resistance to phage infection, response to dissolved oxygen, tolerance of medium components, the production of foam, the production of undesirable by-products and the morphological form of the organism.

## THE SELECTION OF STABLE STRAINS

The ability of the producing strain to maintain its high productivity during both culture maintenance and a fermentation is a very important quality. Yield decay during culture storage may be avoided by the use of maintenance techniques such as those discussed earlier in this chapter, but loss of productivity during the fermentation is far more difficult to control. A decrease in the productivity of a commercial strain is normally due to the occurrence of lower-yielding, spontaneous revertant mutants which frequently have a higher growth rate than the high-producing parent, so that yield decay is especially problematical in long-term fermentations such as fed-batch or continuous culture where the faster-growing, lower producer may predominate, or even replace, the high-producing original strain. This situation is illustrated by the commercial, amino acid-producing organisms, many of which are insufficiently stable to be used in continuous-culture processes. Many workers have attempted to control the stability of amino acid-producing strains. As may be seen from the previous section on the isolation of mutants overproducing primary metabolites the amino acid producers tend to be auxotrophs, analogue-resistant mutants or revertants; such mutations removing the normal mechanisms controlling the production of amino acids. The introduction of more than one mutation giving the same phenotype may give a more stable strain since all the mutations would have to revert for the strain to lose its productivity.

Woodruff and Johnson (1970) selected a double auxotrophic mutant of *Micrococcus glutamicus* requiring both homoserine and threonine and compared its lysine-producing properties with those of a homoserine auxotroph. The authors claimed that the double mutant had a two-fold advantage in that it produced higher levels of lysine compared with the single auxotroph and was also less susceptible to reversion to low productivity, the stability of the double auxotroph being such that it was a suitable organism for the production of lysine by continuous culture.

Nakayama (1972) discussed the problem of strain reversion in the lysine fermentation and cited an example of a fermentation where 87% of the cells from a 60–88-hour culture were revertants. This very high level of revertants was probably due to the faster growth rate of the revertant compared with the single auxotroph in homoserine-limited culture. This situation was controlled, to a certain extent, by the incorporation of antibiotics, such as erythromycin, which are more inhibitory to the rapidly growing revertants than to the homoserine limited auxotrophs. However, the use of mutants which have multiple markers appears to be a better solution, as this considerably reduces the probability of reversion to the wild type as reversion of all the markers must occur, which should be an extremely rare event. Sano and Shiio (1970) cited the use of a *C. glutamicum* mutant for the industrial production of lysine which was auxotrophic for homoserine and leucine and was resistant to *S*-(2-aminoethyl)-L-cysteine and produced 39 g dm$^{-3}$ lysine. This strain would have had to have several reversions before it was restored to anything approaching the wild-type.

The stability of fungal diploids used in commercial fermentations may be controlled by a technique discussed by Ball (1973). Ball claimed that it may be possible to control the degeneration of a diploid into haploids by incorporating non-homologous recessive lethal mutations on separate chromosomes of an homologous pair in the diploid. The deleterious effects of the mutations would be repressed by the dominant alleles in the diploid but would be expressed in a haploid derived from the diploid, resulting in any haploids being non-viable.

A very simple but effective technique for selecting stable strains of *P. chrysogenum* was used in the Panlabs strain development programme (Lein, 1986). Final evaluation of a culture was made using the second slant of a slant-to-slant transfer. If the culture was unstable then the yield of a fermentation from the second slant would be poor, resulting in it being rejected. This procedure was followed sequentially through the programme with the result that the later strains showed less tendency to degenerate after subculturing.

Recombinant plasmids used as vectors in genetic manipulation are susceptible to two types of instability. Segregational instability is due to uneven partitioning of the plasmids at replication resulting in the production of plasmid-free daughter cells. Structural instability is a result of recombination events occurring within the vector, sometimes resulting in disruption of the desired gene. It is essential for acceptable stability that a plasmid vector contains a partitioning locus and that segregation of the plasmid is not simply a random event. The problem has been overcome in some systems by the integration of the vector into the chromosomal DNA, as already seen for filamentous fungal vectors. The disadvantage of this approach is that only one (or a few) copy of the desired gene is present whereas the use of an autonomously replicating plasmid would result in many gene copies.

## THE SELECTION OF STRAINS RESISTANT TO INFECTION

Bacterial fermentations may be affected very seriously by phage infections, which may result in the lysis of the bacteria. A possible method for reducing the risk of failure due to phage contamination is to select bacterial strains which are resistant to the phages isolated in the fermentation plant (Hongo *et al.*, 1972). It is important that the apparent resistant strains isolated are not lysogenic as the carrying of a population of phages in the fermentation is a source of potential lytic phage mutants. The use of phage-resistant mutants does not ensure immunity from phage infections because new host range phages may be introduced on to the plant or phage mutants may appear. Plant hygiene is essential to minimize the risk of contamination and it is also possible to utilize chemical agents in the fermentation which selectively inhibit phage replication (Hongo *et al.*, 1972).

It may be possible to design the host organism of a recombinant fermentation such that it is more resistant to phage infection. Primrose (1990) suggested that the inclusion of one or more host-restriction and modification systems (HRM) may achieve this objective. HRM is a mechanism whereby foreign nucleic acids which enter a cell may be destroyed. Most HRM systems reduce the infectivity of bacteriophage DNA by a factor of $10^2$ to $10^4$. The genes for many HRM systems have been cloned and they can be introduced into host strains provided, of course, that they do not degrade the foreign DNA deliberately introduced into the production strain. Also, recombinant fermentations may be made phage-resistant by changing the host organism. Although this may seem a drastic step it would be feasible if the rationale were built into the development of the process.

Infection of some fermentations with 'wild' microorganisms may be made more easily controlled by selecting commercial strains which are resistant to various antibiotics. The antibiotics to which the commercial strain is resistant may then be used to control the level of contaminants. The danger inherent in this technique is that resistant contaminants will also tend to be selected, but reasonable protection should be given provided that stringent sterilization of a contaminated fermenter is carried out before the vessel is used again so that any antibiotic resistant contaminants are removed.

## THE SELECTION OF NON-FOAMING STRAINS

Foaming during a fermentation may result in the

loss of broth, cells and product via the air outlet as well as putting the fermentation at risk from contamination (see Chapter 9). Thus, foaming is normally controlled by either chemical or mechanical means (see Chapters 4 and 7), but this task may be made easier if a non-foaming strain of the commercial organism can be developed. Foaming which occurs early in the fermentation is usually due to a component in the medium whereas foaming late in the fermentation is normally a property of the growing organism and, therefore, it is only this latter type of foam which may be controlled by strain selection. Both mutant screening and recombination may be used to develop non-foaming types but, obviously, the organisms must be tested for productivity as well. Recombination appears to be a particularly attractive proposition especially if non-foaming mutants have been isolated which are poor producers or good producers which foam. Ancestral crosses using protoplast fusion techniques may then render a strain combining the desirable features of both strains.

## THE SELECTION OF STRAINS WHICH ARE RESISTANT TO COMPONENTS IN THE MEDIUM

Some media components which are required for product formation may interfere with the growth of the organism and, therefore, it may be desirable to select for strains which are resistant to the medium component. Polya and Nyiri (1966) applied this approach to the isolation of mutants of *P. chrysogenum* resistant to phenylacetic acid, a precursor of penicillin and toxic to the organism at high concentrations. Barrios-Gonzalez *et al.* (1993) used a very similar approach in isolating superior penicillin producers. Alikhanian *et al.* (1959) selected strains of the oxytetracycline producer, *S. rimosus*, resistant to high levels of phosphate which prevents product synthesis. This enabled the strain to be used in media containing high levels of corn steep liquor (which is rich in phosphate).

Analogues of repressing media components have been used to select resistant mutants. For example, arsenate and vanadate have been used to isolate phosphate resistant strains (DeWitt *et al.*, 1989) and deoxyglucose to isolate glucose catabolite repression resistant strains.

## THE SELECTION OF MORPHOLOGICALLY FAVOURABLE STRAINS

The morphology of a micro-organism in submerged culture frequently has an effect on the economics or ease of operation of a process. The morphological form of a filamentous micro-organism will affect both the aeration of the system and the ease of filtration of the fermentation broth. As discussed in Chapter 6, the morphology of a fungus in submerged culture may be controlled by the level of a spore inoculum and by the medium, but it is also possible to influence the organism's morphology by altering its genotype. Backus and Stauffer (1955) recognized the influence of the genetics of a strain on the morphology of *P. chrysogenum* in submerged culture and its role in controlling foaming and broth filtration characteristics.

Bartholomew *et al.* (1977) selected strains of *P. chrysogenum* (for penicillin production) which gave a lower viscosity broth which increased the oxygen-transport ability of their fermentation plant equipment. The use of mutation and screening compared with recombination for the production of strains of a particular morphological type has been discussed by Ball (1978). Ball claimed that recombination techniques would be at an advantage over mutation and selection because of the large number of mutants that would have to be screened. A recombination programme would involve the crossing of a strain of the required morphology with a commercial producer, which should involve the screening of relatively few progeny. Hamlyn and Ball (1979) applied the technique of protoplast fusion to the construction of a desirable stain of *C. acremonium* for the production of cephalosporin, as discussed in an earlier section. A cross between an asporulating, slow-growing strain with a sporulating, fast-growing strain which only produced one third of the cephalosporin level of the first strain eventually resulted in the isolation of a recombinant which displayed the desirable properties of both strains, i.e. a strain which demonstrated good sporulation, a high growth rate and produced 40% more antibiotic than the higher-yielding parent. The ability of a strain to sporulate profusely is a very useful characteristic in the development of inoculum, as discussed in Chapter 6.

Rowlands (1992) reported that genetically altering *P. chrysogenum* strains to produce a pelleted rather than a filamentous broth was an important feature in the Panlabs penicillin improvement programme. The pelleted form gives rise to a much lower broth viscosity, resulting in lower power consumption. Although Rowlands did not indicate which genetic technique was used (mutant selection or protoplast fusion), protoplast fusion was certainly used to improve the sporulation and growth of the Panlabs strain, as discussed in a previous section (Table 3.7).

Flocculation of yeasts is the adherence of cells in

clumps resulting in the separation of the cells from the liquid in which they were suspended. Thus, the flocculating property of yeasts may be described as a morphological characteristic. When flocculation occurs in a beer fermentation the yeast will rise to the top of the vessel if it is a 'top-fermenting' yeast or drop to the bottom of the vessel if it is a 'bottom-fermenting' yeast. Thus, the flocculence of a yeast will determine the time of contact of the yeast with the wort (and hence the conversion of the wort to alcohol) and the ease of clarification of the beer. Although mutant selection is rarely applied, the selection of natural variants is a very common practice in many breweries. As discussed in Chapter 6, the yeast produced from one beer fermentation may be used to inoculate a new fermentation, a practice which is very rarely employed in the rest of the fermentation industry. By the careful selection of the yeast produced during a fermentation, strains showing the desired degree of flocculence may be selected. The first cells to rise to the top of a top fermentation, or drop to the bottom of a bottom fermentation, are the most flocculent cells whereas the last to rise or drop (depending on the type of fermentation) are the least flocculent cells. Thus, by selecting those cells that flocculate at an intermediate time the brewer isolates the natural variants which have the most desirable morphological properties for the inoculation of the next fermentation.

## THE SELECTION OF STRAINS WHICH ARE TOLERANT OF LOW OXYGEN TENSION

The provision of oxygen is frequently the limiting factor of many fermentations and it would, therefore, be desirable to select an organism which was capable of producing the product at a lower oxygen tension than normal. This may be achieved by screening for increased production under oxygen limited conditions. For example, Mindlin and Zaitseva (1966) isolated a lysine-producing strain which maintained its productivity under aeration conditions which decreased the parental strain productivity by almost a half.

## THE ELIMINATION OF UNDESIRABLE PRODUCTS FROM A PRODUCTION STRAIN

Athough an industrial micro-organism may produce large quantities of a desirable metabolite it may also produce a large amount of a metabolite which is not required, is toxic or may interfere with the extraction process. Thus, in these circumstances it would be an advantage to alter the strain such that the undesirable product is no longer produced. An example in the

penicillin-producing strains is the elimination of the production of the yellow pigment, chrysogenein, by the selection of non-pigmented mutants which made the extraction of the antibiotic much simpler (Backus and Stauffer, 1955).

Dolezilova *et al.* (1965) considered the production of fungicidin (nystatin) and cycloheximide by mutants of *S. noursei*. These workers demonstrated that mutants could be isolated which produced increased levels of fungicidin but produced no cycloheximide. The use of protoplast fusion techniques for the elimination of *p*-OH penicillin V has been considered earlier in the chapter.

## THE DEVELOPMENT OF STRAINS PRODUCING NEW FERMENTATION PRODUCTS

The isolation of organisms from the natural environment synthesizing commercially useful metabolites is an expensive and laborious process. Therefore, other means of producing novel compounds which may be of some industrial significance have been attempted. Probably the most successful alternative approach has been the semi-synthetic one where microbial products have been chemically modified, for example, the semi-synthetic penicillins. Precursor feeding has also met with some success in this context; by incorporating a precursor of a natural product into the fermentation medium the level of the end product may be increased, for example, the use of phenylacetic acid in the production of penicillin G. The feeding of an analogue of the normal precursor of a natural product frequently results in the production of an analogue of the natural product. Hamill *et al.* (1970) demonstrated that if 5-, 6- or 7-substituted tryptophan replaced tryptophan, the normal precursor of the antifungal agent, pyrrolnitrin, in cultures of *Pseudomonas aureofaciens*, then a series of substituted pyrrolnitrins were obtained, some of which had improved antifungal activity. However, the disadvantages of the analogue-precursor technique are that the end product tends to be very similar to the natural product and the new product will be contaminated with the normal product which the organism may still synthesize from its self-produced natural precursors. Birch (1963) suggested that the problem of mixed products may be overcome by isolating mutants which would not produce the normal precursor but could convert it into the end product. Thus, the analogue precursor could be converted into the novel end product without competition from the normal endogenous precursor. Shier *et al.* (1973) succeeded in applying Birch's idea in the study of neomycin production by *Strep-*

*tomyces fradiae* from the precursor, deoxystreptamine. By using a replica plating technique, the survivors of a mutation treatment were screened for the ability to inhibit the growth of the test organism, *Bacillus subtilis*, only in the presence of deoxystreptamine. By feeding the mutant isolated by this method, different antibiotics were synthesized.

Nagaoka and Demain (1975) described this technique for the isolation of new products as 'mutational biosynthesis' and the mutants isolated as 'idiotrophs'. Besides the aminoglycoside–aminocyclitol antibiotics (of which neomycin is an example), mutational biosynthesis has been applied to the macrolide antibiotics, the novobiocin antibiotics and the β-lactam antibiotics (Daum and Lemk, 1979).

The advent of readily available recombination techniques resulted in attempts to produce novel compounds from recombinants, particularly streptomycetes. The rationale behind these experiments was that by mixing the genotypes of two organisms synthesizing different metabolites then new combinations of biosynthetic genes, and hence pathways, may be produced. Little progress was achieved using protoplast fusion but the exploitation of recombinant DNA technology has yielded some significant successes.

*Streptomyces coelicolor* produces the polyketide actinorhodin (Fig. 3.34) whilst *Streptomyces* sp. AM 7161 produces medermycin. Hopwood *et al.* (1985b) transformed some of the cloned genes coding for actinorhodin into *Streptomyces* sp. AM7161. The recombinant produced another antibiotic, mederrhodin A (Fig. 3.34). The modified strain contained the *act*V gene from *S. coelicolor* coding for the *p*-hydroxylation of actinorhodin; the enzyme was also capable of hydroxylating medermycin. Hopwood *et al.* (1985b) also introduced the entire actinorhodin gene cluster into *Streptomyces violaceoruber* which produces granaticin or dihydrogranaticin (Fig. 3.35). The recombinant synthesized the novel antibiotic, dihydrogranatirhodin which has the same structure as dihydrogranaticin apart from the stereochemistry at one of its chiral centres (Fig. 3.35).

The enzymes responsible for polyketide biosynthesis have been extensively studied in recent years and significant advances have been made in the understanding of both their biochemistry and genetics. Polyketide synthases (PKSs) are multifunctional enzymes which catalyse repeated decarboxylative condensations between coenzyme A thioesters and are very similar to the fatty acid synthases. An enormous range of microbial polyketides are known and the variation is due to the chain length, the nature of the precursors and

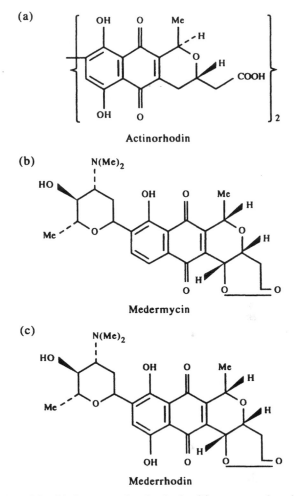

**FIG. 3.34.** (a) Structure of actinorhodin, (b) structure of medermycin, (c) structure of mederrhodin (Hunter and Baumberg, 1989).

the subsequent modification reactions which occur after cyclization. McDaniel *et al.* (1993) have constructed plasmids coding for recombinant PKSs and achieved expression in *Streptomyces coelicolor*. Five novel polyketides were synthesized and the system shows great promise for the production of new, potentially valuable, compounds as well as providing carbon skeletons amenable to derivatization by organic chemists.

A commercially relevant example is provided by the Lilly Research Laboratories group (Beckman *et al.*, 1993) who produced a strain of *Penicillium chrysogenum* capable of accumulating deacetoxycephalosporin C (DAOC), a compound which can be biotransformed into 7-aminodesacetoxycephalosporanic acid

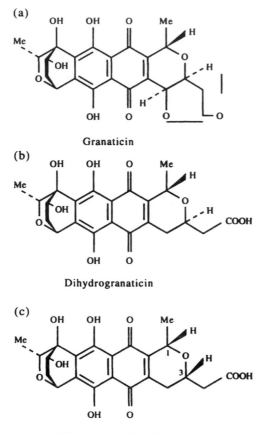

FIG. 3.35. (a) Structure of granaticin, (b) structure of dihydrogranaticin, (c) structure of dihydrogranatirhodin (Hunter and Baumberg, 1989).

(7ADAOC) which is a precursor of at least three chemically synthesized clinically important cephalosporins. The conventional route to 7ADAOC is by the chemical ring expansion of benzylpenicillin which is far more complex than the biotransformation of DAOC.

DAOC is produced as an intermediate in the synthesis of cephalosporin C by *Cephalosporium acremonium*, as shown in Fig. 3.33. However, it would be very difficult to manufacture the compound from *C. acremonium* by blocking the conversion of DAOC to deacetylcephalosporin C because a single bifunctional enzyme catalyses both the conversion of penicillin N to DAOC (expandase activity) and the hydroxylation of DAOC to deacetylcephalosporin C (hydroxylase activity). The bifunctional protein is encoded by the gene *cef* EF. In *Streptomyces clavuligerus* the expandase and hydroxylase enzymes are separate proteins encoded by

the genes *cef* E and *cef* F respectively. Thus, *S. clavuligerus* could be modified for the commercial production of DAOC. However, the Lilly group did not have a strain capable of producing high cephalosporin levels available to them, but they did have a *Penicillium chrysogenum* that overproduced isopenicillin N (IPN) as an intermediate in phenoxymethylpenicillin synthesis as shown in Fig. 3.36. Thus, to enable the *Penicillium* to produce DAOC the insertion of two genes was necessary, i.e. those encoding isopenicillin epimerase and expandase. These workers had already cloned *cef* E (coding for expandase) from *S. clavuligerus* and *cef* D (coding for epimerase) from *S. lipmanii*. Thus, using sophisticated vector technology a *P. chrysogenum* was transformed with a plasmid containing both *cef* D and *cef* E. One of the isolated transformants was capable of synthesizing up to 2.5 g dm$^{-3}$ DAOC and both the *cef* D and *cef* E genes were integrated into the organism's genome.

Thus, the recombinant produced by the Lilly group contained a branched biosynthetic pathway giving rise to the two end products, DAOC and phenoxymethylpenicillin. However, the conversion of IPN to DAOC in the recombinant represented less that 10% of the conversion of IPN to phenoxymethylpenicillin in the parent. Thus, the recombinant may still be developed to improve the productivity. These developments could involve the screening of a larger population of transformants to find one where integration of the inserted genes occurred at sites not associated with penicillin synthesis. Also, the isolation of a strain blocked in the conversion of IPN to phenoxymethylpenicillin should give improved DAOC production.

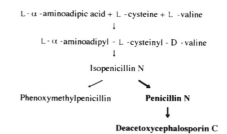

FIG. 3.36. β-Lactam biosynthetic pathway in wild-type *P. chrysogenum* (pathway in medium type) and in *P. chrysogenum* transformed with pBOB13 and pZAZ6 (branched pathway which includes steps in boldface type). Enzyme steps (medium type, top to bottom): ACV synthetase, IPN synthase, acyl coenzyme A:IPN acyltransferase. Enzyme steps (boldface type, top to bottom): IPN epimerase, penicillin N expandase (= DAOC synthase). (Beckman *et al.*, 1993.)

## SUMMARY

The tasks of both discovering new microbial compounds and improving the synthesis of known ones have become more and more challenging. Early work on isolation and improvement relied on a 'blunderbus' approach yet, due to the ingenuity and resourcefulness of the individuals involved, resulted in the establishment of a wide range of highly successful processes. The rational improvement programmes developed in the amino-acid industry pointed the way to the adoption of such approaches for both secondary metabolite discovery and improvement and the development of miniaturized screening sytems allowed the industry to take full advantage of robotic systems to revolutionize screening throughput. However, it is the gigantic developments in molecular biology which have allowed the industry to enter the next stage in its evolution. Recombinant DNA technology has enabled the production of heterologous products and has built on the achievements of directed selection to increase yields of conventional products still further.

## REFERENCES

ABE, S. (1972) Mutants and their isolation. In *The Microbial Production of Amino Acids,* pp. 39–66 (Editors Yamada, K., Kinoshita, S., Tsunoda, T. and Aida, K.). Halstead Press, New York.

AIBA, S., KITAI, K. and IMANAKA, T. (1983) Cloning and expression of thermostable $\alpha$-amylase from *B. subtilis*. *Appl. Env. Micro.* **46**, 109–115.

AJINOMOTO (1983) European patent application 71023.

ALBERTS, A. W., CHEN, J., KURON, G., HUNT, V., HUFF, J., HOFFMAN, C., ROTHTOCK, J., LOPEZ, M., JOSHUA, H., HARRIS, E., PATCHETT, A., MONOGHAN, R., CURRIE, S., STAPLEY, E., ALBERS-SCHONBERG, G., HESENS, O., HIRSHFIELD, J., HOOGSTEEN, K., LIESCH, J. and SPRINGER, J. (1980) Mevinolin, a highly potent inhibitor of HMG-CoA reductase and cholesterol lowering agent. *Proc. Natn. Acad. Sci U.S.A.* **7**, 3957–3961.

ALIKHANIAN, S. I. (1962) Induced mutagenesis in the selection of micro-organisms. *Adv. Appl. Micro.* **4**, 1–5.

ALIKHANIAN, S. I., MINDLIN, S. Z., GOLDAT, S. V. and VLADIMIZOV, A. V. (1959) Genetics of organisms producing tetracyclines. *Ann. N.Y. Acad. Sci.* **81**, 914.

ALLEN, K. E., McNALLY, M. T., LOWENDORF, H. S., SLAYMAN, C. W. and FREE, S. J. (1989) Deoxyglucose resistant mutants of *Neurospora crassa*: Isolation, mapping and biochemical characterisation. *J. Bacteriol.* **171**, 53–58.

ATHERTON, K. T., BYROM, D. and DART, E. C. (1979) Genetic manipulation for industrial processes. *Soc. Gen. Micro. Symp.* **29**, 379–406.

AUNSTRUP, K., OUTTRUP, H., ANDRESEN, O. and DAMBMANN, C. (1972) Proteases from alkalophilic *Bacillus* sp. *Fermentation Technology Today: Proceedings 4th. Int. Fermentation Symposium*, pp. 299–305.

BACKMAN, K., O'CONNOR, M. J., MARUYA, A., RUDD, E., McKAY, D., BALAKRISHNAN, R., RADJAI, M., DiPASQUANTONIO, V., SHODA, D., HATCH, R. and VENKATASUBRAMANIAN, K. (1990) Genetic engineering of metabolic pathways applied to the production of phenylalanine. *Ann. N.Y. Acad. Sci.* **589**, 16–24.

BACKUS, M. P. and STAUFFER, J. F. (1955) The production and selection of a family of strains of *Penicillium chrysogenum*. *Mycologia* **12**, 429–493.

BALL, C. (1973) The genetics of *Penicillium chrysogenum*. *Prog. Ind. Micro.* **12**, 47–72.

BALL, C. (1978) Genetics in the development of the penicillin process. In *Antibiotics and Other Secondary Metabolites, Biosynthesis and Production*, pp. 163–176 (Editors Hutter, R., Leisinger, T., Nuesch, J. and Wehrlin, W.). Academic Press, London.

BALL, C. and McGONAGLE, M. P. (1978) Development and evaluation of a potency index screen for detecting mutants of *P. chrysogenum* having increased penicillin yields. *J. Appl. Bacteriol.* **45**, 67–74.

BALTZ, R. H. (1986) Mutagenesis in *Streptomyces* spp. In *Industrial Microbiology and Biotechnology*, pp. 184–190 (Editors Demain, A. L. and Solomon, N. A.). American Microbiological Society, Washington.

BARRIOS-GONZALEZ, J., MONTENEGRO, E. and MARTIN, J. F. (1993) Penicillin production by mutants resistant to phenylacetic acid. *J. Ferm. Bioeng.* **76** (6), 455–458.

BARTHOLOMEW, W. H., SHEENAN, B. T., SHU, P. and SQUIRES, R. W. (1977) *Abstracts of the 74th National Meeting of the American Chemical Soc.*, Chicago, Illinois. Division of Microbial and Biochemical Technology, Abstract No. 37.

BECKMAN, R., CANTWELL, C., WHITEMAN, P., QUEENER, S. W. and ABRAHAM, E. P. (1993) Production of deacetocephalosporin C by transformants of *Penicillium chrysogenum*: Antibiotic biosynthetic pathway engineering. In *Industrial Microorganisms: Basic and Applied Molecular Genetics*, pp. 177–182 (Editors Baltz, R. H., Hegeman, G. D. and Skatrud, P. L.). American Society for Microbiology, Washington.

BEETON, S. and BULL, A. T. (1989) Biotransformation and detoxification of T-2 toxin by soil and freshwater bacteria. *Appl. Environ. Microbiol.* **56**, 190–197.

BIBIKOVA, M. V., IVANITSKAYA, L. P. and SINGAL, S. M. (1981) Direct screening on selective media with gentamycin of organisms which produce aminoglycoside antibiotics. *Antibiotiki* **26**, 488–492.

BIRCH, A. J. (1963) The biosynthesis of antibiotics. *Pure Appl. Chem.* **7**, 527–537.

BIRGE, E. A. (1988) *Bacterial and Bacteriophage Genetics*, pp. 58–88. Springer Verlag, New York.

BONNER, D. (1947) Studies on the biosynthesis of penicillin. *Arch. Biochem.* **13**, 1–14.

BROWN, A. G., BUTTERWORTH, D., COLE, M., HANSCOMBE, G., HOOD, J. D. and READING, C. (1976) Naturally occurring β-lactamase inhibitors with antibacterial activity. *J. Antibiot.* **29**, 668–669.

BROWN, S. W., SPECK, D., SABATIE, J., GLOECKLER, R., O'REGAN, M., VIRET, J. F., LEMOINE, Y., OSAWA, I., KISOU, T., HAYAKAWA, K. and KAMOGAWA, K. (1991) The production of biotin by recombinant strains of *Escherichia coli*. *J. Chem. Tech. Biotechnol.* **50**, 115–121.

BUCKLAND, B. C. (1992) Reduction to practice. In *Harnessing Biotechnology for the 21st Century*, pp. 215–218 (Editors Ladisch, M. R. and Bose, A.). American Chemical Society, Washington.

BULL, A. T. (1992) Biodiversity as a source of innovation in biotechnology. *Ann. Rev. Microbiol.* **46**, 219–252.

BULL, A. T., ELLWOOD, D. C. and RATLEDGE, C. (1979) The changing scene in microbial technology. *Soc. Gen. Micro. Symp.* **29**, 1–28.

BU'LOCK, J. D. (1979) Process needs and the scope for genetic methods. In *Genetics of Industrial Micro-organisms*, pp. 105–111 (Editors Sebek, O. K. and Laskin, A. I.). American Society for Microbiology, Washington.

BU'LOCK, J. D. (1980) Resistance of a fungus to its own antifungal metabolites and the effectiveness of resistance selection in screening for higher yielding mutants. *Biotechnol. Lett.* **3**, 285–290.

CALAM, C. T., DAGLISH, L. B. and McCANN, E. P. (1976) Penicillin: tactics in strain improvement. In *Second International Symposium on the Genetics of Industrial Micro-organisms*, pp. 273–287 (Editor MacDonald, K. D.). Academic Press, London.

CATCHESIDE, D. G. (1954) Isolation of nutritional mutants of *Neurospora crassa* by filtration enrichment. *J. Gen. Micro.* **11**, 34–36.

CHANG, L. T. and ELANDER, R. P. (1982) Rational selection for improved cephalosporin C production in strains of *Acremonium chrysogenum*. *Dev. Ind. Micro.* **20**, 367–380.

CHENG, S.-C., KIM, R., KING, K., KIM, S.-H. and MODRICH, P. (1984) Isolation of gram quantities of EcoR1 restriction enzymes from an overproducing strain. *J. Biol. Chem.* **259**, 11571–11580.

CHORMONAVA, N. T. (1978) Isolation of *Actinomadura* from soil samples on selective media with kanamycin and rifampicin. *Antibiotiki* **23**, 22–26.

CURRAN, B. P. G. and BUGEJA, V. C. (1993) Yeast cloning and biotechnology. In *Molecular Biology and Biotechnology* (3rd edition), pp. 85–102 (Editors Walker, J. M. and Gingold, E.). The Royal Society of Chemistry, Cambridge.

DAUM, S. J. and LEMKE, J. R. (1979) Mutational biosynthesis of new antibiotics. *Ann. Rev. Micro.* **33**, 241–266.

DAVIES, O. L. (1964) Screening for improved mutants in antibiotic research. *Biometrics*, **20**, 576–591.

DAVIS, B. D. (1949) Nutritionally deficient bacterial mutants isolated by penicillin. *Proc. Natn. Acad. Sci., U.S.A.* **35**, 1–10.

DEBABOV, V. G. (1982) Gene engineering and microbiological industry. In *Overproduction of Microbial Products*, pp.

345–352 (Editors Krumphanzl, V., Sikyta, B. and Vanek, Z.). Academic Press, London.

DEMAIN, A. L. (1957) Inhibition of penicillin formation by lysine. *Arch. Biochem. Biophys.* **67**, 244–245.

DEMAIN, A. L. (1972) Cellular and environmental factors affecting the synthesis and excretion of metabolites. *J. Appl. Chem. Biotechnol.* **22**, 346–362.

DEMAIN, A. L. (1973) Mutation and production of secondary metabolites. *Adv. Appl. Micro.* **16**, 177–202.

DEMAIN, A. L. (1974) How do antibiotic-producing organisms avoid suicide? *Ann. N.Y. Acad. Sci.* **235**, 601–612.

DEMAIN, A. L. (1978) Production of nucleotides by micro-organisms. In *Primary Products of Metabolism. Economic Microbiology*, Vol. 2, pp. 187–209 (Editor Rose, A. H.). Academic Press, London.

DEMAIN, A. L. (1990) Regulation and exploitation of enzyme biosynthesis. In *Microbial Enzymes and Biotechnology* (2nd edition), pp. 331–368 (Editors Fogarty, W. M. and Kelly, C. T.). Elsevier, Barking, UK.

DEMAIN, A. L. and BIRNBAUM, J. (1968) Alteration of permeability for the release of metabolites from the microbial cell. *Curr. Topics Microbiol. Immunol.* **46**, 1–25.

DEMAIN, A. L. and MASUREKAR, P. S. (1974) Lysine inhibition of *in vivo* homocitrate synthesis in *Penicillium chrysogenum*. *J. Gen. Micro.* **82**, 143–151.

DeWITT, J. P., JACKSON, J. V. and PAULUS, T. J. (1989) Actinomycetes. In *Fermentation Process Development of Industrial Organisms*, pp. 1–72 (Editor Neway, J. O.). Marcel Dekker, New York.

DITCHBURN, P., GIDDINGS, B. and MacDONALD, K. D. (1974) Rapid screening for the isolation of mutants in *A. nidulans* with increased penicillin yields. *J. Appl. Bacteriol.* **37**, 515–523.

DOLEZILOVA, L., SPIZEK, J., VONDRACEK, M., PALECKOVA, F. and VANEK, Z. (1965) Cycloheximide producing and fungicidin producing mutants of *Streptomyces noursei*. *J. Gen. Micro.* **39**, 305–310.

DULANEY, E. L. (1954) Induced mutation and strain selection in some industrially important micro-organisms. *Ann. N.Y. Acad. Sci.* **60**, 155–167.

DULANEY, E. L. and DULANEY, D. D. (1967) Mutant populations of *Streptomyces viridifaciens*. *Trans. N.Y. Acad. Sci.* **29**, 782–799.

DUNICAN, L. K. and SHIVNAN, E. (1989) High frequency transformation of whole cells of amino acid producing coryneform bacteria using high voltage electropolation. *Biotechnology* **7**, 1067–1070.

DUNN-COLEMAN, N. S., BLOEBAUM, P., BERKA, R. M., BODIE, E., ROBINSON, N., ARMSTRONG, G., WARD, M., PRZETAK, M., CARTER, G. L., LaCOST, R., WISLON, L. J., KODAMA, K. H., BALIU, E. F., BOWER, B., LAMSA, M. and HEINSOHN, H. (1991) Commercial levels of chymosin production by *Aspergillus*. *Biotechnology* **9** (10), 976–981.

DYKES, C. W. (1993) Molecular biology in the pharmaceutical industry. In *Molecular Biology and Biotechnology* (3rd edition), pp. 155–176 (Editors Walker, J. M. and Gingold, E.). The Royal Society of Chemistry, Cambridge.

ELANDER, R. P. (1966) Two decades of strain development in antibiotic-producing micro-organisms. *Dev. Ind. Micro.* **7**, 61–73.

ELANDER, R. P (1967) Enhanced penicillin synthesis in mutant and recombinant strains of *P. chrysogenum*. In *Induced Mutations and their Utilisation*, pp. 403–423 (Editor Stubble, H.). Academic Verlag, Berlin.

ELANDER, R. P (1980) New genetic approaches to industrially important fungi. *Biotech. Bioeng* **22**, (Supplement 1), 49–61.

ELANDER, R. P. (1989) Bioprocess technology in industrial fungi. In *Fermentation Process Development of Industrial Organisms*. pp. 169–220 (Editor Neway, J. O.). Marcel Dekker, New York.

ELANDER, R. P. and VOURNAKIS, J.N. (1986) Genetic aspects of overproduction of antibiotics and other secondary metabolites. In *Overproduction of Microbial Metabolites. Strain Improvement and Process Control Strategies*, pp. 63–80 (Editors Vanek, Z. and Hostalek, Z.). Butterworths, Boston.

ELANDER, R. P., MABE, J. A., HAMILL, R. L. and GORMAN, M. (1971) Biosynthesis of pyrrolnitrin by analogue resistant mutants of *Pseudomonas fluorescens*. *Fol. Microbiol.* **16**, 157–165.

FANTINI, A. A (1966) Experimental approaches to strain improvement. *Dev. Ind. Micro.* **7**, 79–87.

FERENCZY, L., KEVEI, F. and ZSOLT, J. (1974) Fusion of fungal protoplasts. *Nature (London)* **248**, 793–794.

FLEMING, I. D., NISBET, L. J. and BREWER, S. J. (1982) Target directed antimicrobial screens. In *Bioactive Microbial Products: Search and Discovery*, pp. 107–130 (Editors Bu'Lock, J. D., Nisbet, L. J. and Winstanley, D. J.). Academic Press, London.

FODOR, K. and ALFOLDI, L. (1976) Fusion of protoplasts of *Bacillus megatherium. Proc. Natn. Acad. Sci., U.S.A.* **73**, 2147–2150.

GANJU, P. L. and IYENGAR, M. R. S. (1968) An enrichment technique for isolation of auxotrophic micro-organisms. *Hindustan. Antibiot. Bull.* **11**, 12–21.

GINGOLD, E. B. (1993) An introduction to recombinant DNA technology. In *Molecular Biology and Biotechnology* (3rd edition), pp 23–50. (Editors Walker, J. M. and Gingold, E. B.). The Royal Society of Chemistry, Cambridge, U.K.

GODFREY, O. W. (1973) Isolation of regulatory mutants of the aspartic and pyruvic families and their effects on antibiotic production in *Streptomyces lipmanii*. *Antimicrobiol. Agents Chemother.* **4**, 73–79.

GOODFELLOW, M. and O'DONNELL, A. G. (1989) Search and discovery of industrially significant actinomycetes. In *Microbial Products New Approaches, Soc. Gen. Micro. Symp.* **44**, pp. 343–383 (Editors Baumberg, S., Hunter, I. S. and Rhodes, P. M.). Cambridge University Press, Cambridge.

GORDEE, E. R. and DAY, L. E. (1972) Effect of exogenous penicillin on penicillin synthesis. *Antimicrobiol. Agents Chemother.* **1**, 315–322.

GOULDEN, S. A. and CHATAWAY, F. W. (1969) End product control of acetohydroxy acid synthase by valine in *P. chrysogenum* Q176 and a high yielding mutant. *J. Gen. Micro.* **59**, 111–118.

GUEROLA, N., INGRAHAM, J. L. and CERDA-OLMEDO, E. (1971) Introduction of closely linked multiple mutations by nitrosoguanidine. *Nature New Biol., London* **230**, 122–125.

HAMILL, R. L., ELANDER, R. P., MABE, J. A. and GORMAN, M. (1970) Metabolism of tryptophan by *Pseudomonas aureofaciens. Appl. Micro.* **19**, 721–725.

HAMLYN, P. F. and BALL, C. (1979) Recombination studies with *Cephalosporium acremonium*. In *Genetics of Industrial Micro-organisms* pp. 185–191 (Editors Sebek, O. K. and Laskin, A. I.). American Society for Microbiology, Washington.

HARRISON, D. E. F. (1978) Mixed cultures in industrial fermentation processes. *Adv. Appl. Micro.* **24**, 129–162.

HARRISON, D. E. F, TOPIWALA, H. H. and HAMER, G. (1972) Yield and productivity in SCP production from methane and methanol. *Fermentation Technology Today, Proc. 4th Int. Fermentation Symposium*, pp. 491–495.

HARRISON, D. E. F, WILKINSON, T. G., WREN, S. J. and HARWOOD, J. H. (1976) Mixed bacterial cultures as a basis for continuous production of SCP from $C_1$ compounds. In *Continuous Culture 6, Applications and New Fields* pp. 122–134 (Editors Dean, A. C. R., Ellwood, D. C., Evans, C. G. T. and Melling, J.). Ellis Horwood, Chichester.

HASHIMOTO, S., MURAL, H., EZAKI, M., MORIKAWA, N., HATANAKA, H. *et al.* (1990) Studies on new dehydrogenase inhibitors. 1. Taxonomy, fermentation, isolation and physico-chemical properties. *J. Antibiot.* **43**, 29–35.

HOCKNEY, R. C. (1994) Protein expression in *E. coli*: There and back again. Paper presented at the 127th Meeting of the Society for General Microbiology, Fermentation Group, *Fermentation and Downstream Processing of Recombinant Proteins*, University of Warwick.

HOEKSEMA, H. and SMITH, C. G. (1961) Novobiocin. *Prog. Ind. Micro.* **3**, 93–139.

HONGO, M., OKI, T. and OGATA, S. (1972) Phage contamination and control. In *The Microbial Production of Amino Acids*, pp. 67–90 (Editors Yamada, K., Kinoshita, S., Tsunoda, S. and Aida, K.) Halsted Press, New York.

HOPWOOD, D. A. (1976) Genetics of antibiotic production in Streptomycetes. In *Microbiology—1976*, pp. 558–562 (Editor Schlessinger, D.). American Society for Microbiology, Washington.

HOPWOOD, D. A. (1979) The many faces of recombination. In *Genetics of Industrial Micro-organisms*, pp. 1–9 (Editors Sebek, O. K. and Laskin, A. I.). American Society for Microbiology, Washington.

HOPWOOD, D. A., WRIGHT, H. M., BIBB, M. J. and COHEN, S. N. (1977) Genetic recombination through protoplast fusion in Streptomycetes. *Nature (London)* **268**, 171–174.

HOPWOOD, D. A., BIBB, M. J., CHATER, K. F., KIESER, T., BRUTON, C. J., KIESER, H. M., LYDIATE, D. J., SMITH, C. P., WARD, J. M. and SCHREMPF, H. (1985a) *Genetic Manipulation of Streptomyces: A Laboratory Manual*. The John Innes Foundation, Norwich.

HOPWOOD, D. A., MALPARTIDA, F., KIESER, H. M., IKEDA, H., DUNCAN, J., FUJII, I., RUDD, B. A. M., FLOSS, H. G. and OMURA, S. (1985b) Production of hybrid antibiotics by genetic engineering. *Nature (London)* **314**, 642–646.

HORIUCHI, T., HORIUCHI, S. and NOVICK, A. (1963) *Genetics* **48**, 157–161.

HOSTALEK, Z. (1964) Relation between the carbohydrate metabolism of *Streptomyces aureofaciens* and the biosynthesis of chlortetracycline — The effect of interrupted aeration, inorganic phosphate and benzylthiocyanate on chlortetracycline biosynthesis. *Fol. Microbiol.* **9**, 78–88.

HSIEH, D. P. and MUNNECKE, D. M. (1972) Accelerated microbial degradation of concentrated parathion. *Fermentation Technology Today, Proc. 4th Int. Fermentation Symposium*, pp. 551–554.

HUANG, H. T. (1961) Production of L-threonine by auxotrophic mutants of *E. coli*. *Appl. Micro.* **9**, 419–424.

HUCK, T. A., PORTER, N. and BUSHELL, M. E. (1991) Positive selection of antibiotic producing soil isolates. *J. Gen. Micro.* **137** (10), 2321–2329.

HUNTER, I. S. and BAUMBERG, S. (1989) Molecular genetics of antibiotic formation. In *Microbial Products New Approaches, Soc. Gen. Micro. Symp.* Vol. 44, pp. 121–162 (Editors Baumberg, S., Hunter, I. S. and Rhodes, P. M.). Cambridge University Press, Cambridge.

ICHIKAWA, T., DATE, M., ISHIKURA, T. and OZAKI, A. (1971) Improvement of kasugamycin producing strains by the agar piece method and prototroph method. *Folia Microbiol.* **16**, 218–224.

IKEDA, M. and KATSUMATA, R. (1992) Metabolic engineering to produce tyrosine or phenylalanine in a tryptophan-producing *Corynebacterium glutamicum* strain. *Appl. Environ. Microbiol.* **58** (3), 781–785.

ITAKURA, K., HIROSE, T., CREA, R., RIGGS, A. D., HEYNEKAR, H. L., BOLIVAR, F. and BOYER, H. W. (1977) Expression in *E. coli* of a chemically synthesised gene for the hormone somatostatin. *Science* **198**, 1056–1063.

JOHNSON, M. J. (1972) Techniques for selection and evaluation of culture for biomass production. *Fermentation Technology Today, Proc. 4th Int. Fermentation Symposium*, pp. 473–478.

JONES, A. and VINING, L. C. (1976) Biosynthesis of chloramphenicol in Streptomycetes: Identification of *p*-amino-L-phenylalanine as a product from the action of arylamine synthetase on chorismic acid. *Can. J. Micro.* **22**, 237–244.

KARASAWA, M., TOSAKA, O., IKEDA, S. and YOSHII, H. (1986) Application of protoplast fusion to the development of L-threonine and L-lysine producers. *Agric. Biol. Chem.* **50** (2), 339–346.

KASE, H. and NAKAYAMA, K. (1972) Production of L-threonine by analogue resistant mutants. *Agric. Biol. Chem.* **36**, 1611–1621.

KASE, H., TANAKA, H. and NAKAYAMA, K. (1971) Studies on L-threonine fermentation: Production of L-threonine by auxotrophic mutants. *Agric. Biol. Chem.* **35**, 2089–2096.

KELLEY, W. S., CHALMERS, K. and MURRAY, N. E. (1987)

DNA polymerase from *E. coli*. *Proc. Natn. Acad. Sci. U.S.A.* **74**, 5632–5639.

KINOSHITA, S. and NAKAYAMA, K. (1978) Amino acids. In *Primary Products of Metabolism. Economic Microbiology*, Vol. 2, 210–262 (Editor Rose, A. H.). Academic Press, London.

KINOSHITA, S., UDAKA, S. and SHIMONO, M. (1957a) Studies on the amino acid fermentations, part 1. Production of L-glutamic acid by various micro-organisms. *J. Gen. Appl. Micro.* **3**, 193–205.

KINOSHITA, S., NAKAYAMA, K. and UDAKA, S. (1957b) Fermentative production of L-Orn. *J. Gen. Appl. Micro.* **3**, 276–277.

KIRSOP, B. E. and DOYLE, A. (1991) *Maintenance of Microorganisms and Cultured Cells. A Manual of Laboratory Methods* (2nd edition). Academic Press, London.

KUBOTA, K., ONDA, T., KAMIJO, H., YOSHINAGA, F. and OKAMURA, S. (1973) Microbial production of L-arginine by mutants of glutamic acid-producing bacteria. *J. Gen. Micro.*, **19**, 339–352.

KYOWA HAKKO KOGYO (1983) European patent application 73062.

LEDERBERG, J. and LEDERBERG, E. M. (1952) Replica plating and indirect selection of bacterial mutants. *J. Bacteriol.* **63**, 399–406.

LEIN, J. (1986) The Panlabs strain improvement program. In *Overproduction of Microbial Metabolites. Strain Improvement and Process Control Strategies*, pp. 105–140 (Editors Vanek, Z. and Hostalek, Z.). Butterworths, Boston.

LINCOLN, R. E. (1960) Control of stock culture preservation and inoculum build-up in bacterial fermentations. *J. Biochem. Microbiol. Tech. Eng.* **2**, 481–500.

LIU, C. M., MCDANIEL, L. E. and SCHAFFNER, C. P. (1972) Candicidin biogenesis. *J. Antibiot.* **25**, 116–121.

LUENGO, J. M., REVILLA, G., VILLANUEVA, J. R. and MARTIN, J. F. (1979) Lysine regulation of penicillin biosynthesis in low-producing and industrial strains of *Penicillium chrysogenum*. *J. Gen. Micro.* **115**, 207–211.

MCDANIEL, R., EBERT-KHOSLA, S., HOPWOOD, D. A. and KHOSLA, C. (1993) Engineered biosynthesis of novel polyketides. *Science* **262**, 1546–1550.

MCGUIRE, J. C., GLOTFELTY, G. and WHITE, R. J. (1980) *FEMS Microbiol. Lett.* **9**, 141–143.

MACDONALD, K. D. (1968) The persistence of parental genome segregation in *P. chrysogenum* after nitrogen mustard treatment. *Mutat. Res.* **5**, 302–305.

MACDONALD, K. D. and HOLT, G. (1976) Genetics of biosynthesis and overproduction of penicillin. *Sci. Prog.* **63**, 547–573.

MACDONALD, K. D., HUTCHINSON, J. M. and GILLETT, W. A. (1963) Formation and segregation of heterozygous diploids between a wild strain and derivative of high yield in *Penicillium chrysogenum*. *J. Gen. Micro.* **33**, 385–394.

MACDONALD, K. D., HUTCHINSON, J. M. and GILLETT, W. A. (1965) Heterozygous diploids of *Penicillium chrysogenum* and their segregation patterns. *Genetica* **36**, 378–397.

MALIK, K. A. (1991) Maintenance of microorganisms by simple methods. In *Maintenance of Microorganisms and Cultured Cells. A Manual of Laboratory Methods* (2nd edition), pp. 121–132 (Editors Kirsop, B. E. and Doyle, A.). Academic Press, London.

MARTIN, J. F. (1978) Manipulation of gene expression in the development of antibiotic production. In *Antibiotics and Other Secondary Metabolites. Biosynthesis and Production* (Editors Hutter, R., Leisinger, T., Nuesch, J. and Wehrlin, W.). Academic Press, London.

MARTIN, J. F. (1989) Molecular genetics of amino-acid producing corynebacteria. In *Microbial Products New Approaches, Soc. Gen. Micro. Symp.* Vol. 44, pp. 25–60 (Editors Baumberg, S., Hunter, I. S. and Rhodes, P. M.). Cambridge University Press, Cambridge.

MARTIN, J. F., GILL, J. A., NAHARRO, G., LIRAS, P. and VILLANEUVA, J. R. (1979) Industrial micro-organisms tailor-made by removal of regulatory mechanisms. In *Genetics of Industrial Micro-organisms* pp. 205–209 (Editors Sebek, O. K. and Laskin, A. I.). American Society for Microbiology, Washington.

MINDLIN, S. Z. and ZAITSEVA, Z. M. (1966) Effect of threonine and isoleucine on biosynthesis of L-lysine by a homoserine deficient strain of *Micrococcus glutamicus*. *Prikl. Biokim. i Mikrobiol.* **2** (2), 108–174.

MIWA, K., TSUCHIDA, Y., KURAHASHI, O., NAKAMORI, S., SANO, K. and MOMOSE, H. (1983) Construction of L-threonine overproducing strains of *Escherichia coli* K12 using recombinant DNA techniques. *Agric. Biol. Chem.* **47**, 2329–2334.

MORINAGA, Y., TAKAGI, H., ISHIDA, M., MIWA, K., SATO, T., NAKAMORI, S. and SANO, K. (1987) Threonine production by co-existence of cloned genes coding homoserine dehydrogenase and homoserine kinase in *Brevibacterium lactofermentum*. *Agric. Biol. Chem.* **51**, 93–100.

MURRAY, E. J. (1993) Cloning genes in mammalian cell lines. In *Molecular Biology and Biotechnology* (3rd edition), pp. 103–122 (Editors Walker, J. M. and Gingold, E.). The Royal Society of Chemistry, Cambridge.

NAGAOKA, K. and DEMAIN, A. L. (1975) Mutational biosynthesis of a new antibiotic, streptomutin A, by an idiotroph of *Streptomyces griseus*. *J. Antibiot.* **28**, 627–647.

NAGARAJAN, R., BOECK, L. D., GORMAN, M., HAMILL, R. L., HIGGINS, C. E., HOEHN, M. M., STARK, W. H. and WHITNEY, J. G. (1971) β-Lactam antibiotics from *Streptomyces. J. Am. Chem. Soc.* **93**, 2308–2310.

NAGATA, S., TAIRA, H., HALL, A., JOHNSRUD, L., STREULI, M., ECSODI, J., BOLL, W., CANTELL, K. and WEISSMANN, C (1980) Synthesis in *E. coli* of a polypeptide with human leukocyte interferon activity. *Nature (London)* **284**, 316–320.

NAKAO, Y., IKUICHI, M., SUZUKI, M. and DOI, M. (1970) Microbial production of L-Glu acid from *n*-paraffin by glycerol auxotrophs. *Agr. Biol. Chem.* **34**, 1875–1881.

NAKAO, Y., KIKUCHI, M., SUZUKI, M. and DOI, M. (1972) Microbial production of L-glutamic acid by glycerol auxotrophs and production of L-glutamic acid from *n*-paraffins. *Agric. Biol. Chem.* **36**, 490–496.

NAKAO, Y., KANAMARU, T., KIKUCHI, M. and YMATODANI, S. (1973) Extracellular accumulation of phospholipids, UDP-*N*-acetylhexosamine derivatives and L-glutamic acid by penicillin treated *Corynebacterium alkanolyticum*. *Agric. Biol. Chem.* **37**, 2399–2404.

NAKAYAMA, K. (1972) Lysine and diaminopimelic acid. In *The Microbial Production of Amino Acids*, pp. 369–397 (Editors Yamada, K., Kinoshita, S., Tsunoda, T. and Aida, K.). Halsted Press, New York.

NAKAYAMA, K., KITUDA, S. and KINOSHITA, K. (1961) Induction of nutritional mutants of glutamic acid bacteria and their amino acid accumulation. *J. Gen. Appl. Micro.* **7** (1), 41–51.

NISBET, L. J. (1982) Current strategies in the search for bioactive products. *J. Chem. Tech. Biotech.* **32**, 251–270.

NISBET, L. J. and PORTER, N. (1989) The impact of pharmacology and molecular biology on the exploitation of microbial products. In *Microbial Products New Approaches, Soc. Gen. Micro. Symp.*, Vol. 44, pp. 309–342 (Editors Baumberg, S., Hunter, I. S. and Rhodes, P. M.). Cambridge University Press, Cambridge.

OGATA, K., TOCHIKURA, T., IWAHARA, S., TAKASAWA, S., IKUSHIMA, K., NISHIMURA, A. and KIKUCHI, M. (1965) Studies on biosynthesis of biotin by micro-organisms. Part 1. Accumulation of biotin-vitamers by various micro-organisms. *Agric. Biol. Chem.* **29**, 895–901.

OSAWA, I., SPECK, D., KISOU, T., HAYAKAWA, K., ZINSIUS, M., GLOEKLER, R., LEMOINE, Y. and KAMOGAWA, K. (1989) Cloning and expression in *Escherichia coli* and *Bacillus subtilis* of the biotin synthetase gene from *Bacillus sphaericus. Gene* **80**, 39–48.

O'SULLIVAN, C. Y. and PIRT, S. J. (1973) Penicillin production by lysine auxotrophs of *P. chrysogenum. J. Gen. Micro.* **76**, 65–75.

PEBERDY, J. F., EYSSEN, H. and ANNE, J. (1977) Interspecific hybridisation between *Penicillium chrysogenum* and *Penicillium cyaneofulvum* following protoplast fusion. *Mol. Gen. Genet.* **157**, 281–284.

PERLMAN, D. and KIKUCHI, M. (1977) Culture maintenance. *Annual Reports on Fermentation Processes*, Vol. 1, 41–48 (Editor Perlman, D.). Academic Press, London.

PETTY, T. M. and CRAWFORD, D. L. (1984) Enhancement of lignin degradation in *Streptomyces* spp. by protoplast fusion. *Appl. Environ. Microbiol.* **47**, 439–440.

PICKUP, K., NOLAN, R. D. and BUSHELL, M. E. (1993) A method for increasing the success rate of duplicating antibiotic activity in agar and liquid culture of *Streptomyces* isolates in new antibiotic screens. *J. Ferm. Bioeng.* **76** (2), 89–93.

POLSINELLI, M., ALBERTINI, A., CASSANI, G. and CIFFERI, O. (1965) Relation of biochemical mutations to actinomycin synthesis in *Streptomyces antibioticus. J. Gen. Micro.* **39**, 239–246.

POLYA, K. and NYIRI, L. (1966) *Abstr. 9th Int. Congr. Microbiol.*, p. 172.

PONTECORVO, G., ROPER, J. A., HEMMONS, L. M., MacDo-

NALD, K. D. and BUFTON, A. W. J. (1953) The genetics of *A. nidulans*. *Adv. Genetics*, **5**, 141–238.

PREOBRAZHENSKAYASI, T. P., LAVROVA, N. V. M., UKHOLINA, R. S. and NECHAEVA, N. P. (1975) Isolation of new species of *Actinomadura* on selectice media with streptomycin and bruneomycin. *Antibiotiki* **30**, 404–408.

PRIDHAM, T. G., LYONS, A. J. and PHROMPATIMA, B. (1973) Viability of actinomycetes stored in soil. *Appl. Micro.* **26**, 441–442.

PRIMROSE, S. B. (1990) Controlling bacteriophage infections in industrial bioprocesses. *Adv. Biochem. Eng. Biotechnol.* **43**, 1–10.

PRUESS, D. L. and JOHNSON, M. J. (1967) Penicillin acyltransferases in *P. chrysogenum*. *J. Bacteriol.* **94**, 1502–1509.

RIVIERE, J. (1977) *Industrial Applications of Microbiology* (Translators Moss, M. O. and Smith, J. E.). Surrey University Press, London.

ROWLEY, B. I. and BULL, A. T. (1977) Isolation of a yeast lysing *Arthrobacter* species and the production of lytic enzyme complex in batch and continuous culture. *Biotech. Bioeng.* **19**, 879–900.

ROWLANDS, R. T. (1992) Strain improvement and strain stability. In *Biotechnology of Filamentous Fungi. Technology and Products*, pp. 41–64 (Editors Finkelstein, D. B. and Ball, C.). Butterworth-Heinemann, Stoneham.

RUDGE, R. H. (1991) Maintenance of bacteria by freeze-drying. In *Maintenance of Microorganisms and Cultured Cells. A Manual of Laboratory Methods* (2nd edition), pp. 31–44 (Editors Kirsop, B. E. and Doyle, A.). Academic Press, London.

SAKURAI, N., IMAI, Y., MASUDA, M., KOMATSUBARA, S., and TOSA, T. (1993) Construction of a biotin-overproducing strain of *Serratia marcescens*. *Appl. Environ. Microbiol.* **59** (9), 2857–2863.

SAMSON, R. A., HADLOCK, R. and STOLK, A. C. (1977) A taxonomic study of the *Penicillium chrysogenum* series. *Antonie van Leuwenhoek*, **43**, 169–175.

SAMSON, S. M., DOTZLAV, J. E., BECKER, G. W., VAN FRANK, R. M., VEAL, L. E., YEH, W-K., MILLER, J. R., QUEENER, S. W. and INGOLIA, T. D. (1987) *Biotechnology* **5**, 1207–1214.

SANO, K. and SHIIO, I. (1970) Microbial production of L-lysine III — Production by mutants resistant to S-(L-aminethyl)-L-cysteine. *J. Gen. Appl. Micro.* **16**, 373–391.

SENIOR, E., BULL, A. T. and SLATER, J. H. (1976) Enzyme evolution in a microbial community growing on the herbicide Dalapon. *Nature (London)* **263**, 476–479.

SERMONTI, G. (1969) *Genetics of Antibiotic Producing Organisms*. Wiley-Interscience, London.

SHIBAI, H., ENEI, H. and HIROSE, Y. (1978) Purine nucleoside fermentations. *Process Biochem.* **13** (11), 6–8.

SHIER, W. T., OGAWA, S., HICHENS, M. and RINEHART, K. L. Jr. (1973) Chemistry and biochemistry of the neomycins. XVII. Bioconversion of aminocyclitols to aminocyclitol antibiotics. *J. Antibiot.* **26**, 551–561.

SHIIO, I. and NAKAMORI, S. (1989) Coryneform bacteria. In *Fermentation Process Development of Industrial Organisms*, pp. 133–168 (Editor Neway, J. O.). Marcel Dekker, New York.

SHIIO, I. and SANO, K. (1969) Microbial production of L-lysine II — Production by mutants sensitive to threonine or methionine. *J. Gen. Appl. Micro.* **15**, 267–287.

SHORTLE, D., DIMAIO, D. and NATHANS, D. (1981) Directed mutagenesis. *Ann. Rev. Gen.* **15**, 265–294.

SIBAKOV, M., SARVAS, M. and PALAVA, J. (1983) Increased secretion of $\alpha$-amylase from *B. subtilis*. *FEMS Micro. Lett.* **17**, 81–86.

SIPICZKI, M. and FERENCZY, L. (1977) Protoplast fusion of *Schizosaccharomyces pombe* auxotrophic mutants of identical mating type. *Molec. Gen. Genet.* **157**, 77–83.

SKATRUD, P. L. (1992) Genetic engineering of $\beta$-lactam antibiotic biosynthetic pathways in filamentous fungi. *Trends Biotechnol.* **10**, 324–329.

SLATER, R. J. (1993) The expression of foreign DNA in bacteria. In *Molecular Biology and Biotechnology* (3rd edition), pp. 63–84 (Editors Walker, J. M. and Gingold, E.). The Royal Society of Chemistry, Cambridge.

SNELL, J. J. S. (1991) General introduction to maintenance methods. In *Maintenance of Microorganisms and Cultured Cells. A Manual of Laboratory Methods* (2nd edition), pp. 21–30 (Editors Kirsop, B. E. and Doyle, A.). Academic Press, London.

STUDDARD, C. (1979) Transduction of auxotrophic markers in a chloramphenicol-producing strain of *Streptomyces*. *J. Gen. Micro.* **110**, 479–483.

SVESHNIKOVA, M. A., CHORMONOVA, N. T., LAVROVA, N. V., TREKHOVA, L. P. and PREOBRAZHENSKAYA, T. P. (1976) Isolation of soil actinomycetes on selective media with novobiocin. *Antibiotiki* **21**, 784–787.

SZYBALSKI, W. (1952) Microbial selection. Part 1. Gradient plate technique for study of bacterial resistance. *Science* **116**, 46–48.

TANAKA, M., IZUMI, Y. and YAMADA, H. (1988) Microbial metabolism and production of biotin. *Vitamin (Japan)* **62**, 305–315.

TOMITA, K., HOSHINO, Y., SASHIRA, T., HASEGAWA, K., AKIYAMA, M., TSUKIURA, H. and KAWAGUICHI, H. (1980) Taxonomy of the antibiotic Bu 2313-producing organism *Microtetraspora caesia* sp. nov. *J. Antibiotics* **33**, 1491–1501.

TRENINA, G. A. and TRUTNEVA, E. M. (1966) Use of ristomycin in selection of active variants of *Proactinomyces fructiferi* var. *ristomycini*. *Antibiotiki* **11**, 770–774.

TSUKAGOSHI, N., IRITONI, S., SUSAKI, T., TAKEMURA, T., IHARO, H., IDOTA, Y., YAMAGOTO, H. and UDAKA, S. Efficient synthesis and secretion of a thermophilic $\alpha$-amylase. *J. Bacteriol.* **164**, 1182–1190.

UDAGAWA, K., ABE, S. and KINOSHITA, K. (1962) *J. Ferment. Technol.* (Osaka) (Japanese) **40**, 614.

UPSHALL, A. (1992) The application of molecular genetic methods to filamentous fungi. In *Handbook of Applied Mycology*, Vol. 4, *Fungal Biotechnology*, pp. 81–99 (Editors Arora, D. K., Elander, R. P. and Mukerji, K. G.). Marcel Dekker, New York.

UMEZAWA, H. (1972) *Enzyme Inhibitors of Microbial Origin*, University of Tokyo Press, Tokyo.

VICKERS, J. C., WILLIAMS, S. T. and ROSS, G. W. (1984) A taxonomic approach to selective isolation of streptomycetes from soil. In *Biological, Biochemical and Biomedical Aspects of Actinomycetes*, pp. 553–561 (Editors Oritz-Oritz, L., Bojalil, L. F. and Yakoleff, V.). Academic Press, Orlando.

WAKISAKA, Y., KAWAMURA, Y., YASUDA, Y., KOIZUMA, K. and NISHIMOTO, Y. (1982) A selection procedure for *Micromonospora*. *J. Antibiotics* **35**, 36–82.

WALKER, J. M. and GINGOLD, E. (1993) *Molecular Biology and Biotechnology* (3rd edition). The Royal Society of Chemistry, Cambridge.

WANG, D. I. C., COONEY, C. L., DEMAIN, A. L., DUNNILL, P., HUMPHREY, A. E. and LILLY, M. D. (1979) *Fermentation and Enzyme Technology*. Wiley-Interscience, New York.

WARD, M., WILSON, L. J., KODAMA, K. H., REY, M. W. and BERKA, R. M. (1990) Improved production of chymosin in *Aspergillus* by expression as a glucoamylase-chymosin fusion. *Biotechnology* **8**, 435–440.

WESSELING, A. C. (1982) Protoplast fusion among the Actinomycetes and its indusrial applications. *Dev. Ind. Micro.* **23**, 31–40.

WILLIAMS, S. T. and VICKERS, J. C. (1988) Detection of actinomycetes in natural habitats — problems and perspectives. In *Biology of Actinomycetes '88*, pp. 265–270 (Editors Okami, Y., Beppu, T. and Ogawara, H.). Japan Science Soc., Tokyo.

WILLOUGHBEY, L. G. (1971) Observations on some aquatic actinomycetes of streams and rivers. *Freshwater Biology* **7**, 23–27.

WINDON, J. D., WORSEY, M. J., PIOLI, E. M., PIOLI, D., BARTH, P. T., ATHERTON, K. T., DART, E. C., BYROM, D., POWELL, K. and SENIOR, P. J. (1980) *Nature (London)* **287**, 396.

WOODRUFF, H. B. (1966) The physiology of antibiotic production: role of producing organisms. *Symp. Soc. Gen. Micro.* **16**, 22–46.

WOODRUFF, H. B. and JOHNSON, M. (1970) U.S. Patent 3,524,797.

YOSHIOKOVA, H. (1952) A new rapid isolation procedure of soil streptomycetes. *J. Antibiotics* **5**, 559–561.

ZAHNER, H. (1978) The search for new secondary metabolites. In *Antibiotics and Other Secondary Metabolites. Biosynthesis and Production*, FEMS Symp., Vol. 5, pp. 1–18 (Editors Hutter, R., Leisenger, T., Nuesch and WEHRLI, W.). Academic Press, London.

# CHAPTER 4

# Media for Industrial Fermentations

## INTRODUCTION

DETAILED investigation is needed to establish the most suitable medium for an individual fermentation process, but certain basic requirements must be met by any such medium. All micro-organisms require water, sources of energy, carbon, nitrogen, mineral elements and possibly vitamins plus oxygen if aerobic. On a small scale it is relatively simple to devise a medium containing pure compounds, but the resulting medium, although supporting satisfactory growth, may be unsuitable for use in a large scale process.

On a large scale one must normally use sources of nutrients to create a medium which will meet as many as possible of the following criteria:

1. It will produce the maximum yield of product or biomass per gram of substrate used.
2. It will produce the maximum concentration of product or biomass.
3. It will permit the maximum rate of product formation.
4. There will be the minimum yield of undesired products.
5. It will be of a consistent quality and be readily available throughout the year.
6. It will cause minimal problems during media making and sterilization.
7. It will cause minimal problems in other aspects of the production process particularly aeration and agitation, extraction, purification and waste treatment.

The use of cane molasses, beet molasses, cereal grains, starch, glucose, sucrose and lactose as carbon sources, and ammonium salts, urea, nitrates, corn steep liquor, soya bean meal, slaughter-house waste and fermentation residues as nitrogen sources, have tended to meet most of the above criteria for production media because they are cheap substrates. However, other more expensive pure substrates may be chosen if the overall cost of the complete process can be reduced because it is possible to use simpler procedures. Other criteria are used to select suitable sporulation and inoculation media and these are considered in Chapter 6.

It must be remembered that the medium selected will affect the design of fermenter to be used. For example, the decision to use methanol and ammonia in the single cell protein process developed by ICI plc necessitated the design of a novel fermenter design (MacLennan et al., 1973; Sharp, 1989). The microbial oxidation of hydrocarbons is a highly aerobic and exothermic process. Thus, the production fermenter had to have a very high oxygen transfer capacity coupled with excellent cooling facilities. ICI plc solved these problems by developing an air lift fermenter (see Chapter 7). Equally, if a fermenter is already available this will obviously influence the composition of the medium. Rhodes et al. (1955) observed that the optimum concentrations of available nitrogen for griseofulvin production showed some variation with the type of fermenter used. Some aspects of this topic are considered in Chapter 7.

The problem of developing a process from the laboratory to the pilot scale, and subsequently to the industrial scale, must also be considered. A laboratory medium may not be ideal in a large fermenter with a low gas-transfer pattern. A medium with a high viscosity will also need a higher power input for effective stirring. Besides meeting requirements for growth and product formation, the medium may also influence pH variation, foam formation, the oxidation–reduction potential, and the morphological form of the organism. It may also be necessary to provide precursors or metabolic inhibitors. The medium will also affect product recovery and effluent treatment.

Historically, undefined complex natural materials have been used in fermentation processes because they are much cheaper than pure substrates. However, there is often considerable batch variation because of variable concentrations of the component parts and impurities in natural materials which cause unpredictable biomass and/or product yields. As a consequence of these variations in composition small yield improvements are difficult to detect. Undefined media often make product recovery and effluent treatment more problematical because not all the components of a complex nutrient source will be consumed by the organism. The residual components may interfere with recovery (Chapter 10) and contribute to the BOD of the effluent (Chapter 11).

Thus, although manufacturers have been reluctant to use defined media components because they are more expensive, pure substrates give more predictable yields from batch to batch and recovery, purification and effluent treatment are much simpler and therefore cheaper. Process improvements are also easier to detect when pure substrates are used.

Collins (1990) has given an excellent example of a process producing recombinant protein from *S. cerevisiae* instead of just biomass. The range of growth conditions which can be used is restricted because of factors affecting the stability of the recombinant protein. The control of pH and foam during growth in a fermenter were indentified as two important parameters. Molasses would normally be used as the cheapest carbohydrate to grow yeast biomass in a large scale process. However, this is not acceptable for the recombinant protein production because of the difficulties, and incurred costs caused in subsequent purification which result from using crude undefined media components. Collins and co-workers therefore used a defined medium with glucose, sucrose or another suitable carbon source of reasonable purity plus minimal salts, trace elements, pure vitamins and ammonia as the main nitrogen source and for pH control. Other impurities in molasses might have helped to stabilize foams and led to the need to use antifoams.

Aspects of microbial media have also been reviewed by Suomalainen and Oura (1971), Martin and Demain (1978), Iwai and Omura (1982), DeTilly *et al.* (1983), Kuenzi and Auden (1983), Miller and Churchill (1986), Smith (1986) and Priest and Sharp (1989).

Media for culture of animal cells will be discussed later in this chapter.

## TYPICAL MEDIA

Table 4.1 gives the recipes for some typical media for submerged culture fermentations. These examples are used to illustrate the range of media in use, but are not necessarily the best media in current use.

## MEDIUM FORMULATION

Medium formulation is an essential stage in the design of successful laboratory experiments, pilot-scale development and manufacturing processes. The constituents of a medium must satisfy the elemental requirements for cell biomass and metabolite production and there must be an adequate supply of energy for biosynthesis and cell maintenance. The first step to consider is an equation based on the stoichiometry for growth and product formation. Thus for an aerobic fermentation:

carbon + nitrogen + $O_2$ + other →
and        source           require-
energy                      ments
source

biomass + products + $CO_2$ + $H_2O$ + heat

This equation should be expressed in quantitative terms, which is important in the economical design of media if component wastage is to be minimal. Thus, it should be possible to calculate the minimal quantities of nutrients which will be needed to produce a specific amount of biomass. Knowing that a certain amount of biomass is necessary to produce a defined amount of product, it should be possible to calculate substrate concentrations necessary to produce required product yields. There may be medium components which are needed for product formation which are not required for biomass production. Unfortunately, it is not always easy to quantify all the factors very precisely.

A knowledge of the elemental composition of a process micro-organism is required for the solution of the elemental balance equation. This information may not be available so that data which is given in Table 4.2 will serve as a guide to the absolute minimum quantities of N, S, P, Mg and K to include in an initial medium recipe. Trace elements (Fe, Zn, Cu, Mn, Co, Mo, B) may also be needed in smaller quantities. An analysis of relative concentrations of individual elements in bacterial cells and commonly used cultivation

TABLE 4.1. *Some examples of fermentation media*

**Itaconic acid (Nubel and Ratajak, 1962)**

| | |
|---|---|
| Cane molasses (as sugar) | 150 g dm$^{-3}$ |
| $ZnSO_4$ | 1.0 g dm$^{-3}$ |
| $ZnSO_4 \cdot 7H_2O$ | 3.0 g dm$^{-3}$ |
| $CuSO_4 \cdot 5H_2O$ | 0.01 g gm$^{-3}$ |

**Amylase (Underkofler, 1966)**

| | |
|---|---|
| Ground soybean meal | 1.85% |
| Autolysed Brewers yeast fractions | 1.50% |
| Distillers dried solubles | 0.76% |
| NZ-amine (enzymatic casein hydrolysate) | 0.65% |
| Lactose | 4.75% |
| $MgSO_4 \cdot 7H_2O$ | 0.04% |
| Hodag KG-1 antifoam | 0.05% |

**Avermectin (Stapley and Woodruff, 1982)**

| | |
|---|---|
| Cerelose | 45 g |
| Peptonized milk | 24 g |
| Autolysed yeast | 2.5 g |
| Polyglycol P-2000 | 2.5 cm$^3$ |
| Distilled water | 1 dm$^3$ |
| pH | 7.0 |

**Endotoxin from *Bacillus thuringiensis* (Holmberg et al., 1980)**

| | |
|---|---|
| Molasses | 0–4% |
| Soy flour | 2–6% |
| $KH_2PO_4$ | 0.5% |
| $KH_2PO_4$ | 0.5% |
| $MgSO_4.7H_2O$ | 0.005% |
| $MnSO_4.4H_2O$ | 0.003% |
| $FeSO_4.7H_2O$ | 0.001% |
| $CaCl_2$ | 0.005% |
| $Na(NH_4)_2PO_4.4H_2O$ | 0.15% |

**Lysine (Nakayama, 1972a)**

| | |
|---|---|
| Cane blackstrap molasses | 20% |
| Soybean meal hydrosylate (as weight of meal before hydrolysis with 6N $H_2SO_4$ and neutralized with ammonia water) | 1.8% |
| $CaCO_3$ or $MgSO_4$ added to buffer medium | |
| Antifoam agent | |

**Clavulanic acid (Box, 1980)**

| | |
|---|---|
| Glycerol | 1% |
| Soybean flour | 1.5% |
| $KH_2PO_4$ | 0.1% |
| 10% Pluronic L81 antifoam in soya bean oil | 0.2%(v/v) |

**Oxytetracycline (Anonymous, 1980)**

| | |
|---|---|
| Starch | 12% + 4% (Additional feeding) |
| Technical amylase | 0.1% |
| Yeast (dry wt.) | 1.5% |
| $CaCO_3$ | 2% |
| Ammonium sulphate | 1.5% |
| Lactic acid | 0.13% |
| Lard oil | 2% |
| Total inorganic salts | 0.01% |

**Gibberellic acid (Calam and Nixon, 1960)**

| | |
|---|---|
| Glucose monohydrate | 20 g dm$^{-3}$ |
| $MgSO_4$ | 1 g dm$^{-3}$ |
| $NH_4H_2PO_4$ | 2 g dm$^{-3}$ |
| $KH_2PO_4$ | 5 g dm$^{-4}$ |
| $FeSO_4 \cdot 7H_2O$ | 0.01 g dm$^{-3}$ |
| $MnSO_4 \cdot 4H_2O$ | 0.01 g dm$^{-3}$ |
| $ZnSO_4 \cdot 7H_2O$ | 0.01 g dm$^{-3}$ |
| $CuSO_4 \cdot 5H_2O$ | 0.01 g dm$^{-3}$ |
| Corn steep liquor (as dry solids) | 7.5 g dm$^{-3}$ |

**Glutamic acid (Gore et al., 1968)**

| | |
|---|---|
| Dextrose | 270 g dm$^{-3}$ |
| $NH_4H_2PO_4$ | 2 g dm$^{-3}$ |
| $(NH_4)_2HPO_4$ | 2 g dm$^{-3}$ |
| $K_2SO_4$ | 2 g dm$^{-3}$ |
| $MgSO_4 \cdot 7H_2O$ | 0.5 g dm$^{-3}$ |
| $MnSO_4 \cdot H_2O$ | 0.04 g dm$^{-3}$ |
| $FeSO_4 \cdot 7H_2O$ | 0.02 g dm$^{-3}$ |
| Polyglycol 2000 | 0.3 g dm$^{-3}$ |
| Biotin | 12 $\mu$g dm$^{-3}$ |
| Penicillin | 11 $\mu$g dm$^{-3}$ |

**Penicillin (Perlman, 1970)**

| | |
|---|---|
| Glucose or molasses (by continuous feed) | 10% of total |
| Corn-steep liquor | 4–5% of total |
| Phenylacetic acid (by continuous feed) | 0.5–0.8% of total |
| Lard oil (or vegetable oil) antifoam by continuous addition | 0.5% of total |
| pH to 6.5 to 7.5 by acid or alkali addition | |

Note. The choice of constituents in the ten media is not a haphazard one. The rationale for medium design will be detailed in the remainder of the chapter

TABLE 4.2. *Element composition of bacteria, yeasts and fungi (% by dry weight)*

| Element | Bacteria (Luria, 1960; Herbert, 1976; Aiba *et al.*, 1973 | Yeasts (Aiba *et al.*, 1973; Herbert, 1976) | Fungi (Lilly, 1965; Aiba *et al.*, 1973) |
|---|---|---|---|
| Carbon | 50–53 | 45–50 | 40–63 |
| Hydrogen | 7 | 7 | |
| Nitrogen | 12–15 | 7.5–11 | 7–10 |
| Phosphorus | 2.0–3.0 | 0.8–2.6 | 0.4–4.5 |
| Sulphur | 0.2–1.0 | 0.01–0.24 | 0.1–0.5 |
| Potassium | 1.0–4.5 | 1.0–4.0 | 0.2–2.5 |
| Sodium | 0.5–1.0 | 0.01–0.1 | 0.02–0.5 |
| Calcium | 0.01–1.1 | 0.1–0.3 | 0.1–1.4 |
| Magnesium | 0.1–0.5 | 0.1–0.5 | 0.1–0.5 |
| Chloride | 0.5 | — | — |
| Iron | 0.02–0.2 | 0.01–0.5 | 0.1–0.2 |

media quoted by Cooney (1981) showed that some nutrients are frequently added in substantial excess of that required, e.g. P, K; however, others are often near limiting values, e.g. Zn, Cu. The concentration of P is deliberately raised in many media to increase the buffering capacity. These points emphasize the need for considerable attention to be given to medium design.

Some micro-organisms cannot synthesize specific nutrients, e.g. amino acids, vitamins or nucleotides. Once a specific growth factor has been identified it can be incorporated into a medium in adequate amounts as a pure compound or as a component of a complex mixture.

The carbon substrate has a dual role in biosythesis and energy generation. The carbon requirement for biomass production under aerobic conditions may be estimated from the cellular yield coefficient ($Y$) which is defined as:

$$\frac{\text{Quantity of cell dry matter produced}}{\text{Quantity of carbon substrate utilized}}$$

Some values are given in Table 4.3. Thus for bacteria

TABLE 4.3. *Cellular yield coefficients (Y) of bacteria on different carbon substrates (data from Abbott and Clamen, 1973)*

| Substrate | Cellular yield coefficient (g biomass dry wt. $g^{-1}$ substrate) |
|---|---|
| Methane | 0.62 |
| n-Alkanes | 1.03 |
| Methanol | 0.40 |
| Ethanol | 0.68 |
| Acetate | 0.34 |
| Malate | 0.36 |
| Glucose (molasses) | 0.51 |

with a $Y$ for glucose of 0.5, which is 0.5 g cells $g^{-1}$ glucose, the concentration of glucose needed to obtain 30 g $dm^{-3}$ cells will be $30/0.5 = 60$ g $dm^{-3}$ glucose. One litre of this medium would also need to contain approximately 3.0 g N, 1.0 g P, 1.0 g K, 0.3 g S and 0.1 g Mg. More details of $Y$ values for different micro-organisms and substrates are given by Atkinson and Mavituna (1991b).

An adequate supply of the carbon source is also essential for a product-forming fermentation process. In a critical study, analyses are made to determine how the observed conversion of the carbon source to the product compares with the theoretical maximum yield. This may be difficult because of limited knowledge of the biosynthetic pathways. Cooney (1979) has calculated theoretical yields for penicillin G biosynthesis on the basis of material and energy balances using a biosynthetic pathway based on reaction stoichiometry. The stoichiometry equation for the overall synthesis is:

$$a_2 C_6 H_{12} O_6 + b_2 NH_3 + c_2 O_2 + d_2 H_2 SO_4 + e_2 PAA$$
$$\rightarrow n_2 Pen\ G + p_2 CO_2 + q_2 H_2 O$$

where $a_2$, $b_2$, $c_2$, $d_2$, $e_2$, $n_2$, $p_2$ and $q_2$ are the stoichiometric coefficients and PAA is phenylacetic acid. Solution of this equation yields:

$$10/6\ C_6 H_{12} O_6 + 2NH_3 + 1/2\ O_2 + H_2 SO_4$$
$$+ C_8 H_8 O_2 \rightarrow C_{16} H_{18} O_4 N_2 S + 2CO_2 + 9H_2 O$$

In this instance it was calculated that the theoretical yield was 1.1 g penicillin G $g^{-1}$ glucose (1837 units $mg^{-1}$).

Using a simple model for a batch-culture penicillin fermentation it was estimated that 28, 61 and 11% of the glucose consumed was used for cell mass, maintenance and penicillin respectively. When experimental results of a fed-batch penicillin fermentation were analysed, 26% of the glucose has been used for growth, 70% for maintenance and 6% for penicillin. The maximum experimental conversion yield for penicillin was calculated to be 0.053 g g$^{-1}$ glucose (88.5 units mg$^{-1}$). Thus, the theoretical conversion value is many times higher than the experimental value. Hersbach *et al.* (1984) concuded that there were six possible biosynthetic pathways for penicillin production and two possible mechanisms for ATP production from NADH and FADH$_2$. They calculated that conversion yields by different pathways varied from 638 to 1544 units of penicillin per mg glucose. At that time the best quoted yields were 200 units penicillin per mg glucose. This gives a production of 13 to 29% of the maximum theoretical yield.

The other major nutrient which will be required is oxygen which is provided by aerating the culture, and this aspect is considered in detail in Chapter 9. The design of a medium will influence the oxygen demand of a culture in that the more reduced carbon sources will result in a higher oxygen demand. The amount of oxygen required may be determined stoichiometrically, and this aspect is also considered in Chapter 9. Optimization is dealt with later in this chapter.

## WATER

Water is the major component of all fermentation media, and is needed in many of the ancillary services such as heating, cooling, cleaning and rinsing. Clean water of consistent composition is therefore required in large quantities from reliable permanent sources. When assessing the suitability of a water supply it is important to consider pH, dissolved salts and effluent contamination.

The mineral content of the water is very important in brewing, and most critical in the mashing process, and historically influenced the siting of breweries and the types of beer produced. Hard waters containing high CaSO$_4$ concentrations are better for the English Burton bitter beers and Pilsen type lagers, while waters with a high carbonate content are better for the darker beers such as stouts. Nowadays, the water may be treated by deionization or other techniques and salts added, or the pH adjusted, to favour different beers so that breweries are not so dependent on the local water source. Detailed information is given by Hough *et al.* (1971) and Sentfen (1989).

The reuse or efficient use of water is normally of high priority. When ICI plc and John Brown Engineer-

ing developed a continuous-culture single cell protein (SCP) process at a production scale of 60,000 tonnes per year it was realized that very high costs would be incurred if fresh purified water was used on a once-through basis, since operating at a cell concentration of 30 g biomass (dw) dm$^{-3}$ would require 2700 × 10$^6$ dm$^3$ of water per annum (Ashley and Rodgers, 1986; Sharp, 1989). Laboratory tests to simulate the process showed that the *Methylophilus methylotrophus* could be grown successfully with 86% continuous recycling of supernatant with additions to make up depleted nutrients. This approach was therefore adopted in the full scale process to reduce capital and operating costs and it was estimated that water used on a once through basis without any recycling would have increased water costs by 50% and effluent treatment costs 10-fold.

Water re-usage has also been discussed by Topiwala and Khosrovi (1978), Hamer (1979) and Levi *et al.* (1979).

## ENERGY SOURCES

Energy for growth comes from either the oxidation of medium components or from light. Most industrial micro-organisms are chemo-organotrophs, therefore the commonest source of energy will be the carbon source such as carbohydrates, lipids and proteins. Some micro-organisms can also use hydrocarbons or methanol as carbon and energy sources.

## CARBON SOURCES

### Factors influencing the choice of carbon source

It is now recognized that the rate at which the carbon source is metabolized can often influence the formation of biomass or production of primary or secondary metabolites. Fast growth due to high concentrations of rapidly metabolized sugars is often associated with low productivity of secondary metabolites. This has been demonstrated for a number of processes (Table 4.4). At one time the problem was overcome by using the less readily metabolized sugars such as lactose (Johnson, 1952), but many processes now use semi-continuous or continuous feed of glucose or sucrose, discussed in Chapter 2, and later in this chapter (Table 4.15). Alternatively, carbon catabolite regulation might be overcome by genetic modification of the producer organism (Chapter 3).

TABLE 4.4. *Carbon catabolite regulation of metabolite biosynthesis*

| Metabolite | Micro-organism | Interfering carbon source | Reference |
|---|---|---|---|
| Griseofulvin | *Penicillium griseofulvin* | Glucose | Rhodes (1963); Rhodes *et al.* (1955) |
| Penicillin | *P. chrysogenum* | Glucose | Pirt and Rhigelato (1967) |
| Cephalosporin | *Cephalosporium acremonium* | Glucose | Matsumura *et al.* (1978) |
| Aurantin | *Bacillus aurantinus* | Glycerol | Nishikiori *et al.* (1978) |
| α-Amylase | *B. licheniformis* | Glucose | Priest and Sharp (1989) |
| Bacitracin | *B. licheniformis* | Glucose | Weinberg (1967) |
| Puromycin | *Streptomyces alboniger* | Glucose | Sankaran and Pogell (1975) |
| Actinomycin | *S. antibioticus* | Glucose | Marshall *et al.* (1968) |
| Cephamycin C | *S. clavuligerus* | Glycerol | Aharonowitz and Demain (1978) |
| Neomycin | *S. fradiae* | Glucose | Majumdar and Majumdar (1965) |
| Cycloserine | *S. graphalus* | Glycerol | Svensson *et al.* (1983) |
| Streptomycin | *S. griseus* | Glucose | Inamine *et al.* (1969) |
| Kanamycin | *S. kanamyceticus* | Glucose | Basek and Majumdar (1973) |
| Novobiocin | *S. niveus* | Citrate | Kominek (1972) |
| Siomycin | *S. sioyaensis* | Glucose | Kimura (1967) |

The main product of a fermentation process will often determine the choice of carbon source, particularly if the product results from the direct dissimilation of it. In fermentations such as ethanol or single-cell protein production where raw materials are 60 to 77% of the production cost, the selling price of the product will be determined largely by the cost of the carbon source (Whitaker, 1973; Moo-Young, 1977). It is often part of a company development programme to test a range of alternative carbon sources to determine the yield of product and its influence on the process and the cost of producing biomass and/or metabolite. This enables a company to use alternative substrates, depending on price and availability in different locations, and remain competitive. Up to ten different carbon sources have been or are being used by Pfizer Ltd for an antibiotic production process depending on the geographical location of the production site and prevailing economics (Stowell, 1987).

The purity of the carbon source may also affect the choice of substrate. For example, metallic ions must be removed from carbohydrate sources used in some citric acid processes (Karrow and Waksman, 1947; Woodward *et al.*, 1949; Smith *et al.*, 1974).

The method of media preparation, particularly sterilization, may affect the suitability of carbohydrates for individual fermentation processes. It is often best to sterilize sugars separately because they may react with ammonium ions and amino acids to form black nitrogen containing compounds which will partially inhibit the growth of many micro-organisms. Starch suffers from the handicap that when heated in the sterilization process it gelatinizes, giving rise to very viscous liquids, so that only concentrations of up to 2% can be used without modification (Solomons, 1969).

The choice of substrate may also be influenced by government legislation. Within the European Economic Community (EEC), the use of beet sugar and molasses is encouraged, and the minimum price controlled. The quantity of imported cane sugar and molasses is carefully monitored and their imported prices set so that they will not be competitive with beet sugar. If the world market sugar price is very low then the EEC fermentation industry will be at a disadvantage unless it receives realistic subsidies (Coombs, 1987). Refunds for a defined list of products are available in the EEC when sugar and starch are used as substrates. Legislation for recognition of new products is time consuming and manufacturers may be uncertain as to whether they would benefit from carbon substrate refunds. This uncertainty has meant that some manufacturers might prefer to site factories for new products outside the EEC (Gray, 1987).

Local laws may also dictate the substrates which may be used to make a number of beverages. In the Isle of Man, the Manx Brewers Act (1874) forbids the use of ingredients other than malt, sugar and hops in the brewing of beer. There are similar laws applying to beer production in Germany. Scotch malt whisky may be made only from barley malt, water and yeast. Within France, many wines may be called by a certain name only if the producing vineyard is within a limited geographical locality.

## Examples of commonly used carbon sources

### CARBOHYDRATES

It is common practice to use carbohydrates as the carbon source in microbial fermentation processes. The most widely available carbohydrate is starch obtained from maize grain. It is also obtained from other cereals, potatoes and cassava. Analysis data for these substrates can be obtained from Atkinson and Mavituna (1991a). Maize and other cereals may also be used directly in a partially ground state, e.g. maize chips. Starch may also be readily hydrolysed by dilute acids and enzymes to give a variety of glucose preparations (solids and syrups). Hydrolysed cassava starch is used as a major carbon source for glutamic acid production in Japan (Minoda, 1986). Syrups produced by acid hydrolysis may also contain toxic products which may make them unsuitable for particular processes.

Barley grains may be partially germinated and heat treated to give the material known as malt, which contains a variety of sugars besides starch (Table 4.5). Malt is the main substrate for brewing beer and lager in many countries. Malt extracts may also be prepared from malted grain.

Sucrose is obtained from sugar cane and sugar beet. It is commonly used in fermentation media in a very impure form as beet or cane molasses (Table 4.6), which are the residues left after crystallization of sugar solutions in sugar refining. Molasses is used in the production of high-volume/low-value products such as ethanol, SCP, organic and amino acids and some microbial gums. In 1980, 300,000 tons of cane molasses were used for amino acid production in Japan (Minoda, 1986). Molasses or sucrose also may be used for production of higher value/low-bulk products such as antibiotics, speciality enzymes, vaccines and fine chemicals (Coombs, 1987). The cost of molasses will be very competitive when compared with pure carbohydrates. However, molasses contains many impurities and molasses-based fermentations will often need a more expensive and complicated extraction/purification stage to remove the impurities and effluent treatment will be more expensive because of the unutilized waste materials which are still present. Some new processes may require critical evaluation before the final decision is made to use molasses as the main carbon substrate.

The use of lactose and crude lactose (milk whey powder) in media formulations is now extremely limited since the introduction of continuous-feeding processes utilizing glucose, discussed in a later section of this chapter.

Corn steep liquor (Table 4.7) is a by-product after starch extraction from maize. Although primarily used as a nitrogen source, it does contain lactic acid, small amounts of reducing sugars and complex polysaccharides. Certain other materials of plant origin, usually included as nitrogen sources, such as soyabean meal and Pharmamedia, contain small but significant amounts of carbohydrate.

### OILS AND FATS

Oils were first used as carriers for antifoams in antibiotic processes (Solomons, 1969). Vegetable oils (olive, maize, cotton seed, linseed, soya bean, etc.) may also be used as carbon substrates, particularly for their content of the fatty acids, oleic, linoleic and linolenic acid, because costs are competitive with those of carbohydrates. In an analysis of commodity prices for sugar, soya bean oil and tallow between 1978 and 1985, it would have been cheaper on an available energy basis to use sugar during 1978 to mid 1979 and late 1983 to 1985, whereas oil would have been the chosen substrate in the intervening period (Stowell, 1987).

Bader *et al.* (1984) discussed factors favouring the use of oils instead of carbohydrates. A typical oil contains approximately 2.4 times the energy of glucose on a per weight basis. Oils also have a volume advantage as it would take 1.24 dm$^3$ of soya bean oil to add 10 kcal of energy to a fermenter, whereas it would take 5 dm$^3$ of glucose or sucrose assuming that they are being added as 50% w/w solutions. Ideally, in any fermentation process, the maximum working capacity of a vessel should be used. Oil based fed-batch fermentations permit this procedure to operate more successfully than those using carbohydrate feeds where a larger spare capacity must be catered for to allow for responses to a sudden reduction in the residual nutrient level (Stowell, 1987). Oils also have antifoam properties

TABLE 4.5. *Carbohydrate composition of barley malt* (Harris, 1962) *(expressed as % dry weight of total)*

| | |
|---|---|
| Starch | 58–60 |
| Sucrose | 3–5 |
| Reducing sugars | 3–4 |
| Other sugars | 2 |
| Hemicellulose | 6–8 |
| Cellulose | 5 |

TABLE 4.6. *Analysis of beet and cane molasses* (Rhodes and Fletcher, 1966) *(expressed as % of total w / v)*

|  | Beet | Cane |
|---|---|---|
| Sucrose | 48.5 | 33.4 |
| Raffinose | 1.0 | 0 |
| Invert sugar | 1.0 | 21.2 |

Remainder is non-sugar.

which may make downstream processing simpler, but normally they are not used solely for this purpose.

Stowell (1987) reported the results of a Pfizer antibiotic process operated with a range of oils and fats on a laboratory scale. On a purely technical basis glycerol trioleate was the most suitable substrate. In the UK however, when both technical and economic factors are considered, soyabean oil or rapeseed oil are the preferred substrates. Glycerol trioleate is known to be used in some fermentations where substrate purity is an important consideration. Methyl oleate has been used as the sole carbon substrate in cephalosporin production (Pan *et al.*, 1982).

## HYDROCARBONS AND THEIR DERIVATIVES

There has been considerable interest in hydrocarbons. Development work has been done using n-alkanes for production of organic acids, amino acids, vitamins and co-factors, nucleic acids, antibiotics, enzymes and proteins (Fukui and Tanaka, 1980). Methane, methanol and n-alkanes have all been used as substrates for biomass production (Hamer, 1979; Levi *et al.*, 1979; Drozd, 1987; Sharp. 1989).

Drozd (1987) discussed the advantages and disadvantages of hydrocarbons and their derivatives as fermentation substrates, particularly with reference to cost, process aspects and purity. In processes where the feedstock costs are an appreciable fraction of the total manufacturing cost, cheap carbon sources are important. In the 1960s and early 1970s there was an incentive to consider using oil or natural gas derivatives as carbon substrates as costs were low and sugar prices were high. On a weight basis n-alkanes have approximately twice the carbon and three times the energy content of the same weight of sugar. Although petroleum-type products are initially impure they can be refined to obtain very pure products in bulk quantities which would would reduce the amount of effluent treatment and downstream processing. At this time the view was also held that hydrocarbons would not be subject to the same fluctuations in cost as agriculturally derived feedstocks because it would be a stable priced commodity and might be used to provide a substrate

TABLE 4.7. *Partial analysis of corn-steep liquor* (Belik *et al.*, 1957; Misecka and Zelinka, 1959; Rhodes and Fletcher, 1966)

| | |
|---|---|
| Total solids | 51 %w/v |
| Acidity as lactic acid | 15%w/v |
| Free reducing sugars | 5.6%w/v |
| Free reducing sugars after hydrolysis | 6.8%w/v |
| Total nitrogen | 4%w/v |
| Amino acids as % of nitrogen | |
| Alanine | 25 |
| Arginine | 8 |
| Glutamic acid | 8 |
| Leucine | 6 |
| Proline | 5 |
| Isoleucine | 3.5 |
| Threonine | 3.5 |
| Valine | 3.5 |
| Phenylalanine | 2.0 |
| Methionine | 1.0 |
| Cystine | 1.0 |
| | |
| Ash | 1.25%w/v |
| Potassium | 20% |
| Phosphorus | 1–5% |
| Sodium | 0.3–1% |
| Magnesium | 0.003–0.3% |
| Iron | 0.01–0.3% |
| Copper ⎫ Calcium ⎭ | 0.01–0.03% |
| Zinc | 0.003–0.08% |
| Lead ⎫ Silver ⎬ Chromium ⎭ | 0.001–0.003% |
| | |
| B Vitamins | |
| Aneurine | 41–49 $\mu g\ g^{-1}$ |
| Biotin | 0.34–0.38 $\mu g\ g^{-1}$ |
| Calcium pantothenate | 14.5–21.5 $\mu g\ g^{-1}$ |
| Folic acid | 0.26–0.6 $\mu g\ g^{-1}$ |
| Nicotinamide | 30–40 $\mu g\ g^{-1}$ |
| Riboflavine | 3.9–4.7 $\mu g\ g^{-1}$ |

Also niacin and pyridoxine

for conversion to microbial protein (SCP) for economic animal and/or human consumption. Sharp (1989) gives a very good account of market considerations of changes in price and how this would affect the price of SCP. The SCP would have had to have been cheaper or as cheap as soya meal to be marketed as an animal feed supplement. It is evident that both ICI plc and Shell plc made very careful assessments of likely future prices of soya meal during process evaluation.

SCP processes were developed by BP plc (Toprina from yeast grown on n-alkanes), ICI plc (Pruteen from bacteria grown on methanol), Hoechst / UBHE (Probion from bacteria on methanol) and Shell plc (bacteria on methane). Only BP plc and ICI plc eventually developed SCP at a production scale as an animal feed supplement (Sharp, 1989). BP's product was produced by an Italian subsidiary company, but rapidly withdrawn from manufacture because of Italian government opposition and the price of feed stock quadrupling in 1973. At this time the crude oil exporting nations (OPEC) had collectively raised the price of crude oil sold on the world market. In spite of the significant increase in the cost of crude oil and its derivatives, as well as recognizing the importance of competition from soya bean meal, the ICI plc directorate gave approval to build a full scale plant in 1976. Pruteen was marketed in the 1980s but eventually withdrawn because it could not compete with soya bean meal prices as an animal feed supplement.

Drozd (1987) has made a detailed study of hydrocarbon feedstocks and concluded that the cost of hydrocarbons does not make them economically attractive bulk feedstocks for the production of established products or potential new products where feedstock costs are an appreciable fraction of manufacturing costs of low-value bulk products. In SCP production, raw materials account for three quarters of the operating or variable costs and about half of the total costs of manufacture (Sharp, 1989; see also Chapter 12). It was considered that hydrocarbons and their derivatives might have a potential role as feedstocks in the microbial production of higher value products such as intermediates, pharmaceuticals, fine chemicals and agricultural chemicals (Drozd, 1987).

## NITROGEN SOURCES

### Examples of commonly used nitrogen sources

Most industrially used micro-organisms can utilize inorganic or organic sources of nitrogen. Inorganic nitrogen may be supplied as ammonia gas, ammonium salts or nitrates (Hunter, 1972). Ammonia has been used for pH control and as the major nitrogen source in a defined medium for the commercial production of human serum albumin by *Saccharomyces cerivisiae* (Collins, 1990). Ammonium salts such as ammonium sulphate will usually produce acid conditions as the ammonium ion is utilized and the free acid will be liberated. On the other hand nitrates will normally cause an alkaline drift as they are metabolized. Ammonium nitrate will first cause an acid drift as the ammonium ion is utilized, and nitrate assimilation is repressed. When the ammonium ion has been exhausted, there is an alkaline drift as the nitrate is used as an alternative nitrogen source (Morton and MacMil-

lan, 1954). One exception to this pattern is the metabolism of *Gibberella fujikuroi* (Borrow *et al.*, 1961, 1964). In the presence of nitrate the assimilation of ammonia is inhibited at pH 2.8–3.0. Nitrate assimilation continues until the pH has increased enough to allow the ammonia assimilation mechanism to restart.

Organic nitrogen may be supplied as amino acid, protein or urea. In many instances growth will be faster with a supply of organic nitrogen, and a few micro-organisms have an absolute requirement for amino acids. It might be thought that the main industrial need for pure amino acids would be in the deliberate addition to amino acid requiring mutants used in amino acid production. However, amino acids are more commonly added as complex organic nitrogen sources which are non-homogeneous, cheaper and readily available. In lysine production, methionine and threonine are obtained from soybean hydrolysate since it would be too expensive to use the pure amino acids (Nakayama, 1972a).

Other proteinaceous nitrogen compounds serving as sources of amino acids include corn-steep liquor, soya meal, peanut meal, cotton-seed meal (Pharmamedia, Table 4.8; and Proflo), Distillers' solubles, meal and yeast extract. Analysis of many of these products which include amino acids, vitamins and minerals are given by Miller and Churchill (1986) and Atkinson and Mavituna (1991a). In storage these products may be affected by moisture, temperature changes and ageing.

Chemically defined amino acid media devoid of protein are necessary in the production of certain vaccines when they are intended for human use.

### Factors influencing the choice of nitrogen source

Control mechanisms exist by which nitrate reductase, an enzyme involved in the conversion of nitrate to ammonium ion, is repressed in the presence of ammonia (Brown *et al.*, 1974). For this reason ammonia or ammonium ion is the preferred nitrogen source. In fungi that have been investigated, ammonium ion represses uptake of amino acids by general and specific amino acid permeases (Whitaker, 1976). In *Aspergillus nidulans*, ammonia also regulates the production of alkaline and neutral proteases (Cohen, 1973). Therefore, in mixtures of nitrogen sources, individual nitrogen components may influence metabolic regulation so that there is preferential assimilation of one component until its concentration has diminished.

It has been shown that antibiotic production by many micro-organisms is influenced by the type and

TABLE 4.8. *The composition of Pharmamedia (Traders Protein, Southern Cotton Oil Company, Division of Archer Dariels Midland Co.)*

| Component | Quantity |
|---|---|
| Total solids | 99% |
| Carbohydrate | 24.1% |
| Reducing sugars | 1.2% |
| Non reducing sugars | 1.2% |
| Protein | 57% |
| Amino nitrogen | 4.7% |
| Components of amino nitrogen | |
| Lysine | 4.5% |
| Leucine | 6.1% |
| Isoleucine | 3.3% |
| Threonine | 3.3% |
| Valine | 4.6% |
| Phenylalanine | 5.9% |
| Tryptophan | 1.0% |
| Methionine | 1.5% |
| Cystine | 1.5% |
| Aspartic acid | 9.7% |
| Serine | 4.6% |
| Proline | 3.9% |
| Glycine | 3.8% |
| Alanine | 3.9% |
| Tyrosine | 3.4% |
| Histidine | 3.0% |
| Arginine | 12.3% |
| Mineral components | |
| Calcium | 2 530 ppm |
| Chloride | 685 ppm |
| Phosphorus | 13 100 ppm |
| Iron | 94 ppm |
| Sulphate | 18 000 ppm |
| Magnesium | 7 360 ppm |
| Potassium | 17 200 ppm |
| Fat | 4.5% |
| Vitamins | |
| Ascorbic acid | 32.0 mg kg$^{-1}$ |
| Thiamine | 4.0 mg kg$^{-1}$ |
| Riboflavin | 4.8 mg kg$^{-1}$ |
| Niacin | 83.3 mg kg$^{-1}$ |
| Pantothenic acid | 12.4 mg kg$^{-1}$ |
| Choline | 3 270 mg kg$^{-1}$ |
| Pyidoxine | 16.4 mg kg$^{-1}$ |
| Biotin | 1.5 mg kg$^{-1}$ |
| Folic acid | 1.6 mg kg$^{-1}$ |
| Inositol | 10 800 mg kg$^{-1}$ |

concentration of the nitrogen source in the culture medium (Aharonowitz, 1980). Antibiotic production may be inhibited by a rapidly utilized nitrogen source ($NH_4^+$, $NO_3^-$, certain amino acids). The antibiotic production only begins to increase in the culture broth after most of the nitrogen source has been consumed.

In shake flask media experiments, salts of weak acids (e.g. ammonium succinate) may be used to serve as a nitrogen source and eradicate the source of a strong acid pH change due to chloride or sulphate ions which would be present if ammonium chloride or sulphate were used as the nitrogen source. This procedure makes it possible to use lower concentrations of phosphate to buffer the medium. High phosphate concentrations inhibit production of many secondary metabolites (see Minerals Section).

The use of complex nitrogen sources for antibiotic production has been common practice. They are thought to help create physiological conditions in the trophophase which favour antibiotic production in the idiophase (Martin and McDaniel, 1977). For example, in the production of polyene antibiotics, soybean meal has been considered a good nitrogen source because of the balance of nutrients, the low phosphorus content and slow hydrolysis. It has been suggested that this gradual breakdown prevents the accumulation of ammonium ions and repressive amino acids. These are probably some of the reasons for the selection of ideal nitrogen sources for some secondary metabolites (Table 4.9.).

In gibberellin production the nitrogen source has been shown to have an influence on directing the production of different gibberellins and the relative proportions of each type (Jefferys, 1970).

Other pre-determined aspects of the process can also influence the choice of nitrogen source. Rhodes (1963) has shown that the optimum concentration of available nitrogen for griseofulvin production showed some variation depending on the form of inoculum and the type of fermenter being used. Obviously these factors must be borne in mind in the interpretation of results in media-development programmes.

Some of the complex nitrogenous material may not be utilized by a micro-organism and create problems in downstream processing and effluent treatment. This can be an important factor in the final choice of substrate.

## MINERALS

All micro-organisms require certain mineral elements for growth and metabolism (Hughes and Poole, 1989, 1991). In many media, magnesium, phosphorus, potassium, sulphur, calcium and chlorine are essential components, and because of the concentrations required, they must be added as distinct components. Others such as cobalt, copper, iron, manganese, molyb-

TABLE 4.9. *Best nitrogen sources for some secondary metabolites*

| Product | Main nitrogen source(s) | Reference |
|---|---|---|
| Penicillin | Corn-steep liquor | Moyer and Coghill (1946) |
| Bacitracin | Peanut granules | Inskeep *et al.* (1951) |
| Riboflavin | Pancreatic digest of gelatine | Malzahn *et al.* (1959) |
| Novobiocin | Distillers' solubles | Hoeksema and Smith (1961) |
| Rifomycin | Pharmamedia Soybean meal, $(NH_4)_2SO_4$ | Sensi and Thiemann (1967) |
| Gibberellins | Ammonium salt and natural plant nitrogen source | Jefferys (1970) |
| Butirosin | Dried beef blood or haemo-globin with $(NH_4)_2SO_4$ | Claridge *et al.* (1974) |
| Polyenes | Soybean meal | Martin and MacDaniel (1977) |

denum and zinc are also essential but are usually present as impurities in other major ingredients. There is obviously a need for batch analysis of media components to ensure that this assumption can be justified, otherwise there may be deficiencies or excesses in different batches of media. See Tables 4.7 and 4.8 for analysis of corn steep liquor and Pharmamedia, and Miller and Churchill (1986) for analysis of other media ingredients of plant and animal origin. When synthetic media are used the minor elements will have to be added deliberately. The form in which the minerals are usually supplied and the concentration ranges are given in Table 4.10. As a consequence of product composition analysis, as outlined earlier in this chapter, it is possible to estimate the amount of a specific mineral for medium design, e.g. sulphur in penicillins and cephalosporins, chlorine in chlortetracycline.

The concentration of phosphate in a medium, particularly laboratory media in shake flasks, is often much higher than that of other mineral components. Part of this phosphate is being used as a buffer to minimize pH

TABLE 4.10. *The range of typical concentrations of mineral components* (g dm$^{-3}$)

| Component | Range |
|---|---|
| *$KH_2PO_4$ | 1.0–4.0 (part may be as buffer) |
| $MgSO_4 \cdot 7H_2O$ | 0.25–3.0 |
| KCl | 0.5–12.0 |
| $CaCO_3$ | 5.0–17.0 |
| $FeSO_4 \cdot 4H_2O$ | 0.01–0.1 |
| $ZnSO_4 \cdot 8H_2O$ | 0.1–1.0 |
| $MnSO_4 \cdot H_2O$ | 0.01–0.1 |
| $CuSO_4 \cdot 5H_2O$ | 0.003–0.01 |
| $Na_2MoO_4 \cdot 2H_2O$ | 0.01–0.1 |

*Complex media derived from plant and animal materials normally contain a considerable concentration of inorganic phosphate.

changes when external control of the pH is not being used.

In specific processes the concentration of certain minerals may be very critical. Some secondary metabolic processes have a lower tolerance range to inorganic phosphate than vegetative growth. This phosphate should be sufficiently low as to be assimilated by the end of trophophase. In 1950, Garner *et al.* suggested that an important function of calcium salts in fermention media was to precipitate excess inorganic phosphates, and suggested that the calcium indirectly improved the yield of streptomycin. The inorganic phosphate concentration also influences production of bacitracins, citric acid (surface culture), ergot, monomycin, novobiocin, oxytetracycline, polyenes, ristomycin, rifamycin Y, streptomycin, vancomycin and viomycin (Sensi and Thieman, 1967; Demain, 1968; Liu *et al.*, 1970; Mertz and Doolin, 1973; Weinberg, 1974). However, pyrrolnitrin (Arima *et al.*, 1965), bicyclomycin (Miyoshi *et al.*, 1972), thiopeptin (Miyairi *et al.*, 1970) and methylenomycin (Hobbs *et al.*, 1992) are produced in a medium containing a high concentration of phosphate. Two monomycin antibiotics are selectively produced by *Streptomyces jamaicensis* when the phosphate is 0.1 mM or 0.4 mM (Hall and Hassall, 1970). Phosphate regulation has also been discussed by Weinberg (1974), Aharonowitz and Demain (1977), Martin and Demain (1980), Iwai and Omura (1982) and Demain and Piret (1991).

In a recent review of antibiotic biosynthesis, Liras *et al.* (1990) recognized target enzymes which were (a) repressed by phosphate, (b) inhibited by phosphate, or (c) repression of an enzyme occurs but phosphate repression is not clearly proved. A phosphate control sequence has also been isolated and characterized from the phosphate regulated promoter that controls biosynthesis of candicidin.

Weinberg (1970) has reviewed the nine trace elements of biological interest (Atomic numbers 23–30, 42). Of these nine, the concentrations of manganese, iron and zinc are the most critical in secondary metabolism. In every secondary metabolic system in which sufficient data has been reported, the yield of the product varies linearly with the logarithmic concentration of the 'key' metal. The linear relationship does not apply at concentrations of the metal which are either insufficient, or toxic, to cell growth. Some of the primary and secondary microbial products whose yields are affected by concentrations of trace metals greater than those required for maximum growth are given in Table 4.11.

Chlorine does not appear to play a nutritional role in the metabolism of fungi (Foster, 1949). It is, however, required by some of the halophilic bacteria (Larsen, 1962). Obviously, in those fermentations where a chlorine-containing metabolite is to be produced the synthesis will have to be directed to ensure that the non-chloro-derivative is not formed. The most important compounds are chlortetracycline and griseofulvin. In griseofulvin production, adequate available chloride is provided by the inclusion of at least 0.1% KCl (Rhodes et al., 1955), as well as the chloride provided by the complex organic materials included as nitrogen sources. Other chlorine containing metabolites are caldriomycin, nornidulin and mollisin.

## Chelators

Many media cannot be prepared or autoclaved without the formation of a visible precipitate of insoluble metal phosphates. Gaunt et al. (1984) demonstrated that when the medium of Mandels and Weber (1969) was autoclaved, a white precipitate of metal ions formed, containing all the iron and most of the calcium, manganese and zinc present in the medium.

The problem of insoluble metal phosphate(s) may be eliminated by incorporating low concentrations of chelating agents such as ethylene diamine tetraacetic acid (EDTA), citric acid, polyphosphates, etc., into the medium. These chelating agents preferentially form complexes with the metal ions in a medium. The metal ions then may be gradually utilized by the micro-organism (Hughes and Poole, 1991). Gaunt et al. (1984) were able to show that the precipitate was eliminated from Mandel and Weber's medium by the addition of EDTA at 25 mg dm$^{-3}$. It is important to check that a chelating agent does not cause inhibition of growth of the micro-organism which is being cultured.

In many media, particularly those commonly used in large scale processes, there may not be a need to add a chelating agent as complex ingredients such as yeast extracts or proteose peptones will complex with metal ions and ensure gradual release of them during growth (Ramamoorthy and Kushner, 1975).

## GROWTH FACTORS

Some micro-organisms cannot synthesize a full complement of cell components and therefore require preformed compounds called growth factors. The growth factors most commonly required are vitamins, but there may also be a need for specific amino acids, fatty acids

TABLE 4.11. *Trace elements influencing primary and secondary metabolism*

| Product | Trace element(s) | Reference |
|---|---|---|
| Bacitracin | Mn | Weinberg and Tonnis (1966) |
| Protease | Mn | Mizusawa et al. (1966) |
| Gentamicin | Co | Tilley et al. (1975) |
| Riboflavin | Fe, Co | Hickey (1945) |
| | Fe | Tanner et al. (1945) |
| Mitomycin | Fe | Weinberg (1970) |
| Monensin | Fe | Weinberg (1970) |
| Actinomycin | Fe, Zn | Katz et al. (1958) |
| Candicidin | Fe, Zn | Weinberg (1970) |
| Chloramphenicol | Fe, Zn | Gallicchio and Gottlieb (1958) |
| Neomycin | Fe, Zn | Majumdur and Majumdar (1965) |
| Patulin | Fe, Zn | Brack (1947) |
| Streptomycin | Fe, Zn | Weinberg (1970) |
| Citric acid | Fe, Zn, Cu | Shu and Johnson (1948) |
| Penicillin | Fe, Zn, Cu | Foster et al. (1943) |
| | | Koffler et al. (1947) |
| Griseofulvin | Zn | Grove (1967) |

or sterols. Many of the natural carbon and nitrogen sources used in media formulations contain all or some of the required growth factors (Atkinson and Mavituna, 1991a). When there is a vitamin deficiency it can often be eliminated by careful blending of materials (Rhodes and Fletcher, 1966). It is important to remember that if only one vitamin is required it may be occasionally more economical to add the pure vitamin, instead of using a larger bulk of a cheaper multiple vitamin source. Calcium pantothenate has been used in one medium formulation for vinegar production (Beaman, 1967). In processes used for the production of glutamic acid, limited concentrations of biotin must be present in the medium (see Chapter 3). Some production strains may also require thiamine (Kinoshita and Tanaka, 1972).

## NUTRIENT RECYCLE

The need for water recycling in ICI plc's continuous-culture SCP process has already been discussed in an earlier section of this chapter. It was shown that *M. methylotrophus* could be grown in a medium containing 86% recycled supernatant plus additional fresh nutrients to make up losses. This approach made it possible to reduce the costs of media components, media preparation and storage facilities (Ashley and Rodgers, 1986; Sharp, 1989).

## BUFFERS

The control of pH may be extremely important if optimal productivity is to be achieved. A compound may be added to the medium to serve specifically as a buffer, or may also be used as a nutrient source. Many media are buffered at about pH 7.0 by the incorporation of calcium carbonate (as chalk). If the pH decreases the carbonate is decomposed. Obviously, phosphates which are part of many media also play an important role in buffering. However, high phosphate concentrations are critical in the production of many secondary metabolites (see Minerals section earlier in this chapter).

The balanced use of the carbon and nitrogen sources will also form a basis for pH control as buffering capacity can be provided by the proteins, peptides and amino acids, such as in corn-steep liquor. The pH may also be controlled externally by addition of ammonia or sodium hydroxide and sulphuric acid (Chapter 8).

## THE ADDITION OF PRECURSORS AND METABOLIC REGULATORS TO MEDIA

Some components of a fermentation medium help to regulate the production of the product rather than support the growth of the micro-organism. Such additives include precursors, inhibitors and inducers, all of which may be used to manipulate the progress of the fermentation.

### Precursors

Some chemicals, when added to certain fermentations, are directly incorporated into the desired product. Probably the earliest example is that of improving penicillin yields (Moyer and Coghill, 1946, 1947). A range of different side chains can be incorporated into the penicillin molecule. The significance of the different side chains was first appreciated when it was noted that the addition of corn-steep liquor increased the yield of penicillin from 20 units $cm^{-3}$ to 100 units $cm^{-3}$. Corn-steep liquor was found to contain phenylethylamine which was preferentially incorporated into the penicillin molecule to yield benzyl penicillin (Penicillin G). Having established that the activity of penicillin lay in the side chain, and that the limiting factor was the synthesis of the side chain, it became standard practice to add side-chain precursors to the medium, in particular phenylacetic acid. Smith and Bide (1948) showed that addition of phenylacetic acid and its derivatives to the medium were capable of both increasing penicillin production threefold and to directing biosynthesis towards increasing the proportion of benzyl penicillin from 0 to 93% at the expense of other penicillins. Phenylacetic acid is still the most widely used precursor in penicillin production. Some important examples of precursors are given in Table 4.12.

### Inhibitors

When certain inhibitors are added to fermentations, more of a specific product may be produced, or a metabolic intermediate which is normally metabolized is accumulated. One of the earliest examples is the microbial production of glycerol (Eoff *et al.*, 1919). Glycerol production depends on modifying the ethanol fermentation by removing acetaldehyde. The addition of sodium bisulphite to the broth leads to the formation of the acetaldehyde bisulphite addition compound (sodium hydroxy ethyl sulphite). Since acetaldehyde is

TABLE 4.12. *Precursors used in fermentation processes*

| Precursor | Product | Micro-organism | Reference |
| --- | --- | --- | --- |
| Phenylacetic-acid related compounds | Penicillin G | *Penicillium chrysogenum* | Moyer and Coghill and (1947) |
| Phenoxy acetic acid | Penicillin V | *Penicillium chrysogenum* | Soper *et al.* (1948) |
| Chloride | Chlortetracycline | *Streptomyces aureofaciens* | Van Dyck and de Somer (1952) |
| Chloride | Griseofulvin | *Penicillium griseofulvin* | Rhodes *et al.* (1955) |
| *Propionate | Riboflavin | *Lactobacillus bulgaricus* | Smiley and Stone (1955) |
| Cyanides | Vitamin B12 | *Proprianobacterium, Streptomyces* spp. | Mervyn and Smith (1964) |
| β-Iononones | Carotenoids | *Phycomyces blakesleeanus* | Reyes *et al.* (1964) |
| α-Amino butyric acid | L-Isoleucine | *Bacillus subtilis* | Nakayama (1972b) |
| D-Threonine | L-Isoleucine | *Serratia marcescens* | |
| Anthranilic acid | L-Tryptophan | *Hansenula anomala* | |
| Nucleosides and bases | Nikkomycins | *Streptomyces tendae* | Vecht-Lifshitz and Braun (1989) |
| Dihydronovobionic acid | Dihydronovo-biocin | *Streptomyces* sp. | Walton *et al.* (1962) |
| p-Hydroxycinnamate | Organomycin A and B | *Streptomyces organonensis* | Eiki *et al.* (1992) |
| DL-α-Amino butyric acid | Cyclosporin A | *Tolypocladium inflatum* | Kobel and Traber (1982) |
| L-Threonine | Cyclosporin C | | |
| Tyrosine or p-hydroxy-phenylglycine | Dimethylvanco-mycin | *Nocardia orientalis* | Boeck *et al.* (1984) |

* Yields are not so high as by other techniques

no longer available for re-oxidation of NADH$_2$, its place as hydrogen acceptor is taken by dihydroacetone phosphate, produced during glycolysis. The product of this reaction is glycerol-3-phosphate, which is converted to glycerol.

The application of general and specific inhibitors are illustrated in Table 4.13. In most cases the inhibitor is effective in increasing the yield of the desired product and reducing the yield of undesirable related products. A number of studies have been made with potential chlorination inhibitors, e.g. bromide, to minimize chlortetracycline production during a tetracycline fermentation (Gourevitch *et al.*, 1956; Lepetit S.p.A., 1957; Goodman *et al.*, 1959; Lein *et al.*, 1959; Szumski, 1959).

Inhibitors have also been used to affect cell-wall structure and increase the permeability for release of metabolites. The best example is the use of penicllin and surfactants in glutamic acid production (Phillips and Somerson, 1960).

### Inducers

The majority of enzymes which are of industrial interest are inducible. Induced enzymes are synthesized only in response to the presence in the environment of an inducer. Inducers are often substrates such as starch or dextrins for amylases, maltose for pullulanase and pectin for pectinases. Some inducers are very potent, such as isovaleronitrile inducing nitralase (Kobayashi *et al.*, 1992). Substrate analogues that are not attacked by the enzyme may also serve as enzyme inducers. Most inducers which are included in microbial enzyme media (Table 4.14) are substrates or substrate analogues, but intermediates and products may sometimes be used as inducers. For example, maltodextrins will induce amylase and fatty acids induce lipase. However, the cost may prohibit their use as inducers in a commercial process. Reviews have been published by Aunstrup *et al.* (1979) and Demain (1990).

One unusual application of an inducer is the use of yeast mannan in streptomycin production (Inamine *et al.*, 1969). During the fermentation varying amounts of streptomycin and mannosidostreptomycin are produced. Since mannosidostreptomycin has only 20% of the biological activity of streptomycin, the former is an undesirable product. The production organism *Streptomyces griseus* can be induced by yeast mannan to produce β-mannosidase which will convert mannosidostreptomycin to streptomycin.

It is now possible to produce a number of heterologous proteins in yeasts, fungi and bacteria. These include proteins of viral, human, animal, plant and mi-

TABLE 4.13. *Specific and general inhibitors used in fermentations*

| Product | Inhibitor | Main effect | Micro-organism | Reference |
|---|---|---|---|---|
| Glycerol | Sodium bisulphite | Acetaldehyde production repressed | *Saccharomyces cerivisiae* | Eoff *et al.* (1919) |
| Tetracycline | Bromide | Chlortetracycline formation repressed | *Streptomyces aureofaciens* | Lepetit (1957) |
| Glutamic acid | Penicillin | Cell wall permeability | *Micrococcus glutamicus* | Phillips *et al.* (1960) |
| Citric acid | Alkali metal/phosphate, pH below 2.0 | Oxalic acid repressed | *Aspergillus niger* | Batti (1967) |
| Valine | Various inhibitors | Various effects with different inhibitors | *Brevibacterium roseum* | Uemura *et al.* (1972) |
| Rifamycin B | Di-ethyl barbiturate | Other rifamycins inhibited | *Nocardia mediterranei* | Lancini and White (1973) |
| 7-Chloro-6 de-methyltetracycline | Ethionine | Affects one-carbon transfer reactions | *Streptomyces aureofaciens* | Neidleman *et al.* (1963) |

crobial origin (Peberdy, 1988; Wayne Davies, 1991). However, heterologous proteins may show some degree of toxicity to the host and have a a major influence on the stability of heterologous protein expression. As well as restricting cell growth as biomass the toxicity will provide selective conditions for segregant cells which no longer synthesize the protein at such a high level (Goodey *et al.*, 1987). Therefore, optimum growth conditions may be achieved by not synthesizing a heterologous protein continuously and only inducing it after the host culture has grown up in a vessel to produce sufficient biomass (Piper and Kirk, 1991). In cells of *S. cerevisiae* where the *Gal*1 promoter is part of the gene expression system, product formation may be induced by galactose addition to the growth medium which contains glycerol or low non-repressing levels of glucose as a carbon source.

One commercial system that has been developed is based on the *alc*A promoter in *Aspergillus nidulans* to express human interferon α2 (Wayne Davies, 1991). This can be induced by volatile chemicals, such as ethylmethyl ketone, which are added when biomass has increase to an adequate level and the growth medium contains a non-repressing carbon source or low non-repressing levels of glucose.

Methylotrophic yeasts such as *Hansenula polymorpha* and *Pichia pastoris* may be used as alternative systems because of the presence of an alcohol oxidase

TABLE 4.14. *Some examples of industrially important enzyme inducers*

| Enzyme | Inducer | Micro-organism | Reference |
|---|---|---|---|
| α-Amylase | Starch | *Aspergillus* spp. | Windish and Mhatre (1965) |
| | Maltose | *Bacillus subtilis* | |
| Pullulanase | Maltose | *Aerobacter aerogenes* | Wallenfels *et al.* (1966) |
| α-Mannosidase | Yeast mannans | *Streptomyces griseus* | Inamine *et al.* (1969) |
| Penicillin acylase | Phenylacetic acid | *Escherichia coli* | Carrington (1971) |
| Proteases | Various proteins | *Bacillus* spp. | Keay (1971) |
| | | *Streptococcus* spp. | Aunstrup (1974) |
| | | *Streptomyces* spp. | |
| | | *Asperigillus* spp. | |
| | | *Mucor* spp. | |
| Cellulase | Cellulose | *Trichoderma viride* | Reese (1972) |
| Pectinases | Pectin (beet pulp, apple pomace, citrus peel) | *Aspergillus* spp. | Fogarty and Ward (1974) |
| Nitralase | Isovaleronitrile | *Rhodococcus rhodochrous* | Kobayashi *et al.* (1992) |

promoter (Veale and Sudbery, 1991). During growth on methanol, which also acts as an inducer, the promoter is induced to produce about 30% of the cell protein. In the presence of glucose or ethanol it is undetectable. Expression systems have been developed with *P. pastoris* for tumour necrosis factor, hepatitis B surface antigen and α-galactosidase. Hepatitis B surface antigen and other heterologous proteins can also be expressed by *H. polymorpha*.

## OXYGEN REQUIREMENTS

It is sometimes forgotten that oxygen, although not added to an initial medium as such, is nevertheless a very important component of the medium in many processes, and its availability can be extremely important in controlling growth rate and metabolite production. This will be discussed in detail in Chapter 9.

The medium may influence the oxygen availability in a number of ways including the following:

1. *Fast metabolism.* The culture may become oxygen limited because sufficient oxygen cannot be made available in the fermenter if certain substrates, such as rapidly metabolized sugars which lead to a high oxygen demand, are available in high concentrations.
2. *Rheology.* The individual components of the medium can influence the viscosity of the final medium and its subsequent behaviour with respect to aeration and agitation.

3. *Antifoams.* Many of the antifoams in use will act as surface active agents and reduce the oxygen transfer rate. This topic will be considered in a later section of this chapter.

### Fast metabolism

Nutritional factors can alter the oxygen demand of the culture. *Penicillium chrysogenum* will utilize glucose more rapidly than lactose or sucrose, and it therefore has a higher specific oxygen uptake rate when glucose is the main carbon source (Johnson, 1946). Therefore, when there is the possibility of oxygen limitation due to fast metabolism, it may be overcome by reducing the initial concentration of key substrates in the medium and adding additional quantities of these substrates as a continuous or semi-continuous feed during the fermentation (see Tables 4.1. and 4.15; Chapters 2 and 9). It can also be overcome by changing the composition of the medium, incorporating higher carbohydrates (lactose, starch, etc.) and proteins which are not very rapidly metabolized and do not support such a large specific oxygen uptake rate.

### Rheology

Deindoerfer and West (1960) reported that there can be considerable variation in the viscosity of compounds that may be included in fermentation media.

TABLE 4.15. *Some processes using batch feed or continuous feed or in which they have been tried*

| Product | Additions | Reference |
|---------|-----------|-----------|
| Yeast | Molasses, nitrogen sources, P and Mg | Harrison (1971) |
| | | Reed and Peppler (1973) |
| Glycerol | Sugar, $Na_2CO_3$ | Eoff et al. (1919) |
| Acetone-butyl alcohol | Additions and withdrawals of wort | Soc. Richard et al. (1921) |
| Riboflavin | Carbohydrate | Moss and Klein (1946) |
| Penicillin | Glucose and $NH_3$ | Hosler and Johnson (1953) |
| Novobiocin | Various carbon and nitrogen sources | Smith (1956) |
| Griseofulvin | Carbohydrate | Hockenhull (1956) |
| Rifamycin | Glucose, fatty acids | Pan et al. (1959) |
| Gibberellins | Glucose | Borrow et al. (1960) |
| Vitamin $B_{12}$ | Glucose | Becher et al. (1961) |
| Tetracyclines | Glucose | Avanzini (1963) |
| Citric acid | Carbohydrates, $NH_3$ | Shepherd (1963) |
| Single-cell protein | Methanol | Harrison et al. (1972) |
| Candicidin | Glucose | Martin and McDaniel (1975) |
| Streptomycin | Glucose, ammonium sulphate | Singh et al. (1976) |
| Cephalosporin | Fresh medium addition | Trilli et al. (1977) |

Polymers in solution, particularly starch and other polysaccharides, may contribute to the rheological behaviour of the fermentation broth (Tuffile and Pinho, 1970). As the polysaccharide is degraded, the effects on rheological properties will change. Allowances may also have to be made for polysaccharides being produced by the micro-organism (Banks *et al.*, 1974; Leduy *et al.*, 1974). This aspect is considered in more detail in Chapter 9.

## ANTIFOAMS

In most microbiological processes, foaming is a problem. It may be due to a component in the medium or some factor produced by the micro-organism. The most common cause of foaming is due to proteins in the medium, such as corn-steep liquor, Pharmamedia, peanut meal, soybean meal, yeast extract or meat extract (Schugerl, 1985). These proteins may denature at the air–broth interface and form a skin which does not rupture readily. The foaming can cause removal of cells from the medium which will lead to autolysis and the further release of microbial cell proteins will probably increase the stability of the foam. If uncontrolled, then numerous changes may occur and physical and biological problems may be created. These include reduction in the working volume of the fermenter due to oxygen-exhausted gas bubbles circulating in the system (Lee and Tyman, 1988), changes in bubble size, lower mass and heat transfer rates, invalid process data due to interference at sensing electrodes and incorrect monitoring and control (Vardar-Sukan, 1992). The biological problems include deposition of cells in upper parts of the fermenter, problems of sterile operation with the air filter exits of the fermenter becoming wet, and there is danger of microbial infection and the possibility of siphoning leading to loss of product.

Hall *et al.* (1973) have recognized five patterns of foaming in fermentations:

1. Foaming remains at a constant level through-out the fermentation. Initially it is due to the medium and later due to microbial activity.
2. A steady fall in foaming during the early part of the fermentation, after which it remains constant. Initially it is due to the medium but there are no later effects caused by the micro-organism.
3. The foaming falls slightly in the early stages of the fermentation then rises. There are very slight effects caused by the medium but the major effects are due to microbial activity.

4. The fermentation has a low initial foaming capacity which rises. These effects are due solely to microbial activity.
5. A more complex foaming pattern during the fermentation which may be a combination of two or more of the previously described patterns.

If excessive foaming is encountered there are three ways of approaching the problem:

1. To try and avoid foam formation by using a defined medium and a modification of some of the physical parameters (pH, temperature, aeration and agitation). This assumes that the foam is due to a component in the medium and not a metabolite.
2. The foam is unavoidable and antifoam should be used. This is the more standard approach.
3. To use a mechanical foam breaker. (See Chapter 7.)

Antifoams are surface active agents, reducing the surface tension in the foams and destabilizing protein films by (a) hydrophobic bridges between two surfaces, (b) displacement of the absorbed protein, and (c) rapid spreading on the surface of the film (Van't Riet and Van Sonsbeck, 1992). Other possible mechanisms have been discussed by Ghildyal *et al.* (1988), Lee and Tyman (1988) and Vardar-Sukan (1992).

An ideal antifoam should have the following properties:

1. Should disperse readily and have fast action on an existing foam.
2. Should be active at low concentrations.
3. Should be long acting in preventing new foam formation.
4. Should not be metabolized by the micro-organism.
5. Should be non-toxic to the micro-organism.
6. Should be non-toxic to humans and animals.
7. Should not cause any problems in the extraction and purification of the product.
8. Should not cause any handling hazards.
9. Should be cheap.
10. Should have no effect on oxygen transfer.
11. Should be heat sterilizable.

The following compounds which meet most of these requirements have been found to be most suitable in different fermentation processes (Solomons, 1969; Ghildyal *et al.*, 1988):

1. Alcohols; stearyl and octyl decanol.
2. Esters.
3. Fatty acids and derivatives, particularly glycerides, which include cottonseed oil, linseed oil, soy-bean oil, olive oil, castor oil, sunflower oil, rapeseed oil and cod liver oil.
4. Silicones.
5. Sulphonates.
6. Miscellaneous; Alkaterge C, oxazaline, poly-propylene glycol.

These antifoams are generally added when foaming occurs during the fermentation. Because many antifoams are of low solubility they need a carrier such as lard oil, liquid paraffin or castor oil, which may be metabolized and affect the fermentation process (Solomons, 1967).

Unfortunately, the concentrations of many antifoams which are necessary to control fermentations will reduce the oxygen-transfer rate by as much as 50%; therefore antifoam additions must be kept to an absolute minimum. There are also other antifoams which will increase the oxygen-transfer rate (Ghildyal et al., 1988). If the oxygen-transfer rate is severely affected by antifoam addition then mechanical foam breakers may have to be considered as a possible alternative. Vardar-Sukan (1992) concluded that foam control in industry is still an empirical art. The best method for a particular process in one factory is not necessarily the best for the same process on another site. The design and operating parameters of a fermenter may affect the properties and quantity of foam formed.

## MEDIUM OPTIMIZATION

At this stage it is important to consider the optimization of a medium such that it meets as many as possible of the seven criteria given in the introduction to this chapter. The meaning of optimization in this context does need careful consideration (Winkler, 1991). When considering the biomass growth phase in isolation it must be recognized that efficiently grown biomass produced by an 'optimized' high productivity growth phase is not necessarily best suited for its ultimate purpose, such as synthesizing the desired product. Different combinations and sequences of process conditions need to be investigated to determine the growth conditions which produce the biomass with the physiological state best constituted for product formation. There may be a sequence of phases each with a specific set of optimal conditions.

Medium optimization by the classical method of changing one independent variable (nutrient, antifoam, pH, temperature, etc.) while fixing all the others at a certain level can be extremely time consuming and expensive for a large number of variables. To make a full factorial search which would examine each possible combination of independent variable at appropriate levels could require a large number of experiments, $x^n$, where $x$ is the number of levels and $n$ is the number of variables. This may be quite appropriate for three nutrients at two concentrations ($2^3$ trials) but not for six nutrients at three concentrations. In this instance $3^6$ (729) trials would be needed. Industrially the aim is to perform the minimum number of experiments to determine optimal conditions. Other alternative strategies must therefore be considered which allow more than one variable to be changed at a time. These methods have been discussed by Stowe and Mayer (1966), McDaniel et al. (1976), Hendrix (1980), Nelson (1982), Greasham and Inamine (1986), Bull et al. (1990) and Hicks (1993).

When more than five independent variables are to be investigated, the Plackett-Burman design may be used to find the most important variables in a system, which are then optimized in further studies (Plackett and Burman, 1946). These authors give a series of designs for up to one hundred experiments using an experimental rationale known as balanced incomplete blocks. This technique allows for the evaluation of $X - 1$ variables by $X$ experiments. $X$ must be a multiple of 4, e.g. 8, 12, 16, 20, 24, etc. Normally one determines how many experimental variables need to be included in an investigation and then selects the Plackett-Burman design which meets that requirement most closely in multiples of 4. Any factors not assigned to a variable can be designated as a dummy variable. Alternatively, factors known to not have any effect may be included and designated as dummy variables. As will be shown shortly in a worked example (Table 4.16), the incorporation of dummy variables into an experiment makes it possible to estimate the variance of an effect (experimental error).

Table 4.16 shows a Plackett-Burman design for seven variables ($A-G$) at high and low levels in which two factors, $E$ and $G$, are designated as 'dummy' variables. These can then be used in the design to obtain an estimate of error. Normally three dummy variables will provide an adequate estimate of the error. However, more can be used if fewer real variables need to be studied in an investigation (Stowe and Mayer, 1966). Each horizontal row represents a trial and each vertical

TABLE 4.16. *Plackett–Burman design for seven variables* (Nelson, 1982)

| Trial | Variables | | | | | | | Yield |
|-------|-----------|---|---|---|---|---|---|-------|
|       | *A* | *B* | *C* | *D* | *E* | *F* | *G* |       |
| 1 | H | H | H | L | H | L | H | 1.1 |
| 2 | L | H | H | H | L | H | L | 6.3 |
| 3 | L | L | H | H | H | L | H | 1.2 |
| 4 | H | L | L | H | H | H | L | 0.8 |
| 5 | L | H | L | L | H | H | H | 6.0 |
| 6 | H | L | H | L | L | H | H | 0.9 |
| 7 | H | H | L | H | L | L | H | 1.1 |
| 8 | L | L | L | L | L | L | L | 1.4 |

H denotes a high level value; L denotes a low level value.

column represents the H (high) and L (low) values of one variable in all the trials. This design (Table 4.16) requires that the frequency of each level of a variable in a given column should be equal and that in each test (horizontal row) the number of high and low variables should be equal. Consider the variable *A*; for the trials in which *A* is high, *B* is high in two of the trials and low in the other two. Similarly, *C* will be high in two trials and low in two, as will all the remaining variables. For those trials in which *A* is low, *B* will be high two times and low two times. This will also apply to all the other variables. Thus, the effects of changing the other variables cancel out when determining the effect of *A*. The same logic then applies to each variable. However, no changes are made to the high and low values for the *E* and *G* columns. Greasham and Inamine (1986) state that although the difference between the levels of each variable must be large enough to ensure that the optimum response will be included, caution must be taken when setting the level differential for sensitive variables, since a differential that is too large could mask the other variables. The trials are carried out in a randomized sequence.

The effects of the dummy variables are calculated in the same way as the effects of the experimental variables. If there are no interactions and no errors in measuring the response, the effect shown by a dummy variable should be 0. If the effect is not equal to 0, it is assumed to be a measure of the lack of experimental precision plus any analytical error in measuring the response (Stowe and Mayer, 1966).

This procedure will identify the important variables and allow them to be ranked in order of importance to decide which to investigate in a more detailed study to determine the optimum values to use.

The stages in analysing the data (Tables 4.16 and 4.17) using Nelson's (1982) example are as follows:

1. Determine the difference between the average of the H (high) and L (low) responses for each independent and dummy variable.

Therefore the difference =

$$\Sigma A(H) - \Sigma A(L).$$

The effect of an independent variable on the response is the difference between the average

TABLE 4.17. *Analysis of the yields shown in Table 4.16* (Nelson, 1982)

|  | Factor | | | | | | |
|--|--------|---|---|---|---|---|---|
|  | *A* | *B* | *C* | *D* | *E* | *F* | *G* |
| $\Sigma(H)$ | 3.9 | 14.5 | 9.5 | 9.4 | 9.1 | 14.0 | 9.2 |
| $\Sigma(L)$ | 14.9 | 4.3 | 9.3 | 9.4 | 9.7 | 4.8 | 9.6 |
| Difference | −11.0 | 10.2 | 0.2 | 0.0 | −0.6 | 9.2 | −0.4 |
| Effect | −2.75 | 2.55 | 0.05 | 0.00 | −0.15 | 2.30 | −0.10 |
| Mean square | 15.125 | 13.005 | 0.005 | 0.000 | 0.045 | 10.580 | 0.020 |

Mean square for 'error' = $\dfrac{0.045 + 0.020}{2} = 0.0325$

response for the four experiments at the high level and the average value for four experiments at the low level.

Thus the effect of

$$A = \frac{\Sigma A\,(H)}{4} - \frac{\Sigma A\,(L)}{4}$$

$$= \frac{2(\Sigma A\,(H) - \Sigma A\,(L))}{8}$$

This value should be near zero for the dummy variables.

2. Estimate the mean square of each variable (the variance of effect).

For $A$ the mean square will be $=$

$$\frac{(\Sigma A\,(H) - \Sigma A\,(L))^2}{8}$$

3. The experimental error can be calculated by averaging the mean squares of the dummy effects of $E$ and $G$.

Thus, the mean square for error $=$

$$\frac{0.045 + 0.020}{2} = 0.0325$$

This experimental error is not significant.

4. The final stage is to identify the factors which are showing large effects. In the example this was done using an $F$-test for

$$\frac{\text{Factor mean square}}{\text{error mean square}}.$$

This gives the following values:

$$A = \frac{15.125}{0.0325} = 465.4,$$

$$B = \frac{13.005}{0.0325} = 400.2,$$

$$C = \frac{0.0500}{0.0325} = 1.538$$

$$D = \frac{0.0000}{0.0325} = 0.00$$

$$F = \frac{10.580}{0.0325} = 325.6.$$

When Probability Tables are examined it is found that Factors $A$, $B$ and $F$ show large effects which are very significant, whereas $C$ shows a very low effect which is not significant and $D$ shows no effect. $A$, $B$ and $F$ have been identified as the most important

factors. The next stage would then be the optimization of the concentration of each factor, which will be discussed later.

Nelson (1982) has also referred to the possibility of two factor interactions which might occur when designing Table 4.16. This technique has also been discussed by McDaniel et al. (1976), Greasham and Inamine (1986), Bull et al. (1990) and Hicks (1993).

The next stage in medium optimization would be to determine the optimum level of each key independent variable which has been identified by the Plackett–Burman design. This may be done using response surface optimization techniques which were introduced by Box and Wilson (1951). Hendrix (1980) has given a very readable account of this technique and the way in which it may be applied. Response surfaces are similar to contour plots or topographical maps. Whilst topographical maps show lines of constant elevation, contour plots show lines of constant value. Thus, the contours of a response surface optimization plot show lines of identical response. In this context, response means the result of an experiment carried out at particular values of the variables being investigated.

The axes of the contour plot are the experimental variables and the area within the axes is termed the response surface. To construct a contour plot, the results (responses) of a series of experiments employing different combinations of the variables are inserted on the surface of the plot at the points delineated by the experimental conditions. Points giving the same results (equal responses) are then joined together to make a contour line. In its simplest form two variables are examined and the plot is two dimensional. It is important to appreciate that both variables are changed in the experimental series, rather than one being maintained constant, to ensure that the data are distributed

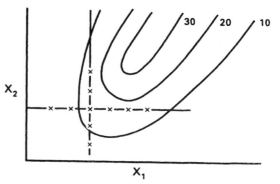

Fig. 4.1. Optimal point of a response surface by one factor at a time.

over the response surface. In Fig. 4.1, the profile generated by fixing $X_1$ and changing $X_2$ and then using the best $X_2$ value and changing $X_1$ constitutes a cross which may not encroach upon the area in which the optimum resides.

The technique may be applied at different levels of sophistication. Hendrix applied the technique at its simplest level to predict the optimum combination of two variables. The values of the variables for the initial experiments are chosen randomly or with the guidance of previous experience of the process. There is little to be gained from using more than 15–20 experiments. The resulting contour map gives an indication of the area in which the optimum combination of variables resides. A new set of experiments may then be designed within the indicated zone. Hendrix proposed the following strategy to arrive at the optimum in an incremental fashion:

1. Define the space on the plot to be explored.
2. Run five random experiments in this space.
3. Define a new space centred upon the best of the five experiments and make the new space smaller than the previous one, perhaps by cutting each dimension by one half.
4. Run five more random experiments in this new space.
5. Continue doing this until no further improvement is observed, or until you cannot afford any more experiments!

The more sophisticated applications of the response surface technique use mathematical models to analyse the first round of experimental data and to predict the relationship between the response and the variables. These calculations then allow predictive contours to be drawn and facilitate a more rapid optimization with fewer experiments. If three or more variables are to be examined then several contour maps will have to be constructed. Hicks (1993) gives an excellent account of the development of equations to model the different interactions which may take place between the variables. Several computer software packages are now available which allow the operator to determine the equations underlying the responses and, thus, to determine the likely area on the surface in which the optimum resides. Some examples of the types of response surface profiles that may be generated are illustrated in Fig. 4.2.

The following examples illustrate the application of the technique:

(i) McDaniel *et al.* (1976), Fig. 4.3. The variables under investigation were cerelose and soybean level, with the analysis indicating the optimum to be 6.2% cerelose and 3.2% soybean.

(ii) Saval *et al.* (1993). The medium for streptomycin production was optimized for four components resulting in a 52% increase in streptomycin yield, a 10% increase in mycelial dry weight and a 48% increase in specific growth rate (Table 4.18).

When further optimization experiments are necessary for medium development in large vessels, the number of experiments will normally be restricted because of the cost and the lack of spare large vessels (Spendley *et al.*, 1962). The simplex search method attempts to optimize $n$ variables by initially performing

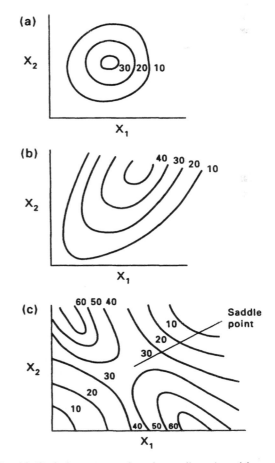

FIG. 4.2. Typical response surfaces in two dimensions; (a) mound, (b) rising ridge, (c) saddle.

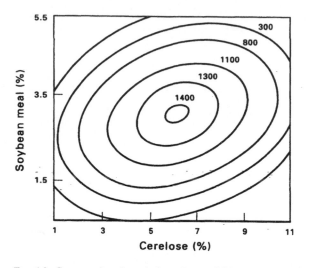

FIG. 4.3. Contour plot of two independent variables, cerelose and soybean meal, for optimization of the candidin fermentation (Redrawn from McDaniel *et al.*, 1976; Bull *et al.*, 1990).

$n + 1$ experimental trials. The results of this initial set of trials are then used to predict the conditions of the next experiment and the situation is repeated until the optimum combination is attained. Thus, after the first set of trials the optimization proceeds as individual experiments. The prediction is achieved using a graphical representation of the trials where the experimental variables are the axes. Using this procedure, the experimental variables are plotted and not the results of the experiments. The initial experimental conditions are chosen such that the points on the graph are equidistant from one another and form the vertices of a polyhedron described as the simplex. Thus, with two variables the simplex will be an equilateral triangle. The results of the initial set of three experiments are then used to predict the next experiment enabling a new simplex to be constructed. The procedure will be explained using an example to optimize the concentra-

TABLE 4.18. *Concentration of nutrients in an original and optimized medium for streptomycin production* (Saval *et al.* 1993)

| Nutrient (g dm$^{-3}$) | Original medium | Optimized medium |
|---|---|---|
| Glucose | 10 | 23 |
| Beer-yeast autolysate | 25 | 27 |
| NaCl | 10 | 8 |
| K$_2$HPO$_4$ | 1 | 1 |

tions of carbon and nitrogen sources in a medium for antibiotic production.

In our example a graph is constructed in which the $x$ axis represents the concentration range of the carbon source (the first variable) and the $y$ axis represents the concentration range of the nitrogen source (the second variable). The first vertex A (experimental point) of the simplex represents the current concentrations of the two variables which are producing the best yield of the antibiotic. The experiment for the second vertex B is planned using a new carbon–nitrogen mixture and the position of the third vertex C can now be plotted on the graph using lengths AC and BC equal to AB (the simplex equilateral triangle, Fig. 4.4a). The concentrations of the carbon and nitrogen sources to use in the third experiment can now be determined graphically and the experiment can be undertaken to determine the yield of antibiotic. The results of the three experiments are assessed and the worst response to antibiotic production indentified. In our example, experiment A was the worst and B the best. The simplex design is

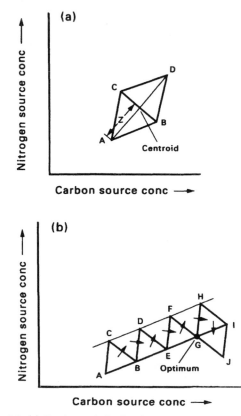

FIG. 4.4. (a) Simplex optimization for a pair of independent variables (with reflection). (b) Simplex optimization of pair of independent variables which has reached the optimum.

now used to design the next experiment. A new simplex (equilateral triangle) BCD is constructed opposite the worst response (i.e. A) using the existing vertices B and C. A line is drawn from A through the centroid (mid point) of BC. D (the next experiment) will be on this line and the sides BD and CD will be the same length as BC. This process of constructing the new simplex is described as reflection. Once the position of D is known, the concentrations of the carbon and nitrogen sources can be determined graphically, the experiment performed and the production of antibiotic assayed. Thus, a series of simplexes can be constructed moving in a crabwise way. The procedure is continued until the optimum is located. At this point the simplex begins to circle on its self, indicating the optimum concentration (Fig 4.4b; Greasham and Inamine, 1986). However, if a new vertex exhibits the lowest response, the simplex would reflect back on to the previous one, halting movement towards the optimum. In this case the new simplex is constructed opposite the second least desirable response using the method previously described.

If it is decided that the supposed optimum should be reached more rapidly then the distance $z$ between the centroid and D may be increased (expanded) by a factor which is often 2. If the optimum is thought to have nearly been reached then the distance $z$ may be decreased by a factor of 0.5 (contraction). This modified simplex optimization was first proposed by Nelder and Mead (1965) and has been discussed by Greasham and Inamine (1986).

The simplex method may also be used in small scale media development experiments to help identify the possible optimum concentration ranges to test in more extensive multifactorial experiments.

### Animal cell media

Mammalian cell lines have been cultured *in vitro* for 40 years. Initially, animal cells were required for vaccine manufacture but they are now also used in the production of monoclonal antibodies, interferon, etc. The media initially used for this purpose contained about 10% serum (foetal calf or calf) plus other organic and inorganic components. Since this pioneering work it has been possible to develop a range of serum-free media (Ham, 1965; Barnes and Sato, 1980). These media contain carbohydrates, amino acids, vitamins, nucleic acids, etc, dissolved in high purity water. Media costs are therefore considerably higher than those for microbial cells. At a 1000 dm³ scale the medium costs

may account for 40% of the unit costs, and serum may be 80% of the medium cost (Wilkinson, 1987).

### Serum

The serum is a very complex mixture containing approximately 1000 components including inorganic salts, amino acids, vitamins, carbon sources, hormones, growth factors, haemoglobin, albumin and other compounds (Brooks, 1975; Glassy *et al.*, 1988). However, most of them do not appear to be needed for growth and differentiation of cell lines which have been tested (Barnes and Sato, 1980; Darfler and Insel, 1982).

Serum is used extensively in production media for animal cell culture to produce recombinant proteins and antibody based products for *in vivo* use in humans. At present the regulations governing the quality of serum which can be used for manufacturing processes vary considerably from country to country (Hodgson, 1993). However, FDA approval of a process will be essential to market a product in the USA and therefore regulate the quality of serum which can be used. Serum tested by approved laboratories should be free of bacterial, viral or BSE (bovine sporangiform encephalitis) contamination and other components should be within strictly defined limits. Serum of this standard is needed for the cell culture media which is used to maintain the cell culture stocks as well as the production media.

The cost of foetal calf serum, US\$190 dm⁻³ in Europe, makes serum free media attractive economic alternatives, but it would take a number of years to develop suitable serum free media. The absence of the many unutilized components in serum will also simplify purification of potential products produced in such media. However, these process changes would need approval by the FDA or other regulatory bodies before a product could be marketed using a modified process.

### Serum-free media supplements

The development of serum-free media was initiated by Ham (1965) who reduced the amount of serum in media and optimized other medium components and Sato (Barnes and Sato, 1980) who investigated a range of components to promote cell growth and differentiation. Some of the more important replacements in serum-free media are albumin, insulin, transferrin, ethanolamine, selenium and β-mercaptoethanol (Glassy *et al.*, 1988).

The advantages of removing serum from media include:

1. More consistent and definable medium composition to reduce batch variation.
2. Reduction in potential contamination to make sterility easier to achieve.
3. Potential cost savings because of cheaper replacement components.
4. Simplifying downstream processing because the total protein content of the medium has been reduced.

### Protein-free media

The elimination of proteins seems an attractive objective. However, the design of such media is difficult and their use may be very limited and not very cost effective. Hamilton and Ham (1977) demonstrated the growth of Chinese hamster cell lines in a protein-free medium formulated from amino acids, vitamins, organic compounds and inorganic salts. Other media have been developed by Cleveland et al. (1983) and Shive et al. (1986).

### Trace elements

The role of trace elements in medium formulation can be significant. Cultured cells normally require Fe, Zn, Cu, Se, Mn, Mo and V (Ham and McKeehan, 1975). These are often present as impurities in other media components. Cleveland et al. (1983) found that if the number of trace elements were increased, insulin, transferrin, albumin and liposomes were not needed in a serum-free hybridoma medium. They included Al, Ag, Ba, Br, Cd, Co, Cr, F, Ge, J, Rb , Zr, Si, Ni and Sn as well as those previously mentioned.

### Osmolality

The optimum range of osmotic pressure for growth is often quite narrow and varies with the type of cell and the species from which it was isolated. It may be necessary to adjust the concentration of NaCl when major additions are made to a medium.

### pH

The normal buffer system in tissue culture media is the $CO_2$-bicarbonate system. This is a weak buffering system and can be improved by the use of a zwitterionic buffer such as Hepes, either in addition to or instead of the $CO_2$-bicarbonate buffer. Continuous pH control is achieved by the addition of sodium bicarbonate or sodium hydroxide (with fast mixing) when too acid. The pH does not normally become too alkaline so acid additions are not required but provision may be made for $CO_2$ additions (Fleischaker, 1987).

### Non-nutritional media supplements

Sodium carboxy methyl cellulose may be added to media at 0.1% to help to minimize mechanical damage caused by the shear force generated by the stirrer impeller. The problems of foam formation and subsequent cell damage and losses can affect animal cell growth. Pluronic F-68 (polyglycol) can provide a protective effect to animal cells in stirred and sparged vessels. In media which are devoid of Pluronic F-68, cells may become more sensitive to direct bubble formation in the presence of an antifoam agent being used to supress foam formation (Zhang et al., 1992).

### REFERENCES

ABBOTT, B. J. and CLAMEN, A. (1973) The relationship of substrate, growth rate and maintenance coefficient to single cell protein production. *Biotech. Bioeng.* **15**. 117–127

AHARONOWITZ, Y. (1980) Nitrogen metabolite regulation of antibiotic biosynthesis. *Ann. Rev. Microbiol.* **34**, 209–233.

AHARONOWITZ, Y. and DEMAIN, A.L. (1977) Influence of inorganic phosphate and organic buffers on cephalosporin production by *Streptomyces clavuligerus*. *Arch. Microbiol.* **115**, 169–173.

AHARONOWITZ, Y. and DEMAIN, A. L. (1978) Carbon catabolite regulation of cephalosporin production in *Streptomyces clavuligerus*. *Antimicrobiol. Agents Chemother.* **14**, 159–164.

AIBA, S., HUMPHREY, A. E. and MILLIS, N. F. (1973) Scale-up. In *Biochemical Engineering* (2nd edition), pp. 195–217. Academic Press, New York.

ANONYMOUS (1980) Research Report. Research Institute of Antibiotics and Biotransformations. Roztoky, Czechoslovakia. Cited by Podojil, M., Blumauerova, M., Vanek, Z. and Culik, K. (1984) The tetracyclines; properties, biosynthesis and fermentation. In *Biotechnology of Industrial Antibiotics*, pp. 259–279 (Editor Vandamme, E. J.). Marcel Dekker, New York.

ARIMA, K., IMANAKA, H., KUSAKA, M., FUKADA, A. and TAMURA, G. (1965) Studies on pyrrolnitrin, a new antibiotic. I. Isolation of pyrrolnitrin. *J. Antibiotics, Ser A.* **18**, 201–204.

ASHLEY, M. H. J. and RODGERS, B. L. F. (1986) The efficient use of water in single cell protein production. In *Perspectives in Biotechnology and Applied Microbiology*, pp. 71–79 (Editors Alani, D. I. and Moo-Young, M.). Elsevier, London.

ATKINSON, B. and MAVITUNA, F. (1991a) Process biotechnology. In *Biochemical Engineering and Biotechnology Handbook* (2nd edition), pp. 43–81. Macmillan, London.

ATKINSON, B. and MAVITUNA, F. (1991b) Stoichiometric aspects of microbial metabolism. In *Biochemical Engineering and Biotechnology Handbook* (2nd edition), pp. 115–167. Macmillan, London.

AUNSTRUP, K. (1974) Industrial production of proteolytic enzymes. In *Industrial Aspects of Biochemistry*, Part A, pp. 23–46 (Editor Spencer, B.). North Holland, Amsterdam.

AUNSTRUP, K.. ANDRESEN, O., FALCH, E. A. and NIELSEN, T. K. (1979) Production of microbial enzymes. In *Microbial Technology* (2nd edition), Vol. 1, pp. 281–309 (Editors Peppler, H. J. and Perlman, D.). Academic Press, New York.

AVANZINI, F. (1963) Preparation of tetracyline antibiotics. Chlortetracycline from *Streptomyces aureofaciens*. British Patent 939,476.

BADER, F. G., BOEKELOO, M. K., GRAHAM, H. E. and CAGLE, J. W. (1984) Sterilization of oils: data to support the use of a continuous point-of-use sterilizer. *Biotech. Bioeng.* **26**, 848–856.

BANKS, G. T., MANTLE, P. G. and SYCZYRBAK, C. A. (1974) Large scale production of clavine alkaloids by *Claviceps fusiformis*. *J. Gen. Microbiol.* **82**, 345–361.

BARNES, D. and SATO, G. (1980) Methods for growth of cultured cells in serum-free medium. *Anal. Biochem.* **102**, 255–270.

BASAK, K. and MAJUMDAR, S. K. (1973) Utilization of carbon and nitrogen sources by *Streptomyces kanamyceticus* for kanamycin production. *Antimicrobiol. Agents Chemother.* **4**, 6–10.

BATTI, M. R. (1967) Process for producing citric acid. U.S. Patent 3,335,067.

BAUCHOP, T. and ELSDEN, S. R. (1960) The growth of microorganisms in relation to their energy supply. *J. Gen. Microbiol.* **23**, 457 466.

BEAMAN, R. G. (l967) Vinegar fermentation. In *Microbial Technology*, pp. 344–359 (Editor Peppler, H. J.). Reinhold, New York.

BECHER, E., BERNHAUER, K. and WILHANN, G. (1961) Process for the conversion of benzimimidazole containing vitamin B12. factors, particularly Factor III to vitamin B12. U.S. Patent 2,976,220.

BELIK, E., HEROLD, M. and DOSKOCIL, J. (1957) The determination of B complex vitamins in corn-steep extracts by microbiological tests. *Chem. Zvesti*, **11**, 51–56 (*Chem. Abs.* **51**, 111508c).

BOECK, L. D., MERTZ, F. D., WOLTER, R. K. and HIGGENS, C. E. (1984) *N*-Diethylvancomycin, a novel antibiotic produced by a strain of *Nocardia orientalis*. Taxonomy and fermentation. *J. Antibiot.* **37**, 446–453.

BORROW, A., JEFFERYS, E. G. and NIXON, I. S. (1960) Gibberellic acid. British Patent 838,033.

BORROW, A., JEFFERYS, E. G., KESSEL, R. H. J., LLOYD, E. C., LLOYD, P. B. and NIXON, I. S. (1961) Metabolism of *Gibberella fujikuroi* in stirred culture. *Can. J. Microbiol.* **7**, 227–276.

BORROW, A., JEFFERYS, E. G., KESSEL, R. H. J., LLOYD, E. C., LLOYD, P. D., ROTHWELL, A., ROTHWELL, B. and SWAIT, J. C. (1964) The kinetics of metabolism of *Gibberella fujikuroi* in stirred culture. *Can. J. Microbiol.* **10**, 407–444.

BOX, G. E. P. and WILSON, K. B. (1951) On the experimental attainment of optimum conditions. *J. R. Stat. Soc. B* **13**, 1–45.

BOX, S. J. (1980) Clavulanic acid and its salts. British Patent 1,563,103.

BRACK, A. (1947) Antibacterial compounds. 1. The isolation of genistyl alcohol in addition to patulin from the filtrate of a penicillin culture. Some derivatives of genistyl alcohol. *Helv. Chim. Acta* **30**, 1–8.

BROOKS, R. F. (1975) Growth regulation *in vitro* and the role of serum. In *Structure and Function of Plasma Proteins*, pp. 239–289 (Editor Allison, A.C.). Plenum, New York.

BROWN, C. M., MacDONALD, D. S. and MEERS, J. F. (1974) Physiological aspects of microbial inorganic nitrogen metabolism. *Adv. Microbial Phys.* **11**, 1–62.

BULL, A. T., HUCK, T. A. and BUSHELL, M. E. (1990) Optimization strategies in microbial process development and operation. In *Microbial Growth Dynamics*, pp. 145–168 (Editors Poole, R. K., Bazin, M. J. and Keevil, C. W.). IRL Press, Oxford.

CALAM, C. T. and NIXON, I. S. (1960) Gibberellic acid. British Patent 839,652.

CARRINGTON, T. R. (1971) The development of commercial processes for the production of 6-aminopenicillanic acid (6-APA). *Proc. R. Soc. London B*, **179**, 321–333.

CLARIDGE, C. A., BUSH, J. A., DEFURIA, M. D. and PRICE, K. E. (1974) Fermentation and mutation studies with a butirosin producing strain of *Bacillus circulans*. *Dev. Industr. Microbiol.* **15**, 101–113.

CLEVELAND, W. L., WOOD, I. and ERLANGER, B. L. (1983) Routine large-scale production of monoclonal antibodies in a protein-free culture medium. *J. Immunol. Meth.* **56**, 221–234.

COHEN, B. L. (1973) The neutral and alkaline proteases of *Aspergillus nidulans*. *J. Gen. Microbiol.* **77**, 21–28.

COLLINS, S. H. (1990) Production of secreted proteins in yeast. In *Protein Production by Biotechnology*, pp. 61–78 (Editor Harris, T. J. R.). Elsevier, London.

COOMBS, J. (1987) Carbohydrate feedstocks: availability and utilization of molasses and whey. In *Carbon Substrates in Biotechnology*, pp. 29–44 (Editors Stowell, J. D., Beardsmore, A. J., Keevil, C.W. and Woodward, J. R.). IRL Press, Oxford.

COONEY, C.L. (1979) Conversion yields in penicillin production: theory vs. practice. *Process Biochem.* **14**(1), 31–33.

COONEY, C. L. (1981) Growth of micro-organisms. In *Biotechnology*, Vol. 1, pp. 73–112 (Editors Rehm, H. J. and Reed, G.). Verlag Chemie, Weinheim.

DARFLER, F. J. and INSEL, P. A. (1982) Serum-free culture of resting, PHA-stimulated, and transformed lymphoid cells, including hybridomas. *Exp. Cell Res.* **138**, 287–295.

DEINDOERFER, F. H. and WEST, J. M. (1960) Rheological properties of fermentation broths. *Adv. Appl. Microbiol.* **2**, 265–273.

DEMAIN, A. L. (1968) Regulatory mechanisms and the industrial production of microbial metabolites. *Lloydia* **31**, 395–418.

DEMAIN, A. L. (1990) Regulation and exploitation of enzyme biosynthesis. In *Microbial Enzymes and Biotechnology* (2nd edition) pp. 331–368 (Editors Fogarty, W. M. and Kelly, K. T.). Elsevier, London.

DEMAIN, A. L. and PIRET, J. M. (1991) Cephamycin production by *Streptomyces clavuligerus*. In *Genetics and Product Formation in Streptomyces*, pp. 87–103 (Editors Baumberg, S., Krugel, H. and Noack, D.). Plenum, New York.

DETILLEY, G., MOU, D. G. and COONEY, C. L. (1983) Optimization and economics of antibiotics production. In *The Filamentous Fungi*, Vol. 4, pp. 190–209 (Editors Smith, J. E., Berry, D. R. and Kristiansen, B.). Arnold, London.

DEY, A. (1985) *Orthogonal Fractional Factorial Designs*, pp. 1–25. Wiley, New York.

DROZD, J. W. (1987) Hydrocarbons as feedstocks for biotechnology. In *Carbon Substrates in Biotechnology*, pp. 119–138 (Editors Stowell, J. D., Beardsmore, A. J., Keevil, C.W. and Woodward, J. R.). IRL Press, Oxford.

EIKI, H., KISHI, I., GOMI, T. and OGAWA, M. (1992) 'Lights out' production of cephamycins in automated fermentation facilities. In *Harnessing Biotechnology for the 21st Century*, pp. 223–227 (Editors Ladisch, M. R. and Bose, A.). American Chemical Society, Washington.

EOFF, J. R., LINDER, W. V. and BEYER, G. F. (1919) Production of glycerine from sugar by fermentation. *Ind. Eng. Chem.* **11**, 82–84.

FLEISCHAKER, H. (1987) Microcarrier cell culture. In *Large Scale Cell Culture Technology*, pp. 59–79 (Editor Lydersen, B. K.). Hanser, Munich.

FOGARTY, W. M. and WARD, O. P. (1974) Pectinases and pectic polysaccharides. *Prog Industr. Microbiol.* **13**, 59–119.

FOSTER, J. W. (1949) *Chemical Activities of the Fungi*. Academic Press, New York.

FOSTER, J. W., WOODRUFF, H. B. and McDANIEL, L. E. (1943) Microbiological aspects of penicillin. 3. Production of penicillin in subsurface cultures of *Penicillium notatum*. *J. Bacteriol.* **46**, 421–432.

FUKUI, S. and TANAKA, A. (1980) Production of useful compounds from alkane media in Japan. *Adv. Biochem. Eng.* **17**, 1–35.

GALLICHIO, V. and GOTTLIEB, D. (1958) The biosynthesis of chloramphenicol. III. Effects of micronutrients on synthesis. *Mycologia* **50**, 490–500.

GARNER, H. R., FAHMY, M., PHILLIPS, R. I., KOFFLER, H., TETRAULT, P. A. and BOHONOS, N. (1950) Chemical

changes in the submerged growth of *Streptomyces griseus*. *Bacteriol. Proc.* pp. 139–140.

GAUNT, D. M., TRINCI, A. P. J. and LYNCH, J. M. (1984) Metal ion composition and physiology of *Trichoderma reesei* grown on a chemically defined medium prepared in two different ways. *Trans. Br. Mycol. Soc.* **83**, 575–581.

GHILDYAL, N. P., LONSANE, B. K. and KARANTH, N. G. (1988) Foam control in submerged fermentation: state of the art. *Adv. Appl. Microbiol.* **33**, 173–221.

GLASSY, M. C., THARAKAN, J. P. and CHAW, P.C. (1988) Serum-free culture media in hybridoma culture and monoclonal antibody production. *Biotech. Bioeng.* **22**, 1015–1028.

GOODEY, A. R., DOEL, S., PIGGOTT, J. R., WATSON, M. E. E. and CARTER, B. L. A. (1987) Expression and secretion of foreign polypeptides in yeast. In *Yeast Biotechnology*, pp. 401–429 (Editors Berry, D. R., Russell, I. and Stewart, G. G.). Allen and Unwin, London.

GOODMAN, J. J., MATRISHIN, M., YOUNG, R. W. and McCORMICK, J. R. D. (1959) Inhibition of the incorporation of chlorine into the tetracycline molecule. *J. Bacteriol.* **78**, 492–499.

GORE, J. H., REISMAN, H. B. and GARDNER, C. H. (1968) L-Glutamic acid by continuous fermentation. U.S. Patent 3,402,102.

GOUREVITCH, A., MISIEK, M. and LEIN, J. (1956) Competitive inhibition by bromide of incorporation of chloride into the tetracycline molecule. *Antibiot. Chemother.* **5**, 448–452.

GRAY, P. S. (1987) Impact of EEC regulations on the economics of fermentation substrates. In *Carbon Substrates in Biotechnology*, pp. 1–27 (Editors Stowell, J. D., Beardsmore, A. J., Keevil, C.W. and Woodward, J. R.). IRL Press, Oxford.

GREASHAM, R. and INAMINE, E. (1986) Nutritional improvement of processes. In *Manual of Industrial Microbiology and Biotechnology*, pp. 41–48 (Editors Demain, A. L. and Solomon, N. A.). American Society for Microbiology, Washington.

GROVE, J. F. (1967) Griseofulvin. In *Antibiotics*, II, pp. 123–13 (Editors Gottlieb, D. and Shaw, P. D.). Springer-Verlag, Berlin.

HALL, M. J. and HASSALL, C. H. (1970) Production of the monomycins, novel depsipeptide antibiotics. *Appl. Microbiol.* **19**, 109–112.

HALL, M. J., DICKINSON, S. D., PRITCHARD, R. and EVANS, J. I. (1973) Foams and foam control in fermentation processes. *Prog. Industr. Microbiol.* **12**, 169–231.

HAM, R. G. (1965) Clonal growth of mammalian cells in a chemically defined synthetic medium. *Proc. Natn. Acad. Sci. U.S.A.* **53**, 288–293.

HAM, R. G. and McKEEHAN, W. (1975) Media and growth requirements. In *Methods in Enzymology*, Vol. 58, pp. 44–93 (Editors Jakoby, W. B. and Pastan, I. H.). Academic Press, New York.

HAMER, G. (1979) Biomass from natural gas. In *Economic*

*Microbiology*, Vol. 4, pp. 315–360 (Editor Rose, A. H.). Academic Press, London.

HAMILTON, W. G. and HAM, R. G. (1977) *In Vitro* **13**, 537.

HARRIS. G. (1962) The structural chemistry of barley and malt. In *Barley and Malt, Biology, Biochemistry and Technology*, pp. 431–582 (Editor Cook, A.H.). Academic Press, London.

HARRISON, D. E. F., TOPIALA, H. and HAMER, G. (1972) Yield and productivity in single cell protein production from methane and methanol. In *Fermentation Technology Today*, pp. 491–495 (Editor Terui, G.). Society of Fermentation Technology, Japan.

HARRISON, J. S. (1971) Yeast production. *Prog. Industr. Microbiol.* **10**, 129–177.

HENDRIX, C. (1980) Through the response surface with test tube and pipe wrench. *Chemtech.* (August), 488–497.

HERBERT, D. (1976) Stoichiometric aspects of microbial growth. In *Continuous Culture 6: Applications and New Fields*. pp. 1–30 (Editors Dean, C. R., Ellwood, D. C ., Evans, C. G. T. and Melling, J.). Ellis Horwood, Chichester.

HERSBACH, G. J. M., VAN DER BEEK, C. P. and VAN DUCK, P. W. M. (1984) The penicillins: properties, biosynthesis and fermentation. In *Biotechnology of Industrial Antibiotics*, pp.45–140 (Editor Vandamme, E. J.). Marcel Dekker, New York.

HICKEY, R. J. (1945) The inactivation of iron by 2,2''-bipyridine and its effect on riboflavine synthesis by *Clostridium acetobutylicum. Arch. Biochem.* **8**, 439–447.

HICKS, C. R. (1993) *Fundamental Concepts in the Design of Experiments* (4th edition). Saunders, New York.

HOBBS, G. *et al.*(1992) An integrated approach to study regulation of production of the antibiotic methylenomycin by *Streptomyces coelicolor* A3(2). *J. Bacteriol.* **174**, 1487–1494.

HOCKENHULL, J. H. (1959) Improvements in or relating to antibiotics. British Patent 868,958.

HODGSON, J. (1993) Bovine serum revisited. *BioTech* **11**, 49–53.

HOEKSEMA, H. and SMITH, C. G. (1961) Novobiocin. *Prog. Industr. Microbiol.* **3**, 91–139.

HOLMBERG, A., SIEVANEN, R. and CARLBERG, G. (1980) Fermentation of *Bacillus thuringiensis* for endotoxin production. *Biotech. Bioeng.* **22**, 1707–1724.

HOSLER, P. and JOHNSON, M. J. (1953) Penicillin from chemically defined media. *Ind. Eng. Chem.* **45**, 871–874.

HOUGH, J. S., BRIGGS, D. E. and STEVENS, R. (1971) *Malting and Brewing Science*, Chapter 7. Chapman and Hall, London.

HUGHES, M. N. and POOLE, R. K. (1989) *Metals and Microorganisms*. Chapman and Hall, London.

HUGHES, M. N. and POOLE, R. K. (1991) Metal speciation and microbial growth — the hard (and soft) facts. *J. Gen. Microbiol.* **137**, 725–734.

HUTNER, S. H. (1972) Inorganic nutrition. *Ann. Rev. Microbiol.* **26**, 313–346.

INAMINE, E., LAGO, B. D. and DEMAIN, A. L. (1969) Regulation of mannosidase, an enzyme of streptomycin biosynthesis. In *Fermentation Advances*, pp. 199–221 (Editor Perlman, D.). Academic Press, New York.

INSKEEP, G. C., BENETT. R. E., DUDLEY, J. F. and SHEPARD, M. W. (1951) Bacitracin, product of biochemical engineering. *Ind. Eng. Chem.* **43**, 1488–1498.

IWAI, Y. and OMURA, S. (1982) Culture conditions for screening of new antibiotics. *J. Antibiot.*, **35**, 123–141.

JEFFERYS, E. G. (1970) The gibberellin fermentation. *Adv. Appl. Microbiol.* **13**, 283–316.

JOHNSON, M. J. (1946) Metabolism of penicillin producing moulds. *Ann. N.Y. Acad. Sci.* **48**, 57–66.

JOHNSON, M. J. (1952) Recent advances in penicillin fermentations. *Bull. World Health Org.* **6**, 99–121

KARROW, E. O. and WAKSMAN, S. A. (1947) Production of citric acid in submerged culture. *Ind. Eng. Chem.* **39**, 821–825.

KATZ, E., PIENTA, P. and SIVAK, A. (1958) The role of nutrition in the synthesis of actinomycin. *Appl. Microbiol.* **6**, 236–241.

KEAY, L. (1971) Microbial proteases. *Process Biochem.* **6**(8), 17–21.

KIMURA, A. (1967) Biochemical studies on *Streptomyces sioyaensis*. 11. Mechanism of the inhibitory effect of glucose on siomycin formation. *Agr. Biol. Chem. (Tokyo)*, **31**, 845–852.

KINOSHITA, S. and TANAKA, K. (1972) Glutamic acid. In *The Microbial Production of Amino Acids*, pp. 263–324 (Editors Yamada, K., Kinoshita, S., Tsunoda, T. and Aida, K.). Halsted Press-Wiley, New York.

KOBAYASHI, M., NAGASAWA, T. and YAMADA, H. (1992) Enzymatic synthesis of acrylamide; a success story not yet over. *Trends Biotech.* **10**, 402–408.

KOBEL, H. and TRABER, R. (1982) Directed synthesis of cyclosporins. *Eur. J. Appl. Microbiol. Biotechnol.* **14**, 237–240.

KOFFLER, H., KNIGHT, S. G. and FRAZIER, W. C. (1947) The effect of certain mineral elements on the production of penicillin in shake flasks. *J. Bacteriol.* **53**, 115–123.

KOMINEK, L. A. (1972) Biosynthesis of novobiocin by *Streptomyces niveus. Antimicrobiol. Agents Chemother.* **1**, 123–134.

KUENZI, M. T. and AUDEN, J. A. L. (1983) Design and control of fermentation processes. In *Bioactive Microbial Products Vol 2, pp. 91–116* (Editors Nisbet, L.J. and Winstanley, D. J.). Academic Press, London.

LANCINI, G. and WHITE, R. J. (1973) Rifamycin fermentation studies. *Process Biochem.* **8**(7), 14–16.

LARSEN, H. (1962) Halophilism. In *The Bacteria*, Vol. 4, pp. 297–342 (Editors Gunsalus, I. C. and Stanier, R. Y.). Academic Press, New York.

LEDUY, J., MARSAN, A. A. and COUPAL, B. (1974) A study of the rheological properties of a non-newtonian fermentation broth. *Biotech. Bioeng.* **16**, 61–76.

LEE, J. C. and TYMAN, K. J. (1988) Antifoams and their effects on coalesence between protein stabilized bubbles. In *2nd International Conference on Bioreactor Fluid Dynamics*, pp. 353–377 (Editor King, R.). Elsevier, London.

LEIN, J., SAWMILLER, L. F. and CHENEY, L. C. (1959) Chlorination inhibitors affecting the biosynthesis of tetracycline. *Appl. Microbiol.* **7**, 149–157.

LEPETIT, S.p.A. (1957) Bromotetracycline. British Patent 772,149.

LEVI, J. D., SHENNAN, J. L. and EBBON, G. P. (1979) Biomass from liquid n-alkanes. In *Economic Microbiology*, Vol. 4, pp. 361–419 (Editor Rose, A. H.). Academic Press, London.

LILLY, V. G. (1965) The chemical environment for growth. 1. Media, macro and micronutrients. In *The Fungi*, Vol. 1, pp. 465–478 (Editors Ainsworth, G. C. and Sussman, A.S.). Academic Press, New York.

LIRAS, P., ASTURIAS, J. A. and MARTIN, J. F. (1990) Phosphate control sequences involved in transcriptional regulation of antibiotic biosynthesis. *Trends Biotech.* **8**, 184–189.

LIU, C.M., MCDANIEL, L.E. and SCHAFFNER, C. P. (1970) Factors affecting the production of candicidin. *Antimicrob. Agents Chemother.* **7**, 196–202.

LURIA, S. E. (1960) The bacterial protoplasm: composition and organisation. In *The Bacteria*, Vol. 1, pp. 1–34 (Editors Gunsalus, I. C. and Stanier, R. Y.). Academic Press, New York.

MACLENNAN, D. G., GOW, J. S. and STRINGER, D. A. (1973) Methanol-bacterium process for SCP. *Process Biochem.* **8**(6), 22–24.

MAJUMDAR, M. K. and MAJUMDAR, S.K. (1965) Effects of minerals on neomycin production by *Streptomyces fradiae*. *Appl. Microbiol.* **13**, 190–193.

MALZAHN, R. C., PHILLIPS, R. F. and HANSON, A. M. (1959) Riboflavin. U.S. Patent 2,876,169.

MANDELS, M. and WEBER, J. (1969) The production of cellulases. In *Cellulases and their Applications. Advances in Chemistry Series*, Vol. 95, pp. 391–414 (Editor Gould, R. F.). American Chemical Society, Washington.

MARSHALL., R., REDFIELD, B., KATZ, E. and WEISSBACK, H. (1968) Changes in phenoxazinone synthetase activity during the growth cycle of *Streptomyces antibioticus*. *Arch. Biochem. Biophys.* **123**, 317–323.

MARTIN, J. F. and DEMAIN, A. L. (1978) Fungal development and metabolite formation. In *The Filamentous Fungi*, Vol. 3, pp.426–450 (Editors Smith, J. E. and Berry, D. R.). Arnold, London.

MARTIN, J. F. and DEMAIN, A. L. (1980) Control of antibiotic biosynthesis. *Microbiol. Rev.* **44**, 230–251.

MARTIN, J. F. and MCDANIEL, L. E. (1975) Kinetics of biosynthesis of polyene macrolide antibiotics by batch cultures: cell maturation time. *Biotech. Bioeng.* **17**, 925–938.

MARTIN, J. F. and MCDANIEL, L. E. (1977) Production of polyene macrolide antibiotics. *Adv. Appl. Microbiol.* **21**, 1–52.

MATSUMURA, M., IMANAKA, T., YOSHIDA, T. and TAGUCHI, H. (1978) Effect of glucose and methionine consumption rates on cephalosporin C production by *Cephalosporium acremonium. J. Ferm. Technol.* **56**, 345–353.

MCDANIEL, L.E., BAILEY, E. G., ETHIRAJ, S. and ANDREWS, H. P. (1976) Application of response surface optimization techniques to polyene macrolide fermentation studies in shake flasks. *Dev. Ind. Microbiol.* **17**, 91–98.

MERTZ, F. P. and DOOLIN, L. E. (1973) The effect of phosphate on the biosynthesis of vancomycin. *Can. J. Microbiol.* **19**, 263–270.

MERVYN, I. and SMITH, E. L. (1964) The biochemistry of vitamin B 12 fermentation. *Prog. Industr. Microbiol.* **5**, 105–201.

MILLER, T. L. and CHURCHILL, B. W. (1986) Substrates for large scale fermentations. In *Manual of Industrial Microbiology*, pp. 122–136 (Editors Demain, A. L. and Solomons, N. A.). American Society for Microbiology, Washington, DC.

MINODA, Y. (1986) Raw materials for amino acid fermentation. *Dev. Indust. Microbiol.* **24**, 51–66.

MISECKA, J. and ZELINKA, J. (1959) The effect of some elements on the quality of corn steep with rererence to the production of penicillin. *Biologia (Bratislava)* **14**, 591–596. *Chem. Abs.* **54**, 23,171c.

MIYAIRI, N., MIYOSHI. T., OKI, H., KOHSAKA, M., IKUSHIMA, H., KUNUGITA, K., SAKAI, H. H., and IMANAKA, H. (1970) Studies on thiopeptin antibiotics. 1. Characteristics of thiopeptin B. *J. Antibiot.* **23**, 113–119.

MIYOSHI, N., MIYAIRI, H., AOKI, H., KOHSAKA, M., SAKAI, H. and IMANAKA, H. (1972) Bicyclomycin, a new antibiotic. I. Taxonomy, isolation and characterization. *J. Antibiot.* **25**, 269–275.

MIZUSAWA, K., ICHISHAWA, E. and YOSHIDA, F. (1966) Proteolytic enzymes of the thermophilic *Streptomyces*. ll. Identification of the organism and some conditions of protease formation. *Agri. Biol. Chem. (Tokyo)*, **30**, 35–41.

MOO-YOUNG, M. (1977) Economics of SCP production. *Process Biochem.* **12**(4), 6–10.

MORTON, A. G. and MACMILLAN, A. (1954) The assimilation of nitrogen from ammonium salts and nitrate by fungi. *J. Exp. Bot. (London)* **5**, 232–252.

MOSS, A. R. and KLEIN, R. (1946) Improvements in or relating to the manufacture of riboflavin. British Patent 615,847.

MOYER, A, J. and COGHILL, R. D. (1946) Penicillin. IX. The laboratory scale production of penicillin by submerged culture by *Penicillium notatum* Westling (NRRL832). *J. Bacteriol.* **51**,79–93.

MOYER, A. J. and COGHILL, R. D. (1947) Penicillin. IX. The effect of phenylacetic acid on penicillin production. *J. Bacteriol.* **53**, 329–341.

NAKAYAMA, K. (1972a) Lysine and diaminopimelic acid. In *The Microbial Production of Amino Acids*, pp. 369–397 (Editors Yamada, K., Kinoshita. S., Tsunoda, T. and Aiba, K.). Halsted Press, New York.

NAKAYAMA, K. (1972b) Micro-organisms in amino acid fermentation. In *Fermentation Technology Today,* pp. 433–438 (Editor Terui, G.). Society of Fermentation Technology, Japan.

NEIDLEMAN, S. L., BIENSTOCK, E. and BENNETT, R. E. (1963) Biosynthesis of 7-chloro-6-demethyltetracycline in the presence of aminopterin and ethionine. *Biochim. Biophys. Acta* **71**, 199–201.

NELDER, J. A. and MEAD, R. (1965) A simplex method for function minimization. *Computer J.* **7**, 308–313.

NELSON, L. S. (1982) Technical aids. *J. Qual. Tech.* **14** (2), 99–100.

NISHIKIORI, T., MASUMA, R., OIWA, R., KATAGIRI, M., AWAYA, J., IWAI, Y. and OMURA, S. (1978) Aurantinin, a new antibiotic of bacterial origin. *J. Antibiot.* **31**, 525–532.

NUBEL, R. D. and RATAJAK, E. J. (1962) Process for producing itaconic acid. U.S. Patent 3,044,941.

PAN, C. H., HEPLER, L. and ELANDER, R. P. (1972) Control of pH and carbohydrate addition in the penicillin fermentation. *Dev. Ind. Microbiol.* **13**, 103–112.

PAN, C.H., SPETH, S. V., McKILLIP, E. and NASH, C. H. (1982) Methyl oleate-based medium for cephalosporin C production. *Dev. Ind. Microbiol.* **23**, 315–323.

PAN, S. C., BONANNO, S. and WAGMAN, G. H. (1959) Efficient utilization of fatty oils as energy source in penicillin fermentation. *Appl. Microbiol.* **7**, 176–180.

PERBERDY, J. F. (1988) Genetic manipulation. In *Physiology of Industrial Fungi*, pp. 187–218 (Editor Berry, D.R.). Blackwell, Oxford.

PERLMAN, D. (1970) The evolution of penicillin manufacturing processes. In *The History of Penicillin Production*, pp. 25–30 (Editor Elder, A. L.). *Chem. Eng. Prog. Symp. Series*, **60** (100).

PHILLIPS, T. and SOMERSON, L. (1960) Glutamic acid. U.S. Patent 3,080,297.

PIPER, P. W. and KIRK, N. (1991) Inducing heterologous gene expression in yeast as fermentations approach maximal biomass. In *Genetically Engineered Proteins and Enzymes from Yeast: Process Control*, pp. 147–184 (Editor Wiseman, A.). Ellis Horwood, Chichester.

PIRT, S.J. and RIGHELATO, R. C. (1967) Effect of growth rate on the synthesis of penicillin by *Penicillium chrysogenum* in batch and chemostat cultures. *Appl. Microbiol.* **15**, 1284–1290.

PLACKETT, R. L. and BURMAN, J. P. (1946) The design of multifactorial experiments. *Biometrika* **33**, 305–325.

PRIEST, F. G. and SHARP, R. J. (1989) Fermentation of bacilli. In *Fermentation Process Development of Industrial Organisms*, pp. 73–112 (Editor Neway, J. O.). Plenum, London.

RAMAMOORTHY, S. and KUSHNER, D. J. (1975) Binding of mecuric and other heavy metal ions by microbial growth media. *Microbial Ecol.* **2**, 162–176.

REED, G. and PEPPLER. H. J. (1973) *Yeast Technology*, Chapter 5. Avi, Westport.

REESE, E. T. (1972). Enzyme production from insoluble substrates. *Biotech. Bioeng. Symp.* **3**, 43–62.

REYES, P., CHICHESTER, C. O. and NAKAYAMA, T. O. M. (1964) Mechanism of β-ionone stimulation of carotenoid and ergosterol biosynthesis in *Phycomyces blakeseeanus*. *Biochim. Biophys. Acta* **90**, 578–592.

RHODES, A. (1963) Griseofulvin: production and biosynthesis. *Prog. Industr. Microbiol.* **4**, 165–187.

RHODES, A. and FLETCHER, D. L. (1966) *Principles of Industrial Microbiology*. Pergamon Press, Oxford.

RHODES, A., CROSSE, R., FERGUSON, T. P. and FLETCHER, D.

L. (1955) Improvements in or relating to the production of the antibiotic griseofulvin. British Patent 784,618.

SANKARAN, L. and POGELL, B. M. (1975) Biosynthesis of puromycin in *Streptomyces alboniger*: Regulation and properties of O-dimethyl puromycin O-methyl transferase. *Antimicrobiol. Agents Chemother.* **8**, 727–732.

SAVAL. S., PABLOS, L., and SANCHEZ, S. (1993) Optimization of a culture medium for streptomycin production using response-surface methodology. *Bioresource Technol.* **43**, 19–25.

SCHUGERL, K. (1985) Foam formation, foam suppression and the effect of foam on growth. *Process Biochem.* **20**(4), 122–123.

SENSI, P. and THIEMANN, J. E. (1967) Production of rifamycins. *Prog. Industr. Microbiol.* **6**, 21–60.

SENTFEN, H. (1989) Brewing liquor: quality requirements and corrective measures. *Brauerei Rundschau* **100**, 53–56.

SHARP, D. H. (1989) *Bioprotein Manufacture: A Critical Assessment*. Ellis Horwood, Chichester.

SHEPHERD, J. (1963) Citric acid. U.S. Patent 3,083,144.

SHIVE, W., PINKERTON, F., HUMPHREYS, J., JOHNSON, M. M., HAMILTON, W. G. and MATTHEWS, K. S. (1986) Development of a chemically defined serum- and protein-free medium for growth of human peripheral lymphocytes. *Proc. Natn. Acad. Sci. U.S.A.* **83**, 9–13.

SHU, P. and JOHNSON, M. J. (1948) Interdependence of medium constituents in citric acid production by submerged fermentation. *J. Bacteriol.* **56**, 577–585.

SINGH, A., BRUZELIUS, E. and HEDING, H. (1976) Streptomycin, a fermentation study. *Eur. J. Appl. Microbiol.* **3**, 97–101.

SMILEY, K. L. and STONE, L. (1955) Production of riboflavine by *Ashbya gossypii*. U.S. Patent 2,702,265.

SMITH, C. G. (1956) Fermentation studies with *Streptomyces niveus*. *Appl. Microbiol.* **4**, 232–236.

SMITH, E. L. and BIDE, A. E. (1948) Penicillin salts. *Biochem. J.* **42**, xvii–xviii.

SMITH, J. E. (1986) Concepts of industrial antibiotic production. In *Perspectives in Biotechnology and Applied Microbiology*, pp. 106–142 (Editors Alani, D. I. and Moo-Young, M.). Elsevier, London.

SMITH, J. E., NOWAKOWSKA-WASZCZUK, K. and ANDERSON, J. G. (1974) Organic acid production by mycelial fungi. In *Industrial Aspects of Biochemistry*, Part A, pp. 297–317 (Editor Spencer, B.). North Holland, Amsterdam.

SOC. RICHARD, ALLENTE ET CIE (1921) Acetone and butyl alcohol. British Patent 176, 284.

SOLOMONS, G. L. (1967) Antifoams. *Process Biochem.* **2**(10), 47–48.

SOLOMONS, G. L. (1969) *Materials and Methods in Fermentation*. Academic Press, London.

SOPER, Q. F., WHITEHEAD, C., BEHRENS, O. K., CORSE, J. J. and JONES, R. G. (1948) Biosynthesis of penicillins. VII. Oxy- and mercapto acetic acids. *J. Chem. Soc.* **70**, 2849–2855.

SPENDLEY, W., HEXT, G. R. and HIMSWORTH, F. R. (1962)

Sequential application of simplex designs in optimization and evolutionary operation. *Technometrics* **4**, 441–461.

STAPLEY, E. O. and WOODRUFF, H. B. (1982) Avermectin, antiparasitic lactones produced by *Streptomyces avermitilis* isolated from a soil in Japan. In *Trends in Antibiotic Research*, pp. 154–170 (Editors Umezawa, H., Demain, A. L., Mata, T. and Hutchinson, C.R.). Japanese Antibiotic Research Association, Tokyo.

STOWE, R. A. and MAYER, R. P. (1966) Efficient screening of process variables. *Ind. Eng. Chem.* **56**, 36–40.

STOWELL, J. D. (1987) The application of oils and fats in antibiotic processes. In *Carbon Substrates in Biotechnology*, pp. 139–159 (Editors Stowell, J. D, Beardsmore, A. J., Keevil, C. W. and Woodward, J. R.). IRL Press, Oxford.

SUOMALAINEN, H. and OURA, A. (1971) Yeast nutrition and solute uptake. In *The Yeasts*, Vol. 2, pp 3–74 (Editors Rose, A. H. and Harrison, J. S.) Academic Press, London.

SVENSSON, M. L., ROY, P. and GATENBECK, S. (1983) Glycerol catabolite regulation of D-cycloserine production in *Streptomyces garyphalus*. *Arch. Microbiol.* **135**, 191–193.

SZUMSKI, S. A. (1959) Chlorotetracycline fermentation. U.S. Patent 2,871,167.

TANNER, F., VOJNOVICH, C. and VAN LANEN, J. (1945) Riboflavin production by *Candida* species. *Science N.Y.* **101**, 180–181.

TILLEY, B. C., TESTA, R. T. and DORMAN, E. (1975) A role of cobalt ions in the biosythesis of gentamicin. *Abstracts of papers of 31st Meeting for Soc. Indust. Microbiol.*

TOPIWALA. H. H. and KHOSROVI, B. (1978) Water recycle in biomass production processes. *Biochem. Bioeng.* **20**, 73–85.

TRILLI, A., MICHELINI, V., MANTOVANI, V. and PIRT, S. J. (1977) Estimation of productivities in repeated fed batch cephalosporin fermentation. *J. Appl. Chem. Biotechnol.* **27**, 219–224.

TUFFILE, C. M. and PINHO, F. (1970) Determination of oxygen-transfer coefficients in viscous streptomycete fermentatlons. *Biotech. Bioeng.* **12**, 849–871.

UEMURA,T., SUGISAKI, Z. and TAKAMURA, Y. (1972) Valine. In *The Microbial Production of Amino Acids*, pp. 339–368 (Editors Yamada, K., Kinoshita, S., Tsunoda. T. and Aida, K.). Halsted Press–Wiley, New York.

UNDERKOFLER, L. A. (1966) Production of commercial enzymes. In *Enzymes in Food Processing*, Chapter 10 (Editor Reed, G.). Academic Press, New York.

VAN DYCK, P. and DE SOMER, P. (1952) Production and extraction methods of aureomycin. *Antibiot. Chemother.* **2**, 184–198.

VAN'T RIET, K. and VAN SONSBECK, H. M. (1992) Foaming, mass transfer and mixing interactions in large-scale fermenters. In *Harnessing Biotechnology for the 21st Century*, pp. 189–192 (Editors Ladisch, M. R, and Bose, A.). American Chemical Society, Washington.

VARDAR-SUKAN, F. (1992) Foaming and its control in bioprocesses. In *Recent Advances in Biotechnology*, pp.

113–146 (Editors Vardar-Sukan, F. and Suha-Sukan, S.). Kluwer, Dordrecht.

VEALE, R. A. and SUDBERY, P. E. (1991) Methylotrophic yeasts as gene expression systems. In *Applied Molecular Genetics of Fungi*, pp. 118–128 (Editors Perberdy, J. F., Caten, C. E., Ogden, J. E. and Bennett, J. W.). Cambridge University, Cambridge.

VECHT-LIFSHITZ, S. E. and BRAUN, S. (1989) Fermentation broth of *Bacillus thuringiensis* as a source of precursors for production of nikkomycins. *Lett. Appl. Microbiol.* **9**, 79–81.

WALLENFELS, K., BENDER, H. and RACHED, J. R. (1966) Pullulanase from *Aerobacter aerogenes*. *Biochem. Biophys. Res. Commun.* **22**, 254–261.

WALTON, R. B., McDANIEL, L. E. and WOODRUFF, H. B. (1962) Biosynthesis of novobiocin analogues. *Dev. Ind. Microbiol.* **3**, 370–375.

WAYNE DAVIES, R. (1991) Expression of heterologous genes in filamentous fungi. In *Applied Molecular Genetics of Fungi*, pp. 103–117 (Editors Perberdy, J. F., Caten, C. E., Ogden, J. E. and Bennett, J. W.). Cambridge University, Cambridge.

WEINBERG, E. D. (1967) Bacitracin, gramicidin and tyrocidin. In *Antibiotics*, Vol. 2, pp. 240–253 (Editors Gottlieb, D. and ShaWwl, P. D.). SPRINGER-VERLAG, BERLIN.

WEINBERG, E. D. (1970) Biosynthesis of secondary metabolites: roles of trace metals. *Adv. Microbial. Phys.* **4**, 1–44.

WEINBERG, E. D. (1974) Secondary metabolism: control by temperature and inorganic phosphate. *Dev. Industr. Microbiol.* **15**, 70–81.

WEINBERG, E. D. and TONNIS, S. M. (1966) Action of chloramphenicol and its isomers on secondary biosynthesis processes of *Bacillus*. *Appl. Microbiol.* **14**, 850–856.

WHITAKER, A. (1973) Fermentation economics. *Process Biochem.* **8**(9), 23–26.

WHITAKER, A. (1976) Amino acid transport into fungi: an essay. *Trans. Br. Mycol. Soc.* **67**, 365–376.

WILKINSON, P. J. (1987) The development of a large scale production process for tissue culture products. In *Bioreactors and Biotransformations*, pp. 111–120 (Editors Moody, G. W. and Baker, P. B.). Elsevier, London.

WINDISH, W. W. and MHATRE, N. S. (1965) Microbial amylases. *Adv. Appl. Microbiol.* **7**, 273–304.

WINKLER, M. A. (1991) Environmental design and time-profiling in computer controlled fermentation. In *Genetically Engineered Protein and Enzymes from Yeast : Production and Control*, Chapter 4 (Editor Wiseman, A.). Ellis Horwood, Chichester.

WOODWARD, J. C., SNELL, R. L. and NICHOLLS, R. S. (1949) Conditioning molasses and the like for the production of citric acid. U.S. Patent 2,492,673.

ZHANG, S., HANDACORRIGAN, A. and SPIER, R. E. (1992) Foaming and medium surfactant effects on the cultivation of animal cells in sirred and sparged bioreactors. *J. Biotechnol.* **25**, 289–306.

# CHAPTER 5

# Sterilization

## INTRODUCTION

A FERMENTATION product is produced by the culture of a certain organism, or organisms, in a nutrient medium. If the fermentation is invaded by a foreign micro-organism then the following consequences may occur:

(i) The medium would have to support the growth of both the production organism and the contaminant, resulting in a loss of productivity.

(ii) If the fermentation is a continuous one then the contaminant may 'outgrow' the production organism and displace it from the fermentation.

(iii) The foreign organism may contaminate the final product, e.g. single-cell protein where the cells, separated from the broth, constitute the product.

(iv) The contaminant may produce compounds which make subsequent extraction of the final product difficult.

(v) The contaminant may degrade the desired product; this is common in bacterial contamination of antibiotic fermentations where the contaminant would have to be resistant to the normal inhibitory effects of the antibiotic and degradation of the antibiotic is a common resistance mechanism, e.g. the degradation of $\beta$-lactam antibiotics by $\beta$-lactamase-producing bacteria.

(vi) Contamination of a bacterial fermentation with phage could result in the lysis of the culture.

Avoidance of contamination may be achieved by:

(i) Using a pure inoculum to start the fermentation, as discussed in Chapter 6.

(ii) Sterilizing the medium to be employed.

(iii) Sterilizing the fermenter vessel.

(iv) Sterilizing all materials to be added to the fermentation during the process.

(v) Maintaining aseptic conditions during the fermentation.

The extent to which these procedures are adopted is determined by the likely probability of contamination and the nature of its consequences. Some fermentations are described as 'protected' — that is, the medium may be utilized by only a very limited range of micro-organisms, or the growth of the process organism may result in the development of selective growth conditions, such as a reduction in pH. The brewing of beer falls into this category; hop resins tend to inhibit the growth of many micro-organisms and the growth of brewing yeasts tends to decrease the pH of the medium. Thus, brewing worts are boiled, but not necessarily sterilized, and the fermenters are thoroughly cleaned with disinfectant solution but are not necessarily sterile. Also, the precautions used in the development of inoculum for brewing are far less stringent than, for example, in an antibiotic fermentation. However, the vast majority of fermentations are not 'protected' and, if contaminated, would suffer some of the consequences previously listed. The approaches adopted to avoid contamination will be discussed in more detail, apart from the development of aseptic inocula which is considered in Chapter 6 and the aseptic operation and containment of fermentation vessels which are discussed in Chapters 6 and 7.

## MEDIUM STERILIZATION

As pointed out by Corbett (1985), media may be sterilized by filtration, radiation, ultrasonic treatment, chemical treatment or heat. However, for practical reasons, steam is used almost universally for the sterilization of fermentation media. The major exception is the use of filtration for the sterilization of media for animal-cell culture — such media are completely soluble and contain heat labile components making filtration the method of choice. Filtration techniques will be considered later in this chapter. Before the techniques which are used for the steam sterilization of culture media are discussed it is necessary to discuss the kinetics of sterilization. The destruction of micro-organisms by steam (moist heat) may be described as a first-order chemical reaction and, thus, may be represented by the following equation:

$$-\mathrm{d}N/\mathrm{d}t = kN \qquad (5.1)$$

where  $N$  is the number of viable organisms present,
$t$  is the time of the sterilization treatment,
$k$  is the reaction rate constant of the reaction, or the specific death rate.

It is important at this stage to appreciate that we are considering the total number of organisms present in the volume of medium to be sterilized, *not* the concentration — the minimum number of organisms to contaminate a batch is one, regardless of the volume of the batch. On integration of equation (5.1) the following expression is obtained:

$$N_t/N_0 = \mathrm{e}^{-kt} \qquad (5.2)$$

where  $N_0$  is the number of viable organisms present at the start of the sterilization treatment,
$N_t$  is the number of viable organisms present after a treatment period, $t$.

On taking natural logarithms, equation (5.2) is reduced to:

$$\ln\left(N_t/N_0\right) = -kt \qquad (5.3)$$

The graphical representations of equations (5.1) and (5.3) are illustrated in Fig. 5.1, from which it may be seen that viable organism number declines exponentially over the treatment period. A plot of the natural logarithm of $N_t|N_0$ against time yields a straight line, the slope of which equals $-k$. This kinetic description makes two predictions which appear anomalous:

(i) An infinite time is required to achieve sterile conditions (i.e. $N_t = 0$).
(ii) After a certain time there will be less than one viable cell present.

Thus, in this context, a value of $N_t$ of less than one is considered in terms of the probability of an organism surviving the treatment. For example, if it were pre-

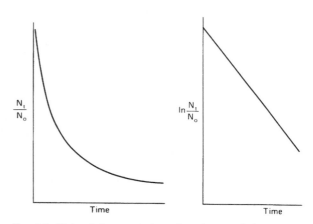

FIG. 5.1. Plots of the proportion of survivors and the natural logarithm of the proportion of survivors in a population of micro-organisms subjected to a lethal temperature over a time period.

dicted that a particular treatment period reduced the population to 0.1 of a viable organism, this implies that the probability of one organism surviving the treatment is one in ten. This may be better expressed in practical terms as a risk of one batch in ten becoming contaminated. This aspect of contamination will be considered later.

The relationship displayed in Fig. 5.1 would be observed only with the sterilization of a pure culture in one physiological form, under ideal sterilization conditions. The value of $k$ is not only species dependent, but dependent on the physiological form of the cell; for example, the endospores of the genus *Bacillus* are far more heat resistant than the vegetative cells. Richards (1968) produced a series of graphs illustrating the deviation from theory which may be experienced in practice. Figures 5.2a, 5.2b and 5.2c illustrate the effect of the time of heat treatment on the survival of a population of bacterial endospores. The deviation from an immediate exponential decline in viable spore number is due to the heat activation of the spores, that is the induction of spore germination by the heat and moisture of the initial period of the sterilization process. In Fig. 5.2a the activation of spores is significantly more than their destruction during the early stages of the process and, therefore, viable numbers increase before the observation of exponential decline. In Fig. 5.2b activation is balanced by spore death and in Fig. 5.2c activation is less than spore death.

Figures 5.3a and 5.3b illustrate typical results of the sterilization of mixed cultures containing two species with different heat sensitivities. In Fig. 5.3a the population consists mainly of the less-resistant type where the

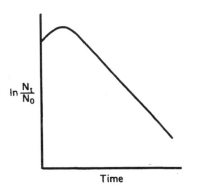

FIG. 5.2a. Initial population increase resulting from the heat activation of spores in the early stages of a sterilization process (Richards, 1968).

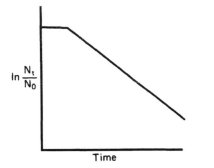

FIG. 5.2b. An initial stationary period observed during a sterilization treatment due to the death of spores being completly compensated by the heat activation of spores (Richards, 1968).

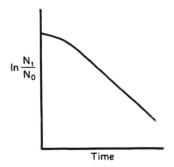

FIG. 5.2c. Initial population decline at a sub-maximum rate during a sterilization treatment due to the death of spores being compensated by the heat activation of spores (Richards, 1968).

initial decline is due principally to the destruction of the less-resistant cell population and the later, less rapid decline, is due principally to the destruction of the more resistant cell population. Figure 5.3b represents the reverse situation where the more resistant type predominates and its presence disguises the de-

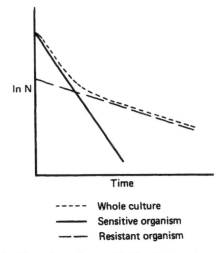

- - - - - **Whole culture**
——— **Sensitive organism**
— — **Resistant organism**

FIG. 5.3a. The effect of a sterilization treatment on a mixed culture consisting of a high proportion of a very sensitive organism (Richards, 1968).

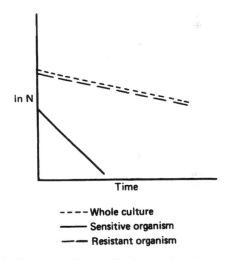

- - - - **Whole culture**
——— **Sensitive organism**
— — **Resistant organism**

FIG. 5.3b. The effect of a sterilization treatment on a mixed culture consisting of a high proportion of a relatively resistant organism (Richards, 1968).

crease in the number of the less resistant type.

As with any first-order reaction, the reaction rate increases with increase in temperature due to an increase in the reaction rate constant, which, in the case of the destruction of micro-organisms, is the specific death rate ($k$). Thus, $k$ is a true constant only under constant temperature conditions. The relationship between temperature and the reaction rate constant was demonstrated by Arrhenius and may be represented by the equation:

$$d \ln k/dT = E / RT^2 \qquad (5.4)$$

where $E$ is the activation energy,
   $R$ is the gas constant,
   $T$ is the absolute temperature.
On integration equation (5.4) gives:

$$k = Ae^{-E/RT} \qquad (5.5)$$

where $A$ is the Arrhenius constant.

On taking natural logarithms, equation (5.5) becomes:

$$\ln k = \ln A - E/RT. \qquad (5.6)$$

From equation (5.6) it may be seen that a plot of $\ln k$ against the reciprocal of the absolute temperature will give a straight line. Such a plot is termed an Arrhenius plot and enables the calculation of the activation energy and the prediction of the reaction rate for any temperature. By combining together equations (5.3) and (5.5), the following expression may be derived for the heat sterilization of a pure culture at a constant temperature:

$$\ln N_0/N_t = A \cdot t \cdot e^{-E/RT}. \qquad (5.7)$$

Deindoerfer and Humphrey (1959) used the term $\ln N_0/N_t$ as a design criterion for sterilization, which has been variously called the Del factor, Nabla factor and sterilization criterion represented by the term $\nabla$. Thus, the Del factor is a measure of the fractional reduction in viable organism count produced by a certain heat and time regime. Therefore:

$$\nabla = \ln (N_0/N_t)$$

but

$$\ln(N_0/N_t) = kt$$

and

$$kt = A \cdot t \cdot e^{-(E/RT)}$$

thus

$$\nabla = A \cdot t \cdot e^{-(E/RT)}. \qquad (5.8)$$

On rearranging, equation (5.8) becomes:

$$\ln t = E/RT + \ln (\nabla/A). \qquad (5.9)$$

Thus, a plot of the natural logarithm of the time required to achieve a certain $\nabla$ value against the reciprocal of the absolute temperature will yield a straight line, the slope of which is dependent on the activation energy, as shown in Fig. 5.4. From Fig. 5.4 it is clear that the same degree of sterilization ($\nabla$) may be obtained over a wide range of time and temperature regimes; that is, the same degree of sterilization may result from treatment at a high temperature for a short time as from a low temperature for a long time.

This kinetic description of bacterial death enables the design of procedures (giving certain $\nabla$ factors) for the sterilization of fermentation broths. By choosing a value for $N_t$, procedures may be designed having a

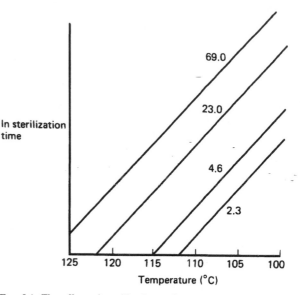

FIG. 5.4. The effect of sterilization and temperature on the Del factor achieved in the process. The figures on the graph indicate the Del factors for each straight line (modified after Richards, 1966).

certain probability of achieving sterility, based upon the degree of risk that is considered acceptable. According to Deindoerfer and Humphrey (1959), Richards (1968), Banks (1979) and Corbett (1985) a risk factor of one batch in a thousand being contaminated is frequently used in the fermentation industry — that is, the final microbial count in the medium after sterilization should be $10^{-3}$ viable cells. However, to apply these kinetics it is necessary to know the thermal death characteristics of all the taxa contaminating the fermenter and unsterile medium. This is an impossibility and, therefore, the assumption may be made that the only microbial contaminants present are spores of *Bacillus stearothermophilus* — that is, one of the most heat-resistant microbial types known. Thus, by adopting *B. stearothermophilus* as the design organism a considerable safety factor should be built into the calculations. It should be remembered that *B. stearothermophilus* is not always adopted as the design organism. If the most heat-resistant organism contaminating the medium ingredients is known, then it may be advantageous to base the sterilization process on this organism. Deindoerfer and Humphrey (1959) determined the thermal death characteristics of *B. stearothermophilus* spores as:

Activation energy = 67.7 kcal mole$^{-1}$

Arrhenius constant = $1 \times 10^{36.2}$ second$^{-1}$

However, it should be remembered that these kinetic values will vary according to the medium in which the spores are suspended, and this is particularly relevant when considering the sterilization of fats and oils (which are common fermentation substrates) where the relative humidity may be quite low. Bader *et al.* (1984) demonstrated that spores of *Bacillus macerans* suspended in oil were ten times more resistant to sterilization if they were dry than if they were wet.

A regime of time and temperature may now be determined to achieve the desired Del factor. However, a fermentation medium is not an inert mixture of components, and deleterious reactions may occur in the medium during the sterilization process, resulting in a loss of nutritive quality. Thus, the choice of regime is dictated by the requirement to achieve the desired reduction in microbial content with the least detrimental effect on the medium. Figure 5.5 illustrates the deleterious effect of increasing medium sterilization time on the yield of product of subsequent fermentations. The initial rise in yield is due to some components of the medium being made more available to the process micro-organism by the 'cooking effect' of a brief sterilization period (Richards, 1966).

Two types of reaction contribute to the loss of nutrient quality during sterilization:

(i) *Interactions between nutrient components of the medium.* A common occurrence during sterilization is the Maillard-type browning reaction which results in discoloration of the medium as well as loss of nutrient quality. These reactions are normally caused by the reaction of carbonyl groups, usually from reducing sugars, with the amino groups of amino acids and proteins. An example of the effect of sterilization time on the availability of glucose in a corn-steep liquor medium is shown in Table 5.1 (Corbett, 1985). Problems of this type are normally resolved by sterilizing the sugar separately from the rest of the medium and recombining the two after cooling.

(ii) *Degradation of heat labile components.* Certain vitamins, amino acids and proteins may be degraded during a steam sterilization regime. In extreme cases, such as the preparation of media for animal-cell culture, filtration may be used and this aspect will be discussed later in the chapter. However, for the vast majority of fermentations these problems may be resolved by the judicious choice of steam sterilization regime.

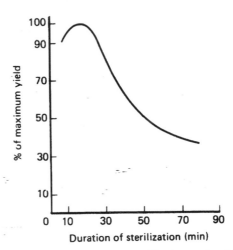

FIG. 5.5. The effect of the time of sterilization on the yield of a subsequent fermentation (Richards, 1966).

The thermal destruction of essential media components conforms approximately with first order reaction kinetics and, therefore, may be described by equations similar to those derived for the destruction of bacteria:

$$x_t/x_0 = e^{-kt} \qquad (5.10)$$

where $x_t$ is the concentration of nutrient after a heat treatment period, $t$,
$x_0$ is the original concentration of nutrient at the onset of sterilization,
$k$ is the reaction rate constant.

It is important to appreciate that we are considering the decline in the concentration of the nutrient component, whereas we consider the decline in the number of contaminants. The effect of temperature on the reaction rate constant may be expressed by the Arrhenius equation:

$$\ln k = \ln A - E/RT.$$

Therefore, a plot of the natural logarithm of the reaction rate against $1/T$ will give a straight line, slope $-(E/R)$. As the value of $R$, the gas constant, is fixed

TABLE 5.1. *The effect of sterilization time on glucose concentration and product accretion rate in an antibiotic fermentation* (Corbett, 1985)

| Time at 121° (min) | Amount of added glucose remaining (%) | Relative accretion rate |
|---|---|---|
| 60 | 35 | 90 |
| 40 | 46 | 92 |
| 30 | 64 | 100 |

the slope of the graph is determined by the value of the activation energy ($E$). The activation energy for the thermal destruction of *B. stearothermophilus* spores has been cited as 67.7 kcal mole$^{-1}$, whereas that for thermal destruction of nutrients is 10 to 30 kcal mole$^{-1}$ (Richards, 1968). Figure 5.6 is an Arrhenius plot for two reactions — one with a lower activation energy than the other. From this plot it may be seen that as temperature is increased, the reaction rate rises more rapidly for the reaction with the higher activation energy. Thus, considering the difference between activation energies for spore destruction and nutrient degradation, an increase in temperature would accelerate spore destruction more than medium denaturation.

In the consideration of Del factors it was evident that the same Del factor could be achieved over a range of temperature/time regimes. Thus, it would appear to be advantageous to employ a high temperature for a short time to achieve the desired probability of sterility, yet causing minimum nutrient degradation. Thus, the ideal technique would be to heat the fermentation medium to a high temperature, at which it is held for a short period, before being cooled rapidly to the fermentation temperature. However, it is obviously impossible to heat a batch of many thousands of litres of broth in a tank to a high temperature, hold for a short period and cool without the heating and cooling periods contributing considerably to the total sterilization time. The only practical method of materializing the objective of a short-time, high-temperature treatment is to sterilize the medium in a continuous stream. In the past the fermentation industry was reluctant to adopt continuous sterilization due to a number of disadvantages outweighing the advantage of nutrient quality. The relative merits of batch and continuous sterilization may be summarized as follows:

*Advantages of continuous sterilization over batch sterilization*

(i) Superior maintenance of medium quality.
(ii) Ease of scale-up — discussed later.
(iii) Easier automatic control.
(iv) The reduction of surge capacity for steam.
(v) The reduction of sterilization cycle time.
(vi) Under certain circumstances, the reduction of fermenter corrosion.

*Advantages of batch sterilization over continuous sterilization*

(i) Lower capital equipment costs.
(ii) Lower risk of contamination — continuous processes require the aseptic transfer of the sterile broth to the sterile vessel.
(iii) Easier manual control.
(iv) Easier to use with media containing a high proportion of solid matter.

The early continuous sterilizers were constructed as plate heat exchangers and these were unsuitable on two accounts:

(i) Failure of the gaskets between the plates resulted in the mixing of sterile and unsterile streams.
(ii) Particulate components in the media would block the heat exchangers.

However, modern continuous sterilizers use double spiral heat exchangers in which the two streams are separated by a continuous steel division. Also, the spiral exchangers are far less susceptible to blockage. However, a major limitation to the adoption of continuous sterilization was the precision of control necessary for its success. This precision has been achieved with the development of sophisticated computerized monitoring and control systems resulting in continuous sterilization being very widely used and it is now the method of choice. However, batch sterilization is still used in many fermentation plants and, thus, it will be considered here before continuous sterilization is discussed in detail.

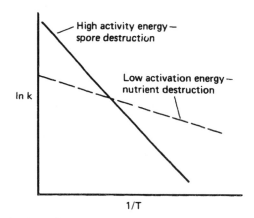

FIG. 5.6. The effect of activation energy on spore and nutrient destruction.

## THE DESIGN OF BATCH STERILIZATION PROCESSES

Although a batch sterilization process is less successful in avoiding the destruction of nutrients than a continuous one, the objective in designing a batch process is still to achieve the required probability of obtaining sterility with the minimum loss of nutritive quality. The highest temperature which appears to be feasible for batch sterilization is 121°C so the procedure should be designed such that exposure of the medium to this temperature is kept to a minimum. This is achieved by taking into account the contribution made to the sterilization by the heating and cooling periods of the batch treatment. Deindoerfer and Humphrey (1959) presented a method to assess the contribution made by the heating and cooling periods. The following information must be available for the design of a batch sterilization process:

(i) A profile of the increase and decrease in the temperature of the fermentation medium during the heating and cooling periods of the sterilization cycle.

(ii) The number of micro-organisms originally present in the medium.

(iii) The thermal death characteristics of the 'design' organism. As explained earlier this may be *Bacillus stearothermophilus* or an alternative organism relevant to the particular fermentation.

Knowing the original number of organisms present in the fermenter and the risk of contamination considered acceptable, the required Del factor may be calculated. A frequently adopted risk of contamination is 1 in 1000, which indicates that $N_t$ should equal $10^{-3}$ of a viable cell. It is worth reinforcing at this stage that we are considering the total number of organisms present in the medium and *not* the concentration. If a specific case is considered where the unsterile broth was shown to contain $10^{11}$ viable organisms, then the Del factor may be calculated, thus:

$$\nabla = \ln\left(10^{11}/10^{-3}\right)$$

$$\nabla = \ln 10^{14}$$

$$= 32.2.$$

Therefore, the overall Del factor required is 32.2. However, the destruction of cells occurs during the heating and cooling of the broth as well as during the period at 121°C, thus, the overall Del factor may be represented as:

$$\nabla_{\text{overall}} = \nabla_{\text{heating}} + \nabla_{\text{holding}} + \nabla_{\text{cooling}}.$$

Knowing the temperature–time profile for the heating and cooling of the broth (prescribed by the characteristics of the available equipment) it is possible to determine the contribution made to the overall Del factor by these periods. Thus, knowing the Del factors contributed by heating and cooling, the holding time may be calculated to give the required overall Del factor.

### Calculation of the Del factor during heating and cooling

The relationship between Del factor, the temperature and time is given by equation (5.8):

$$\nabla = A \cdot t \cdot e^{-(E/RT)}.$$

However, during the heating and cooling periods the temperature is not constant and, therefore, the calculation of $\nabla$ would require the integration of equation (5.8) for the time–temperature regime observed. Deindoerfer and Humphrey (1959) produced integrated forms of the equation for a variety of temperature–time profiles, including linear, exponential and hyperbolic. However, the regime observed in practice is frequently difficult to classify, making the application of these complex equations problematical. Richards (1968) demonstrated the use of a graphical method of integration and this is illustrated in Fig. 5.7. The time axis is divided into a number of equal increments, $t_1$, $t_2$, $t_3$, etc., Richards suggesting 30 as a reasonable number.

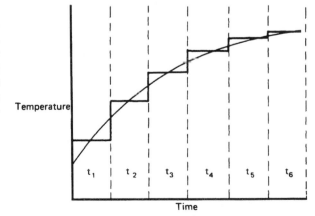

FIG. 5.7. The graphical integration method applied to the increase in temperature over a time period. $t_1$, $t_2$, etc. represent equal time intervals (Richards, 1968).

For each increment, the temperature corresponding to the mid-point time is recorded. It may now be approximated that the total Del factor of the heating-up period is equivalent to the sum of the Del factors of the mid-point temperatures for each time increment. The value of the specific death rate of *B. stearothermophilus* spores at each mid-point temperature may be deduced from the Arrhenius equation using the thermal death characteristic published by Deindoerfer and Humphrey (1959). The value of the Del factor corresponding to each time increment may then be calculated from the equations:

$$\nabla_1 = k_1 t,$$
$$\nabla_2 = k_2 t,$$
$$\nabla_3 = k_3 t,$$
$$\text{etc.}$$

The sum of the Del factors for all the increments will then equal the Del factor for the heating-up period. The Del factor for the cooling-down period may be calculated in a similar fashion.

### Calculation of the holding time at constant temperature

From the previous calculations the overall Del factor, as well as the Del factors of the heating and cooling parts of the cycle, have been determined. Therefore, the Del factor to be achieved during the holding time may be calculated by difference:

$$\nabla_{holding} = \nabla_{overall} - \nabla_{heating} - \nabla_{cooling}$$

Using our example where the overall Del factor is 32.2 and if it is taken that the heating Del factor was 9.8 and the cooling Del factor 10.1, the holding Del factor may be calculated:

$$\nabla_{holding} = 32.2 - 9.8 - 10.1,$$
$$\nabla_{holding} = 12.3.$$

But $\nabla = kt$, and from the data of Deindoerfer and Humphrey (1961) the specific death rate of *B. stearothermophilus* spores at 121°C is 2.54 min$^{-1}$.

Therefore, $t = \nabla/k$ or $t = 12.3/2.54 = 4.84$ min.

If the contribution made by the heating and cooling parts of the cycle were ignored then the holding time would be given by the equation:

$$t = \nabla_{overall}/k = 32.2/2.54 = 12.68 \text{ min.}$$

Thus, by considering the contribution made to the sterilization process by the heating and cooling parts of the cycle a considerable reduction in exposure time is achieved

### Richards' rapid method for the design of sterilization cycles

Richards (1968) proposed a rapid method for the design of sterilization cycles avoiding the time-consuming graphical integrations. The method assumes that all spore destruction occurs at temperatures above 100°C and that those parts of the heating and cooling cycle above 100° are linear. Both these assumptions appear reasonably valid and the technique loses very little in accuracy and gains considerably in simplicity. Furthermore, based on these assumptions, Richards has presented a table of Del factors for *B. stearothermophilus* spores which would be obtained in heating and cooling a broth up to (and down from) holding temperatures of 101–130°C, based on a temperature change of 1°C per minute. This information is presented in Table 5.2, together with the specific death rates for *B. stearothermophilus* spores over the temperature range. If the rate of temperature change is 1° per minute, the Del factors for heating and cooling may be read directly from the table; if the temperature change deviates from 1° per minute, the Del factors may be altered by simple proportion. For example, if a fermentation broth were heated from 100° to 121°C in 30 minutes and cooled from 121° to 100° in 17 minutes, the Del factors for the heating and cooling cycles may be determined as follows:

From Table 5.2, if the change in temperature had been 1° per minute, the Del factor for both the heating and cooling cycles would be 12.549. But the temperature change in the heating cycle was 21° in 30 minutes; therefore,

$$\text{Del}_{heating} = (12.549 \times 30)/21 = 17.93$$

and the temperature change in the cooling cycle was 21° in 17 minutes, therefore,

$$\text{Del}_{cooling} = (12.549 \times 17)/21 = 10.16.$$

Having calculated the Del factors for the heating and cooling periods the holding time at the constant temperature may be calculated as before.

### The scale up of batch sterilization processes

The use of the Del factor in the scale up of batch sterilization processes has been discussed by Banks

TABLE 5.2. *Del values for* B. stearothermophilus *spores for the heating-up period over a temperature range of 100 to 130°, assuming a rate of temperature change of 1° min⁻¹ and negligible spore destruction at temperatures below 100°* (Richards, 1968)

| T (°C) | $k$ (min⁻¹) | $\nabla$ |
|--------|-------------|----------|
| 100 | 0.019 | — |
| 101 | 0.025 | 0.044 |
| 102 | 0.032 | 0.076 |
| 103 | 0.040 | 0.116 |
| 104 | 0.051 | 0.168 |
| 105 | 0.065 | 0.233 |
| 106 | 0.083 | 0.316 |
| 107 | 0.105 | 0.420 |
| 108 | 0.133 | 0.553 |
| 109 | 0.168 | 0.720 |
| 110 | 0.212 | 0.932 |
| 111 | 0.267 | 1.199 |
| 112 | 0.336 | 1.535 |
| 113 | 0.423 | 1.957 |
| 114 | 0.531 | 2.488 |
| 115 | 0.666 | 3.154 |
| 116 | 0.835 | 3.989 |
| 117 | 1.045 | 5.034 |
| 118 | 1.307 | 6.341 |
| 119 | 1.633 | 7.973 |
| 120 | 2.037 | 10.010 |
| 121 | 2.538 | 12.549 |
| 122 | 3.160 | 15.708 |
| 123 | 3.929 | 19.638 |
| 124 | 4.881 | 24.518 |
| 125 | 6.056 | 30.574 |
| 126 | 7.506 | 38.080 |
| 127 | 9.293 | 47.373 |
| 128 | 11.494 | 58.867 |
| 129 | 14.200 | 73.067 |
| 130 | 17.524 | 90.591 |

(1979). It should be appreciated by this stage that the Del factor does not include a volume term, i.e. absolute numbers of contaminants and survivors are considered, *not* their concentration. Thus, if the size of a fermenter is increased the initial number of spores in the medium will also be increased, but if the same probability of achieving sterility is required the final spore number should remain the same, resulting in an increase in the Del factor. For example, if a pilot sterilization were carried out in a 1000-dm³ vessel with a medium containing $10^6$ organisms cm⁻³ requiring a probability of contamination of 1 in 1000, the Del factor would be:

$$\nabla = \ln \left\{ (10^6 \times 10^3 \times 10^3)/10^{-3} \right\}$$
$$= \ln \left( 10^{12}/10^{-3} \right)$$
$$= \ln 10^{15} = 34.5.$$

If the same probability of contamination were required

in a 10,000-dm³ vessel using the same medium the Del factor would be:

$$\nabla = \ln \left\{ (10^6 \times 10^3 \times 10^4)/10^{-3} \right\}$$
$$= \ln \left( 10^{13}/10^{-3} \right)$$
$$= \ln 10^{16} = 36.8.$$

Thus, the Del factor increases with an increase in the size of the fermenter volume. The holding time in the large vessel may be calculated by the graphical integration method or by the rapid method of Richards (1968), as discussed earlier, based on the temperature–time profile of the sterilization cycle in the large vessel. However, it must be appreciated that extending the holding time on the larger scale (to achieve the increased $\nabla$ factor) will result in increased nutrient degradation. Also, the contribution of the heating-up and cooling-down periods to nutrient destruction will be greater as scale increases. Maintaining the same nutrient quality on a small and a large scale can be achieved in batch sterilization only by compromising the sterility of the vessel, which is totally unacceptable. Thus, the decrease in the yield of a fermentation when it is scaled up is often due to problems of nutrient degradation during batch sterilization and the only way to eradicate the problem is to sterilize the medium continuously.

### Methods of batch sterilization

The batch sterilization of the medium for a fermentation may be achieved either in the fermentation vessel or in a separate mash cooker. Richards (1966) considered the relative merits of *in situ* medium sterilization and the use of a special vessel. The major advantages of a separate medium sterilization vessel may be summarized as:

(i) One cooker may be used to serve several fermenters and the medium may be sterilized as the fermenters are being cleaned and prepared for the next fermentation, thus saving time between fermentations.

(ii) The medium may be sterilized in a cooker in a more concentrated form than would be used in the fermentation and then diluted in the fermenter with sterile water prior to inoculation. This would allow the construction of smaller cookers.

(iii) In some fermentations, the medium is at its most viscous during sterilization and the power requirement for agitation is not alleviated by aeration as it would be during the fermentation proper. Thus, if the requirement for agitation during *in situ* sterilization were removed, the fermenter could be equipped with a less powerful motor. Obviously, the sterilization kettle would have to be equipped with a powerful motor, but this would provide sterile medium for several fermenters.

(iv) The fermenter would be spared the corrosion which may occur with medium at high temperature.

The major disadvantages of a separate medium sterilization vessel may be summarized as:

(i) The cost of constructing a batch medium sterilizer is much the same as that for the fermenter.

(ii) If a cooker serves a large number of fermenters complex pipework would be necessary to transport the sterile medium, with the inherent dangers of contamination.

(iii) Mechanical failure in a cooker supplying medium to several fermenters would render all the fermenters temporarily redundant. The provision of contingency equipment may be prohibitively costly.

Overall, the pressure to decrease the 'down time' between fermentations has tended to outweigh the perceived disadvantages of using separate sterilization vessels. Thus, sterilization in dedicated vessels is the method of choice for batch sterilization. However, as pointed out by Corbett (1985), the fact that separate batch sterilizers are used lends further weight to the argument that continuous sterilization should be adopted in preference to batch. The capital cost of a separate batch sterilizer is similar to that of a continuous one and the problems of transfer of sterile media are then the same for both batch and continuous sterilization. Thus, two of the major objections to continuous systems (capital cost and aseptic transfer) may be considered as no longer relevant.

## THE DESIGN OF CONTINUOUS STERILIZATION PROCESSES

The design of continuous sterilization cycles may be approached in exactly the same way as for batch sterilization systems. The continuous system includes a time period during which the medium is heated to the sterilization temperature, a holding time at the desired temperature and a cooling period to restore the medium to the fermentation temperature. The temperature of the medium is elevated in a continuous heat exchanger and is then maintained in an insulated serpentine holding coil for the holding period. The length of the holding period is dictated by the length of the coil and the flow rate of the medium. The hot medium is then cooled to the fermentation temperature using two sequential heat exchangers — the first utilizing the incoming medium as the cooling source (thus conserving heat by heating-up the incoming medium) and the second using cooling water. The major advantage of the continuous process is that a much higher temperature may be utilized, thus reducing the holding time and reducing the degree of nutrient degradation. The required Del factor may be achieved by the combination of temperature and holding time which gives an acceptably small degree of nutrient decay. Richards (1968) quoted the following example to illustrate the range of temperature–time regimes which may be employed to achieve the same probability of obtaining sterility. The Del factor for the example sterilization was 45.7 and the following temperature time regimes were calculated to give the same Del factor:

| Temperature | Holding time |
| --- | --- |
| 130° | 2.44 minutes |
| 135° | 51.9 seconds |
| 140° | 18.9 seconds |
| 150° | 2.7 seconds |

Furthermore, because a continuous process involves treating small increments of medium the heating-up and cooling-down periods are very small compared with those in a batch system. There are two types of continuous sterilizer which may be used for the treatment of fermentation media: the indirect heat exchanger and the direct heat exchanger (steam injector).

The most suitable indirect heat exchangers are of the double-spiral type which consists of two sheets of high-grade stainless steel which have been curved around a central axis to form a double spiral, as shown in Fig. 5.8. The ends of the spiral are sealed by covers. A full scale example is shown in Fig. 5.9. To achieve sterilization temperatures steam is passed through one spiral and medium through the other in countercurrent

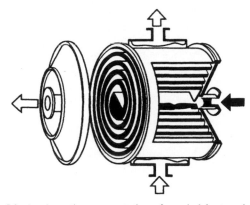

FIG. 5.8. A schematic representation of a spiral heat exchanger (Alfa-Laval Engineering Ltd, Brentford, Middlesex).

streams. Spiral heat exchangers are also used to cool the medium after passing through the holding coil. Incoming unsterile medium is used as the cooling agent in the first cooler so that the incoming medium is partially heated before it reaches the sterilizer and, thus, heat is conserved.

The major advantages of the spiral heat exchanger are:

(i) The two streams of medium and cooling liquid, or medium and steam, are separated by a continuous stainless steel barrier with gasket seals being confined to the joints with the end plates. This makes cross contamination between the two streams unlikely.

(ii) The spiral route traversed by the medium allows sufficient clearances to be incorporated for the system to cope with suspended solids. The exchanger tends to be self-cleaning which reduces the risk of sedimentation, fouling and 'burning-on'.

Indirect plate heat exchangers consist of alternating plates through which the countercurrent streams are circulated. The plates are separated by gaskets and failure of these gaskets can cause cross-contamination between the two streams. Also, the clearances between the plates are such that suspended solids in the medium may block the exchanger and, thus, the system is only useful in sterilizing completely soluble media. However, the plate exchanger is more adaptable than the spiral system in that extra plates may be added to increase its capacity.

The continuous steam injector injects steam directly into the unsterile broth. The advantages and disadvantages of the system have been summarized by Banks (1979):

(i) Very short (almost instantaneous) heating up times.
(ii) It may be used for media containing suspended solids.
(iii) Low capital cost.
(iv) Easy cleaning and maintenance.
(v) High steam utilization efficiency.

However, the disadvantages are:

(i) Foaming may occur during heating.
(ii) Direct contact of the medium with steam requires that allowance be made for condense dilution and requires 'clean' steam, free from anticorrosion additives.

In some cases the injection system is combined with flash cooling, where the sterilized medium is cooled by passing it through an expansion valve into a vacuum chamber. Cooling then occurs virtually instantly. A flow chart of a continuous sterilization system using direct steam injection is shown in Fig. 5.10. In some cases a combination of direct and indirect heat exchangers may be used (Svensson, 1988). This is especially true for starch-containing broths when steam injection is used for the pre-heating step. By raising the temperature virtually instantaneously the critical gelatinization temperature of the starch is passed through very quickly and the increase in viscosity normally associated with heated starch colloids can be reduced.

The most widely used continuous sterilization system is that based on the spiral heat exchangers and a typical layout is shown in Fig. 5.11. The plant is sterilized prior to sterilization of the medium by circulating hot water through the plant in a closed circuit. At the same time, the fermenter and the pipework between the fermenter and the sterilizer are steam sterilized. Heat conservation is achieved by cooling the sterile medium against cold, incoming unsterile medium which will then be partially heated before it reaches the sterilizer.

The Del factor to be achieved in a continuous sterilization process has to be increased with an increase in scale, and this is calculated exactly as described in the consideration of the scale up of batch regimes. Thus, if the volume to be sterilized is increased from 1000 dm$^3$ to 10,000 dm$^3$ and the risk of failure is to remain at 1 in 1000 then the Del factor must be increased from 34.5 to 36.8. However, the advantage of the continuous

FIG. 5.9. Industrial scale spiral heat exchanger (Alfa-Laval Engineering Ltd., Brentford, Middlesex).

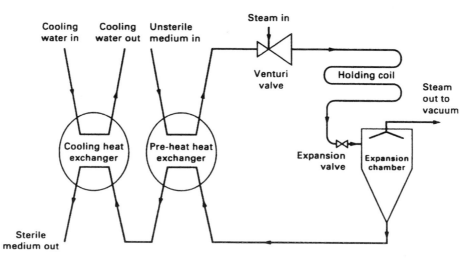

FIG. 5.10. Flow diagram of a typical continuous injector-flash cooler sterilizer.

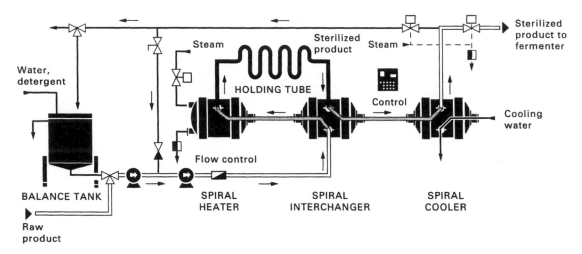

Fig. 5.11. Flow diagram of a typical continuous sterilization system employing spiral heat exchangers (Alfa-Laval Engineering Ltd., Brentford, Middlesex).

process is that temperature may be used as a variable in scaling up a continuous process so that the increased $\nabla$ factor may be achieved whilst maintaining the nutrient quality constant. Deindoerfer and Humphrey (1961) attempted to rationalize the choice of time–temperature regime by the use of a Nutrient Quality Criterion ($Q$), based on similar logic to the Del factor:

$$Q = \ln (x_0/x_t) \qquad (5.11)$$

where $x_0$ is the concentration of essential heat labile nutrient in the original medium,

$x_t$ is the concentration of essential heat labile nutrient in the medium after a sterilization time, $t$.

As considered earlier, the destruction of a nutrient may be considered a first-order reaction:

$$x_t/x_0 = e^{-kt}$$

where $k$ is the reaction rate constant

or $$x_0/x_t = e^{kt}.$$

Thus, taking natural logarithms,

$$\ln (x_0/x_t) = kt.$$

Therefore $$Q = kt.$$

The relationship between $k$ and absolute temperature is described by the Arrhenius equation:

$$k = A \cdot e^{-(E/RT)}$$

therefore, substituting for $k$:

$$Q = A \cdot t \cdot e^{-(E/RT)}.$$

Therefore, as for the Del factor equation, by taking natural logarithms, and rearranging, the following equation is obtained

$$\ln t = E / RT + \ln Q / A. \qquad (5.12)$$

Thus, a plot of the natural logarithms of the time required to achieve a certain $Q$ value against the reciprocal of the absolute temperature will yield a straight line, the slope of which is dependent on the activation energy; that is, a very similar plot to Fig. 5.5 for the Del factor relationship. If both plots were superimposed on the same figure, then a continuous sterilization performance chart is obtained. The example put forward by Deindoerfer and Humphrey (1961) is shown in Fig. 5.12. Thus, in Fig. 5.12 each line of a constant Del factor specifies temperature–time regimes giving the same fractional reduction in spore number and each line of a constant nutrient quality criterion specifies temperature–time regimes giving the same destruction of nutrient. By considering the effect of nutrient destruction on product yield, limits may be imposed on Fig. 5.12 indicating the nutrient quality criterion below which no further increase in yield is achieved (i.e. the nutrient is in excess) and the nutrient quality criterion at which the product yield is at its lowest (i.e. there is no nutrient remaining). Thus, from such a plot a temperature–time regime may be chosen which gives the required Del factor and does not adversely affect the yield of the process.

The adoption of Deindoerfer and Humphrey's approach is possible only if the limiting heat-labile nutrient is identified and the Arrhenius constant and activation energy for its thermal destruction experimentally

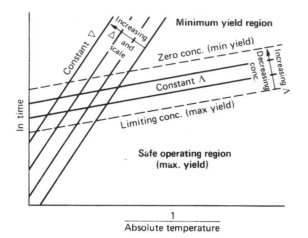

FIG. 5.12. Continuous sterilization performance chart (Deindoerfer and Humphrey, 1961).

determined. As pointed out by Banks (1979), this information may not be available for a complex fermentation medium and, therefore, the technique is fairly limited in its practical application. However, provided this information is available the technique may be used in the scale-up of continuous-sterilization processes. As discussed earlier, the Del factor is scale dependent and therefore as the volume to be sterilized is increased so the Del factor should be increased if the probability of achieving sterility is to remain the same. However, the nutrient-quality criterion is not scale dependent so that by changing the temperature–time regime to accommodate the attainment of sterility the nutrient quality may be adversely affected. Examination of Fig. 5.12 indicates that the only way in which the Del factor may be increased without any change in the nutrient quality criterion is to increase the temperature and to decrease the holding time. Although this scale-up could be performed exactly only if the thermal degradation kinetics of the medium were known, the analysis indicates that even without this information the approach would be to reduce the holding time and increase the temperature.

When designing a continuous sterilization process based on spiral heat exchangers it is important to consider the effect of suspended solids on the sterilization process. Micro-organisms contained within solid particles are given considerable protection against the sterilization treatment. If the residence time in the sterilizer is insufficient for heat to penetrate the particle then the fermentation medium may not be rendered sterile. The routine solution to this problem is to 'overdesign' the process and expose the medium to a

far more severe regime than may be necessary. Armenante and Li (1993) discussed this problem in considerable detail and produced a model to predict the behaviour of a continuous system. Their analysis suggested that the temperature of the particle cores is significantly less than that of the bulk liquid. Furthermore, there is a considerable time lag in heat penetrating to the particle cores, resulting in a very different time–temperature profile for the particles as compared with the liquid medium. Thus, the temperature of the particles may not reach the critical point before they leave the sterilizer and heat penetration into the particles will continue downstream of the sterilizer. Armenante and Li's conclusion is that it is the sterilizer and/or the first cooling exchanger that should be 'overdesigned' rather than the length of the holding coil. Remember that the first cooling exchanger transfers a significant amount of heat from the sterile medium to the incoming medium and increasing its surface area would give more opportunity for the heat to penetrate the particles. This, coupled with increasing the temperature or residence time in the sterilizer, would ensure that the particle cores are up to temperature before the holding coil is reached. Also, this work suggests that it is unwise to use the direct steam injection method to heat a particulate medium because, again, there will be insufficient time for the heat to penetrate the particles.

An example of the scale up of sterilization regimes is given by the work of Jain and Buckland (1988) on the production of efrotomycin by *Nocardia lactamdurans*. In this case a beneficial interaction appeared to be occurring between the protein nitrogen source and glucose during sterilization, thus making the protein less available but resulting in a more controlled fermentation. When glucose was sterilized separately the oxygen demand of the subsequent fermentation was excessive and the fermentation terminated prematurely with very poor product formation. On scaling up the fermentation it was very difficult to attain the correct sterilization conditions using a batch regime. However, continuous sterilization using direct steam injection allowed the design of a precise process producing sterile medium with the required degree of interaction between the ingredients. The identification of this phenomenon was dependent upon careful monitoring of the small scale fermentation and consideration being given to sterilization as an important scale-up factor.

When a fermentation is scaled up it is important to appreciate that the inoculum development process is also increased in scale (see Chapter 6) and a larger seed fermenter may have to be employed to generate sufficient inoculum to start the production scale. Thus, the sterilization regime of the seed fermenter (and its medium) will also have to be scaled up. Therefore, the performance of the seed fermentation should be assessed carefully to ensure that the quality of the inoculum is maintained on the larger scale and that it has not been adversely affected by any increase in the severity of the sterilization regime.

## STERILIZATION OF THE FERMENTER

If the medium is sterilized in a separate batch cooker, or is sterilized continuously, then the fermenter has to be sterilized separately before the sterile medium is added to it. This is normally achieved by heating the jacket or coils (see Chapter 7) of the fermenter with steam and sparging steam into the vessel through all entries, apart from the air outlet from which steam is allowed to exit slowly. Steam pressure is held at 15 psi in the vessel for approximately 20 minutes. It is essential that sterile air is sparged into the fermenter after the cycle is complete and a positive pressure is maintained; otherwise a vacuum may develop and unsterile air be drawn into the vessel.

## STERILIZATION OF THE FEEDS

A variety of additives may be administered to a fermentation during the process and it is essential that these materials are sterile. The sterilization method depends on the nature of the additive, and the volume and feed rate at which it is administered. If the additive is fed in large quantities then continuous sterilization may be desirable, for example, Aunstrup et al. (1979) cited the use of continuous heat sterilization for the feed medium of microbial enzyme fermentations. Aunstrup et al. also referred to the use of filtration for the sterilization of certain feeds. Stowell (1987) described the use of a continuous sterilizer for the addition of oil feeds to industrial scale fermenters, but stressed that each oil has its own characteristics which makes it impossible to predict the performance of a sterilizer for different oil feeds. Batch sterilization of feed liquids normally involves steam injection into the material held in storage vessels. Whatever the sterilization system employed it is essential that all ancillary equipment and feed pipework associated with the additions are sterilizable.

## STERILIZATION OF LIQUID WASTES

Process organisms which have been engineered to produce 'foreign' products and therefore contain heterologous genes are subject to strict containment regulations. Thus, waste biomass of such organisms must be sterilized before disposal. Sterilization may be achieved by either batch or continuous means but the whole process must be carried out under contained conditions. Batch sterilization involves the sparging of steam into holding tanks, whereas continuous processes would employ the type of heat exchangers which have been discussed in the previous section. A holding vessel for the batch sterilization of waste is shown in Fig. 5.13. Whichever method is employed the effluent must be cooled to below 60°C before it is discharged to waste. The sterilization processes have to be validated and are designed using the Del factor approach considered in the previous sections. However, the kinetic characteristics used in the calculations would be those of the process organism rather than of B. stearothermophilus. Also, the $N_t$ value used in the design calculations would be smaller than $10^{-3}$ which is used for medium sterilization and would depend on the assessment of the hazard involved should the organism survive the decontamination process. Thus, the sterilization regime used for destruction of the process organism will be different from that used in sterilizing the medium.

## FILTER STERILIZATION

Suspended solids may be separated from a fluid during filtration by the following mechanisms:

(i)   Inertial impaction.
(ii)  Diffusion.
(iii) Electrostatic attraction.
(iv)  Interception.

(i) *Inertial impaction*
Suspended particles in a fluid stream have momentum. The fluid in which the particles are suspended will flow through the filter by the route of least resistance. However, the particles, because of their momentum, tend to travel in straight lines and may therefore become impacted upon the fibres where they may then remain. Inertial impaction is more significant in the filtration of gases than in the filtration of liquids.

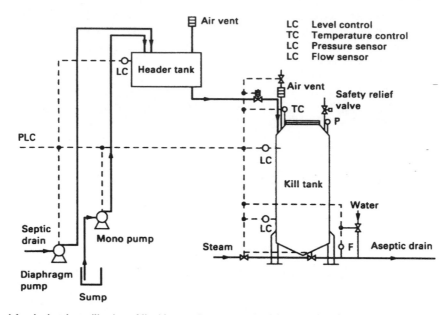

FIG. 5.13. A vessel for the batch sterilization of liquid waste from a contained fermentation (Jansson *et al.*, 1990).

### (ii) *Diffusion*

Extremely small particles suspended in a fluid are subject to Brownian motion which is random movement due to collisions with fluid molecules. Thus, such small particles tend to deviate from the fluid flow pattern and may become impacted upon the filter fibres. Diffusion is more significant in the filtration of gases than in the filtration of liquids.

### (iii) *Electrostatic attraction*

Charged particles may be attracted by opposite charges on the surface of the filtration medium.

### (iv) *Interception*

The fibres comprising a filter are interwoven to define openings of various sizes. Particles which are larger than the filter pores are removed by direct interception. However, a significant number of particles which are smaller than the filter pores are also retained by interception. This may occur by several mechanisms — more than one particle may arrive at a pore simultaneously, an irregularly shaped particle may bridge a pore, once a particle has been trapped by a mechanism other than interception the pore may be partially occluded enabling the entrapment of smaller particles. Interception is equally important a mechanism in the filtration of gases and liquids.

Filters have been classified into two types — those in which the pores in the filter are smaller than the particles which are to be removed and those in which the pores are larger than the particles which are to be removed. The former type may be regarded as an absolute filter, so that filters of this type (provided they are not physically damaged) are claimed to be 100% efficient in removing micro-organisms. Filters of the latter type are frequently referred to as depth filters and are composed of felts, woven yarns, asbestos pads and loosely packed fibreglass. The terms absolute and depth can be misleading as they imply that absolute filtration only occurs at the surface of the filter, whereas absolute filters also have depth and filtration occurs within the filter as well as at the surface. Terms which bear more relationship to the construction of filters are 'non-fixed pore filters' (corresponding with depth filters) and 'fixed pore filters' (corresponding with absolute filters).

Non-fixed pore filters rely on the removal of particles by inertial impaction, diffusion and electrostatic attraction rather than interception. The packing material contains innumerable tortuous routes through the filter but removal is a statistical phenomenon and, thus, sterility of the product is predicted in terms of the probability of failure (similar to the situation for steam sterilization). Thus, in theory, the removal of micro-organisms by a fibrous filter cannot be absolute as there is always the possibility of an organism passing through the filter, regardless of the filter's depth. Also, because the fibres are not cemented into position an increase in the pressure to which the filter is subjected may result in movement of the material, producing larger channels through the filter. Increased pressure may also result in the displacement of previously trapped particles.

Fixed pore filters are constructed so that the filtration medium will not be distorted during operation so that the flow patterns through the filter will not change due to disruption of the material. Pore size is controlled during manufacture so that an absolute rating can be quoted for the filter — i.e. the removal of particles above a certain size can be guaranteed. Thus, interception is the major mechanism by which particles are removed. Because fixed pore filters have depth they are also capable of removing particles which are smaller than the pores by the mechanisms of inertial impaction, diffusion and attraction and these mechanisms do play significant roles in the filtration of gases. Fixed pore filters are superior for most purposes in that they have absolute ratings, are less susceptible to changes in pressure and are less likely to release trapped particles. The major disadvantage associated with absolute filters was the resistance to flow they present and, hence, the large pressure drop across the filters which represents a major operating cost. However, modern absolute filter cartridges which have been developed by many filtration companies contain pleated membranes with very large surface areas, thereby minimizing the pressure drop across the filter.

It is important to realize that filters should be steam sterilized before and after operation (see Chapter 7). Thus, the materials must be stable at high temperatures and the steam must be free of particulate matter because the filter modules are particularly vulnerable to damage at high temperatures. Thus, the steam itself is filtered through stainless steel mesh filters rated at 1 $\mu$m.

### Filter sterilization of fermentation media

Media for animal-cell culture cannot be sterilized by steam because they contain heat-labile proteins. Thus, filtration is the method of choice and as may be appreciated from the previous discussion, fixed pore or absolute filtration is the better system to use. An ideal filtration system for the sterilization of animal cell culture media must fulfil the following criteria:

(i) The filtered medium must be free of fungal, bacterial and mycoplasma contamination.

(ii) There should be minimal adsorption of protein to the filter surface.
(iii) The filtered medium should be free of viruses.
(iv) The filtered medium should be free of endotoxins.

Several filter manufacturers now supply absolute filtration systems for the sterilization of animal cell culture medium. Such systems consist of membrane cartridges which are fitted into stainless steel, steam sterilizable modules. The membranes for media filtration are constructed from steam sterilizable hydrophilic material and are treated to produce a filtrate of particular quality. For example, if minimal protein adsorption is a major criterion then a specially coated filter membrane is used. It would be very difficult to construct a single filtration membrane which would fulfil all four criteria cited above. Thus, a series of filters are used to achieve the desired result. The example shown in Fig. 5.14 is provided by Pall Process Filtration Ltd and illustrates a system to produce sterile, mycoplasma free serum and consists of four filters arranged in sequence. The first filter is a positively charged polypropylene pre-filter with an absolute rating of 5 $\mu$m for the removal of coarse precipitates, clot-like material and other gross contaminants; the second filter is also positively charged polypropylene but with an absolute rating of 0.5 $\mu$m for bulk microbial removal, deformable gels, lipid-based materials and endotoxin reduction; the third filter is a single layered, nylon/polyester positively charged filter with a 0.1-$\mu$m absolute rating for further microbial and endotoxin removal and optimum protection of the final filter; the fourth filter is similar to the third and has the same rating, but is double layered and removes mycoplasmas, gives absolute sterility and final endotoxin control. Thus, the combination of four filters gives a sequential removal of decreasingly small particles and prolongs the life of the final filter. If it is necessary to remove viral contamination then a final 0.04-$\mu$m nylon/polyester filter would be added.

Similar systems may be used in downstream processing of animal cell products where the rating and properties of the filters would be optimized for the particular process. Figure 5.15 illustrates a system for the removal of cells and cell debris from an animal-cell fermentation broth. The pre-filter is a polypropylene 1.0-$\mu$m rated filter to remove the bulk of the cells and debris and the second filter is an hydroxyl modified nylon/polyester 0.2-$\mu$m rated filter giving absolute cell removal with minimal protein adsorption.

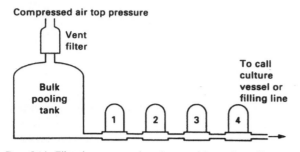

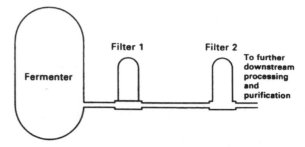

FIG. 5.14. Filtration system for the provision of sterile, mycoplasma free serum.
Filter 1. 5μm absolute rated prefilter for removal of coarse precipitates.
Filter 2. 0.5μm absolute rated prefilter for bulk bioburden removal.
Filter 3. 0.1μm absolute rated single layer prefilter for futher bioburden and endotoxin removal.
Filter 4. 0.1μm absolute rated double layer final filter for absolute sterility, mycoplasma removal and further endotoxin control.
(Pall Process Filtration Ltd., Portsmouth, U.K.)

FIG. 5.15. Filtration system for the removal of cells and cell debris from an animal cell culture fermentation.
Filter 1. 1.0μm absolute rated prefilter for bulk cell and cell debris removal.
Filter 2. 0.2μm absolute rated single layer 'Bio-Inert' filter for final bioburden removal.
(Pall Process Filtration Ltd., Portsmouth, U.K.)

### Filter sterilization of air

Aerobic fermentations require the continuous addition of considerable quantities of sterile air (see Chapter 9). Although it is possible to sterilize air by heat treatment, the most commonly used sterilization process is filtration. Fixed pore filters (which have an absolute rating) are very widely used in the fermentation industry and several manufacturers produce filtration systems for air sterilization. These systems, like those for the sterilization of liquids, consist of pleated membrane cartridges designed to be accommodated in stainless steel modules. A selection of such cartridges and holders is shown in Fig. 5.16 and a sectioned filter unit is shown in Fig. 5.17. The most common construction material used for for the pleated membranes for air sterilization is PTFE, which is hydrophobic and is therefore resistant to wetting. Also, PTFE filters may be steam sterilized and are resistant to ammonia which may be injected into the air stream, prior to the filter, for pH control. As was seen for the filter sterilization of liquids it is essential that a prefilter is incorporated up-stream of the absolute filter. The prefilter traps large particles such as dust, oil and carbon (from the compressor) and pipescale and rust (from the pipework). The use of a coalescing prefilter also ensures the removal of water from the air; entrained water is coalesced in the filter (air flow being from the inside of the filter to the outside) and is discharged via an automatic drain. Figure 5.18 illustrates the layout of such a filtration unit showing the steam sterilization system and Fig. 5.19 is a photograph of the air steriliza-

tion plant for an 85 m$^3$ fermenter. Smith (1981) cited the use of absolute filtration in the sterilization of air for the ICI biomass continuous fermenter.

### Sterilization of fermenter exhaust air

In many traditional fermentations the exhaust gas from the fermenter was vented without sterilization or vented through relatively inefficient depth filters. With the advent of the use of recombinant organisms and a greater awareness of safety and emission levels of allergic compounds the containment of exhaust air is more common (and in the case of recombinant organisms, compulsory). Fixed pore membrane modules are also used for this application but the system must be able to cope with the sterilization of water saturated air, at a relatively high temperature and carrying a large contamination level. Also, foam may overflow from the fermenter into the air exhaust line. Thus, some form of pretreatment of the exhaust gas is necessary before it enters the absolute filter. This pretreatment may be a hydrophobic prefilter or a mechanical separator to remove water, aerosol particles and foam. The pretreated air is then fed to a 0.2-μm hydrophobic filter. Again, it is important to appreciate that the filtration system must be steam sterilizable. Figures 5.18 and 5.19 illustrate the prefilter and mechanical separator systems respectively.

### The theory of depth filters

Although most fermentation companies rely upon the pleated membrane, fixed pore (absolute rated) filter

FIG. 5.16. A selection of absolute membrane filter cartridges and stainless steel housings (Pall Process Filtration, Portsmouth, U.K.).

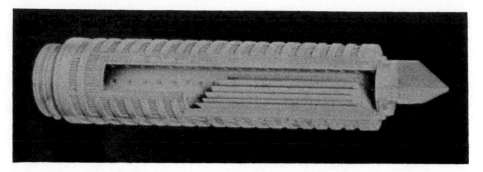

FIG. 5.17. An absolute membrane filter sectioned to show the pleated membrane structure (Pall Process Filtration, Portsmouth, U.K.).

systems it is necessary to consider the theory of depth filtration. Aiba *et al.* (1973) have given detailed quantitative analysis of these mechanisms but this account will be limited to a description of the overall efficiency of operation of fibrous filters. Several workers (Ranz and Wong, 1952; Chen, 1955) have put forward equations relating the collection efficiency of a filter bed to various characteristics of the filter and its components. However, a simpler description cited by Richards (1967) may be used to illustrate the basic principles of filter design.

If it is assumed that if a particle touches a fibre it remains attached to it, and that there is a uniform concentration of particles at any given depth in the filter, then each layer of a unit thickness of the filter should reduce the population entering it by the same proportion; which may be expressed mathematically as:

$$dN/dx = -KN \qquad (5.13)$$

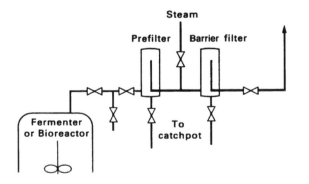

FIG. 5.18. Dual hydrophobic filter system for the sterilization of off-gas from a fermenter (Pall Process Filtration Ltd., Portsmouth, U.K.).

where $N$ is the concentration of particles in the air at a depth, $x$, in the filter and
$K$ is a constant.

On integrating equation (5.13) over the length of the filter it becomes:

$$N/N_0 = e^{-Kx} \qquad (5.14)$$

where $N_0$ is the number of particles entering the filter and $N$ is the number of particles leaving the filter.

On taking natural logarithms, equation (5.14) becomes:

$$\ln(N/N_0) = -Kx. \qquad (5.15)$$

Equation (5.15) is termed the log penetration relationship. The efficiency of the filter is given by the ratio of the number of particles removed to the original number present, thus:

$$E = (N_0 - N)/N_0 \qquad (5.16)$$

where $E$ is the efficiency of the filter.
But:

$$(N_0 - N)/N_0 = 1 - (N/N_0). \qquad (5.17)$$

Substituting $\qquad N/N_0 = e^{-Kx}$

Thus:

$$(N_0 - N)/N_0 = 1 - e^{-Kx} \qquad (5.18)$$

and $\qquad E = 1 - e^{-Kx}.$

The log penetration relationship [equation (5.15)] has been used by Humphrey and Gaden (1955) in filter design, by using the concept $X_{90}$, the depth of filter required to remove 90% of the total number of particles entering the filter; thus:

If $N_0$ were 10 and $x$ were $X_{90}$, then $N$ would be 1:

$$\ln(1/10) = -KX_{90}$$

or $\qquad 2.303 \log_{10}(1/10) = -KX_{90}$

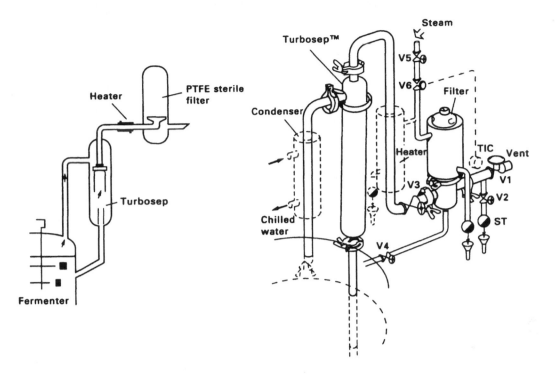

FIG. 5.19. A mechanical separator and hydrophobic filter system for the sterilization of off-gas from a fermenter. Left. Cut-away diagram. Right. Equipment arrangement, showing steam supply. V1–V6, valves; O, steam, traps (Domnick Hunter Ltd., Birtley, Co. Durham, U.K.).

$$2.303(-1) = -KX_{90},$$

therefore, $\quad\quad X_{90} = 2.303/K.$ $\quad\quad$ (5.19)

Consideration of equation (5.15) indicates that a plot of the natural logarithm of $N/N_0$ against $x$, filter length, will yield a straight line of slope $K$. To obtain the information for a plot of this type, data must be gleaned for the removal of organisms from an air stream by filters of increasing length which would involve assessment of microbial levels in the air entering and leaving the filter. Humphrey and Gaden (1955) and Richards (1967) described equipment with which such assessments could be recorded.

The value of $K$ is affected by the nature of the filter material and by the linear velocity of the air passing through the filter. Figure 5.20 is a typical plot of $K$ and $X_{90}$ against linear air velocity from which it may be seen that $K$ increases to an optimum with increasing air velocity, after which any further increase in air velocity results in a decrease in $K$. Table 5.3 (Riviere, 1977) summarizes the effects of linear air velocity on the removal of a range of micro-organisms with a variety of filter materials.

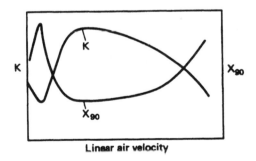

Linear air velocity

FIG. 5.20. The effect of increasing linear air velocity on $K$ and $X_{90}$ of a filtration system (Richards, 1967).

The increase in $K$ with increasing air velocity is probably due to increased impaction, illustrating the important contribution this mechanism makes to the removal of organisms. The decrease in $K$ values at high air velocities is probably due to disruption of the filter, allowing channels to develop and fibres to vibrate, resulting in the release of previously captured organisms.

TABLE 5.3. $X_{90}$ values for the removal of a range of micro-organisms by a variety of filtration materials (Humphrey, 1960; Rivierre, 1977)

| Filter material | Diameter of fibres ($\mu$m) | Micro-organism | Air speed (cm s$^{-1}$) | $X_{90}$ (cm) |
|---|---|---|---|---|
| Glass wool | 16 | Bacillus subtilis spores | 3 | 4 |
| | | | 15 | 9 |
| | | | 30 | 11.5 |
| | | | 150 | 1.5 |
| | | | 300 | 0.4 |
| | 18.5 | Serratia marcescens | 3.9 | 3.1 |
| Glass fibre | 8.5 | Escherichia coli | 3 | 0.4 |
| | | | 15 | 0.6 |
| | | | 30 | 0.7 |
| | | | 150 | 0.8 |
| | | | 300 | 1.1 |
| Norite (15–30 mesh) | — | Bacillus cereus spores | 1.4 | 1.7 |
| | | | 6.4 | 1.5 |
| Activated carbon (4–8 mesh) | | Bacillus cereus spores | 18 | 1.7 |
| | | | 28.5 | 8.7 |

## The design of depth filters

Equation (5.15), the log penetration relationship, is the same form as equation (5.3) in the derivation of thermal-death kinetics. In the case of heat sterilization the theory predicts that an infinite time is required to reduce the population to zero, whereas the theory of filtration predicts that a filter of infinite length is required to remove all organisms from an air stream. Thus, it is not surprising that the same approach is adopted in the design of filters and heat-sterilization cycles, in that an acceptable probability of contamination is determined. The probability of one fermentation in a thousand being contaminated is frequently used in filter design, as it is in the design of heat-sterilization cycles. Having arrived at an acceptable probability of contamination and determined the filtration characteristics (i.e. the value of $K$) of the material to be used, a filter may be designed to filter a certain volume of air containing a certain number of organisms; the following example illustrates the design calculation approach used by Richards (1967):

It is required to provide a 20-m$^3$ fermenter with air at a rate of 10 m$^3$ min$^{-1}$ for a fermentation lasting 100 hours. From an investigation of the filter material to be used, the optimum linear air velocity was shown to be 0.15 m sec$^{-1}$, at which the value of $K$ was 1.535 cm$^{-1}$. The dimensions of the filter may be calculated as follows:

The log penetration relationship states that:

$$\ln (N / N_0) = -Kx.$$

The air in the fermentation plant contained approximately 200 micro-organisms m$^{-3}$. Therefore,

$$N_0 = \text{total amount of air provided} \times 200,$$

$$N_0 = 10 \times 60 \times 100 \times 200$$

$$= 12 \times 10^6 \text{ organisms.}$$

The acceptable degree of contamination is one in a thousand, therefore $N = 10^{-3}$,

$$\ln \{10^{-3}/(12 \times 10^6)\} = -Kx,$$

$$\ln 8.33 \times 10^{-11} = -Kx,$$

$$\ln 8.33 \times 10^{-11} = -1.535x,$$

$$x = -23.21/-1.535 = 15.12 \text{ cm.}$$

Therefore, the filter to be used should be 15.12 cm long.

The cross-sectional area of the filter is given by the volumetric air flow rate divided by the linear air velocity:

$$\pi r^2 = 10/0.15 \times 60$$

where $r$ is the radius of the filter

$$r = 0.59 \text{ m}.$$

Thus the filter to be employed should be 15.12 cm long and 0.59 m radius.

However, as Humphrey (1960) pointed out, the efficient operation of the filter is dependent on the supply of air at the optimum linear velocity. If the air velocity increases or decreases the value of $K$ will decrease, resulting in a loss of filtration efficiency. Considering the example calculation, if the linear air velocity were to drop to 0.03 m sec$^{-1}$, then the value of $K$ would decline to 0.2 cm$^{-1}$. The number of organisms which would enter the fermentation in 1 minute at this reduced air-flow rate would be as calculated below:

$$\ln(N/N_0) = -Kx.$$

At a linear air velocity of 0.03 m sec$^{-1}$, in 1 minute 0.03 $\times$ 60 $\times$ the cross-sectional area of the filter m$^{-3}$ of air would enter the filter, i.e. 1.98 m$^3$. At a microbial contamination level of 200 organisms m$^{-3}$ this means that 396 organisms would enter the filter in 1 minute. Thus:

$$\ln(N/396) = -0.2 \times 15.12,$$

$$N = 19.24.$$

Therefore, 19.24 organisms would have entered the fermenter in 1 minute at the decreased air-flow rate. If the filter had been designed to meet this contingency, then the length would have been:

$$\ln(10^{-3}/396) = -0.2x,$$

$$x = 64.4.$$

Thus a filter length of 64.4 cm would have been required to have maintained the same probability of contamination over the 1 minute of reduced air flow.

This example illustrates the hazards of attempting very precise design and the necessity to consider the reliability of ancillary equipment in any design calculation.

## REFERENCES

AIBA, S., HUMPHREY, A. E. and MILLIS, N. (1973) *Biochemical Engineering* (2nd edition). Academic Press, New York.

ARMENANTE, P. M. and LI, Y. S. (1993) Complete design analysis of a continuous sterilizer for fermentation media containing suspended solids. *Biotech. Bioeng.* **41**, 900–913.

AUNSTRUP, K., ANDRESEN, O., FALCH, E. A. and NIELSEN, T. K. (1979) Production of microbial enzymes. In *Microbial Technology*, Vol. 1 (2nd edition), pp. 282–309 (Editors Peppler, H. J. and Perlman, D.). Academic Press, New York

BADER, F. G., BOEKELOO, M. K., GRAHAM, H. E. and CAGLE, J. W. (1984) Sterilization of oils: data to support the use of a continuous point-of-use sterilizer. *Biotech. Bioeng.* **26**, 848–856.

BANKS, G. T. (1979) Scale-up of fermentation processes. *Topics in Enzyme and Fermentation Biotechnology,* Vol. **3**, 170–266 (Editor Wiseman, A.). Ellis Horwood, Chichester.

CHEN, C. Y. (1955) Filtration of aerosols by fibrous media. *Chem. Rev.* **55**, 595–623.

CORBETT, K. (1985) Design, preparation and sterilization of fermentation media. In *Comprehensive Biotechnology, The Principles of Biotechnology: Engineering Considerations*, pp. 127–139 (Editors Cooney, C. L. and Humphrey, A. E.). Pergamon Press, Toronto.

DEINDOERFER, F. H. and HUMPHREY, A. E. (1959) Analytical method for calculating heat sterilization times. *Appl. Micro.* **7**, 256–264.

DEINDOERFER, F. H. and HUMPHREY, A. E. (1961) Scale-up of heat sterilization operations. *Appl. Micro.* **9**, 134–145.

HUMPHREY, A. E. (1960) Air sterilization. *Adv. Appl. Micro.* **2**, 302–312.

HUMPHREY, A. E. and GADEN, E. L. (1955) Air sterilization by fibrous media. *Ind. Eng. Chem.* **47** 924–930.

JAIN, D. and BUCKLAND, B. C. (1988) Scale-up of the efrotomycin fermentation using a computer-controlled pilot plant. *Bioprocess Eng.* **3**, 31–36.

JANSSON, D. E., LOVEJOY, P. J., SIMPSON, M. T. and KENNEDY, L. D. (1990) Technical problems in large-scale containment of rDNA organisms. In: *Fermentation Technologies: Industrial Applications*, pp. 388–393 (Editor, Yu, R.-L.) Elsevier, London.

RANZ, W. E. and WONG, J. B. (1952) Impaction of dusty smoke particles on surface and body collectors. *Ind. Eng. Chem.* **44**, 1371–1378.

RICHARDS, J. W. (1966) Fermenter mash sterilization. *Proc. Biochem.* **1** (1),41–46.

RICHARDS, J. W. (1967) Air sterilisation with fibrous filters. *Proc. Biochem.* **2** (9), 21–25.

RICHARDS, J. W. (1968) *Introduction to Industrial Sterilisation*. Academic Press, London.

RIVIERE, J. (1977) *Industrial Applications of Microbiology*. (Translated and edited by Moss, M. O. and Smith, J. E.). Surrey University Press.

SMITH, S. R. L. (1981) Some aspects of ICI's single cell protein process. In *Microbial Growth on $C_1$ Compounds*, pp. 342–348 (Editor Dalton, H.). Heyden, London.

STOWELL, J. D. (1987) Application of oils and fats in antibiotic processes. In *Carbon Substrates in Biotechnology*, pp.139–160 (Editors Stowell, J. D., Beardsmore, A. J., Keevil, C. W. and Woodward, J. R.). IRL Press, Oxford.

SVENSSON, R. (1988) Continuous media sterilization in biotechnical fermentation. *Dechema Monographien* **113**, 225–237.

# CHAPTER 6

# The Development of Inocula for Industrial Fermentations

## INTRODUCTION

IT IS ESSENTIAL that the culture used to inoculate a fermentation satisfies the following criteria:

1. It must be in a healthy, active state thus minimizing the length of the lag phase in the subsequent fermentation.
2. It must be available in sufficiently large volumes to provide an inoculum of optimum size.
3. It must be in a suitable morphological form.
4. It must be free of contamination.
5. It must retain its product-forming capabilities.

The process adopted to produce an inoculum meeting these criteria is called inoculum development. Hockenhull is credited with the quotation "once a fermentation has been started it can be made worse but not better" (Calam, 1976). Whereas this is an over-statement it does illustrate the importance of inoculum development. Much of the variation observed in small-scale laboratory fermentations is due to poor inocula being used and, thus, it is essential to appreciate that the establishment of an effective inoculum development programme is equally important regardless of the scale of the fermentation. Such a programme not only aids consistency on a small scale but is invaluable in scaling up the fermentation and forms an essential part in progressing a new process.

A critical factor in obtaining a suitable inoculum is the choice of the culture medium. It must be stressed that the suitability of an inoculum medium is determined by the subsequent performance of the inoculum in the production stage. As discussed elsewhere (Chapter 4), the design of a production medium is determined not only by the nutritional requirements of the organism, but also by the requirements for maximum product formation. The formation of product in the seed culture is not an objective during inoculum development so that the seed medium may be of a different composition from the production medium. However, Lincoln (1960) stated that the lag time in a fermentation is minimized by growing the culture in the 'final-type' medium. Lincoln's argument is an important one, so the inoculum development medium should be sufficiently similar to the production medium to minimize any period of adaptation of the culture to the production medium, thus reducing the lag phase and the fermentation time. Furthermore, Hockenhull (1980) pointed out the dangers of using very different media in consecutive stages. Major differences in pH, osmotic pressure and anion composition may result in very sudden changes in uptake rates which, in turn, may affect viability. Hockenhull also emphasized that for antibiotic fermentations the inoculum medium should contain sufficient carbon and nitrogen to support maximum growth until transfer, so that secondary metabolism remains repressed during growth of the inoculum. If secondary metabolism is derepressed in the seed fermentation, then selection may enrich the culture with non-producing variants having a growth advantage over high-producing types. Hockenhull drew attention to Righelato's (1976) work in which it was shown that chemostat culture of *Penicillium chrysogenum* under carbohydrate-limited conditions led to a loss of penicillin synthesizing ability and an increase in the proportion of non-conidiated variants whereas this did not occur in ammonia-, phosphate- or sulphate-limited conditions. The relevance of this phenomenon is supported by Hockenhull's observation that *P. chrysogenum* inocula produced under non-limiting conditions are remarkably free from variants whereas variants arise relatively frequently during the carbon-limited production phase. Examples of inoculum and production media are given in Table 6.1, from which it may be seen that inoculum media are, generally, less nutritious than production media and contain a lower level of carbon.

The quantity of inoculum normally used is between 3 and 10% of the medium volume (Lincoln, 1960; Meyrath and Suchanek, 1972; Hunt and Stieber, 1986). A relatively large inoculum volume is used to minimize the length of the lag phase and to generate the maximum biomass in the production fermenter in as short a time as possible, thus increasing vessel productivity. Thus, starting from a stock culture, the inoculum must

be built up in a number of stages to produce sufficient biomass to inoculate the production-stage fermenter. This may involve two or three stages in shake flasks and one to three stages in fermenters, depending on the size of the ultimate vessel. Throughout this procedure there is a risk of contamination and strain degeneration necessitating stringent quality-control procedures. The greater the number of stages between the master culture and the production fermenter the greater the risk of contamination and strain degeneration. Therefore, a compromise may be reached regarding the size of the inoculum to be used and the risk of contamination and strain degeneration. Another factor to be considered in the determination of the inoculum volume is the economics of the process. A seed fermenter 10% of the size of the production fermenter represents a considerable financial investment and must be justified in terms of productivity. A large-scale continuous fermentation for the production of biomass

TABLE 6.1. *Inoculum development and production media for a range of processes*

| Process | Inoculum development medium | | Production medium | | References |
|---|---|---|---|---|---|
| Griseofulvin | Whey powder ⎫ to<br>Lactose ⎭ give:<br>  Lactose<br>  Nitrogen<br>Corn-steep<br>  liquor solids<br>  (to give approx.<br>  0.04% N)<br>KH$_2$PO$_4$<br>KCl | <br><br>3.5%<br>0.05%<br><br>0.38%<br><br><br>0.4%<br>0.05% | Lactose<br>Corn-steep<br>  liquor solids<br>  to give:<br>  Nitrogen<br>Limestone<br>KH$_2$PO$_4$<br>KCl | 7%<br><br><br><br>0.2%<br>0.8%<br>0.4%<br>0.1% | Rhodes *et al.*<br>(1957) |
| Clavulanic acid | Soybean flour<br>Dextrin<br>Pluronic L81<br>(Antifoam) | 1.0%<br>2.0%<br>0.03% | Soybean flour<br>Oil<br>KH$_2$PO$_4$ | 1.5%<br>1.0%<br>0.1% | Butterworth<br>(1984) |
| Vitamin B$_{12}$ | <br>Sugar beet<br>Molasses<br>Sucrose<br>Betaine<br>NH$_4$H$_2$PO$_4$<br>(NH$_4$)$_2$SO$_4$<br>MgSO$_4$<br>ZnSO$_4$<br>5-6 Dimethyl-<br>  benzimidazole | (g dm$^{-3}$)<br><br>70<br>—<br>—<br>0.8<br>2<br>0.2<br>0.02<br><br>0.005 | <br><br><br><br><br><br><br><br><br><br> | (g dm$^{-3}$)<br><br>105<br>15<br>3<br>—<br>2.5<br>0.2<br>0.08<br><br>0.025 | Spalla *et al.*<br>(1989) |
| Lysine | Cane molasses<br>Corn-steep liquor<br>CaCO$_3$<br>Soybean meal<br>  hydrolysate | 5%<br>1%<br>1%<br><br>— | <br><br><br><br> | 20%<br>—<br>—<br><br>1.8% | Nakayama<br>(1972) |

would be expected to operate at steady state in excess of 100 days. Thus, the number of times that the fermenter is inoculated should be very few compared with batch or fed-batch systems. In such circumstances it may be more economic to compromise on the size of the inoculum and to tolerate a relatively lengthy period of growth up to maximum biomass than to invest a large seed vessel which would be used on very few occasions. This is particularly relevant for biomass continuous processes where one very large fermenter may be used and, thus, any seed vessel would only be servicing the one production vessel.

A typical inoculum-development programme will now be described in detail. The master culture (see Chapter 3 for an account of master-culture maintenance) is reconstituted and plated on to solid medium; approximately ten colonies of typical morphology of high producers are selected and inoculated on to slopes as the sub-master cultures, each sub-master culture being used for a new production run. At this stage, shake flasks may be inoculated to check the productivity of these cultures, the results of such tests being known before the developing inoculum eventually reaches the production plant. A sub-master culture is used to inoculate a shake flask (250 or 500 cm$^3$ containing 50 or 100 cm$^3$ medium) which, in turn, is used as inoculum for a larger flask, or a laboratory fermenter, which may then be used to inoculate a pilot-scale fermenter. Culture purity checks are carried out at each stage to detect contamination as early as possible. Although the results of these tests may not be available before the culture has reached the production plant, at least it is known at which stage in the procedure contamination occurred. For a sporulating organism the process may be modified to facilitate the use of a spore suspension as inoculum and this will be discussed in more detail later in this chapter.

Lincoln (1960) suggested a more elaborate procedure for the development of inoculum for bacterial fermentations which, with minor modifications, is applicable to any type of culture. The procedure involved the use of one sub-master culture to develop a bulk inoculum which was subdivided, stored in a frozen state and used as inocula for several months. A single colony, derived from a sub-master culture, was inoculated into liquid medium and grown to maximum log phase. This culture was then transferred into nineteen times its volume of medium and incubated again to the maximum log phase, at which point it was dispensed in 20-cm$^3$ volumes, plug frozen and stored at below $-20°$. At least 3% of the samples were tested for purity and productivity in subsequent fermentation and, provided

these were suitable, the remaining samples could be used as initial inocula for subsequent fermentations. To use one of the stored samples as inoculum it was thawed and used as a 5% inoculum for a seed culture which, in turn, was used as a 5% inoculum for the next stage in the programme. This procedure ensured that a proven inoculum was used for the penultimate stage in inoculum development.

## CRITERIA FOR THE TRANSFER OF INOCULUM

The physiological condition of the inoculum when it is transferred to the next culture stage can have a major effect on the performance of the fermentation. The optimum time of transfer must be determined experimentally and then procedures established so that inoculation with an ideal culture may be achieved routinely. These procedures include the standardization of cultural conditions and monitoring the state of an inoculum culture so that it is transferred at the optimum time, i.e. in the correct physiological state. The most widely used criterion for the transfer of vegetative inocula is biomass and such parameters as packed cell volume, dry weight, wet weight, turbidity, respiration, residual nutrient concentration and morphological form have been used (Hockenhull, 1980; Hunt and Stieber, 1986). Ettler (1992) demonstrated that the rheological behaviour of *Streptomyces noursei* could be used as the transfer criterion in the nystatin fermentation. The rheology of the seed fermentation changed from Newtonian to non-Newtonian behaviour and the optimum inoculum transfer time corresponded with this transformation.

Criteria which may be monitored on-line are the most convenient parameters to use as indicators of inoculum quality and these would include dissolved oxygen, pH (although pH would normally be controlled in seed fermentations) and oxygen or carbon dioxide in the effluent gas. Parton and Willis (1990) advocated the use of the carbon dioxide production rate (CPR) as a transfer criterion which requires analysis of the fermenter effluent air (see Chapter 8). This approach is suitable only when transfer is being made from a fermenter, but Parton and Willis stressed the importance of adopting this strategy even for the inoculation of laboratory-scale fermentations despite the fact that an adequate inoculum volume could be produced in shake flask. These workers provide an excellent example of the effect of inoculum transfer time on the productivity of a streptomycete secondary metabolite, as shown in Fig. 6.1a, b and c. The CPR of the

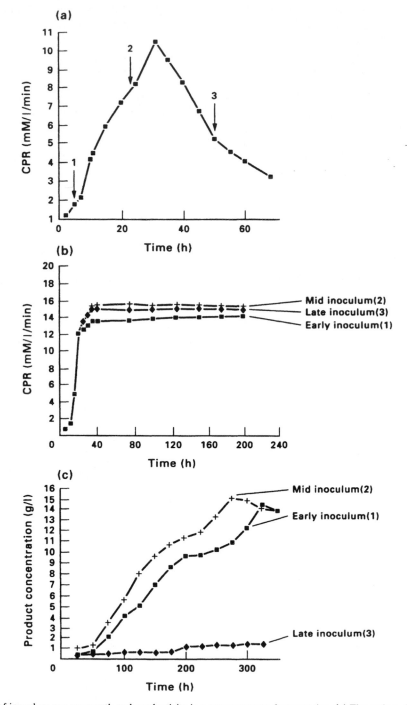

FIG. 6.1. The effect of inoculum age on growth and productivity in a streptomycete fermentation. (a) The carbon dioxide production rate (CPR) profile of the inoculum culture showing the points (1, 2 and 3) at which inocula were removed. (b) The effect of inoculum age on the CPR of the production fermentation. (c) The effect of inoculum age on productivity in the production fermentation (Parton and Willis, 1990).

inoculum fermentation and the points at which inoculum was transferred are shown in Fig. 6.1a. The CPR of the subsequent production fermentations are shown in Fig. 6.1b, from which it may be seen that the three fermentations performed similarly. However, Fig. 6.1c illustrates the very different secondary metabolite production of the three fermentations. Thus, although the time of transfer had only a marginal influence on biomass in the production fermentation the effect on product formation was critical. It should be emphasized that the amount of biomass transferred was standardized for the three fermentations and, thus, the differences in performance were due to the physiological states of the inocula.

In recent years, probes have been developed for on-line assessment of biomass (see Chapter 8) and these could be invaluable in estimating the time of inoculum transfer. Boulton *et al.* (1989) reported the use of a biomass sensor (the Bugmeter) to control the yeast pitching rate (inoculum level) in brewing. The probe measures the dielectric permittivity of viable yeast cells and is unaffected by the presence of dead cells, air bubbles or detritus, making it ideal for the routine monitoring of yeast inoculum. Using the probe, these workers developed an automatic inoculum dispenser allowing a preset viable yeast mass to be transferred from a yeast storage vessel to the brewery fermentation.

Alford *et al.* (1992) reported the use of a real-time expert computer system to predict the time of inoculum transfer for industrial-scale fermentations. The system involves the comparison of on-line fermentation data with detailed historical data of the process. A problem with the interpretation of carbon dioxide production rate figures is that the data are not available continuously because the analyser is not dedicated to any one fermenter, but is analysing process streams from a large number of vessels via a multiplexer system (see Chapter 8). Thus, a fermentation may have passed a critical stage between monitoring times. Also, occasional false readings may be generated. The expert system enabled the verification of data points as well as prediction of the outcome of the fermentation from early information. Data from seed fermentations were analysed by the expert system and the transfer time predicted. As a result of this approach operators were able to plan their work more effectively, the need for manual sampling was reduced and early warning of contamination was provided if the seed-culture profile predicted from early readings was abnormal.

Smith and Calam (1980) compared the quality and enzymic profile of differently prepared inocula *Penicil-lium patulum* (producing griseofulvin) and demonstrated that a low level of glucose 6-phosphate dehydrogenase was indicative of a good quality inoculum. The enzyme profile of good quality inoculum was established early in the growth of the seed culture. Thus, this approach could be used to assess the cultural conditions giving rise to satisfactory inoculum, but would be of less value in determining the time of transfer.

Yeast, unicellular bacterial, fungal and streptomycete fermentations have different requirements for inoculum development and these are dealt with separately.

## THE DEVELOPMENT OF INOCULA FOR YEAST PROCESSES

Whilst the largest industrial fermentations utilizing yeasts are the brewing of beer and the production of biomass, recent processes have also been established for the production of recombinant products.

### Brewing

It is common practice in the British brewing industry to use the yeast from the previous fermentation to inoculate a fresh batch of wort. The brewing terms used to describe this process are 'crop', referring to the harvested yeast from the previous fermentation, and 'pitch', meaning to inoculate. One of the major factors contributing to the continuation of this practice is the wort-based excise laws in the United Kingdom where duty is charged on the sugar consumed rather than the alcohol produced. Thus, dedicated yeast propagation systems are expensive to operate because duty is charged on the sugar consumed by the yeast during growth. It can then be appreciated that the reduced cost of using yeast from a previous fermentation is an attractive proposition (Boulton, 1991). The dangers inherent in this practice are the introduction of contaminants and the degeneration of the strain, the most common degenerations being a change in the degree of flocculence and attenuating abilities of the yeast. In breweries employing top fermentations in open fermenters these dangers are minimized by collecting yeast to be used for future pitching from 'middle skimmings'. During the fermentation the yeast cells flocculate and float to the surface, the first cells to do this being the most flocculent and the last cells the least flocculent. As the head of yeast develops, the surface layer (the most flocculent and highly contaminated yeasts) is removed and discarded and the underlying cells (the 'middle skimmings') are harvested and used for subsequent pitching. Therefore, the 'middle skimmings' contain cells which have the desired flocculence and which have been protected from contamination by the surface layer of the yeast head. The pitching yeast may be treated to reduce the level of contaminating bacteria and remove protein and dead yeast cells by such treatments as reducing the pH of the slurry to 2.5 to 3, washing with water, washing with ammonium persulphate and treatment with antibiotics such as polymixin, penicillin and neomycin (Mandl *et al.*, 1964; Strandskov, 1964; Roessler, 1968; Reed and Nagodawithana, 1991a).

However, traditional open vessels are becoming increasingly rare and the bulk of beer is brewed using cylindro-conical fermenters (see Chapter 7). In these systems the yeast flocculates and collects in the cone at the bottom of the fermenter where it is subject to the stresses of nutrient starvation, high ethanol concentration, low water activity, high carbon dioxide concentration and high pressure (Boulton, 1991). Thus, the viability and physiological state of the yeast crop would not be ideal for an inoculum. The viability of the crop may be assessed using a biomass probe of the type described earlier, thus ensuring that at least the correct amount of viable biomass is used to start the next fermentation. However, the physiological state of the biomass will not have been influenced by such monitoring procedures. The situation is further complicated by the fact that the harvested yeast is stored before it is used as inoculum. Metabolic activity is minimized during this time by chilling rapidly to about 1°, suspending in beer and storing in the absence of oxygen. If oxygen is present during the storage period then the yeast cells consume their stored glycogen which renders them very much less active at the start of the fermentation (Pickerell *et al.*, 1991).

One of the key physiological features of yeast inoculum is the level of sterol in the cells. Sterols are required for membrane synthesis but they are only produced in the presence of oxygen. Thus, we have the anomaly of oxygen being required for sterol synthesis, yet anaerobic conditions are required for ethanol production. This anomaly is resolved traditionally by aerating the wort before inoculation. This oxygen allows sufficient sterol synthesis early in the fermentation to support growth of the cells throughout the process, that is after the oxygen is exhausted and the process is anaerobic. Boulton *et al.* (1991) developed an alternative approach where the pitching yeast was vigorously aerated prior to inoculation. The yeast was then sterol rich and had no requirement for oxygen during the alcohol fermentation.

The difficulties outlined above and the likelihood of strain degeneration and contamination mean that yeasts are rarely used for more than five to ten consecutive fermentations (Thorne, 1970; Reed and Nagodawithana, 1991a) which necessitates the periodical production of a pure inoculum. This would involve developing sufficient biomass from a single colony to pitch a fermentation at a level of approximately 2 grams of pressed yeast per litre. Hansen (1896) pioneered the use of pure inocula and devised a yeast propagation scheme utilizing a 10% inoculum volume at each stage in the programme and employing conditions similar to those used during brewing. However, modern propagation schemes use inoculum volumes of 1% or even lower and may use conditions different from those used during brewing. Therefore, continuous aeration may be used during the propagation stage which seems to have little effect on the beer produced in the subsequent fermentation (Curtis and Clark, 1957). Yeast inoculum produced in this way would also be sterol rich, obviating the need for aerated wort.

A number of yeast propagators (which are basically closed, aerated vessels) have been described in the literature. The simplest type of progagator is a single-stage system resembling an unstirred, aerated fermenter which is inoculated with a shake-flask culture developed from a single colony (Gilliland, 1971). Curtis and Clark (1957) and Thorne (1970) described two-stage systems which could be operated semi-continuously. Thorne's propagator consisted of two linked vessels, 1.5 and 150 dm³ respectively. The smaller vessel was filled with wort, sterilized, cooled, aerated and inoculated with a flask-grown culture. After growth for 3 to 4 days the culture was forced by air pressure into the second vessel which had been filled with sterilized, cooled wort and aerated. An aliquot of 1.5 dm³ was forced back into the first vessel after mixing. In a further 3 to 4 days the larger vessel contained sufficient biomass to pitch a 1000 dm³ fermenter and the first vessel contained sufficient inoculum for another second stage. However, although this procedure should produce a pure inoculum there is a danger of strain degeneration occurring in such a semi-continuous system.

Boulton (1991) speculated that when the UK moves to an alcohol-based tax then the economics of propagating inoculum for each brew would be considerably more attractive. He suggested that the bakers' yeast aerobic fed-batch inoculum programme could be adopted to produce sterol-rich catabolite-derepressed cells. Such an inoculum would remove the necessity for aerating the wort prior to inoculation because sterol synthesis would not be necessary due to the high sterol content of the cells.

### Bakers' yeast

The commercial production of bakers' yeast involves the development of an inoculum through a large number of aerobic stages. Although the production stages of the process may not be operated under strictly aseptic conditions a pure culture is used for the initial inoculum, thereby keeping contamination to a minimum in the early stages of growth. Reed and Nagodawithana (1991b) discussed the development of inoculum for the production of bakers' yeast and quoted a process involving eight stages, the first three being aseptic while the remaining stages were carried out in open vessels. The yeast may be pumped from one stage to the next or the seed cultures may be centrifuged and washed before transfer, which reduces the level of contamination. The yields obtained in the first five stages are relatively low because they are not fed-batch systems, whereas the last three stages are fed-batch (the fed-batch bakers' yeast fermentation is considered in more detail in Chapter 2). A summary of a typical inoculum development programme for the production of bakers' yeast is given in Fig. 6.2.

## THE DEVELOPMENT OF INOCULA FOR BACTERIAL PROCESSES

The main objective of inoculum development for traditional bacterial fermentations is to produce an active inoculum which will give as short a lag phase as possible in subsequent culture. A long lag phase is disadvantageous in that not only is time wasted but also medium is consumed in maintaining a viable culture prior to growth. The length of the lag phase is affected by the size of the inoculum and its physiological condition (Meyrath and Suchanek, 1972). As already stated, the inoculum size normally ranges between 3 and 10% of the culture volume. Lincoln (1960) stressed that bacterial inocula should be transferred in the logarithmic phase of growth, when the cells are still metabolically active. The age of the inoculum is particularly important in the growth of sporulating bacteria, for sporulation is induced at the end of the logarithmic phase and the use of an inoculum containing a high percentage of spores would result in a long lag phase in a subsequent fermentation.

Keay *et al.* (1972) quote the use of a 5% inoculum of a logarithmically growing culture of a thermophilic *Bacillus* for the production of proteases. Aunstrup (1974) described a two-stage inoculum development programme for the production of proteases by *Bacillus subtilis*. Inoculum for a seed fermenter was grown for 1

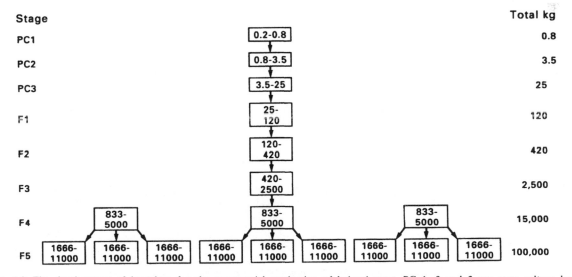

FIG. 6.2. The development of inoculum for the commercial production of bakers' yeast. PC 1, 2 and 3 are pure culture batch fermentations. F1 and 2 are non-aseptic batch fermentations. F3 and 4 are fed-batch fermentations and F5 is the final fed-batch fermentation (Reed and Nagodawithana, 1991a).

to 2 days on a solid or liquid medium and then transferred to a seed vessel where the organism was allowed to grow for a further ten generations before transfer to the production stage. Priest and Sharp (1989) cited the use of a 5% inoculum, still in the exponential phase, for the commercial production of *Bacillus* α-amylase. Underkofler (1976) emphasized that, in the production of bacterial enzymes, the lag phase in plant fermenters could be almost completely eliminated by using inoculum medium of the same composition as used in the production fermenter and employing large inocula of actively growing seed cultures. The inoculum development programme for a pilot-plant scale process for the production of vitamin B₁₂ from *Pseudomonas denitrificans* is shown in Fig. 6.3 (Spalla *et al.*, 1989).

The necessity to use an inoculum in an active physiological state is taken to its extreme in the production of vinegar. The acetic-acid bacteria used in the vinegar process are extremely sensitive to oxygen starvation. Therefore, to avoid disturbing the system, the cells at the end of a fermentation are used as inoculum for the next batch by removing approximately 60% of the culture and restoring the original level with fresh medium (Conner and Allgeier, 1976). The advantage of a highly active inoculum apparently outweighs the disadvantages of possible strain degeneration and contaminant accumulation. However, strain stability is a major concern in inoculum development for fermenta-

tions employing recombinant bacteria. Sabatie *et al.* (1991) demonstrated that plasmid stability and productivity in an *E. coli* biotin fermentation was greatly improved if stationary, rather then exponential, phase cells were used as inoculum. They postulated that the plasmid copy number may be higher in stationary cells than in exponential ones, resulting in a lower plasmid loss in the subsequent fermentation when a stationary culture is used as inoculum. A stationary phase inoculum would result in a lag phase, but this disadvantage was more than compensated for by the considerable improvement in plasmid retention and biotin production compared with that obtained using an exponential inoculum.

In the lactic-acid fermentation the producing organism may be inhibited by lactic acid. Thus, production of lactic acid in the seed fermentation may result in the generation of poor quality inoculum. Yamamoto *et al.* (1993) generated high quality inoculum of *Lactococcus lactis* IO-1 on a laboratory scale using electrodialysis seed culture which reduced the lactate in the inoculum and reduced the length of the lag phase in the production fermentation.

An example of the development of inoculum for an anaerobic bacterial process is provided by the clostridial acetone–butanol fermentation. The process was outcompeted by the petrochemical industry but there is still considerable interest in re-establishing the fermen-

**STOCK CULTURE**

**Lyophilised with skim milk**

↓

**MAINTENANCE CULTURE**

**Agar slope incubated 4 days at 28°**

↓

**SEED CULTURE - FIRST STAGE**

2 dm³ flask containing 0.6 dm³ medium inoculated with culture from one slope; incubated with shaking for 48h at 28°

↓

**SEED CULTURE - SECOND STAGE**

40 - 80 dm³ fermenter containing 25 - 50 dm³ medium inoculated with 1 - 1.2% first stage seed culture. Incubated 25 - 30h at 32°

↓

**PRODUCTION CULTURE**

500 dm³ fermenter with 300 dm³ medium inoculated with 5% second stage seed culture. Incubated at 32° for 140 - 160 h

FIG. 6.3. The inoculum development programme for a vitamin B₁₂ pilot scale fermentation using *Pseudomonas denitrificans* (Spalla *et al.*, 1989).

tation (Gottschalk and Grupe, 1992). The inoculum development programme described by McNeil and Kristiansen (1986) is given in Table 6.2. The stock culture is heat shocked to stimulate spore germination and to eliminate the weaker spores. The production stage is inoculated with a very low volume and this corresponds with Lurie's (1975) description of the South African acetone–butanol fermentation in which a 100,000 $dm^3$ fermenter was inoculated with only 10 $dm^3$ of seed. The use of such small inocula necessitates the achievement of as near perfect conditions as possible to prevent contamination and to avoid an abnormally long lag phase.

## THE DEVELOPMENT OF INOCULA FOR MYCELIAL PROCESSES

The preparation of inocula for fermentations employing mycelial (filamentous) organisms is more involved than that for unicellular bacterial and yeast processes. The majority of industrially important fungi and streptomycetes are capable of asexual sporulation, so it is common practice to use a spore suspension as seed during an inoculum development programme. A major advantage of a spore inoculum is that is contains far more 'propagules' than a vegetative culture. Three basic techniques have been developed to produce a high concentration of spores for use as an inoculum.

### Sporulation on solidified media

Most fungi and streptomycetes will sporulate on suitable agar media but a large surface area must be employed to produce sufficient spores. Parker (1950) described the 'roll-bottle' technique for the production of spores of *Penicillium chrysogenum*. Quantities of medium (300 $cm^3$) containing 3% agar were sterilized in 1 $dm^3$ cylindrical bottles, which were then cooled to 45° and rotated on a roller mill so that the agar set as a cylindrical shell inside the bottle. The bottles were inoculated with a spore suspension from a sub-master slope and incubated at 24° for 6 to 7 days. Parker claimed that although the use of the 'roll-bottle' involved some sacrifice in ease of visual examination, it provided a large surface area for cultivation of spores in a vessel of a convenient size for handling in the laboratory.

Hockenhull (1980) described the production of $10^{10}$ spores of *Penicillium chrysogenum* on a 300-$cm^2$ agar layer in a Roux bottle and El Sayed (1992) quoted the use of spore suspensions derived from agar media containing between $10^7$ and $10^8$ $cm^{-3}$. Butterworth (1984) described the use of a Roux bottle for the production of a spore inoculum of *Streptomyces clavuligerus* for the production of clavulanic acid. The spores produced from one bottle containing 200-$cm^2$ agar surface could be used to inoculate a 75-$dm^3$ seed fermenter which, in turn, was used to inoculate a 1500-$dm^3$ fermenter. The clavulanic acid inoculum development programme is illustrated in Fig. 6.4. Some representative solidified media for the production of streptomycete and fungal spores are given in Tables 6.3 and 6.4 respectively.

### Sporulation on solid media

Many filamentous organisms will sporulate profusely on the surface of cereal grains from which the spores

TABLE 6.2. *The inoculum development programme for the clostridial acetone–butanol fermentation* (Spivey, 1978)

| Stage | Cultural conditions | Incubation time (hours) |
|---|---|---|
| 1. | Heat-shocked spore suspension inoculated into 150 $cm^3$ of potato glucose medium | 12 |
| 2. | Stage 1 culture used as inoculum for 500 $cm^3$ molasses medium | 6 |
| 3. | Stage 2 culture used as inoculum for 9 $dm^3$ molasses medium | 9 |
| 4. | Stage 3 culture used as inoculum for 90,000 $dm^3$ molasses medium | |

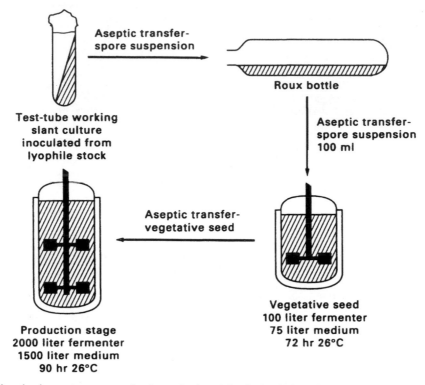

FIG. 6.4. The inoculum development programme for the production of clavulanic acid from *Streptomyces clavuligerus* (Butterworth, 1984).

may be harvested. Substrates such as barley, hard wheat bran, ground maize and rice are all suitable for the sporulation of a wide range of fungi. The sporulation of a given fungus is particularly affected by the amount of water added to the cereal before sterilization and the relative humidity of the atmosphere, which should be as high as possible during sporulation (Vezina and Singh, 1975). Singh *et al.* (1968) have described a system for the sporulation of *Aspergillus ochraces* in which a 2.8-dm$^3$ Fernbach flask containing 200 grams of 'pot' barley or 100 grams of moistened wheat bran produced $5 \times 10^{11}$ conidia after six days at 28° and 98% relative humidity. This was 5 times the number obtainable from a Roux bottle batched with Sabouraud agar and 50 times the number obtainable from such a vessel batched with Difco Nutrient Agar, incubated for the same time period. Vezina *et al.* (1968) have published a list of fungi which are capable of sporulating heavily on cereal grains. El-Sayed (1992) quoted the use of cooked rice for the production of spores of *Penicillium* and *Cephalosporium* in penicillin and cephalosporin inoculum development. Sansing and Cieglem (1973) described the mass production of spores of several *As-*

*pergillus* and *Penicillium* species on whole loaves of white bread and Podojil *et al.* (1984) quoted the use of millet for the sporulation of *Streptomyces aureofaciens* in the development of inoculum for the chlortetracycline fermentation (see Fig. 6.5).

## Sporulation in submerged culture

Many fungi will sporulate in submerged culture provided a suitable medium is employed (Vezina *et al.*, 1965). This technique is more convenient than the use of solid or solidified media because it is easier to operate aseptically and it may be applied on a large scale. The technique was first adopted by Foster *et al.* (1945) who induced submerged sporulation in *Penicillium notatum* by including 2.5% calcium chloride in a defined nitrate–sucrose medium. An example of the use of this technique for the production of inoculum for an industrial fermentation is provided by the griseofulvin process. Rhodes *et al.* (1957) described the conditions necessary for the submerged sporulation of the griseofulvin-producing fungus, *Penicillium patulum*, and

TABLE 6.3. *Solidified media suitable for the sporulation of some representative streptomycetes*

| Organism | Product | Medium | | References |
|---|---|---|---|---|
| S. aureofaciens | Tetracycline* | Malt extract (Difco) | 1.0% | Williams *et al.* |
| | | Yeast extract (Difco) | 0.4% | (1974) |
| | | Glucose | 0.4% | |
| S. erythreus | Erythromycin* | Beef extract (Difco) | 0.1% | Williams *et al.* |
| | | Yeast extract (Difco) | 0.1% | (1974) |
| | | Casamino acids (Difco) | 0.2% | |
| | | Glucose | 0.2% | |
| S. vinaceus | Viomycin* | Corn-steep liquor (50% dry matter) | 1.0% | Williams *et al.* (1974) |
| | | Starch | 1.0% | |
| | | $(NH_4)_2SO_4$ | 0.3% | |
| | | NaCl | 0.3% | |
| | | $CaCO_3$ | 0.3% | |
| S. clavuligerus | Clavulanic* acid | Soluble starch | 1.0% | Butterworth (1984) |
| | | $K_2HPO_4$ | 0.1% | |
| | | $MgSO_4 \cdot 7H_2O$ | 0.1% | |
| | | NaCl | 0.1% | |
| | | $(NH_4)_2SO_4$ | 0.2% | |
| | | $CaCO_3$ | 0.2% | |
| | | $FeSO_4 \cdot 7H_2O$ | 0.001% | |
| | | $MnCl_2 \cdot 4H_2O$ | 0.001% | |
| | | $ZnSO_4 \cdot 7H_2O$ | 0.001% | |
| S. hygroscopicus | Maridomycin | Soluble starch | 1.0% | Miyagawa *et al.* (1979) |
| | | Peptone | 0.04% | |
| | | Meat extract | 0.02% | |
| | | Yeast extract | 0.02% | |
| | | N-Z amine (type A) | 0.02% | |
| | | Agar | 2.0% | |

*Agar content not quoted.

TABLE 6.4. *Solidified media suitable for the sporulation of some representative fungi*

| Fungus | Medium | (g dm$^{-3}$) | References |
|---|---|---|---|
| Penicillium chrysogenum | Glycerol | 7.5 | Booth (1971) |
| | Cane molasses | 7.5 | |
| | Curbay BG | 2.5 | |
| | $MgSO_4 \cdot 7H_2O$ | 0.05 | |
| | $KH_2PO_4$ | 0.06 | |
| | Peptone | 5.0 | |
| | NaCl | 4.0 | |
| | Agar | 20 | |
| Aspergillus niger | Molasses | 300 | Steel *et al.* (1958) |
| | $KH_2PO_4$ | 0.5 | |
| | Agar | 20 | |

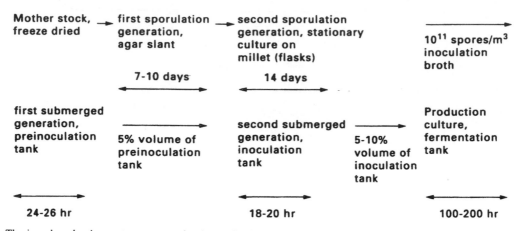

Fig. 6.5. The inoculum development programme for the production of chlortetracycline by *Streptomyces aureofaciens* (Podojil, 1984).

the medium utilized is given in Table 6.5. These authors found that for prolific sporulation the nitrogen level had to be limited to between 0.05 and 0.1% w/v and that good aeration had to be maintained. Also, an interaction was demonstrated between the nitrogen level and aeration in that the lower the degree of aeration the lower the concentration of nitrogen needed to induce sporulation. Submerged sporulation was induced by inoculating 600 cm$^3$ of the above medium, in a 2-dm$^3$ shake flask, with spores from a well-sporulated Czapek-Dox agar culture and incubating at 25° for 7 days. The resulting suspension of spores was then used as a 10% inoculum for a vegetative seed stage in a

stirred fermenter, the seed culture subsequently providing a 10% inoculum for the production fermentation. The vegetative seed and production media are given in Table 6.1.

Most actinomycetes do not sporulate in submerged culture (Whitaker, 1992) and, thus, solid or solidified media tend to be used for the production of spore inocula.

### The use of the spore inoculum

The stage in an inoculum development programme

TABLE 6.5. *Media for the submerged sporulation of selected fungi*

| | | |
|---|---|---|
| Rhodes *et al.* (1957): *Penicillium patulum* | | |
| Whey powder, to give | Lactose 3.5% | |
| | Nitrogen 0.05% | |
| KH$_2$PO$_4$ | | 0.4% |
| KCl | | 0.05% |
| Corn-steep liquor solids to give approx. 0.04% N | | 0.38% |
| | | |
| Foster *et al.* (1945): *Penicillium notatum* | | |
| Sucrose | | 2.0% |
| NaNO$_3$ | | 0.6% |
| KH$_2$PO$_4$ | | 0.15% |
| MgSO$_4 \cdot 7H_2O$ | | 0.05% |
| CaCl$_2$ | | 2.5% |
| | | |
| Vezina *et al.* (1965): *Aspergillus ochraceus* | | |
| Glucose | | 2.5% |
| NaCl | | 2.5% |
| Corn-steep liquor | | 0.5% |
| Molasses | | 5.0% |

at which a large-scale spore inoculum is used varies according to the process; it appears to be common practice that the penultimate stage is so inoculated, but this will depend on the scale of the production fermentation. In the inoculum development programme for the early penicillin fermentation described by Parker (1950) the penultimate stage was inoculated with a spore suspension (from a 'roll-bottle') and this stage may have produced either a vegetative or a submerged spore inoculum for the final fermentation. For the griseofulvin process, Rhodes *et al.* (1957) stated that the spore suspension obtained from the submerged sporulation stage could either be used for direct inoculation of the production fermentation or it could be germinated in an inoculum development medium to yield a vegetative inoculum for the final fermentation. The latter course was preferred and an inoculum volume of 7–10% was used. From Figs 6.4 to 6.6 it can

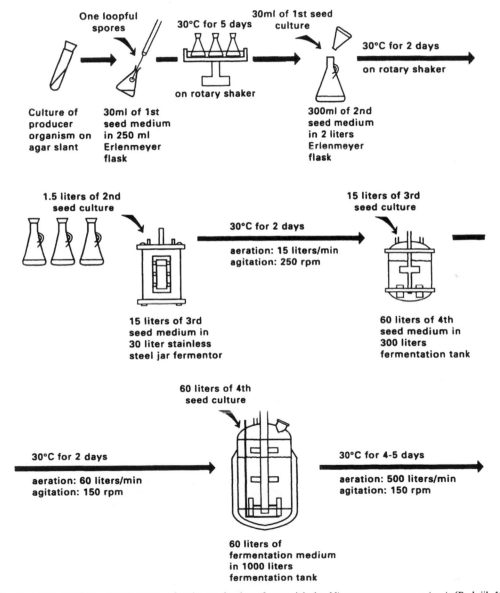

FIG. 6.6. The inoculum development programme for the production of sagamicin by *Micromonospora sagamiensis* (Podojil, 1984).

be seen that in the clavulanic acid process the spore inoculum is used to inoculate the final seed stage, in the chlortetracycline process a vegetative stage is interspersed between the spore inoculated batch and the production fermentation, and in the sagamicin process the spore inoculum is used at a very early stage followed by vegetative growth.

When considering the production of gluconic acid by *Aspergillus niger*, Lockwood (1975) discussed the merits of inoculating the final fermentation directly with a spore suspension as compared with germinating the spores in a seed tank to give a vegetative inoculum. Direct spore inoculation would avoid the cost of installation and operation of the seed tanks whereas the use of germinated spores would reduce the fermentation time of the final stage, thus allowing a greater number of fermentations to be carried out per year. However, labour costs for the production of the vegetative inoculum could be almost as high as for the final fermentation although some of these costs may be recovered, in that gluconic acid produced in the penultimate stage would be recoverable from the final fermentation broth and would contribute to the buffering capacity throughout the fermentation. Thus, Lockwood claimed that the choice of inoculum for the production stage depends on the length of the cycle of the fermentation process, plant size and the availability and cost of labour.

## Inoculum development for vegetative fungi

Some fungi will not produce asexual spores and, therefore, an inoculum of vegetative mycelium must be used. *Gibberella fujikuroi* is such a fungus and is used for the commercial production of gibberellin (Borrow *et al.*, 1961). Hansen (1967) described an inoculum development programme for the gibberellin fermentation. Cultures were grown on long slants ($25 \times 10$ mm test tubes) of potato dextrose agar for 1 week at 24°. Growth from three slants was scraped off and transferred to a 9-dm$^3$ carboy containing 4 dm$^3$ of a liquid medium composed of 2% glucose, 0.3% $MgSO_4 \cdot 7H_2O$, 0.3% $NH_4Cl$ and 0.3% $KH_2PO_4$. The medium was aerated for 75 hours at 28° before transfer to a 100-dm$^3$ seed fermenter containing the same medium.

The major problem in using vegetative mycelium as initial seed is the difficulty of obtaining a uniform, standard inoculum. The procedure may be improved by fragmenting the mycelium in an homogenizer, such as a Waring blender, prior to use as inoculum. This method provides a large number of mycelial particles

and therefore a large number of growing points. Worgan (1968) has given a detailed account of the use of this technique in the preparation of inocula for the submerged culture of the higher fungi.

## The effect of the inoculum on the morphology of filamentous organisms in submerged culture

When filamentous fungi are grown in submerged culture the type of growth varies from the 'pellet' form, consisting of compact discrete masses of hyphae, to the filamentous form in which the hyphae form a homogeneous suspension dispersed through the medium (Whitaker and Long, 1973). The filamentous type of habit gives rise to an extremely viscous broth which may be very difficult to aerate adequately, whereas the pellet type of habit gives rise to a far less viscous, but also less homogeneous, broth (see Chapter 9). In a pelleted culture there is a danger that the mycelium at the centre of the pellet may be starved of nutrients and oxygen due to diffusion limitations. Also, there is considerable evidence that the morphological form of the organism influences the productivity of the culture, but whether this is due to the phenomena already mentioned or to some form of metabolic control is far from clear. Thus, some fermentations are carried out with the fungus in a filamentous habit, whereas others are carried out with the organism growing as pellets. For example, filamentous growth has been claimed to be optimum for penicillin production by *P. chrysogenum* (Smith and Calam, 1980), whereas pelleted growth has been claimed to be optimum for citric-acid production from *Aspergillus niger* (Al Obaidi and Berry, 1980) and lovastatin from *Aspergillus terreus* (Gbewonyo *et al.*, 1992; see also Chapter 9). The necessity for filamentous growth is taken to the extreme in the ICI–Rank Hovis McDougal mycoprotein process where *Fusarium graminearium* is produced for human consumption. A highly filamentous morphology is required to produce the desired texture in the product which resembles the strength and eating texture of white and soft, red meats (Trinci, 1992). Thus, in this process a median hyphal length of 400 $\mu$m is required.

The relevance of this consideration of mycelial morphology to inoculum development is that the morphology may be influenced considerably by both the concentration of spores in a spore inoculum and the inoculum development medium. Usually, a high spore inoculum will tend to produce a dispersed form of growth whilst a low one will favour pellet formation (Foster, 1949). The effect of the concentration of a spore inoculum on the morphology of *P. chrysogenum* is given in Table 6.6. Thus, in the commercial production of fungal products it is critical to grow the organism in the desired morphological form which necessitates the use of an inoculum which achieves this end. If the production fermentation is to be inoculated with a spore suspension then the spore concentration must be such as to produce the production culture in the desired morphological form; if a vegetative inoculum is to be used for the production fermentation then, again, the concentration of its spore inoculum must be such as to produce the vegetative inoculum in the desired morphological form. Although the effects of media on morphological form can be extremely varied dispersed growth is more likely in rich, complex media and pelleted growth tends to occur in chemically defined media (Whitaker and Long, 1973). Thus, the medium used in the spore germination stage must be optimized in terms of the morphology of the inoculum.

An interesting series of experiments on the effects of inoculum conditions on the morphology of *Penicillium citrinum* were reported by Hosobuchi *et al.* (1993). This *Penicillium* species synthesizes compound ML-236B, a precursor of pravastatin which is a cholesterol-lowering drug. Optimum productivity was achieved when the organism grew as compact pellets in the production fermentation. The vegetative inoculum for the production fermentation had to contain an optimum number of short, filamentous propagules in order to initiate pellet formation in the final culture. This was achieved by using a four-stage inoculum development programme (initiated by a spore-inoculated shake flask) with very rich media in the third and fourth cultures. Thus, this system required a dispersed vegetative inoculum to generate a pelleted production fermentation.

The information available on the morphology of actinomycetes in submerged culture is very limited compared with that on fungi. However, Whitaker (1992) has reviewed the area and it is obvious that actinomycetes are capable of producing a wide range of morphological types. Also, it appears to be accepted that a dispersed mycelial morphology is desirable for most industrial actinomycete fermentations. Mycelial forms have been shown to be desirable for the production of streptomycin by *Streptomyces griseus* and turimycin by *S. hygroscopicus*, whereas the pelleted form of *S. nigrificans* was better for glucose isomerase production (Whitaker, 1992). As already discussed for the fungi, the concentration of spores in the inoculum has also been shown to influence the morphology of certain streptomycetes (Lawton *et al.*, 1989). These workers also demonstrated that medium composition and the shear forces operating during culture also affect morphological form. Thus, the principles applied to the optimization of fungal inoculum development regimes are also relevant to actinomycete processes. Hunt and Stieber (1986) described the optimization of the inoculum regime of a small-scale streptomycete cephamycin C fermentation. Pellet formation was observed to be detrimental to product formation and

TABLE 6.6. *The effect of spore concentration and medium on the morphology and penicillin productivity of* Penicillium chrysogenum

| Medium | Spore concentration in the medium | Morphology |
|---|---|---|
| Camici *et al.* (1952): | | |
| Corn-steep dextrin | More than $10 \times 10^5$ dm$^{-3}$ | Filamentous |
| | Less than $10 \times 10^5$ dm$^{-3}$ | Pellets |
| Czapek-dox | More than $3.0 \times 10^5$ dm$^{-3}$ | Filamentous |
| | Less than $3.0 \times 10^5$ dm$^{-3}$ | Pellets |
| Glucose, lactose and ammonium lactate | More than $2.0 \times 10^5$ dm$^{-3}$ | Filamentous |
| | Less than $2.0 \times 10^5$ dm$^{-3}$ | Pellets |

| Spore concentration in the inoculum (cm$^{-3}$) | Penicillin yield (units cm$^{-3}$) | Morphology |
|---|---|---|
| Calam (1976): | | |
| $10^2$ | 500 | Dense pellets |
| $10^3$ | 1800 | Dense pellets |
| $2 \times 10^3$ | 4000 | Open pellets |
| $10^4$ | 5000 | Filamentous |

the key factor in establishing the correct form appeared to be the concentration of iron in the seed medium, a higher iron concentration giving the optimum inoculum.

## THE ASEPTIC INOCULATION OF PLANT FERMENTERS

The inoculation of plant-scale fermenters may involve the transfer of culture from a laboratory fermenter, or spore suspension vessel, to a plant fermenter, or the transfer from one plant fermenter to another. Obviously, it is extremely important that the fermentation is not contaminated during inoculation but if the process is a contained one then it is equally important that the process organism does not escape. Thus, the nature of the inoculation system will be dictated by the containment category of the process (see Chapter 7).

At Containment levels 1 and B2, the addition of inoculum must be carried out in such a way that release of micro-organisms is restricted. This should be done by aseptic piercing of membranes or connections with steam locks. At Containment level 2 and B3/4, no micro-organisms must be released during inoculation or other additions. In order to meet these stringent requirements, all connections must be screwed or clamped and all pipelines must be steam sterilizable (Werner, 1992).

### Inoculation from a laboratory fermenter or a spore suspension vessel

Several systems have been described in the literature which are suitable for inoculating fermentations requiring only Good Industrial Large Scale Practice (GILSP, see Chapter 7). To prevent contamination during the transfer process it is essential that both vessels be maintained under a positive pressure and the inoculation port be equipped with a steam supply. Meyrath and Suchanek (1972) described a system for the inoculation of a plant fermenter from a laboratory vessel. The apparatus is shown in Fig. 6.7. The connecting point A is normally covered with the blank plug a and prior to inoculation this plug is slightly loosened, valve E closed and valve F opened to allow steam to exit at A. Valve F is then closed and E opened so that when a is removed sterile air will be released from the vessel. After removal, plug a is placed in strong disinfectant. Blank plug b is then removed and a coupling made

between B and A. Valve E is closed and an air line attached to point C, establishing a pressure inside the inoculum fermenter greater than in the plant vessel. Valve E is then opened and the inoculum will be forced into the plant fermenter. After closing valve E the inoculum fermenter may be removed, plug a replaced, and the line steamed out by opening valve F. This system may be modified by using the quick connection devices which are now available (see Chapter 7).

Jackson (1958) has described a very similar system for the introduction of a spore suspension into a plant fermenter. The apparatus is shown in Fig. 6.8 and its operation is identical to that described by Meyrath and Suchanek.

Parker (1958) described a more complex system for inoculation of a plant fermenter from a spore-suspension vessel. The apparatus is shown in Fig. 6.9. The sterile spore-suspension vessel is batched with the spore suspension in the sterile room. The plant vessel, containing the medium and with blank plugs screwed on at A and B, is sterilized by steam injection. The plugs A and B are slightly loosened to allow steam to emit for 20 minutes when the whole system is under steam pressure. The blanks are then tightened, the valves E and G shut and sterile air is allowed into the plant fermenter by opening valves D, F and C. The spore suspension vessel is loosely connected at A and B and valves D, H, I and C closed and E, F and G opened. After 20 minutes steaming A and B are tightened and G and E are closed. D is then opened to establish a positive pressure in the pipework. When the pipework has cooled, the pressure in the plant vessel is reduced to approximately 5 psi, valve F is closed and H, I and C are opened. This procedure allows the spore suspension to be forced into the fermenter. Valves D, H, I

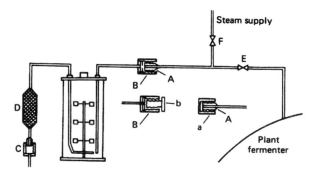

FIG. 6.7. Inoculation of a plant fermenter from a laboratory fermenter (Meyrath and Suchanek, 1972).

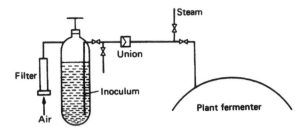

FIG. 6.8. Inoculation of a plant fermenter from a spore suspension vessel (Jackson, 1958).

and C are closed and the suspension vessel replaced by blank plugs at A and B and the pipework steamed by opening valves E and G.

GILSP and category 1 fermentations may also be inoculated by aseptic piercing of a membrane. In this system the inoculum vessel is connected to an inoculating needle assembly (as shown in Fig. 6.10) and the sterile needle pierces a rubber septum set into a fermenter port. However, the use of a needle presents safety problems and many companies prohibit such systems. Also, aerosols may be created on the removal of a needle.

The inoculation of level 2 or B3/4 contained fermentations requires the use of modified systems. It must be possible to steam sterilize all the pipework

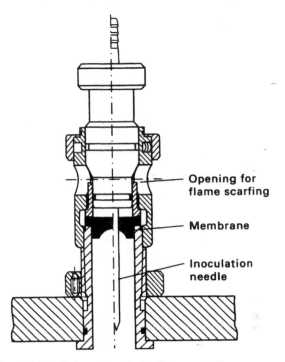

FIG. 6.10. Needle inoculation device (Werner, 1992).

after the inoculation and the condensate from the sterilization must be collected in a kill tank (see Chapters 5 and 7). The inoculation flask is then removed after inoculation and sterilized in an autoclave. One such system is shown in Fig. 6.11 (Vranch, 1990).

### Inoculation from a plant fermenter

Figure 6.12 illustrates the system described by Parker (1958). The two vessels are connected by a flexible pipeline A–B. The batched fermenter is sterilized by steam injection via valves G and J, valves D, I, A, B, H, E and F being open and valve C closed. Valves H and I lead to steam traps for the removal of condensate. After 20 minutes at the desired pressure the steam supply is switched off at J and G and the steam-trap valves I and H are closed. F, E and D are left open so that the connecting pipeline fills with sterile medium. The medium in the fermenter is sparged with sterile air and when it has cooled to the desired temperature the pressure in the seed tank is increased to at least 10 psi whilst the pressure in the fermenter is reduced to about 2 psi. Valve C is opened and the inoculum is forced into the production vessel. After inoculation is

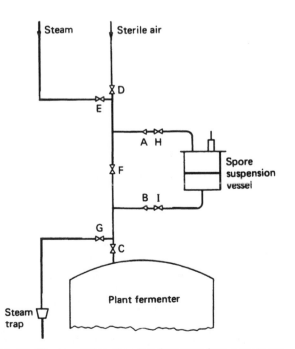

FIG. 6.9. The inoculation of a plant fermenter from a spore suspension vessel (Parker, 1958).

Principles of Fermentation Technology, 2nd Edn.

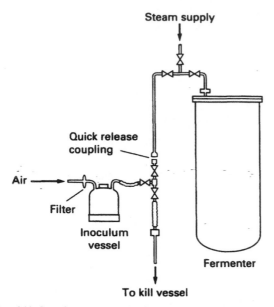

FIG. 6.11. Inoculum system suitable for contained fermentations (Vranch, 1990).

complete the connecting pipeline is resterilized before it is removed. For a contained fermentation the condensate from the two steam traps attached to valves I and H would be directed to a kill tank.

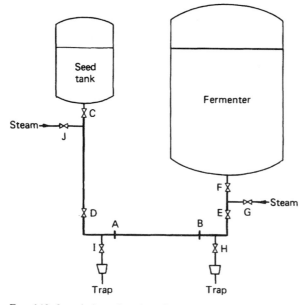

FIG. 6.12. Inoculation of a plant fermenter from another plant fermenter (Parker, 1958).

ALFORD, J. S., FOWLER, G. L., HIGGS, R. E., CLAPP, D. L. and HUBER, F.M. (1992) Development of real-time expert system applications for the on-line analysis of fermentation data. In *Harnessing Biotechnology for the 21st Century*, pp. 375–379 (Editor Ladisch, M. R. and Bose, A). American Chemical Society, Washington, D.C.

AL OBAIDI, Z. S and BERRY, D. R. (1980) cAMP concentration, morphological differentiation and citric acid production in *Aspergillus niger. Biotechnol. Lett.* **2** (1), 5–10.

AUNSTRUP, K. (1974) Industrial production of proteolytic enzymes. In *Industrial Aspects of Biochemistry,* Vol. 30, Part 1, pp. 23–46 (Editor Spencer, B.). North Holland, Amsterdam.

BEECH, S. C. (1952) Acetone–butanol fermentation of sugars. *Ind. Eng. Chem.* **44**, 1677–1682.

BOOTH, C. (1971) Fungal culture media. In *Methods in Microbiology,* Vol. 4, pp. 49–94 (Editor Booth, C.). Academic Press, London.

BORROW, A., JEFFERYS, E. G., KESSELL, R. H. J., LLOYD, E. C., LLOYD, P. B. and NIXON, I. S. (1961) Metabolism of *Gibberella fujikuroi* in stirred culture. *Can. J. Microbiol.* **7**, 227–276.

BOULTON, C. A. (1991) Developments in brewery fermentation. *Biotech. Gen. Eng. Rev.* **9**, 127–182.

BOULTON, C. A., MARYAN, P. S., LOVERIDGE, D. and KELL, D. B. (1989) The application of a novel biomass sensor to the control of yeast pitching rate. *Proc. Eur. Brewing Convention Congress, Zurich,* pp. 653–661.

BOULTON, C. A., JONES, A. R. and HINCHLIFFE, E. (1991) Yeast physiology and fermentation performance. *Proc. Eur. Brewing Convention Congress, Lisbon.*

BUTTERWORTH, D. (1984) Clavulanic acid: properties, biosynthesis and fermentation. In *Biotechnology of Industrial Antibiotics,* pp. 225–236 (Editor, Vandamme, E. J.). Marcel Dekker, New York.

CALAM, C. T. (1976) Starting investigational and production cultures. *Process Biochem.* **11** (3), 7–12.

CAMICI, L., SERMONTI, G. and CHAIN, E. B. (1952) Observations on *Penicillium chrysogenum* in submerged culture. Mycelial growth and autolysis. *Bull. World Health Org.* **6**, 265–272.

CONNER, H. A. and ALLGEIER, R. J. (1976) Vinegar: its history and development. *Adv. Appl. Microbiol.* **20**, 82–127.

CURTIS, N. S. and CLARK, A. G. (1957) Experiments on growing culture yeast for the brewery. Summary of a paper read at European Brewing Conference (Copenhagen 1957). *Brewer's Guardian* **86** (7), 27–28.

EL SAYED A-H, M. M. (1992) Production of penicillins and cephalosporins by fungi. In *Handbook of Applied Mycology,* Vol. 4: *Fungal Biotechnology,* pp. 517–564 (Editors Arora, D. K., Elander, R. P. and Mukerji, K. G.). Marcel Dekker, New York.

ETTLER, P. (1992) The determination of optimum inoculum

quality for submersed fermentation process. *Collect. Czech. Chem. Commun.* **57**, 303–308.

FOSTER, J. W. (1949) *Chemical Activities of Fungi*, p. 62. Academic Press, New York.

FOSTER, J. W. MCDANIEL, L. E., WOODRUFF, H. B. and STOKES, J. L. (1945) Microbiological aspects of penicillin. Production of conidia in submerged culture of *Penicillium notatum. J. Bacteriol* **50**, 365–381.

GBEWONYO, K., HUNT, G. and BUCKLAND, B. (1992) Interactions of cell morphology and transport in the lovastatin fermentation. *Bioprocess Eng.* **8** (1–2), 1–7.

GILLILAND, R. B. (1971) Classification and selection of yeast strains. In *Modern Brewing Technology*, pp. 108–128 (Editor Findlay, W.P.G.). Macmillan, London.

GOTTSCHALK, G. AND GRUPE, H. (1992) Physiological improvements in acetone–butanol fermentation. In *Harnessing Biotechnology for the 21st Century*, pp. 102–105 (Editors Ladisch, M. R. and Bose, A.) American Chemical Society, Washington D.C.

HANSEN, A. M. (1967) Microbial production of pigments and vitamins. In *Microbial Technology*, pp. 222–250 (Editor Peppler, H. J.). Reinhold, New York.

HANSEN, E. C. (1896) Pure culture of systematically selected yeasts in the fermentation industries. In *Practical Studies in Fermentation*, Chapter 1, pp. 1–76 (Translated by A. K. Miller). Spon, London.

HOCKENHULL, D. J. (1980) Inoculum development with particular reference to *Aspergillus* and *Penicillium*. In *Fungal Biotechnology*, pp. 1–24 (Editors Smith, J. E., Berry, D. R. and Christiansen, B.). Academic Press, London.

HOSOBUCHI, M., FUKUI, F., MATSUKAWA, H., SUZUKI, T. and YOSHIKAWA, H. (1993) Morphology control of preculture during production of ML-236B, a precursor of pravastatin sodium, by *Penicillium citrinum. J. Ferm. Bioeng.* **76** (6), 476–481.

HUNT, G. R. and STIEBER, R. W. (1986) Inoculum development. In *Manual of Industrial Microbiology and Biotechnology*, pp. 32–40 (Editors Demain, A. L. and Solomon, N. A.). American Society of Microbiology, Washington, D.C.

JACKSON, T. (1958) Development of aerobic fermentation processes: penicillin. In *Biochemical Engineering*, pp. 183–222 (Editor Steel, R.). Heywood, London.

KEAY, L., MOSELEY, M. H., ANDERSON, R. G., O'CONNOR, R. J. and WILDI, B. S. (1972) Production and isolation of microbial proteases. *Biotech. Bioeng. Symp.* **3**, 63–92.

LAWTON, P., WHITAKER, A., ODELL, D. E. and STOWELL, J. D. (1989) Actinomycete morphology in shaken culture. *Can. J. Microbiol.* **35**, 881–889.

LEAVER, G. (1990) *Integrity of Fermentation Equipment — Containment and Sterility*. State of the Art Report 4. Warren Spring Laboratory, Department of Trade and Industry, Stevenage.

LINCOLN, R. E. (1960) Control of stock culture preservation and inoculum build-up in bacterial fermentations. *J. Biochem. Microbiol. Tech. Eng.* **2**, 481–500.

LOCKWOOD, L. B. (1975) Organic acid production. In *The*

*Filamentous Fungi*, Vol. 1, pp. 140–157. (Editors Smith, J. E. and Berry, D. P.). Arnold, London.

LURIE, J. (1975) In *Large Scale Fermentations for Organic Solvents. Octagon Papers*, Vol. 2, pp. 70–73 (Editors Powell, A. J. and Bu'Lock, J. D.). Dept of Extra Mural Studies, The University, Manchester.

MCNEIL, B. and KRISTIANSEN, K. (1986) The acetone butanol fermentation. *Adv. Appl. Micro.* **31**, 61–92.

MANDL, B., GRUNEWALD, J. and VOERKELIUS, G. A. (1964) The application of time-saving brewing methods. *Brauwelt* **104** (80), 1541–1543.

MEYRATH, J. and SUCHANEK, G. (1972) Inoculation techniques — effects due to quality and quantity of inoculum. In *Methods in Microbiology*, Vol. 7B, pp. 159–209 (Editors Norris, J. R. and Ribbons, D. W.). Academic Press, London.

MIYAGAWA, K., SUZUKI, M., HIGASHIDE, E. and UCHIDA, M. (1979) Effect of aspartate family amino acids on production of maridomycin. III. *Agric. Biol. Chem.* **43**, 1111–1116.

NAKAYAMA, K. (1972) Lysine and diaminopimelic acid. In *The Microbial Production of Amino Acids*, pp. 369–398 (Editors Yamada, K., Kinoshita, S., Tsunoda, T. and Aida, K.). Halsted Press, New York.

PARKER, A. (1958) Sterilization of equipment, air and media. In: *Biochemical Engineering—Unit Processes in Fermentation*, pp. 94–121 (Editor Steel, R.) Heywood, London.

PARTON, C. and WILLIS, P. (1990) Strain preservation, inoculum preparation and inoculum development. In *Fermentation, a Practical Approach*, pp. 39–64 (Editors McNeil, B. and Harvey, L. M.). IRL Press, Oxford.

PICKERELL, A. T. W., HWANG, A. and AXCELL, B. C. (1991) Impact of yeast-handling procedures on beer flavor development during fermentation. *J. Am. Soc. Brew. Chem.* **49**, 87–92.

PODOJIL, M., BLUMAUEROVA, M., CULIK, K. and VANEK, Z. (1984) The tetracyclines: Properties, biosynthesis and fermentation. In *Biotechnology of Industrial Antibiotics*, pp. 259–280 (Editor Vandamme, E. J.). Marcel Dekker, New York.

PRIEST, F. G. and SHARP, R. J. (1989) Fermentation of bacilli. In *Fermentation Process Development of Industrial Organism*, pp. 73–132 (Editor Neway, J. O.). Marcel Dekker, New York.

REED, G. and NAGODAWITHANA, T. W. (1991a) In *Yeast Technology*, pp. 112–116. Van Nostrand Reinhold, New York.

REED, G. and NAGODAWITHANA, T. W. (1991b) In *Yeast Technology*, pp. 288–290. Van Nostrand Reinhold, New York.

RHODES, A., CROSSE, R., FERGUSON, T. P. and FLETCHER, D. L. (1957) Improvements in or relating to the production of the antibiotic griseofulvin. British Patent 784, 618.

RIGHELATO, R. C. (1976) Selection of strains of *Penicillium chrysogenum* with reduced yeild in continuous culture. *J. Appl. Chem. Biotechnol.* **26** (30), 153–159.

ROESSLER, J. G. (1968) Yeast management in the brewery, Part II. Yeast handling and treatment. *Brew. Digest* **43** (9), 94, 96, 98, 102, 115.

SABATIE, J., SPECK, D., REYMUND, J., HEBERT, C., CAUSSIN, L.,

WELTIN, D., GLOECKLER, R., O'REGAN, M., BERNARD, S., LEDOUX, C., OHSAWA, I., KAMOGAWA, K., LEMOINE, Y. and BROWN, S. W. (1991) Biotin formation by recombinant strains of *Escherichia coli*: influence of the host physiology. *J. Biotechnol* **20**, 29–50.

SANSING, G. A. and CIEGLEM, A. (1973) Mass propagation of conidia from several *Aspergillus* and *Penicillium* species. *Appl. Microbiol.* **26**, 830–831.

SINGH, K., SEHGAL, S. N. and VEZINA, C. (1968) Large scale transformation of steroids by fungal spores. *Appl. Microbiol.* **16**, 393–400.

SMITH, G. M. and CALAM, C. T. (1980) Variations in inocula and their influence on the productivity of antibiotic fermentations. *Biotechnol. Lett.* **2** (6), 261–266.

SPALLA, C., GREIN, A., GARAFANO, L. and FERNI, G. (1989) Microbial production of vitamin $B_{12}$. In *Biotechnology of Vitamins, Pigments and Growth Factors*, pp. 257–284 (Editor, Vandamme, E. J.). Elsevier, New York.

SPIVEY, M. J. (1978) The acetone/butanol/ethanol fermentation. *Process Biochem.* **13** (11), 2–3.

STEEL, R., LENZ, C. and MARTIN, S. M. (1954) A standard inoculum for citric acid production in submerged culture. *Can. J. Microbiol.* **1**, 150–157.

STRANDSKOV, F. B. (1964) Yeast handling in a brewery. *Am. Soc. Brewing Chemists Proc.*, pp. 76–79.

THORNE, R. S. W. (1970) Pure yeast cultures in brewing. *Process Biochem.* **5** (4), 15–22.

TRINCI, A. P. J. (1992) Mycoprotein: A twenty year overnight success story. *Mycol. Res.* **96** (1), 1–13.

UNDERKOFLER, L. A. (1976) Microbial enzymes. In *Industrial Microbiology*, pp. 128–164 (Editors Miller, B. M. and Litsky, W.). McGraw-Hill, New York.

VEZINA, C. and SINGH, K. (1975) Transformation of organic compounds by fungal spores. In *The Filamentous Fungi*, Vol. 1, pp. 158–192 (Editors Smith, J. E. and Berry, D. R.). Arnold, London.

VEZINA, C., SINGH, K. and SEHGAL, S. N. (1965) Sporulation of filamentous fungi in submerged culture. *Mycologia* **57**, 722–736.

VEZINA, C., SEHGAL, S. N. and SINGH, K. (1968) Transformation of organic compounds by fungal spores. *Adv. Appl. Microbiol.* **10**, 221–268.

VRANCH, S. P. (1990) Containment and regulations for safe biotechnology. In *Bioprocessing Safety: Worker and Community Safety and Health Considerations*, pp. 39–57 (Editor Hyer, W. C.). American Society for Testing and Materials, Philadelphia, PA.

WERNER, R. G. (1992) Containment in the development and manufacture of recombinant DNA-derived products. In *Safety in Microbiology and Biotechnology*, pp. 190–213 (Editors Collins, C. H. and Beale, A. J.). Butterworth-Heinemann, Oxford.

WHITAKER, A. (1992) Actinomycetes in submerged culture. *Appl. Biochem. Biotechnol.* **32**, 23–35.

WHITAKER, A. and LONG, P. A. (1973) Fungal pelleting. *Process Biochem.* **8** (11), 27–31.

WILLIAMS, S. T., ENTWHISTLE, S. and KURYLOWICZ, W. (1974) The morphology of streptomycetes growing in media used for commercial production of antibiotics. *Microbios.* **11**, 47–60.

WORGAN, J. T. (1968) Culture of the higher fungi. *Prog. Ind. Micro.* **8**, 73–139.

YAMOMOTO, K., ISHIZAKI, A. and STANBURY, P. F. (1993) Reduction in the length of the lag phase of L-lactate fermentation by the use of inocula from electrodialysis seed cultures. *J. Ferm. Bioeng.* **76** (2), 151–152.

# CHAPTER 7

# Design of a Fermenter

## INTRODUCTION

A RESEARCH team led by Chaim Weizmann in Great Britain during the First World War (1914–1918) developed a process for the production of acetone by a deep liquid fermentation using *Clostridium acetobutylicum* which led to the eventual use of the first truly large-scale aseptic fermentation vessels (Hastings, 1978). Contamination, particularly with bacteriophages, was often a serious problem, especially during the early part of a large-scale production stage. Initially, no suitable vessels were available and attempts with alcohol fermenters fitted with lids were not satisfactory as steam sterilization could not be achieved at atmospheric pressure. Large mild-steel cylindrical vessels with hemispherical tops and bottoms were constructed that could be sterilized with steam under pressure. Since the problems of aseptic additions of media or inocula had been recognized, steps were taken to design and construct piping, joints and valves in which sterile conditions could be achieved and maintained when required. Although the smaller seed vessels were stirred mechanically, the large production vessels were not, and the large volumes of gas produced during the fermentation continually agitated the vessel contents. Thus, considerable expertise was built up in the construction and operation of this aseptic anaerobic process for production of acetone–butanol.

The first true large-scale aerobic fermenters were used in Central Europe in the 1930s for the production of compressed yeast (de Becze and Liebmann, 1944). The fermenters consisted of large cylindrical tanks with air introduced at the base via networks of perforated pipes. In later modifications, mechanical impellers were used to increase the rate of mixing and to break up and disperse the air bubbles. This procedure led to the compressed-air requirements being reduced by a factor of 5. Baffles on the walls of the vessels prevented a vortex forming in the liquid. Even at this time it was recognized that the cost of energy necessary to compress air could be 10 to 20% of the total production cost. As early as 1932, Strauch and Schmidt patented a system in which the aeration tubes were provided with water and steam for cleaning and sterilizing.

Prior to 1940, the other important fermentation products besides bakers' yeast were ethanol, glycerol, acetic acid, citric acid, other organic acids, enzymes and sorbose (Johnson, 1971). These processes used highly selective environments such as acidic or anaerobic conditions or the use of an unusual substrate, resulting in contamination being a relatively minor problem compared with the acetone fermentation or the subsequent aerobic antibiotic fermentations.

The decision to use submerged culture techniques for penicillin production, where aseptic conditions, good aeration and agitation were essential, was a very important factor in forcing the development of carefully designed and purpose-built fermentation vessels. In 1943, when the British government decided that surface culture production was inadequate, none of the fermentation plants were immediately suitable for deep fermentation, although the Distillers Company solvent plant at Bromborough only needed aeration equipment to make it suitable for penicillin production (Hastings, 1971). Construction work on the first large-scale plant to produce penicillin by deep fermentation was started on 15th September 1943, at Terre Haute in the United States of America, building steel fermenters with working volumes of 54,000 $dm^3$ (Callahan, 1944). The plant was operational on 30th January 1944. Unfortunately, no other construction details were quoted for the fermenters.

Initial agitation studies in baffled stirred tanks to identify variables were also reported at this time by

Cooper *et al.* (1944), Foust *et al.* (1944) and Miller and Rushton (1944). Cooper's work had a major influence on the design of later fermenters.

## BASIC FUNCTIONS OF A FERMENTER FOR MICROBIAL OR ANIMAL CELL CULTURE

The main function of a fermenter is to provide a controlled environment for the growth of micro-organisms or animal cells, to obtain a desired product. In designing and constructing a fermenter a number of points must be considered:

1. The vessel should be capable of being operated aseptically for a number of days and should be reliable in long-term operation and meet the requirements of containment regulations.
2. Adequate aeration and agitation should be provided to meet the metabolic requirements of the micro-organism. However, the mixing should not cause damage to the organism.
3. Power consumption should be as low as possible.
4. A system of temperature control should be provided.
5. A system of pH control should be provided.
6. Sampling facilities should be provided.
7. Evaporation losses from the fermenter should not be excessive.
8. The vessel should be designed to require the minimal use of labour in operation, harvesting, cleaning and maintenance.
9. Ideally the vessel should be suitable for a range of processes, but this may be restricted because of containment regulations.
10. The vessel should be constructed to ensure smooth internal surfaces, using welds instead of flange joints whenever possible.
11. The vessel should be of similar geometry to both smaller and larger vessels in the pilot plant or plant to facilitate scale-up.
12. The cheapest materials which enable satisfactory results to be achieved should be used.
13. There should be adequate service provisions for individual plants (see Table 7.1).

The first two points are probably the most critical. It is obvious from the above points that the design of a fermenter will involve co-operation between experts in microbiology, biochemistry, chemical engineering, mechanical engineering and costing. Although many dif-

TABLE 7.1. *Service provisions for a fermentation plant*

Compressed air
Sterile compressed air (at 1.5 to 3.0 atmospheres)
Chilled water (12 to 15°)
Cold water (4°)
Hot water
Steam (high pressure)
Steam condensate
Electricity
Stand-by generator
Drainage of effluents
Motors
Storage facilities for media components
Control and monitoring equipment for fermenters
Maintenance facilities
Extraction and recovery equipment
Accessibility for delivery of materials
Appropriate containment facilities

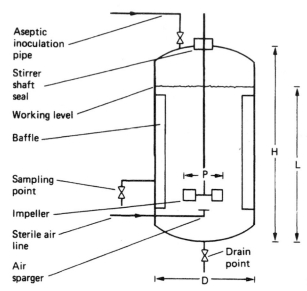

FIG. 7.1. Diagram of a fermenter with one multi-bladed impeller.

ferent types of fermenter have been described in the literature, very few have proved to be satisfactory for industrial aerobic fermentations. The most commonly used ones are based on a stirred upright cylinder with sparger aeration. This type of vessel can be produced in a range of sizes from one $dm^3$ to thousands of $dm^3$. Figures 7.1 and 7.2 are diagrams of typical mechanically agitated and aerated fermenters with one and three multi-bladed impellers respectively. Tables 7.2 and 7.3 give geometrical ratios of various of the dimensions which have been quoted in the literature for a variety of sizes of vessel.

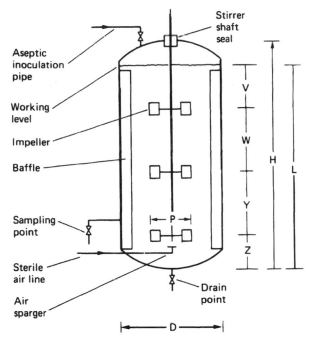

FIG. 7.2. Diagram of a fermenter with three multi-bladed impellers.

At this stage the discussion will be concerned with stirred, aerated vessels for microbial cell culture. More varied shapes are commonly used for alcohol, biomass production, animal cell culture and effluent treatment and will be dealt with later in this chapter.

## ASEPTIC OPERATION AND CONTAINMENT

Aseptic operation involves protection against contamination and it is a well established and understood concept in the fermentation industries, whereas containment involves prevention of escape of viable cells from a fermenter or downstream equipment and is much more recent in origin. Containment guidelines were initiated during the 1970s (East *et al.*, 1984; Flickinger and Sansone, 1984).

To establish the appropriate degree of containment which will be necessary to grow a micro-organism, it, and in fact the entire process, must be carefully assessed for potential hazards that could occur should there be accidental release. Different assessment procedures are used depending on whether or not the organism contains foreign DNA (genetically engineered). Once the hazards are assessed, an organism can be classified into a hazard group for which there is an appropriate level of containment. The procedure which has been adopted within the European Community is outlined in Fig. 7.3. Non-genetically engineered organisms may be placed into a hazard group (1 to 4) using criteria to assess risk such as those given by Collins (1992):

1. The known pathogenicity of the micro-organism.
2. The virulence or level of pathogenicity of the micro-organism — are the diseases it causes mild or serious?
3. The number of organisms required to initiate an infection.
4. The routes of infection.
5. The known incidence of infection in the community and the existence locally of vectors and potential reserves.
6. The amounts or volumes of organisms used in the fermentation process.
7. The techniques or processes used.
8. Ease of prophylaxis and treatment.

A detailed discussion of assessment of risk and ways to reduce it is given by Winkler and Park (1992).

TABLE 7.2. *Details of geometrical ratios of fermenters with single multi-blade impellers (see Fig. 7.1)*

| Dimension | Steel and Maxon (1961) | Wegrich and Shurter (1963) | Blakeborough (1967) |
|---|---|---|---|
| Operating volume | 250 dm$^3$ | 12 dm$^3$ | — |
| Liquid height (L) | 55 cm | 27 cm | — |
| L/D (tank diameter) | 0.72 | 1.1 | 1.0–1.5 |
| Impeller diameter (P/D) | 0.4 | 0.5 | 0.33 |
| Baffle width/D | 0.10 | 0.08 | 0.08–0.10 |
| Impeller height/D | — | — | 0.33 |

TABLE 7.3. *Details of geometrical ratios of fermenters with three multi-bladed impellers*
*(see Fig. 7.2)*

| Dimension | Jackson (1958) | Aiba et al. (1973) | Paca et al. (1976) |
|---|---|---|---|
| Operating volume | — | 100,000 dm$^3$ (total) | 170 dm$^3$ |
| Liquid height (L) | — | — | 150 cm |
| L/D (tank diameter) | — | — | 1.7 |
| Impeller diameter | 0.34–0.5 | 0.4 | 0.33 |
| Baffle width/D | 0.08–0.1 | 0.095 | 0.098 |
| Impeller height/D | 0.5 | 0.24 | 0.37 |
| P/V | 0.5–1.0 | — | 0.74 |
| P/W | 0.5–1.0 | 0.85 | 0.77 |
| P/Y | 0.5–1.0 | 0.85 | 0.77 |
| P/Z | — | 2.1 | 0.91 |
| H/D | 1.0–1.6 | 2.2 | 2.95 |

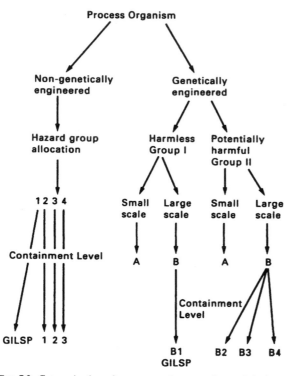

FIG. 7.3. Categorization of a process micro-organism and designation of its appropriate level of containment at research or industrial sites within the European Federation of Biotechnology (GILSP = Good Industrial Large Scale Practice).

Once the organism has been allocated to a hazard group, the appropriate containment requirements can be applied (see Table 7.4). Hazard group 1 organisms used on a large scale only require Good Industrial Large Scale Practice (GILSP). Processes in this cate-gory need to be operated aseptically but no contain-ment steps are necessary, including prevention of es-cape of organisms. If the organism is placed in Hazard group 4 the stringent requirements of level 3 will have to be met before the process can be operated. Details of hazard categories for a range of organisms can be obtained from Frommer *et al.* (1989).

Genetically engineered organisms are classified as either harmless (Group I) or potentially harmful (Group II). The process is then classified as either small scale (A: less than 10 dm$^3$) or large scale (B: more than 10 dm$^3$) according to guidelines which can be found in the Health and Safety Executive document (1993). There-fore large scale processes fall into two categories, IB or IIB. IB processes require containment level B1 and are subject to GILSP, whereas IIB processes are further assessed to determine the most suitable containment level, ranging from B2 to B4 as outlined in Table 7.4. Levels B2 to B4 correspond to levels 1 to 3 for non-genetically engineered organisms.

In future it is possible, under new legislation, that no distinction will be made between non-genetically engi-neered and genetically engineered organisms. The key factor will be whether the organism is harmless or potentially harmful, regardless of its genetic constitu-tion. Containment would then be decided using the scheme which is currently being used for genetically engineered organisms.

Other hazard-assessment systems for classifying or-ganisms have been introduced in many other countries. Production and research workers must abide by ap-propriate local official hazard lists. Problems can occur when different official bodies place the same organism in different hazard categories. In 1989, the European

TABLE 7.4. *Summary of safety precautions for biotechnological operations in the European Federation for Biotechnology (EFB)* (Frommer *et al.*, 1989)

| Procedures | GILSP* | Containment category | | |
|---|---|---|---|---|
| | | 1 | 2 | 3 |
| Written instructions and code of practice | + | + | + | + |
| Biosafety manual | − | + | + | + |
| Good occupational hygiene | + | + | + | + |
| Good Microbiological Techniques (GMT) | − | + | + | + |
| Biohazard sign | − | + | + | + |
| Restricted access | − | + | + | + |
| Accident reporting | + | + | + | + |
| Medical surveillance | − | + | + | + |
| *Primary containment:* | | | | |
| *Operation and equipment* | | | | |
| Work with viable micro-organisms should take place in closed systems (CS), which minimize (m) or prevent (p) the release of cultivated micro-organisms | − | m | p | p |
| Treatment of exhaust air or gas from CS | − | m | p | p |
| Sampling from CS | − | m | p | p |
| Addition of materials to CS, transfer of cultivated cells | − | m | p | p |
| Removal of material, products and effluents from CS | − | m | p | p |
| Penetration of CS by agitator shaft and measuring devices | − | m | p | p |
| Foam-out control | − | m | p | p |
| *Secondary containment:* | | | | |
| *Facilities* | | | | |
| Protective clothing appropriate to the risk category | + | + | + | + |
| Changing/washing facility | + | + | + | + |
| Disinfection facility | − | + | + | + |
| Emergency shower facility | − | − | + | + |
| Airlock and compulsory shower facilities | − | − | − | + |
| Effluents decontaminated | − | − | + | + |
| Controlled negative pressure | − | − | − | + |
| HEPA filters in air ducts | − | − | + | + |
| Tank for spilled fluids | − | − | + | + |
| Area hermetically sealable | − | − | − | + |

*Unless required for product quality, −, not required; +, required.
m, minimize release. The level of contamination of air, working surface and personnel shall not exceed the level found during microbiological work applying Good Microbiological Techniques.
p, prevent release. No detectable contamination during work should be found in the air, working surfaces and personnel.

Federation for Biotechnology were aware of this problem with non-recombinant micro-organisms and produced a consensus list (Frommer *et al.*, 1989).

Most micro-organisms used in industrial processes are in the lowest hazard group which only require GILSP, although some organisms used in bacterial and viral vaccine production and other processes are categorized in higher groups. There is an obvious incentive for industry to use an organism which poses a low risk as this minimizes regulatory restrictions and reduces the need for expensive equipment and associated containment facilities (Schofield, 1992). The costs of some containment design features are discussed in Chapter 12.

In this chapter, where appropriate, a distinction will be made between equipment which is used to maintain aseptic conditions for GILSP and those modifications which are needed to meet physical containment requirements in order to grow specific cultures which have been classified in higher hazard groups. Details on design for containment have been given by East *et al.* (1984), Flickinger and Sansone (1984), Walker *et al.* (1987), Turner (1989), Hambleton *et al.* (1991), Leaver and Hambleton (1992), Vranch (1992) and Werner (1992).

### Overall containment categorization

In this chapter the emphasis will be on containment levels which might be obtained from particular designs of fermenters and associated equipment. However this is only one aspect of assessment. To meet the standards of the specific level of containment it will also be necessary to consider the procedures to be used, staff training, the facilities in the laboratory and factory, downstream processing, effluent treatment, work practice, maintenance, etc. It will be necessary to ensure that **all** these aspects are of a sufficiently high standard to meet the levels of containment deemed necessary for a particular process by a government regulatory body. If these are met, then the process can be operated.

### BODY CONSTRUCTION

#### Construction materials

In fermentations with strict aseptic requirements it is important to select materials that can withstand repeated steam sterilization cycles. On a small scale (1 to 30 dm³) it is possible to use glass and/or stainless steel. Glass is useful because it gives smooth surfaces, is non-toxic, corrosion proof and it is usually easy to examine the interior of the vessel. Two basic types of fermenter are used:

1. A glass vessel with a round or flat bottom and a top flanged carrying plate (Fig. 7.4). The large glass containers originally used were borosilicate battery jars (Brown and Peterson, 1950). All ves-

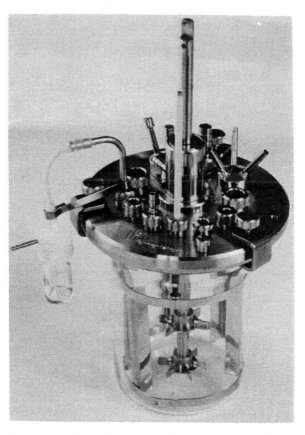

Fig. 7.4. Glass fermenter with a top-flanged carrying plate (Inceltech L.H. Reading, England).

sels of this type have to be sterilized by autoclaving. Cowan and Thomas (1988) state that the largest practical diameter for glass fermenters is 60 cm.

2. A glass cylinder with stainless-steel top and bottom plates (Fig. 7.5). These fermenters may be sterilized *in situ*, but 30 cm diameter is the upper size limit to safely withstand working pressures (Solomons, 1969). Vessels with two stainless steel plates cost approximately 50% more than those with just a top plate.

At pilot and large scale (Figs 7.6 and 7.7), when all fermenters are sterilized *in situ*, any materials used will have to be assessed on their ability to withstand pressure sterilization and corrosion and on their potential toxicity and cost. Walker and Holdsworth (1958), Solomons (1969) and Cowan and Thomas (1988) have discussed the suitability of various materials used in the construction of fermenters. Pilot-scale and industrial-

Fig. 7.5. Three glass fermenters with top and bottom plates (New Brunswick Scientific, Hatfield, England).

scale vessels are normally constructed of stainless steel or at least have a stainless-steel cladding to limit corrosion. The American Iron and Steel Institute (AISI) states that steels containing less than 4% chromium are classified as steel alloys and those containing more than 4% are classified as stainless steels. Mild steel coated with glass or phenolic epoxy materials has occasionally been used (Gordon *et al.*, 1947; Fortune *et al.*, 1950; Buelow and Johnson, 1952; Irving, 1968). Wood, plastic and concrete have been used when contamination was not a problem in a process (Steel and Miller, 1970).

Walker and Holdsworth (1958) stated that the extent of vessel corrosion varied considerably and did not appear to be entirely predictable. Athough stainless steel is often quoted as the only satisfactory material, it has been reported that mild-steel vessels were very satisfactory after 12-years use for penicillin fermentations (Walker and Holdsworth, 1958) and mild steel clad with stainless steel has been used for at least 25 years for acetone–butanol production (Spivey, 1978). Pitting to a depth of 7 mm was found in a mild-steel fermenter after 7-years use for streptomycin production (Walker and Holdsworth, 1958).

The corrosion resistance of stainless steel is thought to depend on the existence of a thin hydrous oxide film on the surface of the metal. The composition of this film varies with different steel alloys and different manufacturing process treatments such as rolling, pickling or heat treatment. The film is stabilized by chromium and is considered to be continuous, non-porous, insoluble and self healing. If damaged, the film will repair itself when exposed to air or an oxidizing agent (Cubberly *et al.*, 1980).

The minimum amount of chromium needed to resist corrosion will depend on the corroding agent in a particular environment, such as acids, alkalis, gases, soil, salt or fresh water. Increasing the chromium content enhances resistance to corrosion, but only grades of steel containing at least 10 to 13% chromium develop an effective film. The inclusion of nickel in high percent chromium steels enhances their resistance and improves their engineering properties. The presence of molybdenum improves the resistance of stainless steels to solutions of halogen salts and pitting by chloride ions in brine or sea water. Corrosion resistance can also be improved by tungsten, silicone and other elements (Cubberly *et al.*, 1980; Duurkoop, 1992).

AISI grade 316 steels which contain 18% chromium, 10% nickel and 2–2.5% molybdenum are now com-

FIG. 7.6. Stainless steel fully automatic 10 dm³ fermenter sterilizable-*in-situ* (LSL, Luton, UK).

monly used in fermenter construction. Table 7.5 (Du-urkoop, 1992) gives the classification systems used to identify stainless steels of AISI 316 grades in the U.S.A. and similar procedures used in Germany, France, the United Kingdom, Japan and Sweden. In a citric acid fermentation where the pH may be 1 to 2 it will be necessary to use a stainless steel with 3–4% molybdenum (AISI grade 317) to prevent leaching of heavy

Fɪɢ. 7.7. Stainless steel pilot plant fermenters (LSL, Luton, UK).

Tᴀʙʟᴇ 7.5 *Classification systems used for stainless steels of 316 grades* (Duurkoop, 1992)

| Germany | DIN | France | Great Britain | Japan | Sweden | U.S.A. | |
|---|---|---|---|---|---|---|---|
| Werkstoff Nummer | | AFNOR | B.S. | JIS | SS | AISI | UNS |
| 1.4401 | X2 CrNiMo 17 12 2 | Z 6 CND 17.11 | 316 S 16<br>316 S 31 | SUS 316 | 2347 | 316 | S31600 |
| 1.4404 | X2 CrNiMo 17 13 2<br>G-X CrNiMo 18 10 | Z 2 CND 18.13<br>Z 2 CND 17 12<br>Z 2 CND 19.10 M | 316 S 11<br>316 S 12 | SUS 316L | 2348 | 316L | S31603 |
| 1.4406 | X2 CrNiMo 17 12 2 | Z 2 CND 17.12 Az | 316 S 61 | SUS 316LN | — | 316LN | S31653 |
| 1.4435 | X2 CrNiMo 18 14 3 | Z 2 CND 17.13 | 316 S 11<br>316 S 12 | SCS 16<br>SUS 316L | 2353 | 316L | S31603 |

metals from the steel which would interfere with the fermentation. AISI Grade 304, which contains 18.5% chromium and 10% nickel, is now used extensively for brewing equipment.

The thickness of the construction material will increase with scale. At 300,000 to 400,000 dm³ capacity, 7-mm plate may be used for the side of a vessel and 10 mm plate for the top and bottom, which should be hemispherical to withstand pressures.

At this stage it is important to consider the ways in which a reliable aseptic seal is made between glass and glass, glass and metal or metal and metal joints such as between a fermenter vessel and a detachable top or base plate. With glass and metal, a seal can be made with a compressible gasket, a lip seal or an 'O' ring (Fig. 7.8). With metal to metal joints only an 'O' ring is suitable. This is placed in a groove, machined in either the end plate, the fermenter body or both. This seal ensures that a good liquid- and/or gas-tight joint is maintained in spite of the glass or metal expanding or contracting at different rates with changes in temperature during a sterilization cycle or an incubation cycle. Nitryl or butyl rubbers are normally used for these seals as they will withstand fermentation process conditions. The properties of different rubbers for seals are discussed by Buchter (1979) and Martini (1984). These rubber seals have a finite life and should be checked regularly for damage or perishing. Before purchasing a fermenter it is important to check that standard sized 'O' rings can be purchased cheaply from local suppliers.

A single 'O' ring seal is adequate for GILSP and levels 1 and B2, a double 'O' ring seal is required for levels 2 and B3 and a double 'O' ring seal with steam between the seals (steam tracing) is necessary for levels 3 and B4 (Chapman, 1989; Hambleton et al., 1991). In the U.S.A., however, simple seals are used at contain-

ment levels comparable with the U.K. B3 (Titchener-Hooker et al., 1993).

Titchener-Hooker et al. (1993) criticized Chapman's proposal to use two seals without a steam trace for a number of reasons including:

a.  Double seals are more difficult to assemble correctly.
b.  It is difficult to detect failure of one seal of a pair during operation or assembly.
c.  Neither of the two seals can be tested independently.
d.  Dead spaces between two seals must be considered to be contaminated.

Leaver (1994) and Titchener-Hooker et al. (1993) consider that correctly fitted single static seals can provide adequate containment for most processes and double static seals with a steam trace should be strictly limited to the small number of processes for which an extreme level of protection may be required.

### Temperature control

Normally in the design and construction of a fermenter there must be adequate provision for temperature control which will affect the design of the vessel body. Heat will be produced by microbial activity and mechanical agitation and if the heat generated by these two processes is not ideal for the particular manufacturing process then heat may have to be added to, or removed from, the system. On a laboratory scale little heat is normally generated and extra heat has to be provided by placing the fermenter in a thermostatically controlled bath, or by the use of internal heating coils or a heating jacket through which water is circulated or by a silicone heating jacket. The silicone jacket consists of a double silicone rubber mat with heating wires between the two mats; it is wrapped around the vessel and held in place by Velcro strips (Applikon, 1989).

Once a certain size has been exceeded, the surface area covered by the jacket becomes too small to remove the heat produced by the fermentation. When this situation occurs internal coils must be used and cold water is circulated to achieve the correct temperature (Jackson, 1990). Different types of fermentation will influence the maximum size of vessel that can be used with jackets alone.

It is impossible to specify accurately the necessary cooling surface of a fermenter since the temperature of the cooling water, the sterilization process, the cultiva-

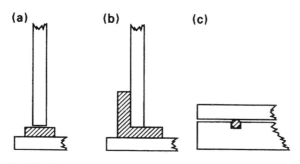

**(a)**  **(b)**  **(c)**

Fig. 7.8. Joint seals for glass–glass, glass–metal and metal–metal; (a) gasket; (b) lip seal; (c) 'O' ring in groove.

tion temperature, the type of micro-organism and the energy supplied by stirring can vary considerably in different processes. A cooling area of 50 to 70 m$^2$ may be taken as average for a 55,000 dm$^3$ fermenter and with a coolant temperature of 14° the fermenter may be cooled from 120° to 30° in 2.5 to 4 hours without stirring. The consumption of cooling water in this size of vessel during a bacterial fermentation ranges from 500 to 2000 dm$^3$ h$^{-1}$, while fungi might need 2000 to 10,000 dm$^3$ h$^{-1}$ (Muller and Kieslich, 1966), due to the lower optimum temperature for growth.

To make an accurate estimate of heating/cooling requirements for a specific process it is important to consider the contributing factors. An overal energy balance for a fermenter during normal operation can be written as:

$$Q_{met} + Q_{ag} + Q_{gas} = Q_{acc} + Q_{exch} + Q_{evap} + Q_{sen}$$
(7.1)

where $Q_{met}$ = heat generation rate due to microbial metabolism,

$Q_{ag}$ = heat generation rate due to mechanical agitation,

$Q_{gas}$ = heat generation rate due to aeration power imput,

$Q_{acc}$ = heat accumulation rate by the system,

$Q_{exch}$ = heat transfer rate to the surroundings and/or heat exchanger,

$Q_{evap}$ = heat loss rate by evaporation,

$Q_{sen}$ = rate of sensible enthalpy gain by the flow streams (exit − inlet).

This equation can be rearranged as:

$$Q_{exch} = Q_{met} + Q_{ag} + Q_{gas} - Q_{acc} - Q_{sen} - Q_{evap}$$
(7.2)

$Q_{exch}$ is the heat which will have to be removed by a cooling system.

Atkinson and Mavituna (1991b) have presented data to estimate $Q_{met}$ for a range of substrates, methods to calculate power input for $Q_{ag}$ and $Q_{gas}$, a formula to calculate the sensible heat loss for flow streams ($Q_{sen}$) and a method to calculate the heat loss due to evaporation ($Q_{evap}$). Cooney et al. (1969) determined representative low and high heats of fermentation values for Bacillus subtilis grown on molasses at 37° (Table 7. 6). They concluded that $Q_{evap}$ and $Q_{sen}$ are small contributory factors and $Q_{acc}$ = 0 in a steady-state system. $Q_{evap}$ can also be eliminated by using a saturated air stream at the temperature of the broth.

TABLE 7.6. Representative low and high values of calculated heats (kcal dm$^{-3}$ h$^{-1}$) of fermentation for Bacillus subtilis on molasses at 37° (Coooney et al., 1969)

| | Heats of fermentation | | | | | |
| | $Q_{acc}$ | $Q_{ag}$ | $Q_{evap}$ | $Q_{sen}$ | $Q_{exch}$ | $Q_{met}$ |
|---|---|---|---|---|---|---|
| Low | 3.81 | 3.32 | 0.023 | 0.005 | 0.61 | 1.12 |
| High | 11.3 | 3.31 | 0.045 | 0.010 | 0.65 | 8.65 |

When designing a large fermenter, the operating temperature and flow conditions will determine $Q_{evap}$ and $Q_{sen}$, the choice of agitator, its speed and the aeration rate will determine $Q_{ag}$ and the sparger design and aeration rate will determine $Q_{gas}$.

The cooling requirements (jacket and/or pipes) to remove the excess heat from a fermenter may be determined by the following formula:

$$Q_{exch} = U \cdot A \cdot \Delta T$$
(7.3)

where $A$ = the heat transfer surface available, m$^2$,

$Q$ = the heat transferred, W,

$U$ = the overall heat transfer coefficient, W/m$^2$K,

$\Delta T$ = the temperature difference between the heating or cooling agent and the mass itself, K.

The coefficient $U$ represents the conductivity of the system and it depends on the vessel geometry, fluid properties, flow velocity, wall material and thickness (Scragg, 1991). $1/U$ is the overall resistance to heat transfer (analogous to $1/K$ for gas–liquid transfer; Chapter 9). It is the reciprocal of the overall heat-transfer coefficient. It is defined as the sum of the individual resistances to heat transfer as heat passes from one fluid to another and can be expressed as:

$$\frac{1}{U} = \frac{1}{h_o} + \frac{1}{h_i} + \frac{1}{h_{of}} + \frac{1}{h_{if}} + \frac{1}{h_w}$$
(7.4)

where $h_o$ = outside film coefficient, W/m$^2$K,

$h_i$ = inside film coefficient, W/m$^2$K,

$h_{of}$ = outside fouling film coefficient, W/m$^2$K,

$h_{if}$ = inside fouling film coefficient, W/m$^2$K,

$h_w$ = wall heat transfer coefficient = $k/x$, W/m$^2$K,

$k$ = thermal conductivity of the wall, W/mK,

$x$ = wall thickness, m.

There is more detailed discussion of $U$ in Bailey and Ollis (1986), Atkinson and Mavituna (1991b) and Scragg (1991).

Atkinson and Mavituna (1991b) have given three methods to determine $\Delta T$ (the temperature driving force) depending on the operating circumstances. If one side of the wall is at a constant temperature, as is often the case in a stirred fermenter, and the coolant temperature rises in the direction of the coolant flow along a cooling coil, an arithmetic mean is appropriate:

$$\Delta T_{am} = \frac{(T_f - T_e) + (T_f - T_i)}{2} \qquad (7.5)$$

$$= \frac{T_f - (T_e - T_i)}{2} \qquad (7.6)$$

where  $T_f$ = the bulk liquid temperature in the vessel,

$T_e$ = the temperature of the coolant entering the system,

$T_i$ = the temperature of the coolant leaving the system.

If the fluids are in counter- or co-current flow and the temperature varies in both fluids then a log mean temperature difference is appropriate:

$$\Delta T_m = \Delta T_e - \Delta T_i / \ln (\Delta T_e / \Delta T_i) \qquad (7.7)$$

where  $\Delta T_e$ = the temperature of the coolant entering,

$\Delta T_i$ = the temperature of the coolant leaving.

If the flow pattern is more complex than either of the two previous situations then the log mean temperature difference defined in equation (7.7) is multiplied by an appropriate dimensionless factor which has been evaluated for a number of heat-exchanger systems by McAdams (1954).

Appropriate techniques have just been discussed to obtain values for $Q_{exch}$, $U$ and $\Delta T$ (or $\Delta T_{am}$ or $\Delta T_m$). If equation (7.3) is now rearranged:

$$A = \frac{Q_{exch}}{U \Delta T}. \qquad (7.8)$$

Substituting in this equation makes it possible to calculate the heat-transfer surface necessary to obtain adequate temperature control.

## AERATION AND AGITATION

Aeration and agitation will be considered in detail in Chapter 9. In this chapter it should be stated that the primary purpose of aeration is to provide micro-organisms in submerged culture with sufficient oxygen for metabolic requirements, while agitation should ensure that a uniform suspension of microbial cells is achieved in a homogeneous nutrient medium. The type of aeration–agitation system used in a particular fermenter depends on the characteristics of the fermentation process under consideration. Although fine bubble aerators without mechanical agitation have the advantage of lower equipment and power costs, agitation may be dispensed with only when aeration provides sufficient agitation, i.e. in processes where broths of low viscosity and low total solids are used (Arnold and Steel, 1958). Thus, mechanical agitation is usually required in fungal and actinomycete fermentations. Non-agitated fermentations are normally carried out in vessels of a height/diameter ratio of 5:1. In such vessels aeration is sufficient to produce high turbulence, but a tall column of liquid does require greater energy input in the production of the compressed air (Muller and Kieslich, 1966; Solomons, 1980).

The structural components of the fermenter involved in aeration and agitation are:

(a) The agitator (impeller).
(b) Stirrer glands and bearings.
(c) Baffles.
(d) The aeration system (sparger).

### The agitator (impeller)

The agitator is required to achieve a number of mixing objectives, e.g. bulk fluid and gas-phase mixing, air dispersion, oxygen transfer, heat transfer, suspension of solid particles and maintaining a uniform environment throughout the vessel contents. It should be possible to design a fermenter to achieve these conditions; this will require knowledge of the most appropriate agitator, air sparger, baffles, the best positions for nutrient feeds, acid or alkali for pH control and antifoam addition. There will also be a need to specify agitator size and number, speed and power imput (see also Chapter 9).

Agitators may be classified as disc turbines, vaned discs, open turbines of variable pitch and propellers, and are illustrated in Fig. 7.9. The disc turbine consists of a disc with a series of rectangular vanes set in a vertical plane around the circumference and the vaned disc has a series of rectangular vanes attached vertically to the underside. Air from the sparger hits the underside of the disc and is displaced towards the vanes where the air bubbles are broken up into smaller bubbles. The vanes of a variable pitch open turbine and

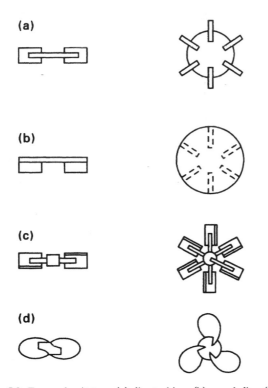

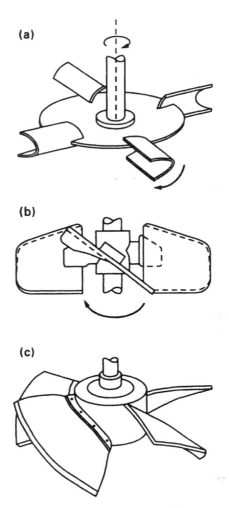

FIG. 7.9. Types of agitators: (a) disc turbine; (b) vaned disc; (c) open turbine, variable pitch; (d) marine propeller (Solomons, 1969).

the blades of a marine propeller are attached directly to a boss on the agitator shaft. In this case the air bubbles do not initially hit any surface before dispersion by the vanes or blades.

Four other modern agitator developments, the Scaba 6SRGT, the Prochem Maxflo T, the Lightning A315 and the Ekato Intermig (Figs 7.10 and 7.11), which are derived from open turbines, will also be discussed for energy conservation and use in high-viscosity broths.

Since the 1940s a Rushton disc turbine of one-third the fermenter diameter has been considered the optimum design for use in many fermentation processes. It had been established experimentally that the disc turbine was most suitable in a fermenter since it could break up a fast air stream without itself becoming flooded in air bubbles (Finn, 1954). This flooding condition is indicated when the bulk flow pattern in the vessel normally associated with the agitator design (radial with the Rushton turbine) is lost and replaced by a centrally flowing air–broth plume up the middle of the vessel with a liquid flow as an annulus (Nienow et al., 1985; see also Chapter 9). The propeller and the open turbine flood when $V_s$ (superficial velocity, i.e. volumet-

FIG. 7.10. Diagrams of (a) Scaba agitator; (b) Lightnin' A315 agitator (four blades) and (c) Prochem Maxflo T agitator (four, five or six blades) (Nienow, 1990).

ric air flow rate/cross-sectional area of fermenter) exceeds 21 m h$^{-1}$, whereas the flat blade turbine can tolerate a $V_s$ of up to 120 m h$^{-1}$ before being flooded, when two sets are used on the same shaft. Besides being flooded at a lower $V_s$ than the disc turbine, the propeller is also less efficient in breaking up a stream of air bubbles and the flow it produces is axial rather than radial (Cooper et al., 1944). The disc turbine was thought to be essential for forcing the sparged air into the agitator tip zone where bubble break up would occur.

In other studies it has been shown that bubble break up occurs in the trailing vortices associated with all agitator types which give rise to gas-filled cavities and

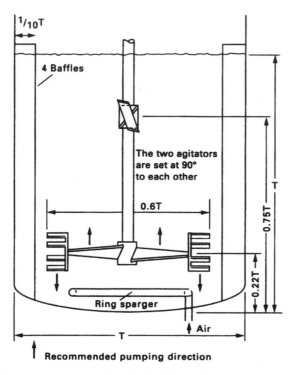

FIG. 7.11. Arrangement for a pair of Intermig agitators. Relative dimensions are given as a proportion of the fermenter vessel diameter (T) (Nienow, 1990).

tions of oxygen in broth away from the agitators and higher concentrations of nutrients, acid or alkali or antifoam near to the feed points.

Another is the Prochem Maxflo agitator. It consists of four, five or six hydrofoil blades set at a critical angle on a central hollow hub (Gbewonyo *et al.*, 1986; Buckland *et al.*, 1988). A high hydrodynamic thrust is created during rotation, increasing the downwards pumping capacity of the blades. This design minimizes the drag forces associated with rotation of the agitator such that the energy losses due to drag are low. This leads to a low power number. The recommended agitator to vessel diameter ratio is greater than 0.4. When the agitator was used with a 800-dm$^3$ *Streptomyces* fermentation, the maximum power requirement at the most viscous stage was about 66% of that with Rushton turbines. The fall in power was also less in a 14,000-dm$^3$ fermentation. The oxygen-transfer efficiency was also significantly improved. It was thought that an improvement in bulk mixing was another contributing factor.

Intermig agitators (Fig. 7.11) made by Ekato of Germany are more complex in design. Two units are used (with agitator diameter to vessel diameter ratios of 0.6 to 0.7) instead of a single Rushton turbine because their power number is so low (Nienow, 1992). A large-diameter air sparger is used to optimize air dispersion (see later section on spargers). The loss in power is less than when aerating with a Rushton turbine. Air dispersion starts from the air cavities which form on the wing tips of the agitator blades. In spite of the downwards pumping direction of the wings, the cavities extend horizontally from the back of the agitator blades, reducing the effectiveness of top to bottom mixing in a vessel.

Cooke *et al.* (1988) and Nienow (1990, 1992) give comparisons of performance of Rushton turbines with some of the newer designs.

These new turbine designs make it possible to replace Rushton turbines by larger low power agitators which do not lose as much power when aerated, are able to handle higher air volumes without flooding and give better bulk blending and heat transfer characteristics in more viscous media (Nienow, 1990). However, there can be mechanical problems which are mostly of a vibrational nature.

Good mixing and aeration in high viscosity broths may also be achieved by a dual impeller combination, where the lower impeller acts as the gas disperser and the upper impeller acts primarily as a device for aiding circulation of vessel contents. This has been discussed by Nienow and Ulbrecht (1985) and in Chapter 9.

provided the agitator speed is high enough, good gas dispersion will occur in low-viscosity broths (Smith, 1985). It has been also shown that similar oxygen-transfer efficiencies are obtained at the same power input per unit volume, regardless of the agitator type.

In high-viscosity broths, gas dispersion also occurs from gas filled cavities trailing behind the rotating blades, but this is not sufficient to ensure satisfactory bulk blending of all the vessel contents. When the cavities are of maximum size, the impeller appears to be rotating in a pocket of gas from which little actual dispersion occurs into the rest of the vessel (Nienow and Ulbrecht, 1985).

Recently, a number of agitators have been developed to overcome problems associated with efficient bulk blending in high-viscosity fermentations (Figs 7.10 and 7.11). The Scaba 6SRGT agitator is one which, at a given power input, can handle a high air flow rate before flooding (Nienow, 1992). This radial-flow agitator is also better for bulk blending than a Rushton turbine, but does not give good top to bottom blending in a large fermenter which leads to lower concentra-

Steel and Maxon (1966) tested a multi-rod mixing impeller. In a 15,000-dm³ vessel, the same novobiocin yield and oxygen availability rate were obtained at about half of the power required by a standard turbine-stirred fermenter, but this type of impeller does not appear to have come into general use.

### Stirrer glands and bearings

The satisfactory sealing of the stirrer shaft assembly has been one of the most difficult problems to overcome in the construction of fermentation equipment which can be operated aseptically for long periods. A number of different designs have been developed to obtain aseptic seals. The stirrer shaft can enter the vessel from the top, side (Richards, 1968) or bottom of the vessel. Top entry is most commonly used, but bottom entry may be advantageous if more space is needed on the top plate for entry ports, and the shorter shaft permits higher stirrer speeds to be used by eliminating the problem of the shaft whipping at high speeds. Originally, bottom entry stirrers were considered undesirable as the bearings would be submerged. Chain *et al.* (1952) successfully operated vessels of this type, and they have since been used by many other workers. Mechanical seals can be used for bottom entry provided that they are routinely maintained and replaced at recommended intervals (Leaver and Hambleton, 1992).

One of the earliest stirrer seals described was that used by Rivett, Johnson and Peterson (1950) in a laboratory fermenter (Fig. 7.12). A porous bronze bearing for a 13-mm shaft was fitted in the centre of the fermenter top and another in a yoke directly above it. The bearings were pressed into steel housings, which screwed into position in the yoke and the fermenter top. The lower bearing and housing were covered with a skirt-like shield having a 6.5 mm overhang which rotated with the shaft and prevented air-borne contaminants from settling on the bearing and working their way through it into the fermenter.

Four basic types of seal assembly have been used: the stuffing box (packed-gland seal), the simple bush seal, the mechanical seal and the magnetic drive. Most modern fermenter stirrer mechanisms now incorporate mechanical seals instead of stuffing boxes and packed glands. Mechanical seals are more expensive, but are more durable and less likely to be an entry point for contaminants or a leakage point for organisms or products which should be contained. Magnetic drives,

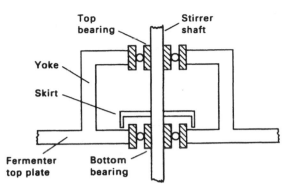

FIG. 7.12. A simple stirrer seal based on a description given by Rivett *et al.* (1950).

which are also quite expensive, are being used in some animal cell culture vessels.

### THE STUFFING BOX (PACKED-GLAND SEAL)

The stuffing box (Fig. 7.13) has been described by Chain *et al.* (1954). The shaft is sealed by several layers of packing rings of asbestos or cotton yarn, pressed against the shaft by a gland follower. At high stirrer speeds the packing wears quickly and excessive pressure may be needed to ensure tightness of fit. The packing may be difficult to sterilize properly because of unsatisfactory heat penetration and it is necessary to check and replace the packing rings regularly. Parker (1950) described a split stuffing box with a lantern ring. Steam under pressure was continually fed into it. Chain *et al.* (1954) used two stuffing boxes on the agitator shaft with a space in between kept filled with steam. Although, at one time, stuffing box-bearings were commonly used in large-scale vessels, operational problems, particularly contamination, have led to their replacement by mechanical seal bearings for many processes. However, these seals are sufficient for the requirements of GILSP containment (Werner, 1992).

### THE MECHANICAL SEAL

The mechanical seal assembly (Figs 7.14 and 7.15) is now commonly used in both small and large fermenters. The seal is composed of two parts, one part is stationary in the bearing housing, the other rotates on the shaft, and the two components are pressed together by springs or expanding bellows. The two meeting surfaces have to be precision machined, the moving surface normally consists of a carbon-faced unit while the stationary unit is of stellite-faced stainless steel. Steam condensate is used to lubricate and cool the seals during operation and serve as a containment

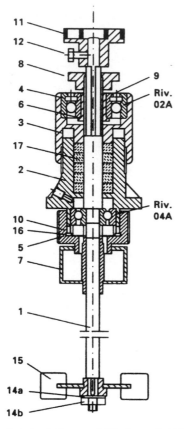

FIG. 7.13. Packed-gland stirrer seal (Chain *et al.*, 1954) (Components: 1, agitator shaft; 2, stuffing box; 3, upper cap; 4, lock ring; 5, lower cap; 6, chuck; 7, greasecup; 8, lock ring; 9, lock nut; 10, distance ring; 11, half coupling; 12, half coupling; 14a, washer; 14b, nut; 15, impeller; 16, shim; 17, packing rings).

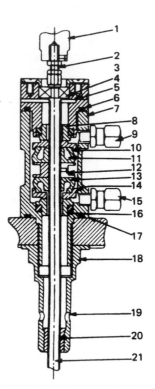

FIG. 7.14. Mechanical seal assembly (Elsworth *et al.*, 1958). (Components; 1, flexible coupling; 2, stirrer shaft; 3, bearing housing; 4, ball journal fit on mating parts; 5, two slots for gland leaks, only one shown; 6, 'O'-ring seal; 7, seal body; 8, stationary counter-face sealed to body with square-section gasket; 9, exit port for condensate, fitted with unequal stud coupling; 10, rotating counter-face; 11, bellows; 12, shaft muff; 13, as 11; 14, as 10; 15, entry port for condensate, as 9; 16, as 8; 17, as 6; 18, shaft bush support; 19, leak holes; 20, Ferobestos bush; 21, ground shaft).

barrier. Single mechanical seals are used with a steam barrier in fermenters for primary containment at level 1 or B2 (Werner, 1992), whereas double mechanical seals are typically used in vessels with the outer seal as a backup for the inner seal for primary containment at level 2 or B3 (Werner, 1992; Leaver and Hambleton, 1992; Fig. 7.15). At level 2 or B3 the condensate is piped to a kill tank. Monitoring of the steam condensate flowing out of the seal is an effective way for checking for seal failure. Disinfectants are alternatives for flushing the seals (Werner, 1992).

MAGNETIC DRIVES

The problems of providing a satisfactory seal when the impeller shaft passes through the top or bottom plate of the fermenter may be solved by the use of a magnetic drive in which the impeller shaft does not pierce the vessel (Cameron and Godfrey, 1969). A magnetic drive (Fig. 7.16) consists of two magnets: one driving and one driven. The driving magnet is held in bearings in a housing on the outside of the head plate and connected to a drive shaft. The internal driven magnet is placed on one end of the impeller shaft and held in bearings in a suitable housing on the inner surface of the headplate. When multiple ceramic magnets have been used, it has been possible to transmit power across a gap of 16 mm. Using this drive, water can be stirred in baffled vessels of up to 300-dm$^3$ capacity at speeds of 300 to 2000 rpm. It would be necessary to establish if adequate power could be transmitted between magnets to stir viscous mould broths or when wanting high oxygen transfer rates in bacterial cultures. Walker *et al.* (1987) have described the development of a magnetic drive suitable for microbial fermentations up to 1500 dm$^3$ which could be used when

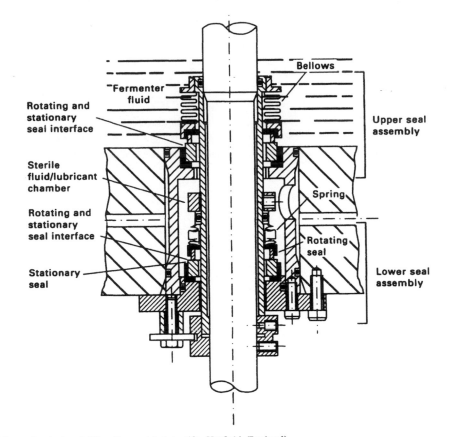

FIG. 7.15. Double mechanical seal (New Brunswick Scientific, Hatfield, England).

higher containment levels are specified. The stirring mechanism is ideal for animal cell culture to minimize the chances of potential contamination. This application is discussed later in this chapter.

## Baffles

Four baffles are normally incorporated into agitated vessels of all sizes to prevent a vortex and to improve aeration efficiency. In vessels over 3-dm³ diameter six or eight baffles may be used (Scragg, 1991). Baffles are metal strips roughly one-tenth of the vessel diameter and attached radially to the wall (see Figs 7.1 and 7.2 and Tables 7.2 and 7.3). The agitation effect is only slightly increased with wider baffles, but drops sharply with narrower baffles (Winkler, 1990). Walker and Holdsworth (1958) recommended that baffles should be installed so that a gap existed between them and the vessel wall, so that there was a scouring action around and behind the baffles thus minimizing microbial growth

on the baffles and the fermenter walls. Extra cooling coils may be attached to baffles to improve the cooling capacity of a fermenter without unduly affecting the geometry.

## The aeration system (sparger)

A sparger may be defined as a device for introducing air into the liquid in a fermenter. Three basic types of sparger have been used and may be described as the porous sparger, the orifice sparger (a perforated pipe) and the nozzle sparger (an open or partially closed pipe). A combined sparger–agitator may be used in laboratory fermenters (Fig. 7.17) and is discussed briefly in a later section.

### POROUS SPARGER

The porous sparger of sintered glass, ceramics or metal, has been used primarily on a laboratory scale in

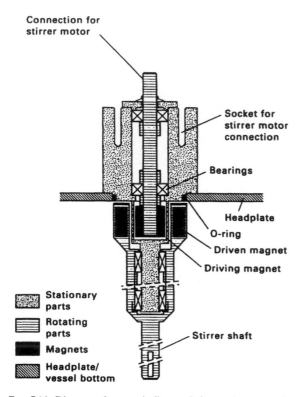

FIG. 7.16. Diagram of magnetically coupled top stirrer assembly (Applikon Dependable Instruments, Schiedam, Netherlands).

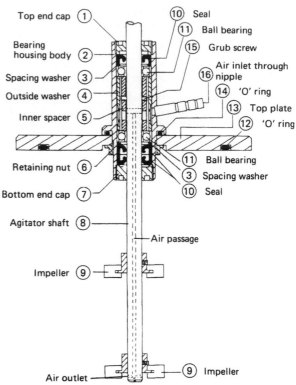

FIG. 7.17. Diagram of bearing housing with combined agitator–sparger (Inceltech L.H. Reading, England).

non-agitated vessels. The bubble size produced from such spargers is always 10 to 100 times larger than the pore size of the aerator block (Finn, 1954). The throughput of air is low because of the pressure drop across the sparger and there is also the problem of the fine holes becoming blocked by growth of the microbial culture.

ORIFICE SPARGER

Various arrangements of perforated pipes have been tried in different types of fermentation vessel with or without impellers. In small stirred fermenters the perforated pipes were arranged below the impeller in the form of crosses or rings (ring sparger), approximately three-quarters of the impeller diameter. In most designs the air holes were drilled on the under surfaces of the tubes making up the ring or cross. Walker and Holdsworth (1958) commented that in production vessels, sparger holes should be at least 6 mm (1/4 inch) diameter because of the tendency of smaller holes to block and to minimize the pressure drop.

In low viscosity fermentations sparged at 1 vvm

(volume of air$^{-1}$ volume of medium$^{-1}$ minute$^{-1}$) with a power input of 1 W kg$^{-1}$, Nienow *et al.* (1988) found that the power often falls to below 50% of its unaerated value when using a single Rushton disc turbine which is one-third the diameter of the vessel and a ring sparger smaller than the diameter of the agitator. If the ring sparger were placed close to the disc turbine and its diameter was 1.2 times that of the disc turbine, a number of benefits could be obtained (Nienow *et al.*, 1988). A 50% higher aeration rate could be obtained before flooding occurred, the power drawn was 75% of the unaerated value, and a higher $K_L a$ could be obtained at the same agitator speed and aeration rate. These advantages were lost at viscosities of about 100 m Pas.

Orifice spargers without agitation have been used to a limited extent in yeast manufacture (Thaysen, 1945), effluent treatment (Abson and Todhunter, 1967) and later in the production of single-cell protein in the air-lift fermenter which are discussed in a later section of this chapter (Taylor and Senior, 1978; Smith, 1980).

## NOZZLE SPARGER

Most modern mechanically stirred fermenter designs from laboratory to industrial scale have a single open or partially closed pipe as a sparger to provide the stream of air bubbles. Ideally the pipe should be positioned centrally below the impeller and as far away as possible from it to ensure that the impeller is not flooded by the air stream (Finn, 1954). The single-nozzle sparger causes a lower pressure loss than any other sparger and normally does not get blocked.

## COMBINED SPARGER–AGITATOR

On a small scale (1 dm$^3$), Herbert *et al.* (1965) developed the combined sparger–agitator design, introducing the air via a hollow agitator shaft and emitting it through holes drilled in the disc between the blades and connected to the base of the main shaft. The design gives good aeration in a baffled vessel when the agitator is operated at a range of rpm (see Fig. 7.17).

## THE ACHIEVEMENT AND MAINTENANCE OF ASEPTIC CONDITIONS

Once the design problems of aeration and agitation have been solved, it is essential that the design meets the requirements of the degree of asepsis and containment demanded by the particular process being considered. It will be necessary to be able to sterilize, and keep sterile, a fermenter and its contents throughout a complete growth cycle. There may be also a need to protect workers and the environment from exposure to hazardous micro-organisms or animal cells (Van Houten, 1992). As has been mentioned earlier, the containment requirements depend on the size of the fermentation vessel.

The following operations may have to be performed according to certain specifications to achieve and maintain aseptic conditions and containment during a fermentation:

1. Sterilization of the fermenter.
2. Sterilization of the air supply and the exhaust gas.
3. Aeration and agitation.
4. The addition of inoculum, nutrients and other supplements.
5. Sampling.
6. Foam control.
7. Monitoring and control of various parameters.

On a small scale, below 10 dm$^3$, the biohazard risk can be controlled by a combination of containment cabinets and work practices (Van Houten, 1992). When the volume of culture exceeds 10 dm$^3$, GILSP is required for those non-pathogenic non-toxigenic agents which have an extended history of large scale use. For this category there should be prevention of contamination of the product, control of aerosols and minimization of the release of micro-organisms during sampling, addition of material, transfer of cells and removal of materials, products and effluents. It should be appreciated that the majority of fermentations fall into this category.

At level 1, B2 or at higher containment levels, the following points need to be considered when designing a fermenter or other vessel, so that it can operate as a contained system (Tubito, 1991):

1. All vessels containing live organisms should be suitable for steam sterilization and have sterile vent filters. This is discussed in Chapter 5.
2. Exhaust gases from vessels should pass through sterile filters. This is discussed in Chapter 5.
3. Seals on flange joints should be fitted with a single 'O'-ring at the lower levels of containment. Flange joints on vessels for Containment levels 3 and B3/4 need double 'O'-rings or double 'O'-rings plus a steam barrier. This has been discussed in an earlier section of this chapter.
4. Appropriate seals should be provided for entry ports for sensor probes, inoculum, sampling, medium addition, acid, alkali and antifoam. This will be discussed later in this chapter.
5. Rotating shafts into a closed system should be sealed with a double acting mechanical seal with steam or condensate between the seals. This has been discussed in an earlier section in this chapter.
6. During operation a steam barrier should be maintained in all fixed piping leading to the 'contained' vessels.
7. Provision of appropriate pressure relief facilities will be discussed later in this chapter.

Further details for containment are given by Giorgio and Wu (1986), Hesselink *et al.* (1990), Kennedy *et al.* (1990), Janssen *et al.* (1990), Hambleton *et al.* (1991), Tubito (1991), Leaver and Hambleton (1992), Van Houten (1992), Vranch (1992) and Werner (1992).

Some of the issues discussed in points 1, 3, 4, 5 and 7

will need to be considered for aseptic operation of GILSP processes.

### Sterilization of the fermenter

The fermenter should be so designed that it may be steam sterilized under pressure. The medium may be sterilized in the vessel or separately, and subsequently added aseptically. If the medium is sterilized *in situ* its temperature should be raised prior to the injection of live steam to prevent the formation of large amounts of condensate. This may be achieved by steam being introduced into the fermenter coils or jacket. As every point of entry to and exit from the fermenter is a potential source of contamination, steam should be introduced through all the entry and exit points except the air outlet from which steam should be allowed to leave.

All pipes should be constructed as simply as possible and slope towards drainage points to make sure that steam reaches all parts of the equipment and is not excluded by siphons or pockets of condensate or mash. Each drainage point in the pipework should be fitted with a steam trap. This will be described in the section on valves and steam traps. Parker (1958), Chain *et al.* (1954) and Muller and Kieslich (1966) and others have all stressed the need to eliminate fine fissures or gaps such as flange seals which might be filled with nutrient solutions and micro-organisms. Hambleton *et al.* (1991) described a high specification pilot scale fermenter with surfaces free of crevices greater than 0.05-mm depth, which is needed if the vessel was to be used for animal cells in suspension culture or on micro-carriers. For long-term aseptic operation welded joints should be used wherever possible, even though sections may have to be cut out and re-welded during maintenance and repair (Smith, 1980).

### Sterilization of the air supply

Sterile air will be required in very large volumes in many aerobic fermentation processes. Although there are a number of ways of sterilizing air, only two have found permanent application. These are heat and filtration. Heat is generally too costly for full-scale operation (see also Chapter 5).

Historically, glass wool, glass fibre or mineral slag wool have been used as filter material, but currently most fermenters are fitted with cartridge-type filters as discussed in Chapter 5. However, before the filter may

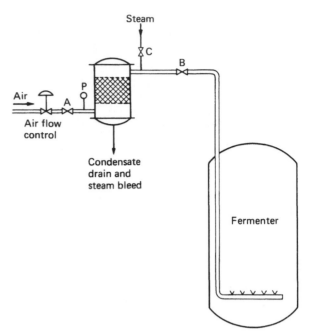

FIG. 7.18. An arrangement of packed air filter and fermenter (Richards, 1968).

be used it, too, must be sterilized in association with the fermenter. Two procedures are commonly followed depending on the construction of the filter unit.

Figure 7.18 shows the simple unit described by Richards (1968). During sterilization the main non-sterile air-inlet valve A is shut, and initially the sterile air valve B is closed. Steam is applied at valve C and air is purged downwards through the filter to a bleed valve at the base. When the steam is issuing freely through the bleed valve, the valve B is opened to allow steam to pass into the fermenter as well as the filter. It is essential to adjust the bleed valve to ensure that the correct sterilization pressure is maintained in the fermenter and filter for the remainder of the sterilization cycle.

An alternative approach is to use a steam-jacketed air filter (Fig. 7.19). At the beginning of a sterilization cycle the valve A will be closed and steam passed through valves B and C, and bled out of D. Simultaneously steam will be passed into the steam jacket through valve F and out of G. When steam is issuing freely from valve D, valve F, may be opened and steam circulated into the fermenter. The bleed valve D will have to be adjusted to ensure that the correct pressure is maintained. Once the sterilization cycle is complete, valves B and E are closed and A is opened to allow air to pass through the heated filter and out of valve D to

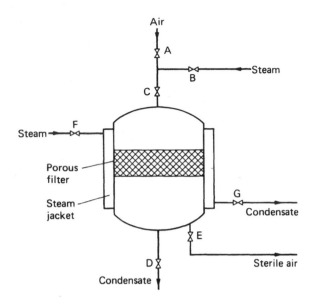

FIG. 7.19. Design for a simple steam-jacketed air-filter.

dry the filter. Finally the steam supply to the steam jacket is stopped. Valve D is closed and valve E opened, thus introducing sterile air into the fermenter to achieve a slight positive pressure in the vessel.

### Sterilization of the exhaust gas from a fermenter

Sterilization of the exhaust gas can be achieved by 0.2-$\mu$m filters on the outlet pipe (Fig. 5.21). Under normal operation aerosol formation may occur in the fermenter and moisture and solid matter may then plug the filter. To ensure satisfactory operation a cyclone separator (for solids) and a coalescer (for liquids) would be included upstream of two filters in series. The filters should be checked regularly to ensure that no viable cells are escaping. A test procedure to ensure integrity has been described by Hesselink et al. (1990).

### The addition of inoculum, nutrients and other supplements

To prevent contamination when operating a fermenter requiring GILSP, it is essential that both the addition vessel and the fermenter should be maintained at a positive pressure and that the addition port is equipped with a steam supply.

At Containment levels 1 and B2, the addition of inoculum, nutrients, etc. must be carried out in such a way that release of micro-organisms is restricted. This should be done by aseptic piercing of membranes or connections with steam locks. At Containment levels 2 and B3/4, no micro-organisms must be released during inoculation or other additions. In order to meet these stringent requirements all connections must be screwed or clamped and all pipelines must be steam sterilizable (Werner, 1992).

Further details of the aseptic inoculation of laboratory, pilot-plant and production fermenters are described in Chapter 6.

### Sampling

The sampling points fitted to larger fermenters also illustrate the principles for maintaining sterility. A sterile barrier must be maintained between the fermenter contents and the exterior when the sample port is not being used and it must be sterilizable after use. A simple design (Fig. 7.20) is described by Parker (1958). In normal operation valves A, B and C are closed and a barrier is formed by submerging the end of the sampling port in 40% formalin or a suitable substitute. A sample is obtained by removing the container of formalin and closing valve A. Valves B and C are then opened until the piping has been sterilized by steam. Valves B and C are then partially closed to allow a slow stream of steam and condensate out of the sampling port. Valve A is then opened slightly to cool the piping. The broth is discarded. Valve C is then closed and a sample is collected. Valve A is then closed and the piping is resterilized and left in the out of use arrangement.

An alternative arrangement for a sampling port is illustrated in Fig. 7.21 where the sterile barrier between

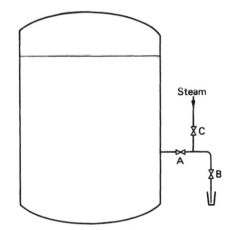

FIG. 7.20. Simple design for a sampling port (Parker, 1958).

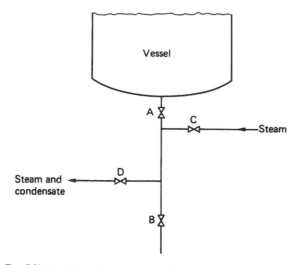

FIG. 7.21. An alternative simple sample port.

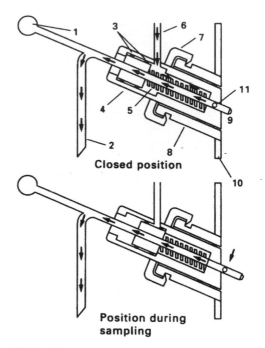

**Closed position**

**Position during sampling**

FIG. 7.22. Sterilizable sampling system for Category 1 (Werner, 1992). Key: (1) handle for opening and closing; (2) piping for sample or condensate, respectively; (3) 'O'-ring; (4) housing; (5) spring; (6) steam inlet; (7) union nut; (8) welding socket; (9) product; (10) wall of the bioreactor; (11) port of sampling tube.

the sample port and the exterior is a condensed flow of steam between valves C and D, valves A and B being closed. Valve D is connected to a steam trap to avoid condensate accumulation. To sterilize valve B prior to sampling, valve C is partially closed, valve D completely closed and valve B partially opened to allow a slow stream of steam and condensate out of the sampling port. Valve A is opened briefly to cool the pipe and the broth is discarded. Valve C is then closed and a sample is collected. Valve A is then closed and the piping is resterilized. In between collecting samples valves C and D are left partially open.

A more modern sterilizable system, as illustrated in Fig. 7.22, has been described by Werner (1992). In the closed position steam enters through the entry point 6 and passes through the sampling hole 11 and into the sampling hole 2. When sampling, the handle is pushed so that the hole in the sampling tube is in the fermenter and the vessel contents may be sampled. When the sampler is closed the unit can be resterilized.

Marshall *et al.* (1990) recognized that sampling requires good mechanical design and good operator practice to ensure sterility. Complex valve sequencing can lead to operator error. The automatic MX-3 rotary valve (Fig. 7.23) was developed by Marshall Biotechnology Ltd and is now marketed by New Brunswick Scientific Ltd. In this device, broth is continuously recycled to and from the fermenter and up to 12 samples can be taken automatically in sample bottles according to a pre-programmed sequence. The storage of the bottles in an integral refrigeration block re-

duces spoilage until they can be removed for analysis. This sampler can be used for GILSP.

In all the sampling devices described above, steam is used merely to maintain aseptic conditions. If the process organism has to be contained then the sample vessel must be designed accordingly. At level 1 or B2, removal of broth should be carried out in such a way that release of micro-organisms is minimized. Such a system (Fig. 7.24) is described by Janssen *et al.* (1990). The sampling bottle assembly is attached to the fermenter by a double-end shut-off quickconnect B and air is vented from the vessel during sampling by a suitable membrane filter. The pipe work is steamed for 15 minutes using 1.3 bar steam with the valve E on the sample bottle closed. With valves C and D closed and A and B open a sample can be taken, then the pipe is resterilized before removing the sample bottle.

At level 2 or B3, sampling should be done in a closed system (Figs 7.25 and 7.26), and all piping coming into contact with a sample must be sterilizable with steam. The sampling vessel should be closed during transport and samples should be examined under containment

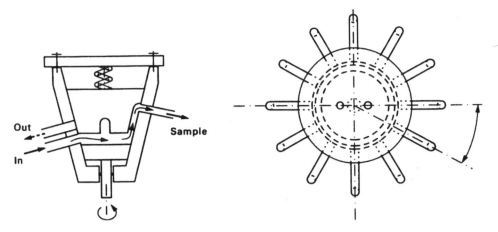

FIG. 7.23. Schematic representation of the MX-3 BioSampler rotary valve system.

conditions corresponding to those specified for the process (Werner, 1992).

### Feed ports

Additions of nutrients and acid/alkali to small fermenters are normally made via silicone tubes which are autoclaved separately and pumped by a peristaltic pump after aseptic connection. In large fermenter units, the nutrient reservoirs and associated piping are usually an integrated part which can be sterilized with the vessel. However, there may also be ports which are used

intermittently (Fig. 7.27). These can be sterilized *in situ* with steam after connection has been completed and before any additions are made.

### Sensor probes

Double 'O' ring seals have been used for many years to provide an aseptic seal for glass electrodes in stainless steel housings in fermenters using GILSP. This system is also suitable for levels 1 and B2, provided that release of micro-organisms is minimized and there are adequate disinfection procedures for dealing with leakages (Werner, 1992).

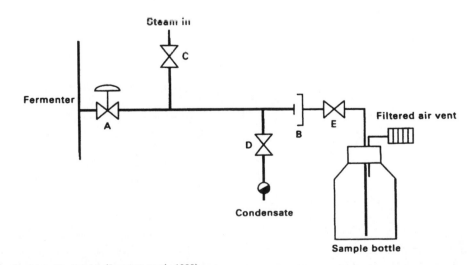

FIG. 7.24. Sample system for level 1 (Jannsen *et al.*, 1990).

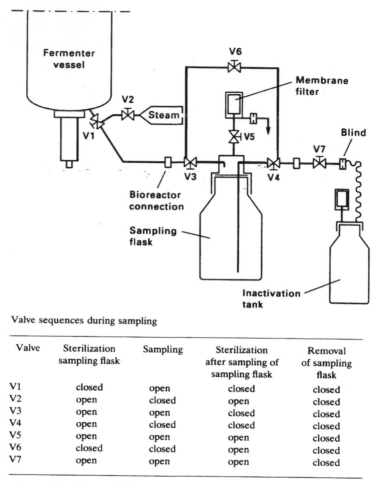

Valve sequences during sampling

| Valve | Sterilization sampling flask | Sampling | Sterilization after sampling of sampling flask | Removal of sampling flask |
|---|---|---|---|---|
| V1 | closed | open | closed | closed |
| V2 | open | closed | open | closed |
| V3 | open | open | closed | closed |
| V4 | open | closed | closed | closed |
| V5 | open | open | open | closed |
| V6 | closed | closed | open | closed |
| V7 | open | open | open | closed |

FIG. 7.25. Containment sampling unit for Chapter 2 (Werner, 1992).

At Containment levels 2 and B3/4, probes are fitted with triple 'O'-ring seals (Hambleton *et al.*, 1991). Although double 'O'-ring seals with steam tracing have been described they are not normally considered to be feasible (Leaver and Hambleton, 1992). The use of pre-inserted back-up probes is recommended as a means for dealing with probe failure rather than using a retractable electrode housing during a fermentation cycle because of the danger of leakage of broth (Hambleton *et al.*, 1991).

#### Foam control

In any fermentation it is important to minimize foaming. When foaming becomes excessive, there is a danger that filters become wet resulting in contamina-

tion. There is also the possibility that siphoning will develop, leading to the loss of all or part of the contents of the fermenter. Methods for foam control are considered in Chapter 8 and antifoams are discussed in Chapter 4. In certain circumstances antifoams may cause problems with aeration or downstream processing. Foam breakers which break down foam by an impact mechanism created by some type of rotating mechanism inside the fermenter are manufactured by Bioengineering AG (Switzerland), Chemap AG (Switzerland), Electrolux (Sweden), Frings (Germany) and New Brunswick Scientific Ltd (U.S.A.). Mechanical foam breakers used in Acetators for vinegar production are described later in this chapter. Domnick Hunter (U.K.) manufacture the 'Turbosep', in which foam is directed over stationary turbine blades in a separator and the liquid fraction is returned to the fermenter.

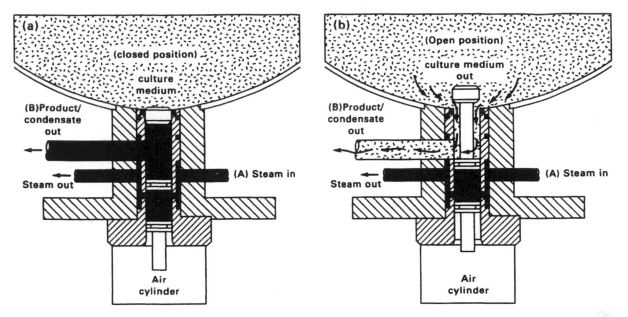

Fɪɢ. 7.26. Resterilizable harvest-sampling valve for Level 2 containment (New Brunswick, Hatfield, England). This spool-type valve is connected to the bottom of the fermenter vessel in the closed position (a), pressurized steam circulates throughout the entire valve body and through the product condensate line (B) via a steam inlet line (A). Aseptic withdrawal of samples is achieved with the valve in the open position (b). To prevent possible contamination when the plunger is raised, steam is circulated to the lower valve area. Action of the plunger is controlled by an air cyclinder.

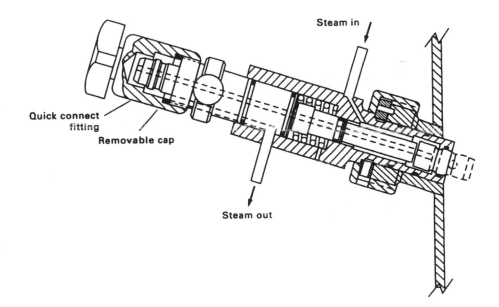

Fɪɢ. 7.27. Steam sterilizable liquid-feed port (Beck *et al.*, 1987).

At Containment level 1 or B2, the exhaust air system should shut down automatically if the foam control system fails (Werner, 1992).

### Monitoring and control of various parameters

These factors are considered in detail in Chapter 8.

## VALVES AND STEAM TRAPS

Valves attached to fermenters and ancillary equipment are used for controlling the flow of liquids and gases in a variety of ways. The valves may be:

1. Simple ON/OFF valves which are either fully open or fully closed.
2. Valves which provide coarse control of flow rates.
3. Valves which may be adjusted very precisely so that flow rates may be accurately controlled.
4. Safety valves which are constructed in such a way that liquids or gases will flow in only one direction.

When making the decision as to which valves to use in the design and construction of a fermenter it is essential to consider the following points:

1. Will the valve serve its chosen purpose? Is it suitable for aseptic operation or contained processes, and of the correct dimensions? Is the pressure drop across the valve tolerable?
2. Will the valve withstand the rigours of the process? The materials used to construct the valve should be suited to the process. It is also important to know whether corrosive liquids are used or synthesized during the process. The maximum operating temperature and pressure of the process should be known.
3. Are there welds or flanges in the valve?
4. Is the valve one which can be operated by remote control?
5. The cost and availability of suitable valves.
6. Can the valve be used for containment purposes?

A wide range of valves is available, but not all of them are suitable for use in fermenter construction (Solomons, 1969; Kemplay, 1980).

The valves described in this section open and close by (a) raising or lowering the blocking unit with a screw

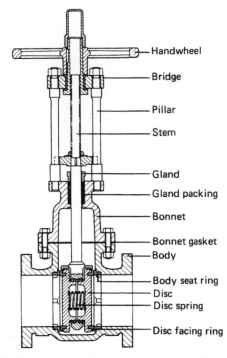

FIG. 7.28. Sectional view of a two-piece gate valve (British Valve Manufacturers Association, 1972).

Handwheel
Bridge
Pillar
Stem
Gland
Gland packing
Bonnet
Bonnet gasket
Body
Body seat ring
Disc
Disc spring
Disc facing ring

thread (rising stem), (b) a drilled sphere or plug, or a disc rotating in between two bearings or (c) a rubber diaphragm or tube which is pinched.

### Gate valves

In this valve (Fig. 7.28), a sliding disc is moved in or out of the flow path by turning the stem of the valve. It is suitable for general purposes on a steam or a water line for use when fully open or fully closed and therefore should not be used for regulating flow. The flow path is such that the pressure drop is minimal, but unfortunately it is not suitable for aseptic conditions as mash solids can pack in the groove where the gate slides, and there may be leakage round the stem of the valve which is sealed by a simple stuffing box. This means that the nut around the stem and the packing must be checked regularly.

### Globe valves

In this valve (Fig. 7.29), a horizontal disc or plug is raised or lowered in its seating to control the rate of

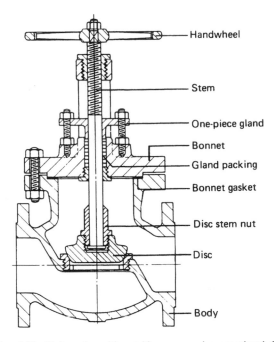

FIG. 7.29. Globe valve with outside screw and conventional disc (Kemplay, 1980).

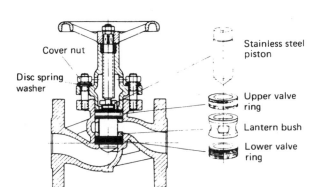

FIG. 7.30. Piston valve (Kemplay, 1980).

### Needle valves

The needle valve (Fig. 7.31) is similar to the globe valve, except that the disc is replaced by a tapered plug or needle fitting into a tapered valve seat. The valve can be used to give fine control of steam or liquid flow. Accurate control of flow is possible because of the variable orifice formed between the tapered plug and the tapered seat. The aseptic applications are very limited.

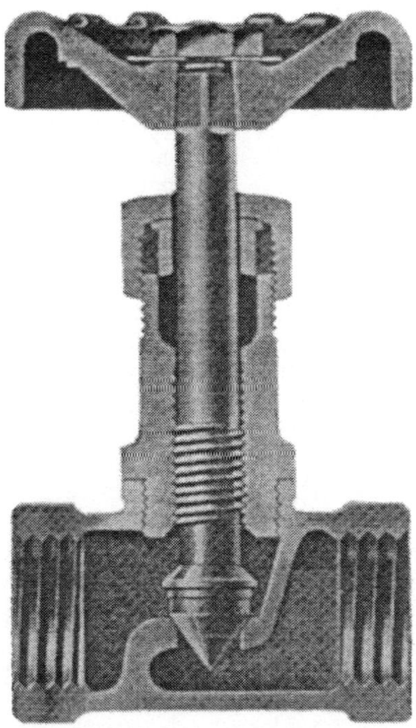

flow. This type of valve is very commonly used for regulating the flow of water or steam since it may be adjusted rapidly. It is not suitable for aseptic operation because of potential leakage round the valve stem, which is similar in design to that of the gate valve. There is a high-pressure drop across the valve because of the flow path.

In both the gate and globe valves it is possible to incorporate a flexible metallic membrane around the stem of the valve, to replace the standard packing. This modified type of valve can be operated aseptically, but is bigger and more expensive. Valves with non-rising stems have been used, but they are still potential sources of contamination.

### Piston valves

The piston valve (Fig. 7.30) is similar to a globe valve except that flow is controlled by a piston passing between two packing rings. This design has proved in practice to be very efficient under aseptic operation. It is important to sterilize them partly open so that steam can reach as far as possible into the valve body. There may be blockage problems with mycelial cultures. The pressure drop is similar to that of a globe valve.

FIG. 7.31. Needle valve for accurate control of flow rate (Kemplay, 1980).

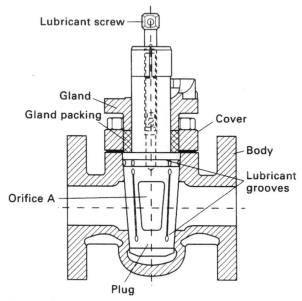

FIG. 7.32. Sectional view of lubricated taper plug valve (British Valve Manufacturers Association, 1972).

## Plug valves

In this valve there is a parallel or tapered plug sitting in a housing through which an orifice, A, has been been machined. The valve shown in Fig. 7.32 is in the closed position. When the plug is turned through 90° the valve is fully open and the flow path is determined by the cross-sectional area of A, which may not be as large as that of the pipeline. This type of valve has a tendency to leak or seize up, but the use of lubricants and/or sealants may overcome these problems. If suitable packing sleeves (e.g. compressed asbestos) are incorporated into the valve it will be suitable for use in a steam line as it is quick to operate, has protected seals, a minimal pressure drop and a positive closure. It can also provide good flow control.

## Ball valves

This valve (Fig. 7.33) has been developed from the plug valve. The valve element is a stainless-steel ball through which an orifice is machined. The ball is sealed between two wiping surfaces which wipe the surface and prevent deposition of matter at this point. The orifice in the ball can be of the same diameter as the pipeline, giving an excellent flow path. The valve is suitable for aseptic operation, can handle mycelial broths and can be operated under high temperatures and pressures. The pressure and temperature range is normally limited by the PTFE seat and stem seals.

## Butterfly valves

The butterfly valve (Fig. 7.34) consists of a disc which rotates about a shaft in a housing. The disc closes against a seal to stop the flow of liquid. This type of valve is normally used in large diameter pipes operating under low pressure where absolute closure is not essential. It is not suitable for aseptic operation.

## Pinch valves

In the pinch valve (Fig. 7.35) a flexible sleeve is closed by a pair of pinch bars or some other mechanism which can be operated by compressed air remotely or automatically. The flow rate can be controlled from 10 to 95% of rated flow capacity (Pikulik, 1976). The valve is suitable for aseptic operation with fermentation broths, even when mycelial, as there are no dead spaces in the valve structure, and the closing mechanism is isolated from the contents of the piping. Obviously, the sleeve of rubber, neoprene, etc., must be checked regularly for signs of wear.

## Diaphragm valves

Like the pinch valve, the diaphragm valve (Fig. 7.36) makes use of a flexible closure, with or without a weir. It may also be fitted with a quick action lever. This

194

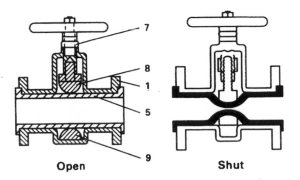

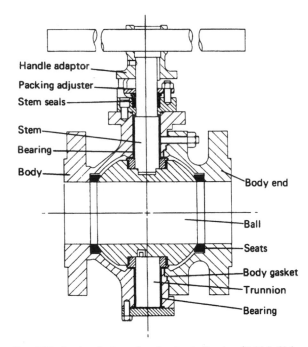

Handle adaptor
Packing adjuster
Stem seals
Stem
Bearing
Body
Body end
Ball
Seats
Body gasket
Trunnion
Bearing

FIG. 7.33. Sectional view of end-entry ball valve (British Valve Manufacturers Association, 1972).

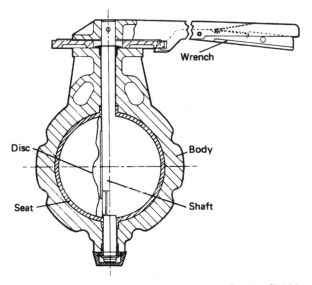

Wrench
Disc
Body
Seat
Shaft

FIG. 7.34. Sectional view of wafer-pattern butterfly valve (British Valve Manufacturers Association, 1972).

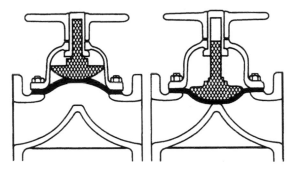

**Open**  **Shut**

FIG. 7.35. Sectional view of pinch valve in open and shut position: (1) body; (5) flexible tube; (7) spindle; (8) top pinch bar; (9) lower pinch bar (British Valve Manufacturers Association, 1966; Kemplay, 1980).

FIG. 7.36. Sectional views of weir-type diaphragm valves in open and closed positions (Thielsch, 1967).

pressure limits. Diaphragm failure, which is often due to excessive handling, is the primary fault of the valve. Ethylene propylene diene modified (EPDM) is now the preferred material. Hambleton *et al.* (1991) consider that a diaphragm valve with a steam seal on the 'clean' side (APV Ltd, Crawley, U.K.) is a potentially safer valve. However their widescale use would make a fermentation plant much more complex and expensive. Steam barriers on valves have been used by ICI plc for the 'Pruteen' air-lift fermenter (Smith, 1980; Sharp, 1989).

### The most suitable valve

Among these group of valves which have just been described, globe and butterfly valves are most commonly used for ON/OFF applications, gate valves for crude flow control, needle valves for accurate flow control and ball, pinch or diaphragm valves for all sterile uses.

valve is very suitable for aseptic operation provided that the diaphragm is of a material which will withstand repeated sterilization. The valve can be used for ON/OFF, flow regulation, and for steam services within

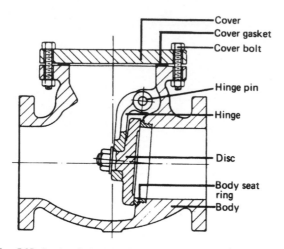

FIG. 7.37. Sectional view of swing check valve (Kemplay, 1980).

Ball and diaphragm valves are now the most widely used designs in fermenter and other biotechnology equipment (Leaver and Hambleton, 1992). Although ball valves are more robust, they contain crevices which make sterilization more difficult.

### Check valves

The purpose of the check valve is to prevent accidental reversal of flow of liquid or gas in a pipe due to breakdown in some part of the equipment. There are three basic types of valve: swing check, lift check and combined stop and check with a number of variants. The swing check valve (Fig. 7.37) is most commonly used in fermenter designs. The functional part is a hinged disc which closes against a seat ring when the intended direction of flow is accidentally reversed.

### Pressure-control valves

When planning the design of a plant for a specific process, the water, steam and air should be at different, but specified pressures and flow rates in different parts of the equipment. For this reason it is essential to control pressures precisely and this can be done using reduction or retaining valves.

PRESSURE-REDUCTION VALVES

Pressure-reduction valves are incorporated into pipelines when it is necessary to reduce from a higher to a lower pressure, and be able to maintain the lower pressure in the downstream side within defined limits irrespective of changes in the inlet pressure or changes in demand for gas, steam or water.

PRESSURE-RETAINING VALVES

A pressure-retaining valve will maintain pressure in the pipeline upstream of itself, and the valve is designed to open with a rising upstream pressure. It is constructed with a reverse action of the pressure-reducing valve.

### Safety valves

Safety valves must be incorporated into every air or steam line and vessel which is subject to pressure to ensure that the pressure will never exceed the safe upper limit recommended by the manufacturer or a code of practice and to satisfy government legislation and insurance companies. They must also be of the correct type and size to suit the operating conditions and be in sufficient numbers to protect the plant. The reliability of such valves is crucial.

In the simplest valves (Fig. 7.38), a spindle is lifted from its seating against the pressure of gas, steam or

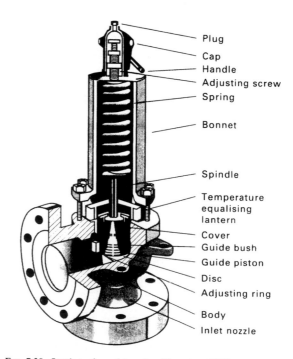

FIG. 7.38. Section of a safety valve (Kemplay, 1980).

liquid. Once the pressure falls below the value set by the tensioned spring, the spindle should return to its original position. However, the valve may stick open if waste material lodges on the valve seat and plant operators may interfere with the release pressure setting. Bursting/rupture discs may be used as an alternative and are of a more hygienic design than some valves (Leaver and Hambleton, 1992).

For GILSP operational categories, venting the escaping gas through the factory roof in emergencies would be satisfactory. At higher containment levels it would be necessary to treat the escaping gas. A kill tank with a HEPA venting filter is one possible solution, but not fully satisfactory.

It is also important to ensure that vessels do not collapse when vacuums occur. Vacuum release valves have been designed to cope with cold rinsing of a hot vessel or absorption of $CO_2$ when cleaning with caustic solutions (Maule, 1986). Shuttlewood (1984) has discussed design pressure and vacuum pressure in relation to vessel thickness requirements for design codes for American and British standards of safety.

### Steam traps

In all steam lines it is essential to remove any steam condensate which accumulates in the piping to ensure optimum process conditions. This may be achieved by incorporating steam traps, which will collect and remove automatically any condensate at appropriate points in steam lines. A steam trap has two elements. One is a valve and seat assembly which provides an opening, which may be of variable size, to ensure effective removal of any condensate. This opening may operate on an open/close basis so that the average discharge rate matches the steam condensation rate or the dimensions of this opening may be varied continually to provide a continuous flow of condensate The second element is a device which will open or close the valve by measuring some parameter of the condensate reaching it to determine whether it should be discharged (Armer, 1991).

The steam trap may be designed to operate automatically on the basis of:

1. The density of the fluid by using a float (ball or bucket) which will float in water or sink in steam (Fig. 7.39).
2. By measuring the temperature of the fluid, closing the valve at or near steam temperature and opening it when the fluid has cooled to a temperature, say 8° below the saturated steam temperature of the steam. The sensing element is often a stainless-steel capsule filled with a water–alcohol mixture which expands when steam is present and presses a ball into a valve seat or contracts when cooler condensate is present and lifts the ball from the seating. These are used in thermostatic and balanced pressure steam traps (Figs 7.40 and 7.41).
3. By measuring the kinetic effects of the fluid in motion. At a given pressure drop, low-density steam will move at a much greater velocity than higher density condensate. The conversion of pressure energy into kinetic energy can be used to control the degree of opening of a valve. This type of steam trap is not used very widely.

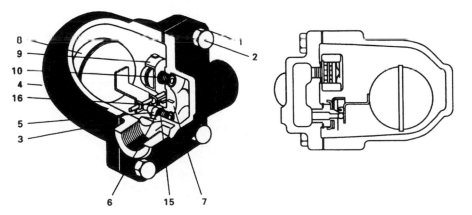

FIG. 7.39. Ball float steam trap with thermostatic air vent (Spirax/Sarco, Cheltenham, England). (5) valve seat; (8) ball float and lever; (9) air vent.

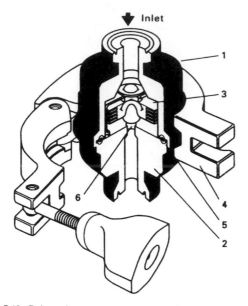

FIG. 7.40. Balanced pressure steam trap (Spirax/Sarco, Cheltenham, England). (1) body; (2) end connection; (3) element; (4) Tri-Clover clamp; (5) Tri-Clover joint gasket; (6) ball seat.

Armer (1991) thinks that balanced pressure thermostatic traps are the most suitable for autoclaves and sterilizers. Some can operate close to steam-saturation temperature with little back up of condensate. Sarco Ltd (Cheltenham, U.K.) make a hermetically sealed balanced pressure steam trap (Model No. SBP 30; Fig. 7.42) which has been used with a high containment level pilot-plant fermenter (Hambleton et al., 1991).

Any steam condensate may be (a) returned to the

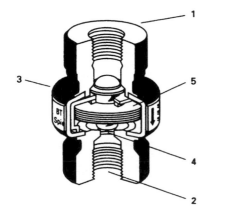

FIG. 7.41. Stainless steel thermostatic steel trap (Spirax/Sarco, Cheltenham, England). (1) inlet; (2) outlet; (3) body; (4) ball seat; (5) element.

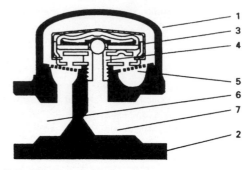

FIG. 7.42. SBP 30 Sealed balanced pressure thermostatic steam trap (Spirax/Sarco, Cheltenham, England). (1) cover; (2) body; (3) element/capsule; (4) ball seat assembly; (5) strainer screen; (6) inlet; (7) outlet.

boiler, (b) used in a steam condensate line, (c) vented to waste or (d) piped to a kill tank in a containment location

### Complete loss of contents from a fermenter

Leaver and Hambleton (1991) have suggested building a wall or dyke around a fermenter which would retain the entire contents were they to leak out. The escaping fluid would then be pumped into a suitable vessel for sterilization. The floor should be covered with an impervious epoxy based surface and coved up the walls (Kennedy et al., 1990).

### Testing new fermenters

When a new fermenter (10 dm$^3$ or larger) has finally been assembled in a factory it must be hydraulically pressure tested and checked by independent inspectors using nationally approved test procedures. A vessel which has passed the approved tests is given a certificate which is recognized by insurance companies which will allow it to be operated in a laboratory or factory. If any subsequent modifications are made to a vessel in this size range it must be retested and certificated before it can be legally used. Thus a number of fermenter manufacturers cut extra 'O' ring grooves and steam traces in head plates or ports to satisfy higher containment-level needs so that the vessels will not have to be modified subsequently and to avoid extra pressure checks.

## OTHER FERMENTATION VESSELS

Some of the other forms of fermentation vesels will now be considered. These vessels have more limited applications and have been developed for specific purposes or closely related processes. Some are historical developments, such as the earliest forms of packed tower, others were being developed in parallel with the standard mechanically stirred fermenter during the 1940s, while other approaches are more recent. Aspects of this topic have been reviewed by Prokop and Votruba (1976), Katinger (1977), Hamer (1979), Levi *et al.* (1979), Solomons (1980), Schugerl (1982, 1985), Sittig (1982), Winkler (1990) and Atkinson and Mavituna (1991a).

### The Waldhof-type fermenter

The investigations on yeast growth in sulphite waste liquor in Germany, Japan and the United States of America led to the development of the Waldhof-type fermenter (Inskeep *et al.*, 1951; Watanabe, 1976). Inskeep *et al.* (1951) have given a description of a production vessel based on a modification of the original design of Zellstofffabrik Waldhof. The fermenter was of carbon steel, clad in stainless steel, 7.9 m in diameter and 4.3-m high with a centre draught tube 1.2 m in diameter. A draught tube was held by tie rods attached to the fermenter walls. The operating volume was 225,000 dm$^3$ of emulsion (broth and air) or 100,000 dm$^3$ of broth without air. Non-sterile air was introduced into the fermenter through a rotating pin-wheel type of aerator, composed of open-ended tubes rotating at 300 rpm (Fig. 7.43). The broth passed down the draught tube from the outer compartment and reduced the foaming.

### Acetators and cavitators

Fundamental studies by Hromatka and Ebner (1949) on vinegar production showed that if *Acetobacter* cells were to remain active in a stirred aerated fermenter, the distribution of air had to be almost perfect within the entire contents of the vessel. They solved the full-scale problem by the use of a self-aspirating rotor (Ebner *et al.*, 1967). In this design (Fig. 7.44), the turning rotor sucked in air and broth and dispersed the mixture through the rotating stator (d). The aerator also worked without a compressor and was self-priming.

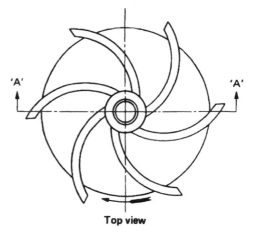

**Top view**

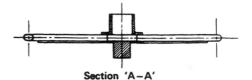

**Section 'A–A'**

Fig. 7.43. Top view and section of a Waldhof aeration wheel (Inskeep *et al.*, 1951).

Vinegar fermentations often foam and chemical antifoams were not thought feasible because they would decrease aeration efficiency (Chapter 9) and additives were not desirable in vinegar. A mechanical defoamer therefore had to be incorporated into the vessel and as foam builds up it is forced into a chamber in which a rotor turns at 1,000 to 1,450 rpm. The centrifugal force breaks the foam and separates it into gas and liquid. The liquid is pumped back into the fermenter and the gas escapes by a venting mechanism. Descriptions of the design and various sizes of model have been given by Ebner *et al.* (1967). Fermenters of this design are manufactured by Heindrich Frings, Bonn, Germany. An illustration of the basic components is given in Fig. 7. 45. In 1981, 440 acetators were in operation all over the world with a total production of 767 × 10$^6$ dm$^3$ year$^{-1}$. The major vinegar producers were the U.S.A. (152 × 10$^6$ dm$^3$), France (90 × 10$^6$ dm$^3$) and Japan (46 × 10$^6$ dm$^3$) while the remainder was produced by over 50 countries (Ebner and Follmann, 1983).

Chemap AG of Switzerland manufacture the Vinegator. A self-aspirating stirrer and a central suction tube aerates a good recirculation of liquid. Additional air is provided by a compressor. Foam is broken down by a mechanical defoamer (Ebner and Follmann, 1983).

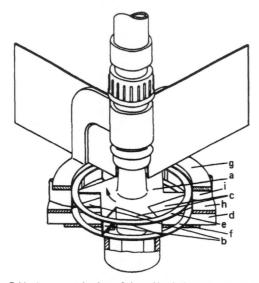

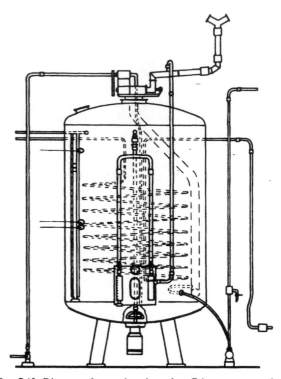

FIG. 7.44. Axonometric view of the self-priming aerator used with the Frings generator (Ebner *et al.*, 1967). The turbine is designed as a hollow body (a) with openings which are arranged radially and open against the direction of rotation (b). The openings are shielded by vertical sheets (c). The turbine sucks liquid from above and below and mixes it with air sucked in through the openings. The suspension is thrown through the stator (d) towards the circumference of the tank. An upper and lower ring on the turbine (e,f) helps to direct and regulate the air–liquid suspension. The stator (d) consists of an upper and lower ring (g,h) which are connected by vertical sheets (i) inclined at about 30° towards the radius.

FIG. 7.45. Diagram of a section through a Frings generator fermenter used for the manufacture of vinegar. The fermenter, which can be used semicontinuously or continuously, employs vortex stirring (Greenshields, 1978).

At least three other vinegar fermenters are no longer manufactured. The Bourgeois process was sold in Europe between 1955 and 1980 and the Fardon process between 1960 and 1975. The Yeomans cavitator was sold in the U.S.A. between 1959 and 1970 (Cohee and Steffen, 1959; Mayer, 1961; Ebner and Follmann, 1983). The fermenter had an agitator of different design, but similar operating principles to the acetator. Uniform distribution of air bubbles was obtained by means of the circulation pattern created by the centrally located draught (draft) tube. The agitator withdrew liquid from the draught tube and pushed liquid into the main part of the vessel. The outer level rose and overflow occurred back into the top of the draught tube.

### The tower fermenter

It is difficult to formulate a single definition which encompasses all the types of tower fermenter. Their main common feature appears to be their height:diameter ratio or aspect ratio. Such a definition has been given by Greenshields *et al.* (1971) who described a tower fermenter as an elongated non-mechanically stirred fermenter with an aspect ratio of at least 6:1 for the tubular section or 10:1 overall, through which there is a unidirectional flow of gases. Several different types of tower fermenter exist and these will be examined in broad groups based on their design.

The simplest types of fermenter are those that consist of a tube which is air sparged at the base (bubble columns). This type of fermenter was first described for citric acid production on a laboratory scale (Snell and Schweiger, 1949). This batch fermenter was in the form of a glass column having a height:diameter ratio of 16:1 with a volume of 3 dm³. Humid sterile air was supplied through a sinter at the base. Steel *et al.* (1955) reported an increase in scale to 36 dm³ for a fermenter of this type. Pfizer Ltd has always used non-agitated tower vessels for a range of mycelial fermentation processes including citric acid and tetracyclines (Solomons, 1980; Carrington *et al.*, 1992). Recently Pfizer Ltd sold their citric acid interests to Arthur Daniels Midland who are operating such vessels up to 23 m high (Burnett, 1993).

Volumes of between 200 m³ and 950 m³ have been reported elsewhere (Rohr *et al.*, 1983).

In 1965 the brewing industry began to use tower fermenters which were more complex in design and could be operated continuously. Hall and Howard (1965) described small-scale fermenters that consisted of water jacketed tubes of various dimensions which were inclined at angles of 9 to 90° to the horizontal. Air and mash were passed in at the base and effluent beer was removed at the top.

A vertical-tower beer fermenter design (Chapter 2) was patented by Shore *et al.* (1964). Perforated plates were positioned at intervals in the tower to maintain maximum yeast production. The settling zone which could be of various designs, was to provide a zone free of rising gas so that the cells could settle and return to the main body of the tower and the clear beer could be removed. This design must be considered as an intermediate between single- and multistage systems. Towers of up to 20,000 dm³ capacity and capable of producing up to 90,000 dm³ day⁻¹ have been installed. Greenshields and Smith (1971) commented that it was difficult to predict the upper operating limits for these fermenters. Experiments with particular yeast strains in pilot-size towers were essential to establish optimum full-scale operating conditions.

The next group of tower fermenters are the multistage systems, first described by Owen (1948) and Victorero (1948) for brewing beer, although these systems were not used on an industrial scale. Later work reported using these systems includes continuous cultures of *E. coli* (Kitai *et al.*, 1969), bakers' yeast (Prokop *et al.*, 1969) and activated sludge (Lee *et al.*, 1971; Besik, 1973). The fermenters used by all these workers were basically similar. Each consisted of a column forming the body of the vessel and a number of perforated plates which were positioned across the fermenter, dividing it into compartments. Approximately 10% of the horizontal plate area was perforated. The possibility of introducing media into individual stages independently was discussed by Lee *et al.* (1971). Besik (1973) decribed a down-flow tower in which substrate was fed in at the top and overflowed through down spouts to the next section while air was supplied from the base. Schugerl, Lucke and Oels (1977) have written a comprehensive general review.

### Cylindro-conical vessels

The use of cylindro-conical vessels in the brewing of lager was first proposed by Nathan (1930), but his ideas were not adopted for the brewing of lagers and beers until the 1960s (Hoggan, 1977). Breweries throughout the world have now adopted this method of brewing. The vessel (Fig. 7.46) consists of a stainless-steel vertical tube with a hemispherical top and a conical base with an included angle of approximately 70° (Boulton, 1991).

Aspect ratios are usually 3:1 and fermenter heights are 10 to 20 m. Operating volumes are chosen to suit the individual brewery requirements, but are often 150,000 to 200,000 dm³. Vessels are not normally agitated unless a particularly flocculant yeast is used, but small impellers may be used to ensure homogeneity when filling with wort (Boulton, 1991). In the vessel, the wort is pitched (inoculated) with yeast and the fermentation proceeds for 40 to 48 hours. Mixing is achieved by the generation of carbon dioxide bubbles that rise rapidly in the vessel. Temperature control is monitored by probes positioned at suitable points within the vessel. A number of cooling jackets are fitted to the vessel wall to regulate and cause flocculation and settling of the yeast (Ulenberg *et al.*, 1972; Maule, 1986;

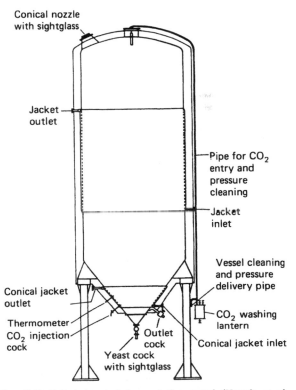

FIG. 7.46. Cylindro-conical fermentation vessel (Hough *et al.*, 1971).

Boulton, 1991). The fermentation is terminated by the circulation of chilled water via the cooling jackets which results in yeast flocculation. Thus, it is necessary to select a yeast strain which will flocculate readily in the period of chilling. Part of this yeast may be withdrawn and used for repitching another vessel. The partially cleared beer may be left to allow a secondary fermentation and conditioning. Some of the adantages of this vessel in brewing are:

1. Reduced process times may be achieved due to increased movement within the vessel.
2. Primary fermentation and conditioning may be carried out in the same vessel.
3. The sedimented yeast may be easily removed since yeast separation is good.
4. The maturing time may be reduced by gas washing with carbon dioxide.

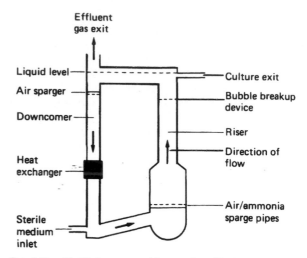

FIG. 7.47a. Air-lift fermenter with outer loop (Taylor and Senior, 1978).

### Air-lift fermenters

An air-lift fermenter (Fig. 7.47) is essentially a gas-tight baffled riser tube (liquid ascending) connected to a downcomer tube (liquid descending). Figure 7.47a shows an external riser and Fig. 7.47b an internal riser. Air or gas mixtures are introduced into the base of the riser by a sparger during normal operating conditions. The driving force for circulation of medium in the vessel is produced by the difference in density between the liquid column in the riser (excess air bubbles in the medium) and the liquid column in the downcomer (depleted in air bubbles after release at the top of the loop). Circulation times in loops of 45-m height may be 120 seconds. More details on liquid circulation and mixing characteristics are discussed by Chen (1990). This type of vessel can be used for continuous culture. The first patent for this vessel was obtained by Scholler and Seidel (1940).

It would be uneconomical to use a mechanically stirred fermenter to produce SCP (single-cell protein) from methanol as a carbon substrate, as heat removal would be needed in external cooling loops because of the high rate of aeration and agitation required to operate the process. To overcome these problems, particularly that of cooling the medium when mechanical agitation is used, air-lift fermenters with outer or inner loops (Fig. 7.47) were chosen. Development work for operational processes for SCP has been done by ICI plc in Great Britain (Taylor and Senior, 1978; Smith, 1980), Hoechst AG-Uhde GmbH in Germany (Faust et

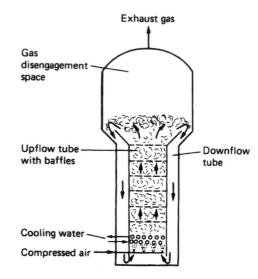

FIG. 7.47b. Air-lift fermenter with inner loop (Smith, 1980).

al., 1977) and Mitsubishi Gas Chemical Co. Inc. in Japan (Kuraishi et al., 1978). Although ICI plc initially used an outer-loop system in their pilot plant, all three companies preferred an inner-loop design for large-scale operation. Hamer (1979) and Sharp (1989) have reviewed these fermenters. In the ICI plc continuous process, air and gaseous ammonia were introduced at the base of the fermenter. Sterilized methanol, other nutrients and recycled spent medium were also introduced into the downcomer. Heat from this exothermic fermentation was removed by surrounding part of the downcomer with a cooling jacket in the pilot plant, while at full scale ($2.3 \times 10^6$ dm$^3$) it was found necessary to insert cooling coils at the base of the riser.

Unfortunately the production of SCP for animal feed has not proved an economic proposition because of the price of methanol and the competition from animal feeds based on arable protein crops. ICI plc's vessel at Billingham, U.K. has now been dismantled.

In 1964, Rank Hovis McDougall decided to develop a protein-rich food primarily for human consumption (Trinci, 1992). They have grown *Fusarium graminareum* on a wheat starch based medium using a modified ICI plc 40 m$^3$ air-lift fermenter (Fig. 7.48) to produce the myco-protein Quorn. The use of an air-lift fermenter for culture of a mycelial fungus would seem unusual as lower rates of oxygen transfer occur in a viscous culture which give rise to lower biomass yields. Because low shear conditions are present in the vessel, long fungal hyphae can be cultured (the preferred product

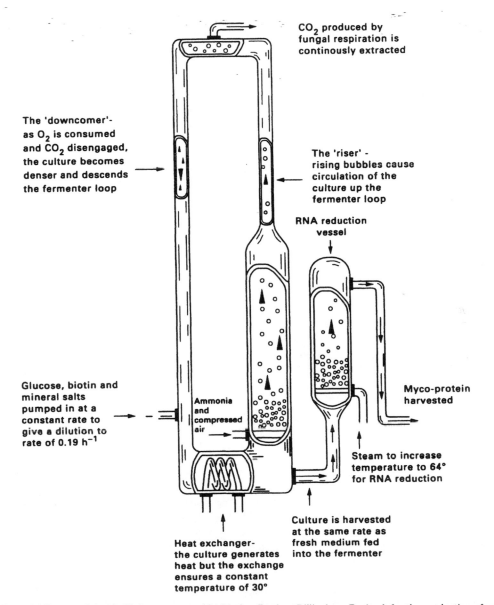

FIG. 7.48. Schematic diagram of the air-lift fermenter used by Marlow Foods at Billingham, England, for the production of myco-protein in continuous flow culture (Trinci, 1992).

form) even though production yields are only 20 g dm$^{-3}$. At the present time a fermenter to produce 10,000 tonnes per annum of myco-protein is considered economically feasible.

Okabe *et al.* (1993) modified a 3-dm$^3$ air-lift fermenter by putting stainless steel four-mesh sieves at the top and bottom of the draught tube to manipulate the morphology of *Aspergillus terreus* for optimum production of itaconic acid as the culture circulates in the vessel flow path. The fungal morphology was an intermediate state between pellets and pulp. Using this vessel, the itaconic acid production rate (g dm$^{-1}$ h$^{-1}$) was double that obtained with a stirred fermenter or an air-lift fermenter with a conventional draught tube.

Work has also begun to examine oxygen transfer rates with modified draught tubes. Carrington *et al.* (1992) used a 20 m$^3$ pilot-scale bubble column fermenter fitted with an internal helical cooling coil (Fig. 7.49) or a solid draught tube. The fermentation studied was a commercial *Streptomyces* antibiotic fermentation in a complex medium which produced a viscous non-Newtonian broth. Tracer studies indicated that the vessel fitted with only the cooling coil behaved like an air-lift fermenter with a region of good mixing in the zone above the cooling coil. The coil acted as a leaky draught tube with back mixing taking place between the coils into the riser section. No poorly oxygenated zones were observed. Liquid velocities of 1 m sec$^{-1}$ were measured giving circulation times of 9 to 12 seconds and mixing times of 14 to 18 seconds. The $K_L a$ at different power inputs and viscosities was found to increase almost linearly with increasing power input and decreased exponentially with increasing viscosity. When a solid draught tube was installed inside the cooling coil the circulation time was similar, but the mixing time increased to 18 to 24 seconds. The $K_L a$ was also determined at different power inputs and viscosities. This gave a reduction of 5 to 25% in $K_L a$ with the biggest reduction at a high viscosity.

Wu and Wu (1990) compared $K_L a$s in a range of mesh draught tubes and a solid draught tube in an air-lift vessel of 15 dm$^3$ working volume. When a 24-mesh tube was used at high superficial gas velocities the $K_L a$ was double that of the same vessel with a solid draught tube.

Bakker *et al.* (1993) have developed a multiple air-lift fermenter in which three air-lift fermenters with internal loops are incorporated into one vessel (Fig. 7.50). Fresh medium is fed into the central compartment, depleted medium overflows into the middle compartment, from here to the outer compartment where medium is eventually discharged. The hydrodynamics

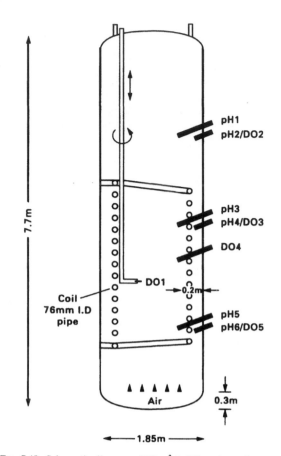

FIG. 7.49. Schematic diagram of 20 m$^3$ bubble column fermenter fitted with internal cooling coil (Carrington *et al.*, 1992). (pH and DO indicate positions of pH and oxygen electrodes.)

and mixing in the middle compartment were found to be comparable with those obtained with conventional internal loop air-lift vessels, although some tangential liquid flow was observed in this compartment.

## The deep-jet fermenter

Some designs of continuous culture fermenter achieve the necessary mechanical power input with a pump to circulate the liquid medium from the fermenter through a gas entrainer and back to the fermenter (Fig. 7.51; Hamer, 1979; Meyrath and Bayer, 1979). Two basic construction principles have been used for the gas entrainer nozzles. The injector and the ejector (Fig. 7.52 ; Reuss, 1992). In an injector a jet of medium is surrounded by a jet of compressed air. The gas from the outlet enters the larger tube with a nozzle velocity of 5 to 100 m s$^{-1}$ and expands in the tube to

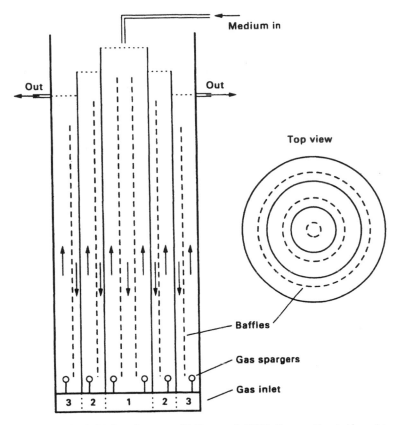

FIG. 7.50. Three-compartment multiple air-lift loop fermenter (Bakker *et al.*, 1993). Cross-sectional side and top view.

form large air bubbles which are dispersed by the shear of the water jet. In an ejector the liquid jet enters into a larger converging–diverging nozzle and entrains the gas around the jet. The gas which is sucked into the converging–diverging jet is dispersed in that zone. One of the industrial-scale fermenters using the ejector design is marketed by Vogelbusch (Vogelbusch AG, Vienna, Austria). Partially aerated medium is pumped by a multiphase pump through a broth cooler to an air entrainer above the fermenter. The air–medium mixture falls down a slightly conical shaft at a high velocity and creates a turbulence in the fermenter. Two-thirds of the exhaust gas is vented from the fermenter headspace and the remainder via the multiphase pump. Oxygen-transfer rates of 4.5 g dm$^{-3}$ h$^{-1}$ with an energy consumption of 1 kW h$^{-1}$ kg$^{-1}$ have been achieved for industrial-scale yeast production from whey using such a fermentation system.

### The cyclone column

Dawson (1974) developed the cyclone column, par-

ticularly for the growth of filamentous cultures (Fig. 7.53). The culture liquid was pumped from the bottom to the top of the cyclone column through a closed loop. The descending liquid ran down the walls of the column in a relatively thin film. Nutrients and air were fed in near the base of the column whilst the exhaust gases left at the top of the column. Good gas exchange, lack of foaming and limited wall growth have been claimed with this fermenter. Dawson (1974, 1988) has listed a number of potential bacterial, fungal and yeast applications including the batch production of a vaccine for scours in calves with the vessel being operated as batch, continuous or fed-batch.

### The packed tower

The packed tower is a well established application of immobilized cells. A vertical cylindrical column is packed with pieces of some relatively inert material, e.g. wood shavings, twigs, coke, an aggregate or polythene. Initially both medium and cells are fed into the top of the packed bed. Once the cells have adhered to

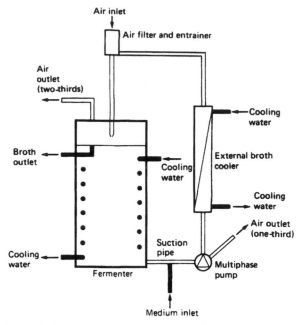

FIG. 7.51. Diagram of the Vogelbusch deep-jet fermenter system (from Schreier, 1975; Hamer, 1979).

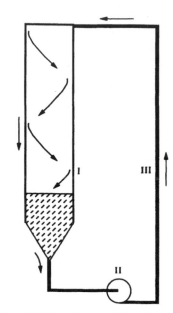

FIG. 7.53. Schematic diagram of a cyclone column fermenter. (I) cyclone column; (II) circulating pump; (III) recirculating limb (Dawson, 1974).

the support and are growing well as a thin film, fresh medium is added at the top of the column and the fermented medium is removed from the bottom of the column. The best known example is the vinegar generator, in which ethanol was oxidized to acetic acid by strains of *Acetobacter* supported on beech shavings; the first recorded use was in 1670 (Mitchell, 1926). More recently, packed towers have been used for

sewage and effluent treatment (Noble *et al.,* 1964). In treatment of gas liquor, a column was packed to a height 7.9 m with 'Dowpac', a polystyrene derivative. The main advantages compared with other methods of effluent treatment being its simplicity of operation and a saving in land because of the increased surface areas within the column. Other possible applications with immobilized cells are now being investigated.

### Rotating-disc fermenters

Rotating-disc contactors have been used in effluent treatment (Chapter 11). They utilize a growing microbial film on slow rotating discs to oxidize the effluent. Anderson and Blain (1980) have used the same principle to construct small fermenters of up to 40-dm$^3$ working volume. A range of filamentous fungi, including species of *Aspergillus*, *Rhizopus*, *Mucor* and *Penicillium*, could be grown on the polypropylene discs. It has been possible to obtain yields of 80 g dm$^{-3}$ of citric acid from *A. niger* using this design of fermenter.

## ANIMAL CELL CULTURE

Interest in the *in vitro* cultivation of animal cells has developed because of the need for large scale produc-

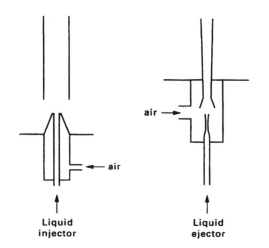

FIG. 7.52. Gas entrainer nozzles of deep-jet fermenters (Reuss, 1992).

tion of monoclonal antibodies, hormones, vaccines and other products which are difficult or impossible to produce synthetically or by using other culture techniques.

Animal cells are usually more nutritionally demanding than microbial cells. They lack cell walls which makes them sensitive to shear and extremes of osmolarity. The doubling times are normally 12 to 48 hours and cell densities in suspension cultures rarely exceed $10^6$ to $10^7$ cells $cm^{-3}$. Two distinct modes of growth can be recognized:

1. *Anchorage dependent cells*. These cells require a solid support for their replication. They produce cellular protrusions (pseudopodia) which allow them to adhere to positively charged surfaces and often grow as monolayers.
2. *Anchorage independent cells*. These cells do not require a support and can grow as a suspension in submerged culture. Established and transformed cell lines are normally in this category.

There are also intermediate categories of cells which may be grown as anchorage dependent or suspension cells. It is also possible to grow some anchorage dependent cells in suspension, provided the cells can be grown on suitable microcarriers.

A range of free and immobilized culture systems have therefore been developed for the culture of different lines of animal cells and their products. The range of available equipment and techniques may be confusing to scientists not familiar with animal cells who may have to decide on the most appropriate culture systems for laboratory, pilot or production scale. Because of the relatively small quantities of product required, the volume at production scale may only be 100 to 10,000 $dm^3$. Some useful introductions include Griffiths (1986, 1988), Propst *et al.* (1989), Lavery (1990) and Bliem *et al.* (1991).

Shear is a phenomenon recognized as being critical to the scale up of animal cell culture processes, irrespective of the cell line or reactor configuration (Bliem and Katinger, 1988). Shear may influence the cell culture causing damage which may result in cell death or metabolic changes. The shear sensitivity of animal cells may vary between cell lines, the phase of growth or with a change to fresh medium

Mijnbeek (1991) has reviewed research on shear stress of free and immobilized animal cells in stirred fermenters and air-lift fermenters. It was concluded that the predominant damage mechanism in both types

of vessel was due to sparging and the break up of bubbles on the medium surface. This type of damage causing cell death might be reduced by increasing the height to diameter ratio in the vessel, increasing the bubble size, decreasing the gas flow rate or by adding protective agents. Other damage mechanisms in stirred fermenters are caused by cell–microcarrier eddies and microcarrier–microcarrier interactions. Damage of this type may be reduced by reducing the impeller speed, impeller diameter, microcarrier size and concentration or by increasing the viscosity of the medium.

The maximum cell densities obtainable in stirred and air lift fermenters are often only $10^6$ cells $cm^{-3}$. This is not ideal if secreted proteins are present only in very low concentrations and mixed with other proteins present in the original medium. A number of modified fermenters and reactors have been developed to grow cells at higher concentrations using microcarriers, encapsulation, perfusion, glass beads or hollow fibres to obtain the required product at higher concentrations.

### Stirred fermenters

Unmodified stirred fermenters have been used for the batch production of some virus vaccines (Propst *et al.*, 1989), but modified vessels are used for most cultures (Propst *et al.*, 1989; Lavery, 1990). The modifications made to fermenters are to reduce the possibility of cell damage due to shear, heat or contamination. Marine propellers revolving at a slow speed (10 to 100 rpm) will normally provide adequate mixing. Hemispherical bottoms on the vessels will ensure better mixing of the broth at slow stirrer speeds. Water jacket heating is often preferred since heating probes may give rise to localized zones of high temperature which might damage some of the cells. Magnetic driven stirrers may be used to reduce the risk of contamination. A novel sparger–impeller design which improves aeration at slow speeds is incorporated into the Celligen system manufactured by New Brunswick, U.S.A. (Fig. 7.54; Beck *et al.*, 1987). When the impeller rotates the swept-back ports produce a negative pressure inside the hollow impeller complex. This creates a suction lift that produces highly efficient circulation and gas transfer at low rpm without damaging the cells. Gases are introduced through a ring sparger into the medium as it circulates through a fine stainless steel mesh jacket which excludes cells and microcarriers. Because the gas sparging is restricted to a relatively small zone, foaming

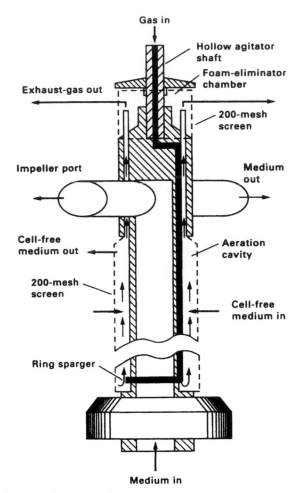

FIG. 7.54. Sparger-impeller in the Celligen cell-culture fermenter (New Brunswick Ltd, Hatfield, England).

is reduced, and any foam formed is broken up by the mesh in the foam eliminator chamber.

## Air-lift fermenters

Air-lift fermenters have proved ideal for growth of some cell lines because of the gentle mixing action and reduced shear forces when compared with those in stirred vessels. The absence of a stirrer and associated seals excludes a potential source of contamination (Griffiths, 1988; Propst *et al.*, 1989; Lavery, 1990).

Vessels of 1000 to 2000 dm$^3$ are commercially available. Celltech Ltd, U.K., have used such vessels to produce monoclonal antibodies from hybridoma cells (Wilkinson, 1987).

## Microcarriers

Microcarriers may provide a solution to the problem of growing anchorage-dependent cultures in suspension culture in fermenters, by providing the necessary surface for attachment. Animal cells normally have a net negative surface charge and will attach to a positively charged surface by electrostatic forces. Van Wezel (1967) made use of this property and attached anchorage dependent cells to chromatographic grade DEAE Sephadex A-50 resin beads and suspended the coated beads in a slowly stirred liquid medium. The density of the electrostatic charges on the microcarrier surface is critical if cell growth at high bead concentrations is to be achieved. If the net charge density on the bead surface is too low, cell attachment will be restricted. When the charge density is too high, apparent toxic effects will limit cell growth (Fleischaker, 1987). A number of microcarrier beads, manufactured from dextran, cellulose, gelatin, plastic or glass, are now commercially available (Fleischaker, 1987; Butler, 1988). Dextran microcarriers have been used for large scale production of viral vaccines and interferon. Unfortunately some of the microcarriers cost £1200 to £1500 kg$^{-1}$.

## Encapsulation

At least three methods of encapsulation have been developed (Griffiths, 1988; Lavery, 1990). Encapsel, a technique developed by Damon Biotechnology, U.S.A., traps the animal cells in sodium alginate spheres which are then coated with polylysine to form a semi-permeable membrane. The enclosed alginate gel is solubilized with sodium citrate to release the cells into free suspension within the capsules which are usually 50 to 500-$\mu$m diameter. After a few weeks growth it is possible to obtain cell concentrations of $5 \times 10^8$ cm$^{-3}$ and product levels 100 times higher than with free cells can be achieved. The high molecular weight products are retained within the capsules. This technique has been used commercially for monoclonal antibody production.

In a second method the cells are entrapped in calcium alginate which will allow high molecular weight products to diffuse into the medium. Unfortunately, the spheres tend to be 0.5 to 1.0 mm diameter and slow diffusion into the spheres may cause nutrient limitations. Alternatively the cells can be entrapped in agarose beads in which the cells are contained in a honeycomb matrix within the gel. These capsules have a wide size distribution and a low mechanical strength compared with alginate ones.

### Hollow fibre chambers

Anchorage dependent cells can be cultured at densities of $10^8$ cells $cm^{-3}$ using bundles of hollow fibres held together in cartridge chambers (Hirschel and Gruenberg, 1987; Griffiths, 1988). The cells are grown in the extra capillary spaces (ECS) within the cartridge. Medium and gases diffuse through from the capillary lumea to the ECS. The molecular weight cut-off of the fibre walls may be selected so that the product is retained in the ECS or released into the perfusing medium. Many chambers will be needed for scale-up because of diffusion limitations in larger chambers. These have been used to study production of monoclonal antibodies, viruses, gonadotropin, insulin and antigens.

### Packed glass bead reactors

Packed glass bead reactors have proved to be useful for long term culture of attached dependent cell lines. It is possible to obtain cell densities of $10^{10}$ viable cells in a 1-$dm^3$ vessel with moderate medium flow rates through the vessel (Fig. 7.55; Propst *et al.*, 1989). Increasing the size of vessels causes problems with mass transfer of oxygen and nutrients and scale up can be achieved by increasing the number of small vessels. This technique is available only as a contract production service from Bioresponse Inc. of California, U.S.A.

### Perfusion cultures

Perfusion culture is a technique where modified fermenters of up to 100 $dm^3$ are gently stirred and broth is withdrawn continuously from the vessel and passed through a stainless steel or ceramic filter. This type of culture is sometimes referred to as spin culture, since the filter is spun to prevent blocking with cells. The filtered medium is pumped to a product reservoir and fresh medium is pumped into the culture vessel. The rate of addition and removal can be regulated depending on the cell concentration in the vessel. With this method it is possible to obtain cell densities 10 to 30 times higher than the maximum cell density in an unmodified fermenter (Lydersen, 1987; Tolbert *et al.*, 1988; Lavery, 1990). This procedure has been used commercially by Invitron Corp. U.S.A. This company has patented a gentle 'sail' agitator to prevent cell damage in a 100-$dm^3$ vessel. Attachment dependent cells have been grown on microcarriers using sail agitators rotating at 8 to 12 rpm.

At Hybridtech Inc. USA, animal cells have been immobilized on a ceramic matrix and medium is perfused through this matrix. In this way high cell densities can be maintained. The apparatus is marketed as the 'Opticell' (Lydersen, 1987).

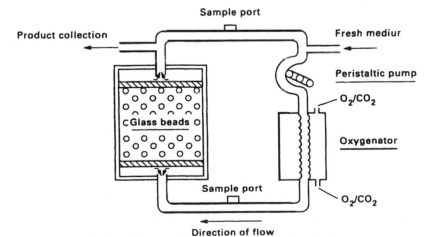

FIG. 7.55. Schematic diagram of a glass bead reactor (Browne *et al.*, 1988).

# REFERENCES

ABSON, J. W. and TODHUNTER, K. H. (1967) Effluent disposal. In *Biochemical and Biological Engineering Science*, Vol.1, pp. 309–343 (Editor Blakeborough, N.). Academic Press, London.

AIBA, S., HUMPHREY, A. E. and MILLIS, N. F. (1973) *Biochemical Engineering* (2nd Edition), Chapter 11, pp. 303–316. Academic Press, New York.

ANDERSON, J. G. and BLAIN, J. A. (1980) Novel developments in microbial film reactors. In *Fungal Biotechnology*, pp. 125–152 (Editors Smith, J. E., Berry, D. R. and Kristiansen, B). Academic Press, London.

APPLIKON (1991) Silicone heating jacket for temperature control. *Bioteknowledge* (Applikon), (1), 9–11.

ARMER, A. (1991) Steam utilization. In *Plant Engineer's Reference Book*, Section 13 (Editor Snow, D. A.). Butterworth-Heinemann, London.

ARNOLD, R. H. and STEEL, R. (1958) Oxygen supply and demand in aerobic fermentations. In *Biochemical Engineering*, pp. 149–181 (Editor Steel, R.). Heywood, London.

ATKINSON, B. and MAVITUNA, F. (1991a) Reactors. *Biochemical Engineering and Biotechnology Handbook* (2nd edition), pp. 607–668. Macmillan, London.

ATKINSON, B. and MAVITUNA, F. (1991b) Heat transfer. *Biochemical Engingineering and Bioengineering Handbook* (2nd edition), pp. 793–829. Macmillan, London.

BAILEY, J. E. and OLLIS, D. F. (1986) *Biochemical Engineering Fundamentals* (2nd edition), pp. 512–532. McGraw-Hill, New York.

BAKKER, W. A. M., VAN CAN, H. J. L., TRAMPER, J. and de GOOIJER, C. D. (1993) Hydrodynamics and mixing in a multiple air-lift loop reactor. *Biotech. Bioeng.* **42**, 994–1001.

BECL, C., STIEFEL, H. and STINNETT, T. (1987) Cell-culture bioreactors. *Chem. Eng.* **94**, 121–129.

DE BECZE, G. and LIEBMANN, A. J. (1944) Aeration in the production of compressed yeast. *Ind. Eng. Chem.* **36**, 882–890.

BESIK, F. (1973) Multi-stage tower-type activated sludge process for complete treatment of sewage. *Water Sewage Works* (September), pp.122–127.

BLAKEBOROUGH, N. (1967) Industrial fermentations. In *Biochemical and Biological Engineering Science*, Vol. 1, pp. 25–48 (Editor Blakeborough, N.). Academic Press, London.

BLIEM, R. and KATINGER, H. (1988) Scale-up engineering in animal cell technology: Part II. *Trends Biotechnol.* **6**, 224–230.

BLIEM, R., KONOPITZKY, K. and KATINGER, H. (1991) Industrial animal cell reactor systems: aspects of selection and evaluation. *Adv. Biochem. Eng. Biotechnol.* **44**, 1–26.

BOULTON, C. A. (1991) Developments in brewery fermentation. *Biotechnol. Genetic Eng. Rev.* **9**, 127–181.

BRITISH VALVE MANUFACTURERS ASSOCIATION (1972) *Technical Reference Book on Values for the Control of Fluids* (3rd edition).

BROWN, W. E. and PETERSON, W. H. (1950) Factors affecting production of penicillin in semi-pilot plant equipment. *Ind. Eng. Chem.* **42**, 1769–1774.

BROWNE, P. C., FIGUEROA, C., COSTELLO, M. A. C., OAKLEY, R. and MACIUKAS, S. M. (1988) Protein production from mammalian cells grown on glass beads. In *Animal Cell Biotechnology*, Vol. 3, pp. 251–262 (Editors, R. E. and Griffiths, J. B.) Academic Press, London.

BUCHTER, H. H. (1979) *Industrial Sealing Technology*. Wiley, New York.

BUCKLAND, B. C., GBEWONYO, K., DiMASSI, D., HUNT, G., WESTERFIELD, G. and NIENOW, A. W. (1988) Improved performance in viscous mycelial fermentations by agitator retrofitting. *Biotech. Bioeng.* **31**, 737–742.

BUELOW, G. H. and JOHNSON, M. J. (1952) Effect of separation on citric acid production in 50 gallon tanks. *Ind. Eng. Chem.* **44**, 2945–2946.

BURNETT, J. M. (1993) The industrial production of citric acid. *124th Meeting of the Society for General Microbiology*. Canterbury, U.K.

BUTLER, M. (1988) A comparative review of microcarriers available for the growth of anchorage dependent animal cells. In *Animal Cell Biotechnology*, Vol. 3, pp. 284–323 (Editors Spiers, R. E. and Griffiths, J. B.). Academic Press, London.

CALLAHAN, J. R. (1944) Large scale production by deep fermentation. *Chem. Metal. Eng.* **51**, 94–98.

CAMERON, J. and GODFREY, E. I. (1969) The design and operation of high-power magnetic drives. *Biotech. Bioeng.* **11**, 967–985.

CARRINGTON, R., DIXON, K., HARROP, A. J. and MACALONEY, G. (1992) Oxygen transfer in industrial air agitated fermentations. In *Harnessing Biotechnology for the 21st Century*, pp. 183–188 (Editors Ladisch, M. R. and Bose, A.). American Chemical Society, Washington, DC.

CHAIN, E. B., PALADINO, S., UGOLINI, F., CALLOW, D. S. and VAN DER SLUIS, J. (1952) Studies on aeration. 1. *Bull. World Health Org.* **6**, 83–98.

CHAIN, E. B., PALADINO, S., UGOLINO, F.,CALLOW, D. S. and VAN DER SLUIS, J. (1954) A laboratory fermenter for vortex and sparger aeration. *Rep. Inst. Sup. Sanita. Roma (Engl. Ed.)*, **17**, 61–120.

CHAPMAN, C. (1989) Client requirements for supply of contained bioreactors and associated equipment. In *Proceedings of DTI / HSE / SCI Symposium on Large Scale Bioprocessing Safety, Laboratory Report LR 746 (BT)*, pp. 58–62 (Editor Salisbury, T.). Warren Spring Laboratories, Stevenage.

CHEN, N. Y. (1990) The design of airlift fermenters for use in biotechnology. *Biotechnol. Genetic Eng. Rev.* **8**, 379–396.

COHEE, R. F. and STEFFEN, G. (1959) Make vinegar continuously. *Food Eng.* **3**, 58–59.

COLLINS, C. H. (1992) Hazard groups and containment categories in microbiology and biotechnology. In *Safety in*

*Industrial Microbiology and Biotechnology*, pp. 23–33 (Editors Collins, C. H. and Beale, A. J.). Butterworth-Heinemann, London.

COOKE, M., MIDDELTON, J. C. and BUSH, J. R. (1988) Mixing and mass transfer in filamentous fermentations. In *2nd International Conference on Bioreactor Fluid Dynamics*, pp. 37–64 (Editor King, R.). Elsevier, London.

COONEY, C. L., WANG, D. I. C. and MATELES, R. I. (1969) Measurement of heat evolution and correlation with oxygen consumption during microbial growth. *Biotech. Bioeng.* **11**, 269–280.

COOPER, F. M., FERNSTROM, G. A. and MILLER, S. A. (1944) Performance of gas–liquid contactors. *Ind. Eng. Chem.* **36**, 504–509.

COWAN, C. T. and THOMAS, C. R. (1988) Materials of construction in the biological process industries. *Process Biochem.* **23**(1), 5–11.

CUBBERLY, W. H., UNTERWEISER, P. M. BENJAMIN, D., KIRKPATRICK, C. W., KNOLL, V. and NIEMAN, K. (1980) Stainless steels in corrosion service. In *Metals Handbook* (9th edition), Vol. 3, pp. 56–93. American Society for Metals, Ohio.

DAWSON, P. S. S. (1974) The cyclone column fermenter. *Biotech. Bioeng. Symp.* **4**, 809–819.

DAWSON, P. S. S. (1988) The Cyclone Column and continuous phased culture. In *Biotechnology Research and Applications*, pp. 141–154 (Editors Gavora, J., Gerson, D. F., Luong, J., Storer, A. and Woodley, J. H.). Elsevier, London.

DUURKOOP, A. (1992) Stainless steel reflects the quality of Applikon's bioreactors. *Bioteknowledge* (Applikon, Holland), **12**(1), 4–6.

EAST, D., STINNETT, T. and THOMA, R. W. (1984) Reduction of biological risk in fermentation processes. by physical containment. *Dev. Ind. Microbiol.* **25**, 89–105.

EBNER, H. and FOLLMAN, H. (1983) Vinegar. In *Biotechnology*, Vol. 5, pp. 425–446 (Editor Reed, G.) Verlag Chemie, Weinheim.

EBNER, H., POHL, K. and ENEKEL, A. (1967) Self-priming aerator and mechanical defoamer for microbiological process. *Biotech. Bioeng.* **9**, 357–364.

ELSWORTH, R., CAPELL, G. A. and TELLING, R.C. (1958) Improvements in the design of a laboratory culture vessel. *J. Appl. Bacteriol.* **21**, 80–85.

FAUST, U., PRAVE, P. and SUKATSCH, D. A. (1977) Continuous biomass production from methanol by *Methylomonas clara*. *J. Ferm. Technol.* **55**, 605–614.

FINN, R. F. (1954) Agitation-aeration in the laboratory and in industry. *Bacteriol. Rev.* **18**, 254–274.

FLEISCHAKER, R. (1987) Microcarrier cell culture. In *Large Scale Cell Culture Technology*, pp. 59–79 (Editor Lydersen, B. K.). Hanser, Munich.

FLICKINGER, M. C. and SANSONE, E. B. (1984) Pilot-and production-scale containment of cytotoxic and oncogenic fermentation processes. *Biotech. Bioeng.* **26**, 860–870.

FORTUNE, W. B., McCORMICK, S. L., RHODEHAMEL, H. W and

STEFANIAK, J. J. (1950) Antibiotics development. *Ind. Eng. Chem.* **42**, 191–198.

FOUST, H. G., MACK, D. E. and RUSHTON, J. H. (1944) Gas–liquid contacting by mixers. *Ind. Eng. Chem.* **36**, 517–522.

FROMMER, W. *et al.* (1989) Safety biotechnology. III. Safety precautions for handling micro-organisms of different risk classes. *Appl. Microbiol. Biotechnol.* **30**, 541–552.

GBEWONYO, K, DiMASI, D. and BUCKLAND, B. C. (1986) The use of hydrofoil impellers to improve oxygen transfer efficiency in viscous mycelial fermentaions. In *International Conference on Biorector Fluid Dynamics*. pp. 281–299. BHRA, Cranfield, U.K.

GIORGIO, R. J. and WU, J. J. (1986) Design of large scale containment facilities for recombinant DNA fermentations. *Trends Biotechnol.* **4**, 60–65.

GORDON, J. J., GRENFELL, E., KNOWLES, E., LEGG, B. J. McALLISTER, R. C. A. and WHITE, T. (1947) Methods for penicillin production in submerged culture on a pilot scale. *J. Gen. Microbiol.* **1**, 171–186.

GREENSHIELDS, R. N. (1978) Acetic acid: vinegar. In *Economic Microbiology*, Vol. 2, pp. 121–186 (Editor Rose, A. H.). Academic Press, London.

GREENSHIELDS, R. N. and SMITH, E. L. (1971) Tower fermentation systems and their applications. *Chem. Eng.* (London), (249), 182–190.

GRIFFITHS, J. B. (1986) Scaling up of animal cell cultures. In *Animal Cell Culture — A Practical Approach*, pp. 33–69 (Editor Freshney, R. I.). IRL Press, Oxford.

GRIFFITHS, J. B. (1988) Overview of cell culture systems and their scale-up. In *Animal Cell Biotechnology*, Vol. 3, pp. 179–220 (Editor Spier, R. E. and Griffiths, J. B.). Academic Press, London.

HALL, R. D. and HOWARD, G. A. (1965) Improvements in or relating to brewing of beer. British Patent 979,491.

HAMBLETON, P., GRIFFITHS, B., CAMERON, D. R. and MELLING, J. (1991) A high containment polymodal pilot-plant- design concepts. *J. Chem. Tech. Biotechnol.* **50**, 167–180.

HAMER, G. (1979) Biomass from natural gas. In *Economic Microbiology*, Vol. 4, pp. 315–360 (Editor Rose, A. H.). Academic Press, London.

HASTINGS, J. J. H. (1971) Development of the fermentation industries in Great Britain. *Adv. Appl. Microbiol.* **16**, 1–45.

HASTINGS, J. J. H. (1978) Acetone-butanol fermentation. In *Economic Microbiology*, Vol. 2, pp. 31–45 (Editor Rose, A. H.). Academic Press, London.

HEALTH and SAFETY EXECUTIVE (1993) A guide to the Genetically Modified Organisms (Contained Use) Regulations, 1992. H.M.S.O., London.

HERBERT, D., PHIPPS, P. J. and TEMPEST, D. W. (1965) The chemostat: design and instrumentaion. *Lab. Practice* **14**, 1150–1161.

HESSELINK, P. G. M., KASTELIEM, J. and LOGTENBERG, M. T. (1990) Biosafety aspects of biotechnological processes: testing and evaluating equipment and components. In

*Fermentation Technology: Industrial Applications*, pp. 378–382 (Editor Yu, P. L.). Elsevier, London.

HIRSCHEL, M. D. and GRUENBERG, M. L. (1987) An automated hollow fiber system for the large scale manufacture of mammalian cell secreted product. In *Large Scale Cell Culture Technology*, pp. 113–144 (Editor Lydersen, B. K.). Hanser, Munich.

HOGGAN, J. (1977) Aspects of fermentation in conical vessels. *J. Inst. Brewing*, **83**, 133–138.

HOUGH, J. S., BRIGGS, D. E. and STEVENS, R. (1971) *Malting and Brewing Science*. Chapman and Hall, London.

HROMATKA, O. and EBNER, H. (1949) Untersuchungen uber sie Essiggerung. 1. Fesselgarung und Durchlufrungsverfahren. *Enzymologia* **13**, 369–387.

INSKEEP, G. C., WILEY, A. J., HOLDERBY, J. M. and HUGHES, L. P. (1951) Food yeast from sulphite liquor. *Ind. Eng. Chem.* **43**, 1702–1711.

IRVING, G. M. (1968) Construction materials for breweries. *Chem. Eng. (N.Y.)* **75**(14), 100, 102–104.

JACKSON, A. T. (1990) Basic heat transfer. In *Process Engineering in Biotechnology*, pp. 58–71. Open University Press, Milton Keynes.

JACKSON, T. (1958) Development of aerobic fermentation processes: penicillin. In *Biochemical Engineering*, pp. 185–221 (Editor Steel, R.). Heywood, London.

JANSSEN, D. E., LOVEJOY, P. J., SIMPSON, M. T. and KENNEDY, L. D. (1990) Technical problems in containment of rDNA organisms. In *Fermentation Technologies: Industrial Applications*, pp. 388–393 (Editor Yu, P. L.). Elsevier, London.

JOHNSON, M. J. (1971) Fermentation — yesterday and tomorrow. *Chem. Tech.* **1**, 338–341.

KATINGER, H. W. D. (1977) New fermenter configuration. In *Biotechnology and Fungal Differentiation, FEMS Symp.* **4**, 137–155 (Editors Meyrath, J. and Bu'Lock, J. D.). Academic Press, London.

KEMPLAY, J. (1980) *Valve Users Manual*. Mechanical Engineering Publications, London.

KENNEDY, L. D., BOLAND, M. J., JANNSEN, D. E. and FRUDE, M. J. (1990) Designing a pilot-scale fermentation facility for use with genetically modified micro-organisms. In *Fermentation Technologies: Industrial Applications*, pp. 383–387 (Editor Yu, P. L.). Elsevier, London.

KITAI, A., TONE, H. and OZAKI, A. (1969). The performance of a perforated plate column as a multistage continuous fermenter. 1. Washout and growth phase differentiation in the column. *J. Ferm. Technol.* **47**, 333–339.

KURAISHI, M., TEROA, I., OHKOUCHI, H., MATSUDA, N. and NAGAI, I. (1977) SCP-process development with methanol as substrate. *DECHEMA Monograph* (1978), **83**, (1704–1723), 111–124.

LAVERY, M. (1990) Animal cell fermentation. In *Fermentation — A Practical Approach*, pp. 205–220 (Editor McNiel, B. and Harvey, L. M.). IRL Press, Oxford.

LEAVER, G. (1994) Interpretation of regulatory requirements to large scale biosafety — The role of the industrial biosafety project. In *International Safety Aspects in Biotechnology* (Editors Hambleton, P., Salusbury, T. and Melling, J.). Elsevier, Amsterdam.

LEAVER, G. and HAMBLETON, P. (1992) Designing bioreactors to minimize or prevent inadvertant release into the workplace and natural environment. *Pharmac. Technol.* (April), 18–26.

LEE, S. S., ERIKSON, L. E. and FAN, L. T. (1971) Modeling and optimization of a tower type activated sludge system. *Biotech. Bioeng. Symp.* **2**, 141–173.

LEVI, J. D., SHENNAN, J. L. and EBBON, G. P. (1979) Biomass from liquid n-alkanes. In *Economic Microbiology*, Vol. 4, pp. 362–419 (Editor Rose, A. H.). Academic Press, London.

LYDERSEN, B. K. (1987) Perfusion cell culture system based on ceramic matrices. In *Large Scale Cell Culture Technology*, pp. 169–192 (Editor Lydersen, B. K.). Hanser, Munich.

MARSHALL, R. D., WEBB, C., MATTHEWS, T. M. and DEAN, J. F. (1990) Automatic aseptic sampling of fermentation broth. In *Practical Advances in Fermentation Technology*, pp. 6.1–6.11. Inst. Chem. Eng. North Western Branch Symposium Papers No. 3. University of Manchester Institute of Science and Technology.

MARTINI, L. J. (1984) *Practical Seal Design*. Dekker, New York.

MAULE, D. R. (1986) A century of fermenter design. *J. Inst. Brew.* **92**, 137–147.

MAYER, E. (1961) Vinegar by oxidative fermentation of alcohol. U.S. Patent 2,997,424.

McADAMS, W. H. (1954) Heating and cooling inside tubes. *Heat Transmission* (3rd edition), pp. 202–251. McGraw-Hill, New York.

MEYRATH, J. and BEYER, K. (1979) Biomass from whey. In *Economic Microbiology*, Vol. 4, pp. 207–269 (Editor Rose, A. H.). Academic Press, London.

MIJNBEEK, G. (1991) Sheer stress effects on cultured animal cells. *Bioteknowledge* (Applikon), (1), 3–7.

MILLER, F. D. and RUSHTON, J. H. (1944) A mass velocity theory for liquid agitation. *Ind. Eng. Chem.* **36**, 499–503.

MITCHELL, C. A. (1926) *Vinegar: Its Manufacture and Examination* (2nd edition). Griffin, London.

MULLER, R. and KIESLICH, K. (1966) Technology of the microbiological preparation of organic substances. *Angwante Chem. (Int. Ed.)* **5**, 653–662.

NATHAN, L. (1930) Improvements in the fermentation and maturation of beer. *J. Inst. Brewing*, **36**, 538–550.

NIENOW, A. W. (1990) Agitators for mycelial fermentations. *Trends Biotechnol.* **8**, 224–233.

NIENOW, A. W. (1992) New agitators versus Rushton turbines: A critical comparison of transport phenomena. In *Harnessing Biotechnology for the 21st Century*, pp. 193–196 (Editors Ladisch, M. R. and Bose, A.). American Chemical Society, Washington, DC.

NIENOW, A. W., ALLSFORD, K. V., CRONIN, D., HUOXING, L., HAOZHUNG, W. and HUDCOVA, V. (1988) The use of large ring spargers to improve the performance of fermenters agitated by single and multiple standard Rushton turbines. In *2nd International Conference on Bioreactor Fluid Dynamics*, pp. 159–177 (Editor King, R.). Elsevier, London.

NIENOW, A. W. and ULBRECHT, J. J. (1985) Fermentation Broths. In *Mixing of Liquids by Mechanical Agitation*, pp. 203–235 (Editord Ulbrecht, J. J. and Patterson, G. E.). Gordon and Breach, New York.

NIENOW, A. W., WAROESKERKEN, M. M. C. G., SMITH, J. M. and KONNO, M. (1985) On the flooding/loading transition and the complete dispersal conditions in aerated vessels agitated by a Rushton turbine. *Proc. 5th European Mixing Conference*, pp. 143–154. BHRA, Cranfield.

NOBLE, T. G., JACKMAN, M. I. and BADGER, E. M. H. (1964) The biological treatment of gas liquor in aeration tanks and packed towers. *J. Inst. Sewage Purification* **5**, 440–463.

OKABE, M., OHTA, N. and PARK, Y. S. (1993) Itaconic acid production in an air-lift bioreactor using a modified draft tube. *J. Ferm. Bioeng.* **76**, 117–122.

OWEN, W. L. (1948) Continuous fermentation. *Sugar*, **43**, 36–38.

PACA, J., ETTLER, P. and GREGR, V. (1976) Hydrodynamic behaviour and oxygen transfer rate in a pilot plant fermenter. *J. Appl. Chem. Biotechnol.* **26**, 310–317.

PARKER, A. (1958) Sterilization of equipment, air and media. In *Biochemical Engineering—Unit Processes in Fermentation*, pp. 94–121 (Editor Steel, R.) Heywood, London.

PIKULIK, A. (1976) Selecting and specifying valves for new plants. *Chem. Eng. N.Y.* **83**, 168–190.

PROKOP, A. and VOTRUBA, J. (1976) Bioengineering problems connected with the use of conventional and unconventional raw materials in fermentation. *Folia Microbiol.* **21**, 58–69.

PROKOP, A., ERIKSON, L. E., FERNANDEZ, J. and HUMPHREY, A. E. (1969) Design and physical characteristics of a multistage continuous tower fermenter. *Biotech. Bioeng.* **11**, 945–966.

PROPST, C. L., VON WEDEL, R. J. and LUBINIECKI, A. S. (1989) Using mammalian cells to produce products. In *Fermentation Process Development of Industrial Organisms*, pp. 221–276 (Editor Neway, J. O.). Dekker, New York.

RICHARDS, J. W. (1968) Design and operation of aseptic fermenters. In *Introduction to Industrial Sterilization.*, pp. 107–122. Academic Press, London.

RIVETT, R. W., JOHNSON, M. J. and PETERSON, W. H. (1950) Laboratory fermenter for aerobic fermentations. *Ind. Eng. Chem.* **42**, 188–190.

ROHR, M., KUBICEK, C. P. and KOMINEK, J. (1983) Citric acid. In *Biotechnology*, Vol. 3, pp. 419–454 (Editor Dallweg, H.). Verlag Chemie, Weinheim.

SCHOFIELD, G. M. (1992) Current legislation and regulatory frameworks. In *Safety in Industrial Microbiology and Biotechnology*, pp. 6–22 (Editors Collins, C. H. and Beale, A. J.). Buttterworth-Heinemann, London.

SCHOLLER, M. and SEIDEL, M. (1940) Yeast production and fermentation. U.S. Patent 2,188,192.

SCHREIER, K. (1975) High-efficiency fermenter with deep-jet aerators. *Chemiker Zeitung* **99**, 328–331.

SCHUGERL, K. (1982) New bioreactors for aerobic processes. *Int. Eng. Chem.* **22**, 591–610.

SCHUGERL, K. (1985) Nonmechanically agitated bioreactor systems. In *Comprehensive Biotechnology*, Vol. 2, pp. 99–118 (Editors Cooney, C. L. and Humphrey, A. E.). Pergamon Press, Oxford.

SCHUGERL, K., LUCKE, J. and OELS, U. (1977) Bubble column bioreactors. *Adv. Biochem. Eng.* **7**, 1–84.

SCRAGG, A. H. (1991) *Bioreactors in Biotechnology — A Practical Approach*, pp. 112–125. Ellis Horwood, Chichester.

SHARP, D. H. (1989) The development of Pruteen by ICI. In *Bioprotein Manufacture*, pp. 53–78. Ellis Horwood, Chichester.

SHORE, D. T., ROYSTON, M. G. and WATSON, E. G. (1964) Improvements in or relating to the production of beer. British Patent 959,049.

SHUTTLEWOOD, J. R. (1984) *Brew Distilling Int.* (August), p.22.

SITTIG, W. (1982) The present state of fermentation reactors. *J. Chem. Tech. Biotechnol.* **32**, 47–58.

SMITH, J. M. (1985) Dispersion of gases in liquids: the hydrodynamics of gas dispersion in low viscosity liquids. In *Mixing of Liquids by Mechanical Agitation*, pp.139–201 (Editors Ulbrecht, J. J. and Patterson, G. K.). Gordon and Breach, New York.

SMITH, S. R. L. (1980) Single cell protein. *Phil. Trans. Roy. Soc. (London) B* **290**, 341–354.

SNELL, R. L. and SCHWEIGER, L. B. (1949) Production of citric acid by fermentation. U.S. Patent 2,492,667.

SOLOMONS, G. L. (1969) *Materials and Methods in Fermentation*. Academic Press, London.

SOLOMONS, G. L. (1980) Fermenter design and fungal growth. In *Fungal Biotechnology*, pp. 55–80 (Editors Smith, J. E., Berry, D. R. and Kristiansen, B). Academic Press, London.

SPIVEY, M. J. (1978) The acetone–butanol–ethanol fermentation. *Process Biochem.* **13**(11), 2–4, 25.

STEEL, R. and MAXON, W. D. (1961) Power requirements of a typical actinomycete fermentation. *Ind. Eng. Chem.* **53**, 739–742.

STEEL, R. and MAXON, W. D. (1966) Studies with a multiple-rod mixing impeller. *Biotech. Bioeng.* **8**, 109–115.

STEEL, R. and MILLER, T. L. (1970) Fermenter design. *Adv. Appl. Microbiol.* **12**, 153–188.

STEEL, R., LENTZ, C. P. and MARTIN, S. M. (1955) Submerged citric acid production of sugar beet molasses: increase in scale. *Can. J. Microbiol.* **1**, 299–311.

STRAUCH and SCHMIDT (1932) German Patent 552,241. Cited by de Becze and Liebmann (1944).

TAYLOR, I. J. and SENIOR, P. J. (1978) Single cell proteins: a new source. *Endeavour (N.S.)*, **2**, 31–34.

THAYSEN, A. C. (1945) Production of food yeast. *Food* (May), pp. 116–119.

THIELSCH, H. (1967) Manufacture, fabrication and joining of commercial piping. In *Piping Handbook* (5th edition), pp. 7.1–7.300 (Editor King, R. C.). McGraw-Hill, New York.

TITCHENER-HOOKER, N. J., SINCLAIR, P. A., HOARE, M., VRANCH, S. P., COTTAM, A. and TURNER, M. K. (1993) The specification of static seals for contained operations; an engineering appraisal. *Pharmaceutical Technol. (Europe)* October, 26–30.

TOLBERT, W. R., SRIGLEY, W. R. and PRIOR, C. P. (1988) Perfusion culture systems for large-scale pharmaceutical production. In *Animal Cell Biotechnology*, Vol. 3, pp. 374–393 (Editors Spiers, R. E. and Griffiths, J. B.). Academic Press, London.

TRINCI, A. P. J. (1992) Mycoprotein: A twenty year overnight sucess story. *Mycol. Res.* **96**, 1–13.

TUBITO, P. J. (1991) Contamination control facilities for the biotechnology industry. In *Cleanroom Design*, pp. 85–120 (Editor Whyte, W.). Wiley, Chichester.

TURNER, M. K. (1989) Categories of large-scale containment for manufacturing processes with recombinant organisms. *Biotechnol. Gen. Eng. Rev.* **7**, 1–43.

ULENBERG, G. H., GERRITSON, H. and HUISMAN, J. (1972) Experiences with a giant cylindro-conical tank. *Master Brew. Assoc. Amer. Q.* **9**, 117–122.

VAN HOUTEN, J. (1992) Containment of fermentations: Comprehensive assessment and integrated control. In *Harnessing Biotechnology for the 21st Century*, pp. 415–418 (Editors Ladisch, M. R. and Bose, A.). American Chemical Society, Washington, DC.

VAN WEZEL, A. L. (1967) Growth of cell strains and primary cells on microcarriers in homogenous cultures. *Nature (London)* **216**, 64–65.

VICTORERO, F. A. (1948) Apparatus for continuous fermentation. U.S. Patent 2,450,218.

VRANCH, S. P. (1992) Engineering for safe bioprocessing. In *Safety in Industrial Microbiology and Biotechnology*, pp. 176–189 (Editors Collins, C. H. and Beale, A. J.). Butterworth-Heinemann, London.

WALKER, J. A. H. and HOLDSWORTH, H. (1958) Equipment design. In *Biochemical Engineering*, pp. 223–273 (Editor Steel, R.). Heywood, London.

WALKER, P. D., NARENDRANATHAN, T. J., BROWN, D. C., WOODHOUSE, F. and VRANCH, S. P. (1987) Containment of micro-organisms during fermentation and downstream processing. In *Separation for Biotechnology*, pp. 469–479 (Editors Verrall, M. S. and Hudson, M. J.). Ellis Horwood, Chichester.

WATANABE, K. (1976) Production of RNA. In *Microbial Production of Nucleic Acid Related Substances*, pp. 55–65 (Editors Ogata, K., Kinoshita, S., Tsonoda, T. and Aida, K.). Kodansha, Tokyo and Halsted, New York.

WEGRICH, R. H. and SHURTER, R. A. (1953) Development of a typical aerobic fermentation. *Ind. Eng. Chem.* **45**, 1153–1160.

WERNER, R. G. (1992) Containment in the development and manufacture of recombinant DNA-derived products. In *Safety in Industrial Microbiology and Biotechnology*, pp. 190–213 (Editors Collins, C. H. and Beale, A. J.). Butterworth-Heinemann, London.

WILKINSON, P. J. (1987) The development of a large scale production process for tissue culture products. In *Bioreactors and Biotransformation*, pp. 111–120 (Editors Moody, G. W. and Baker, P. B.). Elsevier, London.

WINKLER, M. A. (1990) Problems in fermenter design and operation. In *Chemical Engineering Problems in Biotechnology*, pp. 215–350 (Editor Winkler, M. A.). SCI/Elsevier, London.

WU, W. T. and WU, J. Y. (1990) Air lift reactor with draught tube. *J. Ferm. Bioeng.* **70**, 359–361.

# CHAPTER 8

# Instrumentation and Control

**INTRODUCTION**

THE SUCCESS of a fermentation depends upon the existence of defined environmental conditions for biomass and product formation. To achieve this goal it is important to understand what is happening to a fermentation process and how to control it to obtain optimal operating conditions. Thus, temperature, pH, degree of agitation, oxygen concentration in the medium and other factors may have to be kept constant during the process. The provision of such conditions requires careful monitoring (data acquisition and analysis) of the fermentation so that any deviation from the specified optimum might be corrected by a control system. Criteria which are monitored frequently are listed in Table 8.1, along with the control processes with which they are associated. As well as aiding the maintenance of constant conditions, the monitoring of a process may provide information on the progress of the fermentation. Such information may indicate the optimum time to harvest or that the fermentation is progressing abnormally which may be indicative of contamination or strain degeneration. Thus, monitoring equipment produces information indicating fermentation progress as well as being linked to a suitable control system.

In initial studies the number of functions which are to be controlled may be restricted in order to gain more knowledge about a particular fermentation. Thus, the pH may be measured and recorded but not maintained at a specified pH or the dissolved oxygen concentration may be determined but no attempt will be made to prevent oxygen depletion.

Also, it is important to consider the need for a sensor and its associated control system to interface with a computer (to be discussed in a later section). This chapter will consider the general types of control systems which are available, specific monitoring and control systems and the role of computers. More information on intrumentation and control has been written by Flynn (1983, 1984), Armiger (1985), Bull (1985), Rolf and Lim (1985), Bailey and Ollis (1986), Kristiansen (1987), Montague *et al.* (1988), Dusseljee and Feijen (1990), Atkinson and Mavituna (1991) and Royce (1993).

It is apparent from Table 8.1 that a considerable number of process variables may need to be monitored during a fermentation. Methods for measuring these variables, the sensors or other equipment available and possible control procedures are outlined below.

There are three main classes of sensor:

1. Sensors which penetrate into the interior of the fermenter, e.g. pH electrodes, dissolved-oxygen electrodes.
2. Sensors which operate on samples which are continuously withdrawn from the fermenter, e.g. exhaust-gas analysers.
3. Sensors which do not come into contact with the fermentation broth or gases, e.g. tachometers, load cells.

It is also possible to characterize a sensor in relation to its application for process control:

1. *In-line sensor*. The sensor is an integrated part of the fermentation equipment and the measured value obtained from it is used directly for process control.
2. *On-line sensor*. Although the sensor is an integral part of the fermentation equipment, the measured value cannot be used directly for control.

TABLE 8.1. *Process sensors and their possible control functions*

| Category | Sensor | Possible control function |
|---|---|---|
| Physical | Temperature | Heat/cool |
| | Pressure | |
| | Agitator shaft power | |
| | rpm | |
| | Foam | Foam control |
| | Weight | Change flow rate |
| | Flow rate | Change flow rate |
| Chemical | pH | Acid or alkali addition, carbon source feed rate |
| | Redox | Additives to change redox potential |
| | Oxygen | Change feed rate |
| | Exit-gas analysis | Change feed rate |
| | Medium analysis | Change in medium composition |

An operator must enter measured values into the control system if the data is to be used in process control.

3. *Off-line sensor.* The sensor is not part of the fermentation equipment. The measured value cannot be used directly for process control. An operator is needed for the actual measurement (e.g. medium analysis or dry weight sample) and for entering the measured values into the control system for process control.

When evaluating sensors to use in measurement and control it is important to consider response time, gain, sensitivity, accuracy, ease and speed of calibration, stability, reliability, output signal (continuous or discontinuous), materials of construction, robustness, sterilization, maintenance, availability to purchase and cost (Flynn, 1983, 1984; Royce, 1993).

## METHODS OF MEASURING PROCESS VARIABLES

### Temperature

The temperature in a vessel or pipe is one of the most important parameters to monitor and control in any process. It may be measured by mercury-in-glass thermometers, bimetallic thermometers, pressure bulb thermometers, thermocouples, metal-resistance thermometers or thermistors. Metal-resistance thermometers and thermistors are used in most fermentation applications. Accurate mercury-in-glass thermometers

are used to check and calibrate the other forms of temperature sensors, while cheaper thermometers are still used with laboratory fermenters.

## MERCURY-IN-GLASS THERMOMETERS

A mercury-in-glass thermometer may be used directly in small bench fermenters, but its fragility restricts its use. In larger fermenters it would be necessary to insert it into a thermometer pocket in the vessel, which introduces a time lag in registering the vessel temperature. This type of thermometer can be used solely for indication, not for automatic control or recording.

## ELECTRICAL RESISTANCE THERMOMETERS

It is well known that the electrical resistance of metals changes with temperature variation. This property has been utilized in the design of resistance thermometers. The bulb of the instrument contains the resistance element, a mica framework (for very accurate measurement) or a ceramic framework (robust, but for less accurate measurement) around which the sensing element is wound. A platinum wire of 100-$\Omega$ resistance is normally used. Leads emerging from the bulb are connected to the measuring element. The reading is normally obtained by the use of a Wheatstone bridge circuit and is a measure of the average temperature of the sensing element. This type of thermometer does have a greater accuracy ($\pm 0.25\%$) than some of the other measuring devices and is more sensitive to small temperature changes. There is a fast response to detectable changes (1 to 10 seconds), and there is no restriction on distance between the very compact sensing point ($30 \times 5$ mm) and the display point of reproducible readings. These thermometers are normally enclosed in stainless-steel sheaths if they are to be used in large vessels and ancillary equipment.

## THERMISTORS

Thermistors are semiconductors made from specific mixtures of pure oxides of iron, nickel and other metals. Their main characteristic is a large change in resistance with a small temperature change. The change in resistance is a function of absolute temperature. The temperature reading is obtained with a Wheatstone bridge or a simpler or more complex circuit depending on the application. Thermistors are relatively cheap and have proved to be very stable, give reproducible readings, and can be sited remotely from the read-out point. Their main disadvantage is the marked non-linear temperature versus resistance curve.

## TEMPERATURE CONTROL

The use of water jackets or pipe coils within a fermenter as a means of temperature control has been described in Chapter 7. In many small systems there is a heating element, 300 to 400 W capacity being adequate for a 10-dm³ fermenter, and a cooling water supply; these are on or off depending on the need for heating or cooling. The heating element should be as small as possible to reduce the size of the 'heat sink' and resulting overshoot when heating is no longer required. In some cases it may be better to run the cooling water continuously at a steady rate and to have the heating element only connected to the control unit. This can be an expensive mode of operation if the water flows directly to waste. For small-scale use, Harvard Apparatus Ltd (Fircroft Way, Edenbridge, Kent, U.K.) make a unit, the Thermocirculator, which will pump recirculating thermostatically heated water through fermenters of up to 10 dm³ capacity and give temperature control of ±0.1°.

In large fermenters, where heating during the fermentation is not normally required, a regulatory valve at the cooling-water inlet may be sufficient to control the temperature. There may be provision for circulation of refrigerated brine if excessive cooling is required. Steam inlets to the coil and jacket must be present if a fermenter is being used for batch sterilization of media.

Low agitation speeds are often essential in animal cell culture vessels to minimize shear damage. In these vessels, heating fingers can create local 'hot-spots' which may cause damage to cells very close to them. Heating jackets which have a lower heat output proportional to the surface area (water or silicone rubber covered electrical heating elements — Chapter 7) are used to overcome this problem.

## Flow measurement and control

Flow measurement and control of both gases and liquids is important in process management.

### GASES

One of the simplest methods for measuring gas flow to a fermenter is by means of a variable area meter. The most commonly used example is a rotameter, which consists of a vertically mounted glass tube with an increasing bore and enclosing a free-moving float which may be a ball or a hollow thimble. The position of the float in the graduated glass tube is indicative of flow rate. Different sizes can cater for a wide range of flow rates. The accuracy depends on having the gas at a constant pressure, but errors of up to ±10% of full-scale deflection are quoted (Howe *et al.*, 1969). The errors are greatest at low flow rates. Ideally, rotameters should not be sterilized and are therefore normally placed between a gas inlet and a sterile filter. There is no provision for on-line data logging with the simple rotameters. Metal tubes can be used in situations where glass is not satisfactory. In these cases the float position is determined by magnetic or electrical techniques, but this provision has not been normally utilized for fermentation work. Rotameters can also be used to measure liquid flow rates, provided that abrasive particles or fibrous matter are not present.

The use of oxygen and carbon dioxide gas analysers for effluent gas analysis requires the provision of very accurate gas-flow measurement if the analysers are to be used effectively. For this reason thermal mass flowmeters have been utilized for the range 0 to 500 dm³ min⁻¹. These instruments have a ±1% full-scale accuracy and work on the principle of measuring a temperature difference across a heating device placed in the path of the gas flow (Fig. 8.1). Temperature probes such as thermistors are placed upstream and downstream of the heat source, which may be inside or outside the piping.

The mass flow rate of the gas, $Q$, can be calculated from the specific heat equation:

$$H = QC_p(T_2 - T_1)$$

where $H$ = heat transferred,

$Q$ = mass flow rate of the gas,

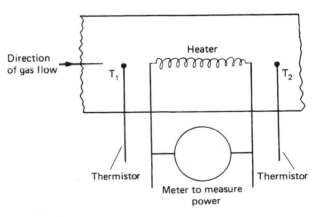

FIG. 8.1. Thermal mass flowmeter.

$C_p$ = specific heat of the gas,
$T_1$ = temperature of gas before heat is transferred to it,
$T_2$ = temperature of gas after heat is transferred to it.

This equation can then be rearranged for $Q$:

$$Q = \frac{H}{C_p(T_2 - T_1)}.$$

A voltage signal can be obtained by this method of measurement which can be utilized in data logging.

Control of gas flow is usually by needle valves. Often this method of control is not sufficient, and it is necessary to incorporate a self-acting flow-control valve. At a small scale, such valves as the 'flowstat' are available (G.A. Platon, Ltd, Wella Road, Basingstoke, Hampshire, U.K.). Fluctuations in pressure in a flow-measuring orifice cause a valve or piston pressing against a spring to gradually open or close so that the original, preselected flow rate is restored. In a gas 'flowstat', the orifice should be upstream when the gas supply is at a regulated pressure and downstream when the supply pressure fluctuates and the back pressure is constant. Valves operating by a similar mechanism are available for larger scale applications.

## LIQUIDS

The flow of non-sterile liquids can be monitored by a number of techniques (Howe *et al.*, 1969), but measurement of flow rates of sterile liquids presents a number of problems which have to be overcome. On a laboratory scale flow rates may be measured manually using a sterile burette connected to the feed pipe and timing the exit of a measured volume. The possible use of rotameters has already been mentioned in the previous section. A more expensive method is to use an electrical flow transducer (Howe *et al.*, 1969) which can cope with particulate matter in suspension and measure a range of flow rates from very low to high (50 cm³ min⁻¹ to 500,000 dm³ min⁻¹) with an accuracy of ±1%. In this flowmeter (Fig. 8.2) there are two windings outside the tube, supplied with an alternating current to create a magnetic field. The voltage induced in the field is proportional to the relative velocity of the fluid and the magnetic field. The potential difference in the fluid can be measured by a pair of electrodes, and is directly proportional to the velocity of the fluid.

In batch and fed-batch culture fermenters, a cheaper alternative is to measure flow rates indirectly by load cells (see Weight section). The fermenter and all ancillary reservoirs are attached to load cells which monitor

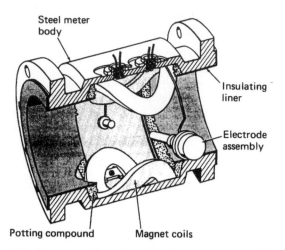

FIG. 8.2. A cut-away view of a short-form magnetic flowmeter (Howe *et al.*, 1969).

the increases and decreases in weight of the various vessels at regular time intervals. Provided the specific gravities of the liquids are known it is possible to estimate flow rates fairly accurately in different feed pipes. This is another technique which may be used with particulate suspensions.

Another indirect method of measuring flow rates aseptically is to use a metering pump which pumps liquid continuously at a predetermined and accurate rate. A variety of metering pumps are commercially available including motorized syringes, peristaltic pumps, piston pumps and diaphragm pumps. Motorized syringes are used only when very small quantities of liquid have to be added slowly to a vessel. In a peristaltic pump, liquid is moved forwards gradually by squeezing a tubing held in a semicircular housing. A variety of sizes of tubes can be used in different pumps to produce different known flow rates over a very wide range. Suspensions can be handled since the liquid has no direct contact with moving parts.

A piston pump contains an accurately machined ceramic or stainless-steel piston moving in a cylinder normally fitted with double ball inlet and outlet valves. The piston is driven by a constant-speed motor. Flow rates can be varied within a defined range by changing the stroke rate, the length of the piston stroke and by using a different piston size. Sizes are available from cm³ h⁻¹ to thousands of dm³ h⁻¹ and all can be operated at relatively high working pressures. Unfortunately, they cannot be used to pump fibrous or particulate suspensions. Piston pumps are more expensive than comparable sized peristaltic pumps but do not suffer from tube failure.

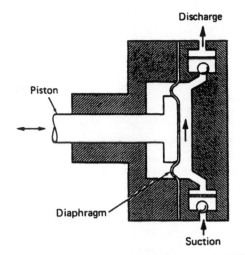

FIG. 8.3. A direct-driven diaphragm pump (Howe *et al.*, 1969).

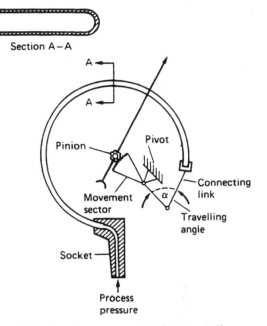

FIG. 8.4. 'C' Bourdon tube pressure gauge (Liptak, 1969).

Leakage can occur via the shaft housing of a piston pump. The problem can be prevented by the use of a diaphragm pump. This pump uses a flexible diaphragm to pump fluid through a housing (Fig. 8.3) with ball valves to control the direction of flow. The diaphragm may be made of, e.g., teflon, neoprene, stainless steel, and is actuated by a piston. A range of sizes of pumps is available for flow rates up to thousands of $dm^3 h^{-1}$.

Liquid flow from a nutrient feed tank or to or out of a fermenter may be monitored by continuous weighing on a balance or load cell(s). This will be discussed in the Weight section.

### Pressure measurement

Pressure is one of the crucial measurements that must be made when operating many processes. Pressure measurements may be needed for several reasons, the most important of which is safety. Industrial and laboratory equipment is designed to withstand a specified working pressure plus a factor of safety. It is therefore important to fit the equipment with devices that will sense, indicate, record and control the pressure. The measurement of pressure is also important in media sterilization. In a fermenter, pressure will influence the solubility of gases and contribute to the maintenance of sterility when a positive pressure is present.

One of the standard pressure measuring sensors is the Bourdon tube pressure gauge (Fig. 8.4), which is used as a direct indicating gauge. The partial coil has an elliptical cross-section (A–A) which tends to be-

come circular with increasing pressure, and because of the difference between the internal and external radii, gradually straightens out. The process pressure is connected to the fixed socket end of the tube while the sealed tip of the other end is connected by a geared sector and pinion movement which actuates an indicator pointer to show linear rotational response (Liptak, 1969).

When a vessel or pipe is to be operated under aseptic conditions a diaphragm gauge can be used (Fig. 8.5). Changes in pressure cause movements of the diaphragm capsule which are monitored by a mechanically levered pointer.

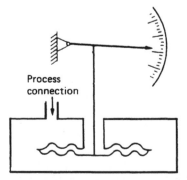

FIG. 8.5. Nested diaphragm-type pressure sensor (Liptak, 1969).

Alternatively, the pressure could be measured remotely using pressure bellows connected to the core of a variable transformer. The movement of the core generates a corresponding output. It is also possible to use pressure sensors incorporating strain gauges. If a wire is subject to strain its electrical resistance changes; this is due, in part, to the changed dimensions of the wire and the change in resistivity which occurs due to the stress in the wire. The output can then be measured over long distances. Another electrical method is to use a piezoelectric transducer. Certain solid crystals such as quartz have an asymmetrical electrical charge distribution. Any change in shape of the crystal produces equal, external, unlike electric charges on the opposite faces of the crystal. This is the piezoelectric effect. Pressure can therefore be measured by means of electrodes attached to the opposite surfaces of the crystal. Bioengineering AG (Wald, Switzerland) have made a piezoelectrical transducer with integral temperature compensation, to overcome pyroelectric effects, and built into a housing which can be put into a fermenter port.

It will also be necessary to monitor and record atmospheric pressure if oxygen concentrations in inlet and/or exit gases are to be determined using oxygen gas analysers (see later section). Paramagnetic gas analysers are susceptible to changes in barometric pressure. A change of 1% in pressure may cause a 1% change in oxygen concentration reading. This size of error may be very significant in a vessel where the oxygen consumption rate is very low and there is very little difference between the inlet and exit gas compositions. The pressure changes should be constantly monitored to enable the appropriate corrections to be made.

### Pressure control

Different working pressures are required in different parts of a fermentation plant. During normal operation a positive head pressure of 1.2 atmospheres ($161 \text{ kN}^{-1}$) absolute is maintained in a fermenter to assist in the maintenance of aseptic conditions. This pressure will obviously be raised during a steam-sterilization cycle (Chapter 5). The correct pressure in different components should be maintained by regulatory valves (Chapter 7) controlled by associated pressure gauges.

### Safety valves

Safety valves (Chapter 7) should be incorporated at various suitable places in all vessels and pipe layouts which are likely to be operated under pressure. The valve should be set to release the pressure as soon as it increases markedly above a specified working pressure. Other provisions will be necessary to meet any containment requirements.

### Agitator shaft power

A variety of sensors can be used to measure the power consumption of a fermenter. On a large scale, a watt meter attached to the agitator motor will give a fairly good indication of power uptake. This measuring technique becomes less accurate as there is a decrease in scale to pilot scale and finally to laboratory fermenters, the main contributing factor being friction in the agitator shaft bearing (Chapter 7). Torsion dynamometers can be used in small-scale applications. Since the dynamometer has to be placed on the shaft outside the fermenter the measurement will once again include the friction in the bearings. For this reason strain gauges mounted on the shaft within the fermenter are the most accurate method of measurement and overcome frictional problems (Aiba *et al.*, 1965; Brodgesell, 1969). Aiba *et al.* (1965) mounted four identical strain gauges at 45° to the axis in a hollow shaft. Lead wires from the gauges passed out of the shaft via an axial hole and electrical signals were then picked up by an electrical slip-ring arrangement. Theoretical treatment of the strain gauge measurements has been covered by Aiba *et al.* (1973).

### Rate of stirring

In all fermenters it is important to monitor the rate of rotation (rpm) of the stirrer shaft. The tachometer used for this purpose may employ electromagnetic induction voltage generation, light sensing or magnetic force as detection mechanisms (Brodgesell, 1969). Obviously, the final choice of tachometer will be determined by the type of signal which is required for recording and/or process control for regulating the motor speed and other ancillary equipment. Provision is often made on small laboratory fermenters to vary the rate of stirring. In most cases it is now standard practice to use an a.c. slip motor that has an acceptable torque curve that is coupled to a thyristor control. At pilot or full scale, the need to change rates of stirring is normally reduced. When necessary it can be done using gear boxes, modifying the sizes of wheels and drive belts, or by changing the drive motor, the most expensive alternative.

### Foam sensing and control

The formation of foam is a difficulty in many types of microbial fermentation which can create serious problems if not controlled. It is common practice to add an antifoam to a fermenter when the culture starts foaming above a certain predetermined level. The methods used for foam sensing and antifoam additions will depend on process and economic considerations. The properties of antifoams have been discussed elsewhere (Chapters 4 and 7), as has their influence on dissolved oxygen concentrations (Chapter 9).

A foam sensing and control unit is shown in Fig. 8.6. A probe is inserted through the top plate of the fermenter. Normally the probe is a stainless-steel rod, which is insulated except at the tip, and set at a defined level above the broth surface. When the foam rises and touches the probe tip, a current is passed through the circuit of the probe, with the foam acting as an electrolyte and the vessel acting as an earth. The current actuates a pump or valve and antifoam is released into the fermenter for a few seconds. Process timers are routinely included in the circuit to ensure that the antifoam has time to mix into the medium and break down the foam before the probe is programmed after a preset time interval to sense the foam level again and possibly actuate the pump or valve. Alternatively antifoam may be added slowly at a predetermined rate by a small pump so that foaming never occurs and there is therefore no need for a sensing system.

A number of mechanical antifoam devices have been described including discs, propellers, brushes or hollow cones attached to the agitator shaft above the surface of the broth. The foam is broken down when it is thrown against the walls of the fermenter. Other devices which have been manufactured include horizontal rotating shafts, centrifugal separators and jets spraying on to deflector plates (Hall *et al.*, 1973; Viesturs *et al.*, 1982). Unfortunately most of these devices have to be used in conjunction with an antifoam.

### Weight

A load cell offers a convenient method of determining the weight of a fermenter or feed vessel. This is done by placing compression load cells in or at the foot of the vessel supports. When designing the support system for a fermenter or other vessel, the weight of which is to be measured by load cells, the principle of the three-legged stool should be remembered. Three feet will always rest in stable equilibrium even though the supporting surface is uneven. If more feet are provided, the additional feet must each be fitted with means of adjustment or precision packing to ensure load bearing on all the feet.

A load cell is essentially an elastic body, usually a solid or tubular steel cylinder, the compressive strain of which under axial load may be measured by a series of electrical resistance strain gauges which are cemented to the surface of the cylinder. The load cell is assembled in a suitable housing with electrical cable connecting points. The cell is calibrated by measuring compressive strain over the appropriate range of loading. Changes of resistance with strain which are proportional to load are determined by appropriate electrical apparatus.

It is therefore possible to use appropriately sized load cells to monitor feed rates from medium reservoirs, acid and base utilization for pH control and the use of antifoam for foam control. The change in weight in a known time interval can be used indirectly as a measure of liquid flow rates.

### Microbial biomass

Real-time estimation of microbial biomass in a fermenter is an obvious requirement, yet it has proved very difficult to develop a satisfactory sensor. Most monitoring has been done indirectly by dry weight samples (made quicker with microwave ovens), cell density (spectrophotometers), cell numbers (Coulter counters) or by the use of gateway sensors which will be discussed later in this chapter. Other alternative approaches are real-time estimation of a cell component which remains at a constant concentration, such

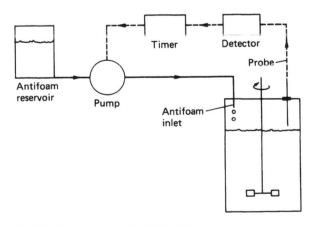

FIG. 8.6. Foam sensing and control unit.

as nicotinamide adenine dinucleotide (NAD), by fluorimetry or measurement of a cell property which is proportional to the concentration of viable cells, such as radio frequency capacitance.

It is well established that fluorimetric measurements are very specific and rapid, but their use in fermentation studies is limited. The measurement of NAD, provided that it remains at a constant concentration in cells, would be an ideal indirect method for continuous measurement of microbial biomass. In pioneer studies, Harrison and Chance (1970) used a fluorescence technique to determine NAD–NADH levels inside microbial cells growing in continuous culture. Einsele *et al.* (1978) mounted a fluorimeter on a fermenter observation port located beneath the culture surface which enabled the measurement of NADH fluorescence *in situ*, making it possible to determine bulk mixing times in the broth and to follow glucose uptake by monitoring NADH levels. Beyeler *et al.* (1981) were able to develop a small sterilizable probe for fitting into a fermenter to monitor NADH, which had high specificity, high sensitivity, high stability and could be calibrated *in situ*. In batch culture of *Candida tropicalis*, the NAD(P)H-dependent fluorescence signal correlated well with biomass, so that it could be used for on-line estimation of biomass. Changes in the growth conditions, such as substrate exhaustion or the absence of oxygen, were also very quickly detected.

Schneckenburger *et al.* (1985) used this technique to study the growth of methanogenic bacteria in anaerobic fermentations. They thought cost was a problem, fluorescence equipment being too expensive for routine biotechnology applications when the minimum price was about US$10,000. Ingold (Switzerland) have developed the Fluorosensor, a probe which can be integrated with a small computer or any data transformation device (Gary, Meier and Ludwig, 1988).

Dielectric spectroscopy can be used on-line to monitor biomass. Details of the theory and principles of this technique have been described by Kell (1987). At low radio frequencies (0.1 to 1.0 MHz), a microbial cell membrane will act as a capacitor, and become charged by the so-called $\beta$-dispersion effect (Schwan, 1957), making it possible to discriminate between microbial cells, gas bubbles and insoluble media particles. The size of this $\beta$-dispersion is linearly proportional to the membrane enclosed volume fraction up to high cell densities. Kell *et al.* (1987) were able to show that the capacitance (dielectric permittivity) was linearly proportional to the biomass concentration. The output for unicellular organisms is proportional to the mean cell radius whereas with mycelial suspensions the output

remains linear for increases in biomass of a particular cell morphology. The capacitance has been shown to give a linear response with biomass using a number of strains of bacteria, yeasts, mycelial fungi, plant and animal cells.

The sterilizable probe can be inserted directly into a fermenter using a 25-mm diameter port. Fouling of the gold electrodes in the probe can be avoided by the automatic application of electrolytic cleaning pulses. The sensor ('Bug meter') is manufactured by Aber Instruments (Aberystwyth, Wales) and marketed by Applikon (Schiedam, The Netherlands). It has a capacitance range of 0.1 to 200 pF (picoFarads), which is equivalent to 0.1 to 200 mg dry weight $cm^{-3}$ (approximately $10^6$ to $2 \times 10^9$ cells $cm^{-3}$ of *Saccharomyces cerivisiae*). The resolution depends on the type of cells and the conductivity of the medium, but is normally 0.1 mg dry weight $cm^{-3}$. In order for the sensor to work effectively the suspending medium must have a minimum conductance. Yeast slurries after acid washing (Chapter 6) are satisfactory, but before such washing there may be a need for extra salts in the medium in order to make measurements. This sensor has proved ideal for yeast cells and is now being used by the brewing industry to control yeast pitching rates (Boulton *et al.*, 1989).

### Measurement and control of dissolved oxygen

In most aerobic fermentations it is essential to ensure that the dissolved oxygen concentration does not fall below a specified minimal level. Since the 1970s steam sterilizable oxygen electrodes have become avail-

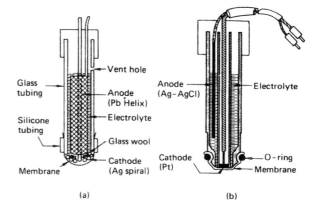

FIG. 8.7. Construction of dissolved-oxygen electrodes: (a) galvanic, (b) polarographic (Lee and Tsao, 1979).

able for this monitoring (Fig. 8.7). Details of electrodes are given by Lee and Tsao (1979).

These electrodes measure the partial pressure of the dissolved oxygen and not the dissolved oxygen concentration. Thus at equilibrium, the probe signal of an electrode will be determined by:

$$P(O_2) = C(O_2) \times P_T$$

where $P(O_2)$   is the partial pressure of dissolved oxygen sensed by the probe,

$C(O_2)$   is the volume or mole fraction of oxygen in the gas phase,

$P_T$   is the total pressure.

The actual reading is normally expressed as percentage saturation with air at atmospheric pressure, so that 100% dissolved oxygen means a partial pressure of approximately 160 mmHg.

Pressure changes can have a significant effect on readings. If the total pressure of the gas equilibrating with the fermentation broth varies, the electrode reading will change even though there is no change in the gas composition. Changes in atmospheric pressure can often cause 5% changes and back pressure due to the exit filters can also cause increases in readings. Allowance must also be made for temperature. The output from an electrode increases by approximately 2.5% per °C at a given oxygen tension. This effect is due mainly to increases in permeability in the electrode membrane. Many electrodes have built-in temperature sensors which allow automatic compensation of the output signal. It is also important to remember that the solubility of oxygen in aqueous media is influenced by the composition. Thus, water at 25°C and 760 mmHg pressure saturated with air will contain 8.4 mg $O_2$ dm$^{-3}$, while 25% NaCl in identical conditions will have an oxygen solubility of 2.0 mg $O_2$ dm$^{-3}$. However, the measured partial pressure outputs for $O_2$ would be the same even though the oxygen concentrations would be very different. Therefore it is best to calibrate the electrode in percentage oxygen saturation. More details on oxygen electrodes and their calibration has been given by Halling (1990).

In small fermenters (1 dm$^3$), the commonest electrodes are galvanic and have a lead anode, silver cathode and employ potassium hydroxide, chloride, bicarbonate or acetate as an electrolyte. The sensing tip of the electrode is a teflon, polyethylene or polystyrene membrane which allows passage of the gas phase so that an equilibrium is established between the gas phases inside and outside the electrode. Because of the relatively slow movement of oxygen across the membrane, this type of electrode has a slow response of the

order of 60 seconds to achieve a 90% reading of true value (Johnson et al., 1964). Buhler and Ingold (1976) quote 50 seconds for 98% response for a later version. These electrodes are therefore suitable for monitoring very slow changes in oxygen concentration and are normally chosen because of their compact size and relatively low cost. Unfortunately, this type of electrode is very sensitive to temperature fluctuations, which should be compensated for by using a thermistor circuit. The electrodes also have a limited life because of corrosion of the anode.

Polarographic electrodes, which are bulkier than galvanic electrodes, are more commonly used in pilot and production fermenters, needing instrument ports of 12, 19 or 25 mm diameter. Removable ones need a 25 mm port. They have silver anodes which are negatively polarized with respect to reference cathodes of platinum or gold, using aqueous potassium chloride as the electrolyte. Response times of 0.05 to 15 seconds to achieve a 95% reading have been reported (Lee and Tsao, 1979). The electrodes which can be very precise may be both pressure and temperature compensated. Although a polarographic electrode may initially cost 600% more than the galvanic equivalent, the maintenance costs are considerably lower as only the membrane should need replacing.

Prototypes of a fast response phase fluorometric sterilizable oxygen sensor are now being developed (Bambot et al., 1994). The sensor utilizes the differential quenching of a fluorescence lifetime of a chromophore, tris(4,7-diphenyl-1,10-phenanthroline)ruthenium(II) complex, in response to the partial pressure of oxygen. The fluorescence of this complex is quenched by oxygen molecules resulting in a reduction of fluorescence lifetime. Thus, it is possible to obtain a correlation between fluorescence lifetime and the partial pressure of oxygen. However, at room temperature when a Clark type oxygen electrode shows a linear calibration the optical sensor shows a hyperbolic response. The sensitivity of the optical sensor when compared with an oxygen electrode is significantly higher at low oxygen tensions whereas the sensitivity is low at high oxygen tensions. The sensor is autoclavable, free of maintenance requirements, stable over long periods and gives reliable measurements of low oxygen tensions in dense microbial cultures.

Dissolved oxygen concentrations may also be determined by a tubing method, described by Phillips and Johnson (1961) and Roberts and Shepherd (1968). The probe consists of a coil of permeable teflon or propylene tubing within the fermenter through which is passed a stream of helium or nitrogen. The oxygen, which diffuses from the fermentation medium through the tubing wall into the inert gas stream, is then determined using a paramagnetic gas analyser (see next section). Times of 2 to 10 minutes are required before taking readings. The tubing will withstand repeated sterilization and has been used continuously for up to 1000 hours at pilot scale.

If it is necessary to increase the dissolved oxygen concentration in a medium this may be achieved by increasing the air flow rate or the rpm of the impeller or a combination of both processes. Another way is to increase the ratio of oxygen to nitrogen in the input gas using a variable proportionating valve while maintaining a constant gas flow rate (Siegall and Gaden, 1962). The cost of this technique would normally restrict its use to laboratories and pilot plants.

### Inlet and exit-gas analysis

The measurement and recording of the inlet and/or exit gas composition is important in many fermentation studies. By observing the concentrations of carbon dioxide and oxygen in the entry and exit gases in the fermenter and knowing the gas flow rate it is possible to determine the oxygen uptake of the system, the carbon dioxide evolution rate and the respiration rate of the microbial culture.

The oxygen concentration can be determined by a paramagnetic gas analyser. Oxygen has a strong affinity for a magnetic field, a property which is shared with only nitrous and nitric oxides. The analysers may be of a deflection or thermal type (Brown *et al.*, 1969).

In the deflection analyser, the magnetic force acts on a dumb-bell test body that is free to rotate about an axis (Fig. 8.8). The magnetic force which is created around the test body is proportional to the oxygen concentration. When the test body swings out of the magnetic field a corrective electrostatic force must be applied to return it to the original position. Electrostatic force readings can therefore be used as a measure of oxygen concentrations.

In a thermal analyser, a flow through 'ring' element is the detector component (Fig. 8.9). After entering the ring, the paramagnetic oxygen content of the sample is attracted by the magnetic field to the central glass tube where resistors heat the gases. The resistors are connected into a Wheatstone bridge circuit to detect variations in resistance due to flow-rate changes. The oxygen in the heated sample loses a high proportion of its paramagnetism. Cool oxygen in the incoming gas flow

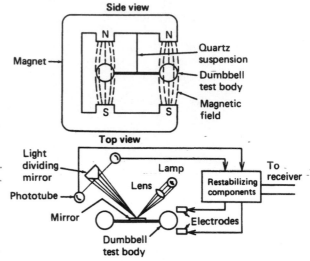

FIG. 8.8. A deflection-type paramagnetic oxygen analyser (Brown *et al.*, 1969).

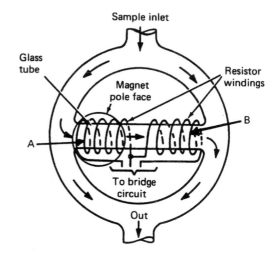

FIG. 8.9. The measuring element in a thermal-type paramagnetic oxygen analyser (Brown *et al.*, 1969).

will now be attracted and displace the hot oxygen. This displacement action produces a convection current. The flow rate of the convection current is a function of the oxygen concentration and can be detected by resistors. The resulting gas flow cools the winding A and heats the winding B, and it is the resulting temperature difference that imbalances the Wheatstone bridge.

Carbon dioxide is commonly monitored by infrared analysis using a positive filtering method. The unit (Fig. 8.10) consists of a source of infrared energy, a 'chopper'

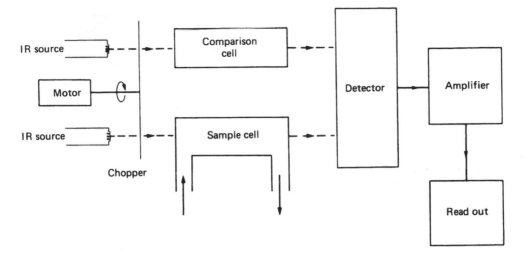

FIG. 8.10. Simple positive filtering infrared analyser.

to ensure that energy passes through each side of the optical system, a sample cell, a comparison (or reference) cell, and an infrared detector sensitized at a wavelength at which the gas of interest absorbs infrared energy. In this case the detector will be filled with carbon dioxide. This optical system senses the reduced radiation energy of the measuring beam reaching the detector, which is due to the absorption in the carbon dioxide in the sample cell.

It is expensive to have separate carbon dioxide and oxygen analysers for each separate fermenter. Therefore it may be possible to couple up a group of fermenters via a multiplexer to a single pair of gas analysers (Meiners, 1982). Gas analysis readings can then be taken in rotation for each fermenter every 30 to 60 minutes. In many cases this will be adequate. Alternatively the gas analysers can be replaced by a mass spectrometer which can analyse a number of components as well as oxygen and carbon dioxide (see later section).

### pH measurement and control

In batch culture the pH of an actively growing culture will not remain constant for very long. In most processes there is a need for pH measurement and control during the fermentation if maximum yield of a product is to be obtained. Rapid changes in pH can often be reduced by the careful design of media, particularly in the choice of carbon and nitrogen sources, and also in the incorporation of buffers or by batch feeding (Chapters 2 and 4). The pH may be further controlled by the addition of appropriate quantities of ammonia or sodium hydroxide if too acidic, or sulphuric acid if the change is to an alkaline condition. Normally the pH drift is only in one direction.

pH measurement is now routinely carried out using a combined glass reference electrode that will withstand repeated sterilization at temperatures of $121°$ and pressures of $138$ kN m$^{-2}$. The electrodes may be silver/silver chloride with potassium chloride or special formulations (e.g. Friscolyt by Ingold) as an electrolyte. Occasionally calomel/mercury electrodes are used. The electrode is connected via leads to a pH meter/controller. If the electrode and its fermenter have to be sterilized in an autoclave then the associated leads and plugs to a pH meter must be able to withstand autoclaving and retain their electrical resistance. Repeated sterilization may gradually change the performance of the electrode. The long culture times associated with animal cell culture or continuous culture of any cells makes a withdrawal option highly desirable to allow for servicing the electrode or troubleshooting without interrupting the fermentation. The housing for this option needs to be carefully designed to ensure that the fermentation does not become contaminated when the electrode is withdrawn, serviced, resterilized and inserted into the housing (Gary et al., 1988).

Ingold electrodes contain a ceramic housing in the reference half cell which has pore dimensions capable of preventing fungal or bacterial infections. However, it is often desirable in animal-cell culture to sterilize the reference electrolyte as well as the electrode surfaces and seals. Both liquid and gel filled electrodes are available. The liquid system gives a faster response, is most stable and accurate. When the electrode is pressurized above the operating pressure in a fermenter the liquid electolyte will gradually flow out of the ceramic diaphragm and prevent fouling, particularly from proteins in the fermentation broth precipitating on the membrane after contact with the electrolyte (Gary *et al.*, 1988).

Readers should consult Halling (1990) for more information on calibration and checking, sterilization, routine maintenance and problems in use.

Control units, to be discussed later in this chapter, may be simple ON/OFF or more complex. In the case of the ON/OFF controller, the controller is set to a predetermined pH value. When a signal actuates a relay, a pinch valve is opened or a pump started, and acid or alkali is pumped into the fermenter for a short time which is governed by a process timer (0 to 5 seconds). The addition cycle is followed by a mixing cycle which is governed by another process timer (0 to 60 seconds) during which time no further acid or alkali can be added. At the end of the mixing cycle another pH reading will indicate whether or not there has been adequate correction of the pH drift. In small volumes the likelihood of overshoot is minimal. A recording unit may be wired to the pH meter to monitor the pH pattern throughout a process cycle.

Shinskey (1973) has discussed pH control of batch processes using proportional and proportional plus derivative control (see later section) when overshoot of the set point is to be avoided. In the case of proportional action, the controller must be adjusted so that a valve on an acid feed-line is shut when the error is zero. However, overshoot is possible as there may be a delay in closing the valve once the set point is achieved. In some cases the overshoot cannot be corrected because of lack of alkali nor may it be desirable. Therefore to preclude an overshoot, the valve must be closed before the controlled variable reaches the set point. This may be done using proportional plus derivative control. The derivative action will need careful adjustment. Too little derivative action will cause some overshoot while too much will lead to the premature closure of the valve. This premature closure may be only for a short time before the valve opens again to give a response pattern as shown in Fig. 8.11.

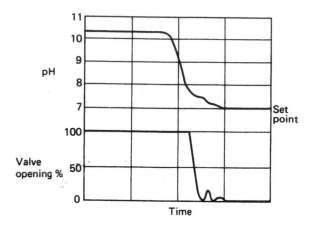

FIG. 8.11. Valve opening and pH changes with proportional plus derivative control (Shinskey, 1973).

is a measure of the oxidation–reduction potential of a biological system and can be determined as a voltage (mV), the value in any system depending on the equilibrium of:

$$\text{Reduced form} \rightleftharpoons \text{Oxidized form} + \text{electron(s)}$$
$$\text{(negative value)} \qquad \text{(positive value)}$$

The measuring electrode consists of gold, platinum or iridium which is welded to a copper lead. The interpretation of results presents difficulties and is confusing (Halling, 1990). The culture is not at redox equilibrium until possibly the end of a growth cycle. During the cycle, although some redox half-reactions may be in equilibrium they cannot all be in a dynamic system. The micro-organisms can also be at a different redox potential from the broth. It has been speculated that the probe signal is indicating something about the relative concentrations of uncertain and probably varying chemical species! This is far from the ideal for a sensor which should only be measuring a specific factor. If the broth contains traces of oxygen this will probably dominate the signal. It is the ability to detect low concentrations of oxygen in media (1 ppm) where redox electrodes may have a good application for determining oxygen availability in anaerobic or microaerophilic processes operated on a small scale (Kjaergaard and Joergensen, 1979).

Halling (1990) gives further details on routine handling, sterilization, testing and calibration.

## Redox

Aspects of redox potential have been reviewed by Jacob (1970), Kjaergaard (1977) and Halling (1990). It

## Carbon dioxide electrodes

The measurement of dissolved carbon dioxide is possible with an electrode, since a pH or voltage change can be detected as the gas goes into solution. The first available electrode consisted of a combined pH electrode with a bicarbonate buffer (pH 5) surrounding the bulb and ceramic plug, with the solution being retained by a PTFE membrane held by an O-ring. Unfortunately, this electrode was not steam sterilizable. This basic design has been modified so that dissolved $CO_2$ from the sample permeates a stainless-steel reinforced silicone bi-layer membrane and dissolves in the internal bicarbonate electrolyte of the electrode. The subsequent pH shift is determined by an internal pH element. This system is extremely sensitive to shifts in the pH element as one pH unit change represents a tenfold increase in $pCO_2$, but precautions have been taken in the design to ensure minimal drift of the sensor. It is possible to calibrate the electrode on-line using specially formulated buffer solutions. This new version can be steam sterilized (Gary *et al.*, 1988).

## ON-LINE ANALYSIS OF OTHER CHEMICAL FACTORS

If good control of a fermentatation is to be obtained, then all chemical factors which can influence growth and product formation ought to be continuously monitored. This ideal situation has not yet been achieved but a number of techniques are currently being developed.

### Ion-specific sensors

Ion-specific sensors have been developed by Orion Research and Radiometer to measure $NH_4^+$, $Ca^{2+}$, $K^+$, $Mg^{2+}$, $PO^{3-}$, $SO^{2-}$, etc. (Orion Research Inc., 11 Blackstone Street, Cambridge, MA, U.S.A.; Radiometer — Aagard Nielson and Schroder, Emdrupvej 72, Copenhagen-NV, Denmark). The response time of these electrodes varies from 10 seconds to several minutes depending on the concentration of the ion species, the composition of the sample, etc. However, none of these probes is steam sterilizable.

### Enzyme and microbial electrodes

Enzyme or microbial cell electrodes can be used in some analyses. A suitable enzyme or microbial cell which produces a change in pH or forms oxygen in the enzyme reaction is chosen and immobilized on a membrane held in close contact to a pH or oxygen electrode. Unfortunately, the oxygen demand of an enzyme may restrict the maximum substrate concentration which might be detected in a medium. Enfors (1981) overcame this problem for glucose determination by co-immobilizing glucose oxidase and catalase. More recently, it has been possible to use a ferrocene derivative as an artificial redox carrier to shuttle electrons from glucose oxidase to a carbon electrode, thus making the device largely independent of oxygen concentrations (Higgins and Boldot, 1992). Enzyme electrodes are also commercially available to monitor cholesterol, triglycerides, lactate, acetate, oxalate, methanol, ethanol, creatine, ammonia, urea, amino acids, carbohydrates and penicillin (Higgins *et al.*, 1987; Luong *et al.*, 1988; Higgins and Boldot, 1992).

Ideally, an electrode that can be inserted into a fermenter and steam sterilized is required, but none is yet available which can be used in this manner, even when using enzymes which are stable at high temperatures which can be obtained from thermophilic microorganisms. Hewetson *et al.* (1979) prepared sterilized penicillinase electrodes by assembling sterile components or by standing assembled components in chloroform before placing in a fermenter.

### Near infra-red spectroscopy

Hammond and Brookes (1992) have described the development of near infra-red spectroscopy (NIR; 460–1200 nm) for rapid, continuous and batch analysis of components of fermentation broths. In samples from an antibiotic fermentation, they used NIR absorbance bands to simultaneously estimate fat (in the medium), techoic acid (biomass) and antibiotic (the product). Fat analysis has been made possible with a fibre optic sensor placed *in situ* through a port in the fermenter wall. The assay time for an antibiotic has been reduced from 2 hours to 2 minutes. A method has also been developed to measure alkaline protease production in broths. Vaccari *et al.* (1994) have used this technique to measure glucose, lactic acid and biomass in a lactic acid fermentation.

### Mass spectrometers

The mass spectrometer can be used for on-line analysis since it is very versatile and has a response time of less than 5 seconds for full-scale response and taking about 12 seconds for a sample stream. It allows for monitoring of gas partial pressures ($O_2$, $CO_2$, $CH_4$, etc.), dissolved gases ($O_2$, $CO_2$, $CH_4$, etc.) and volatiles (methanol, ethanol, acetone, simple organic acids, etc.). Heinzle *et al.* (1981) combined a data processor with a mass spectrometer equipped with a capillary gas inlet to measure gas partial pressures and a membrane inlet to detect dissolved gases and volatiles. It is therefore possible to multiplex this type of system to analyse several fermenters sequentially. Merck and Company Inc. have installed an integrated gas- and liquid-phase analysis system incorporating a mass spectrometer to sample a number of fermenters (Omstead and Greasham, 1989). The availability of low cost mass spectrometers at approximately $15,000 does make their use in a laboratory or pilot plant financially feasible as an alternative to oxygen and carbon dioxide analysers, as well as analysing for a range of other gaseous and volatile compounds.

## CONTROL SYSTEMS

The process parameters which are measured using probes described in the previous sections may be controlled using control loops.

A control loop consists of four basic components:

1. A measuring element.
2. A controller.
3. A final control element.
4. The process to be controlled.

In the simplest type of control loop, known as feedback control (Fig. 8.12), the measuring element senses a process property such as flow, pressure, temperature, etc., and generates a corresponding output signal. The controller compares the measurement signal with a predetermined desired value (set point) and produces an output signal to counteract any differences between the two. The final control element receives the control signal and adjusts the process by changing a valve opening or pump speed and causing the controlled process property to return to the set point.

### Manual control

A simple example of control is manual control of a steam valve to regulate the temperature of water flowing through a pipe (Fig. 8.13). Throughout the time of operation a plant operative is instructed to monitor the temperature in the pipe. Immediately the tempera-

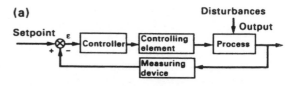

FIG. 8.12. A feedback control loop (Rolf and Lim, 1985).

ture changes from the set point on the thermometer, the operative will take appropriate action and adjust the steam valve to correct the temperature deviation. Should the temperature not return to the set point within a reasonable time, further action may be necessary. Much depends on the skill of individual operatives in knowing when and how much adjustment to make. This approach with manual control may be very costly in terms of labour and should always be kept to a strict minimum when automatic control could be used instead. A justifiable use of manual control may be in the adjustment of minor infrequent deviations.

### Automatic control

When an automatic control loop is used, certain modifications are necessary. The measuring element must generate an output signal which can be monitored by an instrument. In the case of temperature control, the thermometer is replaced by a thermocouple, which is connected to a controller which in turn will produce a signal to operate the steam valve (Fig. 8.14).

Automatic control systems can be classified into four main types:

1. Two-position controllers (ON/OFF).
2. Proportional controllers.
3. Integral controllers.
4. Derivative controllers.

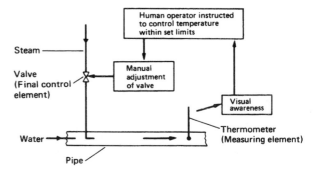

FIG. 8.13. Simple manual-control loop for temperature control.

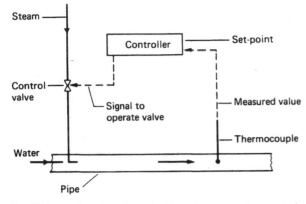

FIG. 8.14. Simple automatic control loop for temperature control.

## TWO-POSITION CONTROLLERS (ON / OFF)

The two-position controller, which is the simplest automatic controller, has a final control unit (valve, switch, etc.) which is either fully open (ON) or fully closed (OFF). The response pattern to such a change will be oscillatory. If there is instant response then the pattern will be as shown in Fig. 8.15.

If one considers the example of the heating of a simple domestic water tank controlled by a thermostat operating with ON/OFF, then there will be a delay in response when the temperature reaches the set point and the temperature will continue rising above this point before the heating source is switched off. At the other extreme, the water will continue cooling after the heating source has been switched on. With this mode of operation an oscillatory pattern will be obtained with a repeating pattern of maximum and minimum temperature oscillating about the set point, provided that all the other process conditions are maintained at a steady level (Fig. 8.16).

If this type of controller is to be used in process control then it is important to establish that the maxi-

mum and minimum values are acceptable for the specific process, and to ensure that the oscillation cycle time does not cause excessive use of valves or switches.

ON/OFF control is not satisfactory for controlling any process parameter where there are likely to be large sudden changes from the equilibrium. In these cases alternative forms of automatic control must be used.

In more complex automatic control systems three different methods are commonly used in making error corrections. They are: proportional, integral and derivative. These control methods may be used singly or in combinations in applying automatic control to a process, depending upon the complexity of the process and the extent of control required. Since many of the controllers used in the chemical industries are pneumatic, the response to an error by the controller will be represented by a change in output pressure. Pneumatic controllers are still widely used because they are robust and reliable. In other cases, when the controller is electronic, the response to an error will be represented as a change in output current or voltage.

## PROPORTIONAL CONTROL

Proportional control can be explained as follows: the change in output of the controller is proportional to the input signal produced by the environmental change (commonly referred to as error) which has been detected by a sensor.

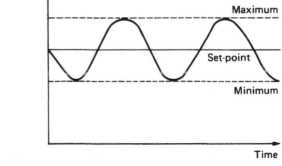

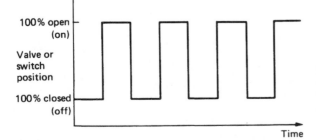

FIG. 8.15. Oscillatory pattern of a simple two-position valve or switch.

FIG. 8.16. Oscillatory pattern of the temperature of a domestic water tank (no water being drawn off) using ON/OFF control of the heating element.
_____ Output without control
........... Proportional action
– – – – Integral action
--------- Proportional + integral action
–·–·–· Proportional + derivative action
--------- Proportional + integral + derivative action

Mathematically it can be expressed by the following equation:

$$M = M_0 + K_c \Sigma$$

where $M$ = output signal,
$M_0$ = controller output signal when there is no error,
$K_c$ = controller gain or sensitivity,
$\Sigma$ = the error signal.

Hence, the greater the error (environmental change) the larger is the initial corrective action which will be applied. The response to proportional control is shown in Fig. 8.17, from which it may be observed that there is a time of oscillation which is reduced fairly quickly. It should also be noted that the controlled variable attains a new equilibrium value. The difference between the original and the new equilibrium value is termed the offset.

The term $K_c$ (controller gain) is the multiplying factor (which may be dimensioned) which relates a change in input to the change in output.

$$\Delta I \rightarrow \boxed{\text{Controller}} \rightarrow \Delta O$$

(Change in input)  (Change in output)

Then

$$\Delta I = K_c \Delta O.$$

$K_c$ may contain conversion units if there is an electrical input and a pressure output or vice versa.

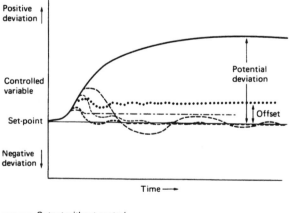

———— Output without control
• • • • • • Proportional action
– – – – – Integral action
- - - - - - Proportional + integral action
–·–·–·– Proportional + derivative action
– – – – – Proportional + integral + derivative action

FIG. 8.17. Typical controlled plant responses.

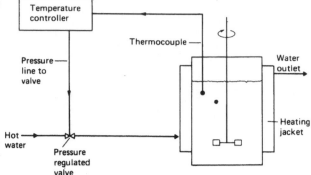

FIG. 8.18. A fermenter with a temperature-controlled heating jacket.

If the input to the controller is 1 unit of change, then:

(a) with a controller gain of 1, the output will be 1 unit,
(b) with a controller gain of 2, the output will be 2 units, etc.

On many controllers $K_c$ is graduated in terms of proportional band instead of controller gain.

Now
$$K_c \propto \frac{1}{PB} \text{ (proportional band)}$$

or
$$K_c = c' \frac{1}{PB} \text{ where } c' \text{ is a constant.}$$

This quantity $PB$ is defined as the error required to move the final control element over the whole of its range (e.g. from fully open to fully shut) and is expressed as a percentage of the total range of the measured variable (e.g. two extremes of temperature).

A fermenter with a heating jacket will be used as an example (Fig. 8.18). A thermocouple is connected to a temperature controller which has a span of 10° covering the range 25° to 35°, with a set point at 30°. The controller valve which is controlled by a pressure regulator, is fully open at 5 psig (46 kN m$^{-2}$) and fully closed at 15 psig (138 kN m$^{-2}$), while the set point of 30° corresponds to a control pressure on the valve of 10 psig (92 kN m$^{-2}$). When the controller gain is 1, a change of 10° will cause a pressure change of 10 psig when the valve will be fully open at 25° and fully closed at 35°. Thus, the proportional band is 100%. If the controller gain is 2, a 5° change will cause the valve to go from fully open to fully closed, i.e. 27.5° to 32.5°. In this case the proportional band width will be:

$$\frac{\text{Actual band width}}{\text{Total band range}} = \frac{5°}{10°} \times 100 = 50\%.$$

Either side of this band the pressure will be constant. In the case of a controller gain of 4, a 2.5° change will cause a pressure change of 10 psig (92 kN m$^{-2}$), which will cause the control valve to go from fully open to fully closed (28.75° to 31.25°). In this case the proportional band will be

$$= \frac{2.5°}{10°} \times 100 = 25\%.$$

These results are summarized in Table 8.2 for controller gains of 1, 2 and 4.

When the proportional band is very small (the controller gain is high) the control mode can be likened to simple ON/OFF, with a high degree of oscillation but no offset. As the proportional band is increased (low controller gain) the oscillations are reduced but the offset is increased. Settings for proportional band width are normally a compromise between degree of oscillation and offset. If the offset is not desirable it can be eliminated by the use of proportional control in association with integral control (see Fig. 8.17 and later section).

## INTEGRAL CONTROL

The output signal of an integral controller is determined by the integral of the error input over the time of operation. Thus:

$$M = M_0 + \frac{1}{T_i} \int \Sigma \, dt$$

where $T_i$ = integral time.

It is important to remember that the controller output signal changes relatively slowly at first as time is required for the controller action to integrate the error. It is evident from Fig. 8.17 that the maximum deviation from the set point is significant when compared with the use of proportional control for control of the chosen parameter, and the system takes longer to settle down. There is, however, no offset which is advantageous in many control processes.

## DERIVATIVE CONTROL

When derivative control is applied the controller senses the rate of change of the error signal and contributes a component of the output signal that is proportional to a derivative of the error signal. Thus:

$$M = M_0 + T_d \frac{d\Sigma}{dt}$$

where $T_d$ is a time rate constant.

It is important to remember that if the error is constant there is no corrective action with derivative control. In practice, derivative control is never used on its own. The response curve has therefore been deliberately omitted from Fig. 8.17.

Figure 8.19 demonstrates the response of derivative control to sinusoidal error inputs. The output is always in a direction to oppose changes in error, both away from and towards the set point, which in this example results in a 90° phase shift. This opposition to a change has a fast damping effect and this property is very useful in combination with other modes of control which will be discussed later.

TABLE 8.2. *The effect of controller gain* ($K_c$) *on band width of a proportional temperature controller*

| | Measured temperature | | Pressure output (psig) at $K_c = 1$ psig/1° | | Pressure output (psig) at $K_c = 2$ psig/1° | | Pressure output (psig) at $K_c = 4$ psig/1° | |
|---|---|---|---|---|---|---|---|---|
| | 35° | 15 | | | 15 | | 15 | |
| | 34° | 14 | | | 15 | | 15 | |
| | 33° | 13 | | | 15 | | 15 | |
| | 32.5° | 12.5 | | | 15 | | 15 | |
| | 32° | 12 | | | 14 | | 15 | |
| | 31.25° | 11.25 | | | 12.5 | | 15 | |
| | 31° | 11 | | | 12 | | 14 | |
| Set → | 30° | 10 | Prop. band 100% | | 10 | Prop. band 50% | 10 | Prop. band 25% |
| point | 29° | 9 | | | 8 | | 6 | |
| | 28.75° | 8.75 | | | 7.5 | | 5 | |
| | 28° | 8 | | | 6 | | 5 | |
| | 27.5° | 7.5 | | | 5 | | 5 | |
| | 27° | 7 | | | 5 | | 5 | |
| | 26° | 6 | | | 5 | | 5 | |
| | 25° | 5 | | | 5 | | 5 | |

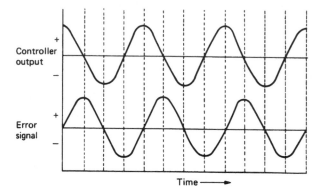

FIG. 8.19. Response of a derivative controller to sinusoidal error inputs.

## Combinations of methods of control

Three combinations of control systems are used in practice:

(a) Proportional plus integral.
(b) Proportional plus derivative.
(c) Proportional plus integral plus derivative.

### PROPORTIONAL PLUS INTEGRAL CONTROL

When proportional plus integral control is used, the output response to an error gives rise to a slightly higher initial deviation in the output signal compared with that which would be obtained with proportional control on its own (Fig. 8.17). This is due to a contribution in the signal from integral control. However, the oscillations are soon reduced and there is finally no offset. This mode of control finds wide applications since the proportional component is ideal in a process where there are moderate changes, whereas the integral component will allow for large load changes and eliminate the offset that would have occurred.

### PROPORTIONAL PLUS DERIVATIVE CONTROL

If proportional plus derivative control is used, the output response to an error will lead to reduced deviations, faster stabilization and a reduced offset (Fig. 8.17) compared with proportional control alone. Because the derivative component has a rapid stabilizing influence, the controller can cope with rapid load changes.

### PROPORTIONAL PLUS INTEGRAL PLUS DERIVATIVE CONTROL

The combination of proportional plus integral plus derivative normally provides the best control possibilities (Fig. 8.17). The advantages of each system are retained. The maximum deviation and settling time are similar to those for a proportional plus derivative controller whilst the integral action ensures that there is no offset. This method of control finds the widest applications because of its ability to cope with wide variations of patterns of changes which might be encountered in different processes.

### Controllers

The first primary automatic controllers were electronic control units which were adjusted manually to set up desired PID response patterns to a disturbance in a control loop. These controllers are relatively expensive and some knowledge of control engineering is necessary to make the correct adjustments to obtain the required control responses.

The availability of cheap computers and suitable computer programs to mimic PID control and handle a number of control loops simultaneously has made it possible to use some very complex control techniques for process optimization. Some aspects of this work on computer applications will be discussed later in this chapter.

### More complex control systems

In certain situations PID control is not adequate to control a disturbance. Control may be difficult when there is a long time lag between a change in a manipulated variable and its effect on the measured variable. Consider the example of a heat exchanger where the water temperature is regulated by the flow rate of steam from a steam valve (Hall et al., 1974). By the time the effects of a change in steam flow influence the hot water temperature, a considerable energy change has occurred in the heat exchanger which will continue to drive the hot water temperature away from the set-point after a correction has been made to the steam-flow valve. This lag will lead to cycling of the measured variable about the set point. Cascade control can solve this problem. When cascade control is used (Fig. 8.20), the output of one controller is the set-point for another. Each controller has its own measured variable with only the primary controller having an

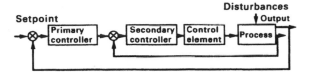

FIG. 8.20. A cascade feedback control loop (Rolf and Lim, 1985).

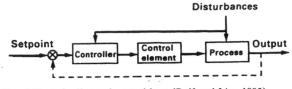

FIG. 8.22. A feedforward control loop (Rolf and Lim, 1985).

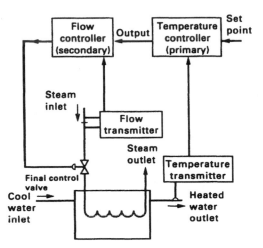

FIG. 8.21. Cascade control of a heat-exchange process (Hall *et al.*, 1974).

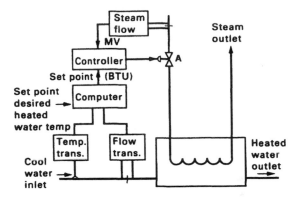

FIG. 8.23. Feedforward control of a heat-exchange process (Hall *et al.*, 1974).

individual set-point and only the secondary controller providing an output to the process. In more complex cascade systems more loops may be included.

When cascade control is used to control water temperature in a heat exchanger process (Fig. 8.21) there is the primary loop (slow or outer loop), consisting of the temperature sensor and the primary temperature controller of the process water temperature, and the secondary loop (fast or inner loop) consisting of the steam flow sensor, the steam flow controller, its process variable (steam flow), the control valve and the process. The addition of the secondary controller, whose measured variable is steam flow, allows steam flow variations to be corrected immediately before they can affect the hot water temperature. This extra control loop should help to minimize temperature cycling.

Feed-forward control (Fig. 8.22) makes it possible to utilize other disturbances besides the measured values of the process that have to be controlled and enable fast control of a process. The cascade control system (Fig. 8.21) which has just been discussed would be adequate provided that the water inlet pressure and temperature do not fluctuate. When fluctuations occur, feed-forward control may be used, but this will require the measurement of both inlet water temperature and flow rate with both variables combined to control the

steam flow rate (Fig. 8.23). In the example shown (Hall *et al.*, 1974) a computer is programmed to determine the heat energy required to be added to the cool inlet water to bring it to the appropriate temperature at the outlet and allow the correct amount of steam to enter the heating coils. Inputs for temperature and flow are made from the inlet pipe to the computer. Calculations can then be made to determine the required heat input to raise the water temperature in the heat exchanger. The output control signal to the valve A is computed to allow the correct amount of steam to enter the heating coils.

When one or more of the process variables or characteristics is not known and cannot be measured directly, then adaptive control should be considered. The on-line identification of process characteristics and the subsequent use of this information to improve the process constitute adaptive control (Hall *et al.*, 1974). The sequences and the interactions in the adaptive control loop are outlined in Fig. 8.24. Adaptive control is useful in circumstances where the process dynamics are not well defined or change with time. This may be most useful in controlling a batch fermentation where considerable and often complex changes may occur (Bull, 1985; Dusseljee and Feijen, 1990). Adaptive control strategies have been used in fed-batch yeast cultivation (Montague *et al.*, 1988), amino acid production (Radjai *et al.*, 1984) and penicillin production (Lorenz *et al.*, 1985).

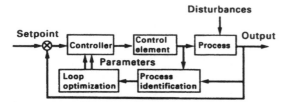

FIG. 8.24. Adaptive control (Rolf and Lim, 1985).

It is also possible to prepare sequential programs using a computer based control system. Consider control of the dissolved oxygen concentration in a fermentation broth. This may be changed by altering the agitation rate, the air or gas flow rate, the partial pressure of oxygen in the inlet gas or the total fermenter pressure. In practice, combinations of these variables may be used either sequentially or simultaneously using suitable computer programs. Initially the agitation rate can be increased to respond to decreases in dissolved oxygen concentration. When a predetermined maximum agitation rate is reached, the air flow can be steadily increased to a preset maximum, followed by the third and subsequent stages.

## COMPUTER APPLICATIONS IN FERMENTATION TECHNOLOGY

Since the initial use of computers in the 1960s for modelling fermentation processes (Yamashita and Murao, 1978) and in process control for production of glutamic acid (Yamashita et al., 1969) and penicillin (Grayson, 1969), there have been numerous publications on computer applications in fermentation technology (Rolf and Lim, 1985; Bushell, 1988; Whiteside and Morgan, 1989; Fish et al., 1990). Initially, the use of large computers was restricted because of their cost but reductions in costs and the availability of cheaper small computers has widened interest in their possible applications. The availability of efficient small computers has led to their use for pilot plants and laboratory systems since the financial investment for the on-line computer amounts to a relatively insignificant part of the whole system.

Three distinct areas of computer function were recognized by Nyiri (1972):

1. *Logging of process data.* Data logging is performed by the data acquisition system which has both hardware and software components. There is an interface between the sensors and the computer. The software should include the computer program for sequential scanning of the sensor signals and the procedure of data storage.

2. *Data analysis (Reduction of logged data).* Data reduction is performed by the data-analysis system, which is a computer program based on a series of selected mathematical equations. The analysed information may then be put on a print out, fed into a data bank or utilized for process control.

3. *Process control.* Process control is also performed using a computer program. Signals from the computer are fed to pumps, valves or switches via the interface. In addition the computer program may contain instructions to display devices or teletypes, to indicate alarms, etc.

At this point it is necessary to be aware that there are two distinct fundamental approaches to computer control of fermenters. The first is when the fermenter is under the direct control of the computer software. This is termed Direct Digital Control (DDC) and will be discussed in the next section. The second approach involves the use of independent controllers to manage all control functions of a fermenter and the computer communicates with the controller only to exchange information. This is termed Supervisory Set-Point Control (SSC) and will be discussed in more detail in the Process Control section.

It is possible to analyse data, compare it with model systems in a data store, and use control programs which will lead to process optimization. However, process optimization by this method is not a widely used procedure in the fermentation industries at present. It is important to be aware of these different applications, since this will influence the size and type of computer system which will be appropriate for the precise role that it is intended to perform, whether in a laboratory, a pilot plant, or manufacturing plant, or a combination of these three.

### Components of a computer-linked system

When a computer is linked to a fermenter to operate as a control and recording system, a number of factors must be considered to ensure that all the components interact and function satisfactorily for control and data logging. A DDC system will be used as an example to explain computer controlled addition of a liquid from a reservoir to a fermenter. A simple outline of the main components is given in Fig. 8.25. A sensor S in the fermenter produces a signal which may need to be amplified and conditioned in the correct analogue form. At this stage it is necessary to convert the signal to a digital form which can be subsequently transmitted to the computer. An interface is placed in the circuit at this point. This interface serves as the junction point for the inputs from the fermenter sensors to the computer and the output signals from the computer to the fermenter controls such as a pump T attached to an additive reservoir. Digital to analogue conversion is necessary between the interface and the pump T.

A sensor will generate a small voltage proportional to the parameter it is measuring. For example, a temperature probe might generate 1 V at 10°C and 5 V at 50°C. Unfortunately, the signal cannot be understood by the computer and must be converted by an analogue to digital converter (ADC) into a digital form.

The accuracy of an ADC will depend on the number of bits (the unit of binary information) it sends to the computer. An 8-bit converter will work in the range 0–255 and it is therefore able to divide a signal voltage into 256 steps. This will give a maximum accuracy of 100/256, which is approximately 0.4%. However, a 10-bit converter can give 1024 steps with an accuracy of 100/1024, which is approximately 0.1%. Therefore when a parameter is to be monitored very accurately a converter of the appropriate degree of accuracy will be required. The time taken for an ADC to convert voltage signals to a digital output will vary with accuracy, but improved accuracy leads to slower conversion and hence slower control responses. However, cycle times of about 1 second may be adequate in many fermentation systems. It is also important to ensure that the voltage ranges of the sensors are matched to the ADC input range. More detailed discussion is given by Whiteside and Morgan (1989).

A digital to analogue converter (DAC) converts a

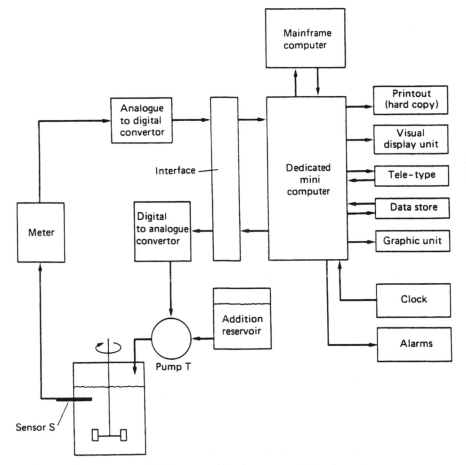

FIG. 8.25. Simplified layout of computer-controlled fermenter with only one control loop shown.

digital signal from the computer into an electrical voltage which can be used to drive electrical equipment, e.g. a stirrer motor. Like the ADC, the accuracy of the DAC will be determined by whether it is 8-bit, 10-bit, 12-bit, etc., and will for example determine the size of steps in the control of rpm of a stirrer motor.

The small computer itself is dedicated solely to one or more fermenters. This computer is coupled to a real-time clock, which determines how frequently readings from the sensor(s) should be taken and possibly recorded. The other ancillary equipment linked directly to the computer might include a visual display unit, a data store, a teletype, a graphic display unit, a print out, alarms and a barometer.

The small computer is often connected to a large main frame computer for random access, not on a real-time scale, but for long-term data storage and retrieval and for complex data analysis which will not be utilized subsequently in real-time control.

It is also possible to develop programs so that on-line instruments can be checked regularly and recalibrated when necessary. Swartz and Cooney (1979) were able to routinely recalibrate a paramagnetic oxygen analyser and an infrared carbon dioxide analyser every 12 hours utilizing a program which connected a gas of known composition to the analysers and subsequently monitored the analyser outputs.

### Data logging

The simplest task for a computer is data logging. Parameters such as those listed in Table 8.1 can be measured by sensors which produce a signal which is compatible with the computer system.

Programs have been developed so that by reference to the real-time clock, the signals from the appropriate sensors will be scanned sequentially in a predetermined pattern and logged in a data store. Typically, this may be 2- to 60-second intervals, and the data is printed out on a visual display unit. In preliminary scanning cycles the values are compared with predefined limit values, and deviations from these values result in an error print out, or if more extreme then an alarm may be activated. In the final cycle of a sequence, say every 5 to 60 minutes, the program instructs that the sensor readings are permanently recorded on a print out or in a data store.

At the same time as on-line data is being recorded from sensors, analytical data for broth viscosity, microbial growth, substrate and precursor utilization and product formation, which have to be determined sepa-

rately may be logged into the data store for specific known times. Carleysmith (1989) has described a data handling system being used by Smith Kline Beecham in the United Kingdom.

Thus, it is now possible to record data continuously for a range of parameters from a number of fermenters simultaneously using minimal manpower, provided that the capital outlay is made for fermenters with suitable instrumentation coupled with adequate computer facilities.

### Data analysis

Because a computer can undertake so many calculations very rapidly, it is possible to design programs to analyse fermentation data in a number of ways. A linked main-frame computer may be used for part of this analysis as well as the dedicated small computer.

A number of the monitoring systems were described as 'Gateway Sensors' by Aiba *et al.* (1973) and are given in Table 8.3. Gateway sensors are so called because the information they yield can be processed to give further information about the fermentation. More details of analysis of direct measurements, indirect measurements and estimated variables have been discussed by Zabriskie (1985) and Royce (1993).

The respiratory quotient of a culture may be calculated from the metered gas-flow rates and analyses for oxygen and carbon dioxide leaving a known volume of culture in the fermenter. This procedure was used to monitor growth of *Candida utilis* in a 250-$dm^3$ fermenter, to follow or forecast events during operation (Nyiri *et al.*, 1975).

If one defines the fraction of substrate which is converted to product then it is possible to write mass

TABLE 8.3. *Gateway sensors* (Aiba *et al.*, 1973)

| Sensor | Information that may be determined from the sensor signal |
|---|---|
| pH | Acid product formation |
| Dissolved oxygen | Oxygen-transferrate |
| Oxygen in exit gas ⎫ Gas-flow rate ⎭ | Oxygen-uptake rate |
| Carbon dioxide in exit gas ⎫ Gas-flow rate ⎭ | Carbon dioxide evolution rate |
| Oxygen-uptake rate ⎫ Carbon dioxide evolution rate ⎭ | Respiratory quotient |
| Sugar-level and feed rate ⎫ Carbon dioxide evolution rate ⎭ | Yield and cell density |

balances for C, H, O and N with the measurement of only a few quantities ($O_2$, $CO_2$, $NH_3$, etc.). All the other quantities can be calculated, including biomass and yield, if the biomass elemental composition is known (Chapter 2). This procedure was used for the analysis of a bakers' yeast fermentation (Cooney *et al.*, 1977). Biomass production can be regarded as a stoichiometric relationship in which substrate is converted, in the presence of oxygen and ammonia to biomass, carbon dioxide and water:

Carbon source-energy + oxygen + ammonium →

cells + water + carbon dioxide.

Thus, the equation can be written in the form:

$$aC_xH_yO_z + bO_2 + cNH_3 \rightarrow dC_rH_sO_tN_u$$
$$+ H_2O + fCO_2$$

where *a, b, c, d, e* and *f* are moles of the respective reactants and products. $C_xH_yO_z$ is the molecular formula of the substrate where *x, y* and *z* are the specific carbon, hydrogen and oxygen atom numbers. Biomass is represented by $C_rH_sO_tN_u$ where *r, s, t* and *u* are the corresponding numbers of each element in the cell. This technique was developed to use with bakers' yeast fermentations (Cooney *et al.*, 1977; Wang *et al.*, 1977).

### Process control

Arminger and Moran (1979) recognized three levels of process control that might be incorporated into a system. Each higher level involves more complex programs and needs a greater overall understanding of the

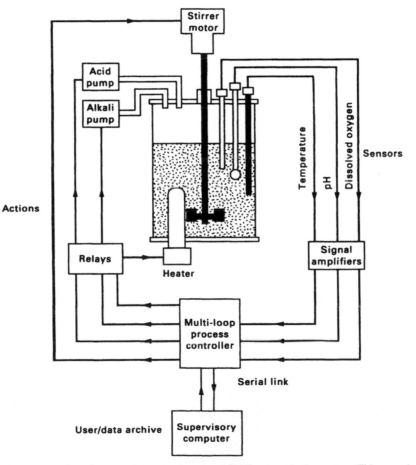

FIG. 8.26. Diagrammatic representation of a supervisory setpoint control (SSC) system for fermenters. This example illustrates a system controlling temperature by means of heating only, dissolved oxygen tension by stirrer speed and pH by the addition of acid and alkali. All control functions are performed by the intelligent process controller and the computer only communicates with this in order to log data and send new setpoints when instructed to do so by the user (Whiteside and Morgan, 1989).

process. The first level of control, which is already routinely used in the chemical industries, involves sequencing operations, such as manipulating valves or starting or stopping pumps, instrument recalibration, on-line maintenance and fail-safe shut-down procedures. In most of these operations the time base is at least in the order of minutes, so that high-speed manipulations are not vital. Two applications in fermentation processes are sterilization cycles and medium batching.

The next level of computer control involves process control of temperature, pH, foam control, etc. where the sensors are directly interfaced to a computer (Direct Digital Control (DDC); Fig. 8.25). When this is done separate controller units are not needed. The computer program determines the set point values and the control algorithms, such as PID, are part of the computer software package. Better control is possible as the control algorithms are mathematically stored functions rather than electrical functions. This procedure allows for greater flexibility and more precise representation of a process control policy. The system is not very expensive as separate electronic controllers are no longer needed, but computer failure can cause major problems unless there is some manual back-up facility.

The alternative approach is to use a computer in a purely supervisory role. All control functions are performed by an electronic controller using a system illustrated in Fig. 8.26 where the linked computer only logs data from sensors and sends signals to alter set points when instructed by a computer program or manually. This system is known as Supervisory Set-Point Control (SSC) or Digital Set-Point Control (DSC). When SSC is used, the modes of control are limited to proportional, integral and derivative because the direct control of the fermenter is by an electronic controller. However, in the event of computer failure the process controller can be operated independently.

Whiteside and Morgan (1989) have discussed some of the relative merits of DDC and SSC systems and given case histories of the installation and operation of both systems.

The most advanced level of control is concerned with process optimization. This will involve understanding a process, being able to monitor what is happening and being able to control it to achieve and maintain optimum conditions. Firstly, there is a need for suitable on-line sensors to monitor the process continuously. A number are now available for dissolved oxygen, dissolved carbon dioxide, pH, temperature, biomass (the bug meter, NADH fluorescence, near infra-red spec-

troscopy) and some metabolites (mass spectroscopy and near infra-red spectroscopy). All these sensors have been discussed earlier in this chapter. Secondly, it is important to develop a mathematical model that adequately describes the dynamic behaviour of a process. Shimizu (1993) has stressed the vital role which these models play in optimization and reviewed the use of this approach in batch, fed-batch and continuous processes for biomass and metabolites. This approach with appropriate on-line sensors and suitable model programs has been used to optimize bakers' yeast production (Ramirez et al., 1981; Shi et al., 1989), an industrial antibiotic process and lactic acid production (Shi et al., 1989).

Although much progress has been made in the ability to control a process, few sensors are yet available to monitor on-line for many metabolites or other parameters in a fermentation broth thus delaying or making a fast response difficult for on-line control action. Also, it is possible that not all the important parameters in a process have been identified and the mathematical model derived to describe a process may be inadequate. Because of these limitations, an artificial neural network may be used to achieve better control (Karim and Rivera, 1992). These are highly interconnected networks of non-linear processing units arranged in layers with adjustable connecting strengths (weights).

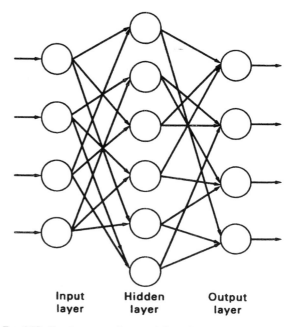

FIG. 8.27. Two-layer neural network (not all the possible interconnections are shown).

In simpler neural networks there is one input layer, one hidden layer and one output layer (Fig. 8.27). Unlike recognized knowledge-based systems, neural networks do not need information in the form of a series of rules, but learn from process examples from which they derive their own rules. This makes it possible to deal with non-linear systems and approximate or limited data.

When training a neural network the aim is to adjust the strengths of the interconnections (neurons) so that a set of inputs produces a desired set of outputs. The inputs may be process variables such as temperature, pH, flow rates, pressure and other direct or indirect measurements which give information about the state of the process. The process outputs obtained (biomass, product, etc.) produce the teacher signal(s) which trains the network. The difference between the desired output and the value predicted by the network is the prediction error. Adjustments are made to minimize the total prediction error by modifying the interconnection strengths until no further decrease in error is achieved. Commercial computer packages are now available to help to determine which of the input variables to use for training and to determine the optimum number of interconnections and hidden layers (Glassey et al., 1994). Readers requiring more detail of the theory of neural networks should consult Karim and Rivera (1992).

This method of control is still at an early stage of development, but it has already been used in a case study on ethanol production by Zymomonas mobilis (Karim and Rivera, 1992), in real-time variable estimation and control of a glucoamylase fermentation (Linko and Zhu, 1992) and recombinant Escherichia coli fermentations (Glassey et al., 1994).

In industrial systems where a significant amount of on-line and off-line process data may be available, but there are tight time restraints imposed on process optimization, the potential for developing a relatively accurate neural network model within short time scales becomes very attractive (Glassey et al., 1994).

# REFERENCES

AIBA, S., OKAMOTO, R. and SATOH, K. (1965). Two sorts of measurements with a jar type of fermenter-power requirements of agitations and capacity coefficient of mass transfer in bubble aeration. J. Ferm. Technol. **43**, 137–145.

AIBA, S., HUMPHREY, A. E. and MILLIS, N. F. (1973) Biochemical Engineering (2nd edition), Chaper 12. Academic Press, New York.

ARMIGER, W. B. (1985) Instrumentation for monitoring and controlling reactors. In Comprehensive Biotechnology, Vol. 2, pp. 133–148 (Editors Cooney, C.L. and Humphrey, A. E.). Pergamon Press, Oxford.

ARMIGER, W. B. and MORAN, D. M. (1979) Review of alternatives and rationale for computer interfacing and system configuration. Biotech. Bioeng. Symp. **9**, 215–225.

ATKINSON, B. and MAVITUNA, F. (1991) Measurement and instrumentation. In Biochemical Engineering and Biotechnology Handbook, pp. 1023–1057. Macmillan, Basingstoke.

BAILEY, J. E. and OLLIS, D. F. (1986) Instrumentation and control. In Biochemical Engineering Fundamentals, pp. 658–725. McGraw-Hill, New York.

BAMBOT, S. B., HOLAVANAHALI, R., LAKOWICZ, J. R., CARTER, G. M. and RAO, G. (1994) Phase fluorometric sterilizable optical oxygen sensor. Biotech. Bioeng. **43**, 1139–1145.

BEYELER, W., EINSELE, A. and FIECHTER, A. (1981) Fluorometric studies in bioreactors: methods and applications. Abstracts of Communications. Second European Congress of Biotechnology, Eastbourne, p. 142. Society of Chemical Industry, London.

BOULTON, C. A., MARYAN, P. S., LOVERIDGE, D. and KELL, D. B. (1989) The application of a novel biomass sensor to the control of yeast pitching rate. Proceedings of the 22nd European Brewing Convention, Zurich, pp. 653–661. European Brewing Convention.

BRODGESELL, A. (1969) Miscellaneous sensors. In Instrument Engineer's Handbook, Vol. 1, pp. 914–943 (Editor Liptak, B. G.). Chilton, Philadelphia, PA.

BROWN, J, E., KAMINSKI, R. K, BLAKE, A. C. and BRODGESELL, A. (1969) Analytical measurements. In Instrument Engineer's Handbook, Vol. 1, pp. 713–918 (Editor Liptak, B. G.). Chilton, Philadelphia.

BUHLER, H. and INGOLD, W. (1976) Measuring pH and oxygen in fermenters. Proc. Biochem. **11**(3), 19–22, 24.

BULL, D. N. (1985) Instrumentation for fermentation process control. In Comprehensive Biotechnology, Vol. 2, pp. 149–163 (Editor Cooney, C. L. and Humphrey, A. E.). Pergamon Press, Oxford.

BUSHELL, M. E. (1988) Computers in Fermentation Technology. Progress in Industrial Microbiology, Vol. 25. Elsevier, Amsterdam.

CARLEYSMITH, S. W. (1989) Data handling for fermentation development — An industrial approach. In Computer Applications in Fermentation Technology — Modelling and Control of Biotechnological Processes, pp. 393–400 (Editors Fish, N. M., Fox, R. I. and Thornhill, N. F.). Society for Chemical Industry/Elsevier Science Publishers B.V., London.

COONEY, C. L., WANG, H. Y. and WANG, D. I. C. (1977) Computer-aided balancing for prediction of fermentation parameters. Biotech. Bioeng. **19**, 55–67.

DUSSELJEE, P. J. B and FEIJEN, J. (1990) Instrumentation and control. In Fermentation: A Practical Approach, pp. 149–172 (Editors McNeil, B. and Harvey, L. M.). IRL Press, Oxford.

EINSELE, A., RISTROPH, D. I. and HUMPHREY, A. E. (1978) Mixing times and glucose uptake measured with a fluorimeter. *Biotech. Bioeng.* **20**, 1487–1492.

ENFORS, S.O. (1981). An enzyme electrode for control of glucose concentration in fermentation broths. In *Abstracts of Communications. Second Congress of Biotechnology, Eastbourne*, p. 141. Society for Chemical Industry, London.

FISH, N. M., FOX, R. I. and THORNHILL, N. F. (1989) *Computer Applications in Fermentation Technology — Modelling and Control of Biotechnological Processes*. Society for Chemical Industry/Elsevier, London.

FLYNN, D. S. (1983) Instrumentation for fermentation processes. In *Modelling and Control of Fermentation Processes*, pp. 5–12 (Editor Halme, A.). Pergamon Press, Oxford.

FLYNN, D. S. (1984) Instrumentation and control of fermenters. In *Filamentous Fungi*, Vol. 4, pp. 77–100 (Editors Smith, J. E., Berry, D. R. and Kristiansen, B.). Arnold, London.

GARY, K., MEIER, P. and LUDWIG, K. (1988) General aspects of the use of sensors in biotechnology with special emphasis on cell cultivation. In *Biotechnology Research and Applications*, pp. 155–164 (Editors GAVORA, J., GERSON, D. F., LUONG, J., STOVER, A. and WOODLEY, J. H.). Elsevier, London.

GLASSEY, J., MONTAGUE, G. A., WARD, A. C. and KARA, B. V. (1994) Enhanced supervision of recombinant *E. coli* fermentations via artifical neural networks. *Process Biochem.* **29**, 387–398.

GRAYSON, P. (1969) Computer control of batch fermentations. *Proc. Biochem.* **4**(3), 43–44.

HALL, M. J., DICKINSON, S. D., PRITCHARD, R. and EVANS, J. I. (1973) Foams and foam control in fermentation processes. *Prog. Ind. Microbiol.* **12**, 169–231.

HALL, G. A., HIGGINS, S. P., KENNEDY, R. H. and NELSON, J. M. (1974) Principles of automatic control. In *Process Instrumentation and Control Handbook* (2nd edition), Section 18 (Editor Considine, D.M.). McGraw-Hill, New York.

HALLING, P. J. (1990) pH, dissolved oxygen and related sensors. In *Fermentation — A Practical Approach*, pp. 131–147 (Editors McNeil, B. and Harvey, L. M.). IRL, Oxford.

HAMMOND, S. V. and BROOKES, I. K. (1992) Near infrared spectroscopy — A powerful technique for at-line and on-line analysis of fermentation. In *Harnessing Biotechnology for the 21st Century*, pp. 325–333 (Editors Ladisch, M. R. and Bose, A.). American Chemical Society, Washington, DC.

HARRISON, D. E. F. and CHANCE. B. (1970) Fluorimetric technique for monitoring changes in the level of reduced nicotinamide nucleotides in continuous cultures of micro-organisms. *Appl. Microbiol.* **19**, 446–450.

HEINZLE, E., DUNN, I. J. and BOURNE, J. R. (1981) Continuous measurement of gases and volatiles during fermentations using mass spectrometry. In *Abstracts of Communications.*

*Second European Congress of Biotechnology, Eastbourne*, p. 24. Society of Chemical Industry, London.

HEWETSON, J. W., JONG, T. H. and GRAY, P. P.(1979) Use of an immobilized penicillinase electrode in the monitoring of the penicillin fermentation. *Biotech. Bioeng. Symp.* **9**, 125–135.

HIGGINS, I. J. and BOLDOT, J. (1992) Development of practical electrochemical biological sensing devices. In *Harnessing Biotechnology for the 21st Century*, pp. 316–318 (Editors Ladisch, M. R. and Bose, A.). American Chemical Society, Washington, DC.

HIGGINS, I. J., SWAIN, A. and TURNER, A. P. F. (1987) Principles and applications of biosensors in microbiology. *J. Appl. Bacteriol. Symp. Suppl.* 95S–105S.

HOWE, W. H., KOP, J. G., SIEV, R. and LIPTAK, B. G. (1969) Flow measurement. In *Instrument Engineer's Handbook*, Vol. 1, pp. 411–567 (Editor Liptak, B. G.). Chilton. Philadelphia, PA.

JACOB, H. E. (1970) Redox potential. In *Methods in Microbiology*, Vol. 2, pp. 91–123 (Editors Morris, J. R. and Ribbons, D. W.). Academic Press, London.

JOHNSON, M. J., BORKOWSKI, J. and ENGBLOM, C. (1964) Steam sterilisable probes for dissolved oxygen measurement. *Biotech. Bioeng.* **6**, 457–468.

KARIM, M. N. and RIVERA, S. L. (1992) Artificial neural networks in bioprocess state estimation. *Adv. Biochem. Eng. Biotechnol.* **46**, 1–33.

KELL, D. B. (1987) The principles and potential of electrical admittance spectroscopy; an introduction. In *Biosensors: Fundamentals and Applications*, pp. 427–468 (Editors Turner, A. P. F., Karube, I. and Wilson, G. S.). University Press, Oxford.

KELL, D. B., SAMWORTH, C. M., TODD, R. W., BUNGARD, S. J. and MORRIS, G. J. (1987) Real-time estimation of microbial biomass during fermentations, using a dielectric probe. *Studia Biophysica* **119**, 153–156.

KJAERGAARD, L. (1977) The redox potential: its use and control in biotechnology. *Adv. Biochem. Eng.* **7**, 131–150.

KJAERGAARD, L. and JOERGENSEN, B. B. (1979) Redox potential as a state variable in fermentation systems. *Biotech. Bioeng. Symp.* **9**, 85–94.

KRISTIANSEN, B. (1987) Instrumentation. In *Basic Biotechnology*, pp. 253–280 (Editors Bu'Lock, J. D. and Kristiansen, B.). Academic Press, London.

LEE, Y. H . and TSAO, G. T. (1979) Dissolved oxygen electrodes. *Adv. Biochem. Eng.* **13**, 36–86.

LINKO, P. and ZHU, Y.-H. (1992) Neural network modelling for real-time variable estimation and prediction in the control of glucoamylase fermentation. *Proc. Biochem.* **27**, 275–283.

LIPTAK, B. G. (1969) Pressure mesurement. In *Instrument Engineer's Handbook*, Vol. 1, pp. 171–263 (Editor Liptak, B. G.). Chilton, Philadelphia, PA.

LORENZ, T., FRUEH, K., DIEKMANN, J., HIDDESSEN, R., LUEBBERT, A. and SCHUEGERL, K (1985) On-line measurement and control in a bubble column reactor with an external loop. *Chem. Ing. Tech.* **57**, 116–117.

LUONG, J. H. T., MULCHANDANI, A. and GUILBAULT, G. G. (1988) Developments and applications of biosensors. *Trends Biotechnol.* **8**, 310–316.

MEINERS, M. (1982) Computer applications in fermentation. In *Overproduction of Microbial Products*, pp. 637–649 (Editors Krumphanzl, V., Sikyta, B. and Vanek, Z.). Academic Press, London.

MONTAGUE, G., MORRIS, A. and WARD, A. (1988) Fermentation monitoring and control: a perspective. *Biotechnol. Gen. Eng. Res.* **7**, 147–188.

NYIRI, L. K. (1972) A philosophy of data acquisition, analysis and computer control of fermentation processes. *Dev. Ind. Microbiol.* **13**, 136–145.

NYIRI, L. K., TOTH, G. M. and CHARLES, M. (1975) Measurement of gas-exchange conditions in fermentation processes. *Biotech. Bioeng.* **17**, 1663–1678.

OMSTEAD, D. R. and GREASHAM, R. H. (1989) Integrated fermenter sampling and analysis. In *Computer Applications in Fermentation Technology: Modelling and Control of Technological Processes*, pp. 5–13 (Editors Fish, N. M., Fox, R. I. and Thornhill, N. F.). SCI-Elsevier, London.

PHILLIPS, D. H. and JOHNSON, M. J. (1961) Measurement of dissolved oxygen in fermentations. *J. Biochem. Microbiol. Technol. Eng.* **3**, 261–275.

RADJAI, M. K., HATCH, R. T. and CADMAN, T. W. (1984) Optimization of amino acid production by automatic self-tuning digital control of redox potential. *Biotech. Bioeng. Symp.* **14**, 657–679.

RAMIREZ, A., DURAND, A. and BLACHERE, H. T. (1981) Optimal baker's yeast production in extended fed-batch culture using a computer coupled pilot-fermentor. In *Abstracts of Communications. Second European Congress of Biotechnology, Eastbourne*, p. 26. Society of Chemical Industry, London.

ROBERTS, A. and SHEPHERD, A. S. (1968) Dissolved oxygen measurement in continuous aseptic fermentation. *Proc. Biochem.* **3**(2), 23–24.

ROLF, M. J. and LIM, H. C. (1985) Systems for fermentation process control. In *Comprehensive Biotechnology*, Vol. 2, pp. 165–174 (Editors Cooney, C.L. and Humphrey, A. E.). Pergamon Press, Oxford.

ROYCE, P. N. (1993) A discussion of recent developments in fermentation monitoring from a practical perspective. *Crit. Rev. Biotechnol.* **13**, 117–149.

SCHNECKENBURGER, H., REUTER, B. W. and SCHOBERTH, S. M. (1985) Fluorescence techniques in biotechnology. *Trends in Biotechnol.* **3**, 257–261.

SCHWAN, H. P. (1957) Electrical properties of tissue and cell suspensions. In *Advances in Biological and Medical Physics*, Vol. 5, pp. 147–209 (Editors Lawrence, J. H. and Tobias, C. A.). Academic Press, New York.

SHI, Z., SHIMIZU, K., WATANABE, N. and KOBAYASHI, T. (1989) Adaptive on-line optimizating control of bioreactor systems. *Biotech. Bioeng.* **33**, 999–1009.

SHIMIZU, K. (1993) An overview on the control system design of bioreactors. *Adv. Biochem. Eng./Biotechnol.* **50**, 65–84.

SHINSKEY, F. G. (1973) *pH and pIon Control in Process and Waste Streams*, Chapter 9. Wiley, New York.

SIEGALL, S. O. and GADEN, E. L. (1962) Automatic control of dissolved oxygen levels in fermentations. *Biotech. Bioeng.* **4**, 345–351.

SWARTZ, J. R. and COONEY, C. L. (1979) Indirect fermentation measurements as a basis for control. *Biotech. Bioeng. Symp.* **9**, 95–101.

VACCARI, G., DOSI, E., CAMPSI, A. L., GONZALEZ-VARAY, R. A., MATLEUZZI, D. and MANTOVANI, C. (1994) A near infrared spectroscopy technique for the control of fermentation processes: An application to lactic acid fermentation. *Biotech. Bioeng.* **43**, 913–917.

VIESTURS, U. E., KRISTAPSONS, M. Z. and LEVITANS, E. S. (1982) Foam in microbiological processes. *Adv. Biochem. Eng.* **21**, 169–224.

WANG, H. Y., COONEY, C. L. and WANG, D. I. C. (1977) Computer-aided baker's yeast fermentations *Biotech. Bioeng.* **19**, 69–86.

WHITESIDE, M. C. and MORGAN, P. (1989) Computers in fermentation. In *Computers in Microbiology — A Practical Approach*, pp. 175–190 (Editors Bryant, T. N. and Wimpenny, J. W. T.). IRL Press, Oxford.

YAMASHITA, S. and MURAO, C. (1978) Fermentation Process. *J. Soc. Instr. Control Eng.* **6** (10), 735–740.

YAMASHITA, S., HISOSHI, H. and INAGAKI, T. (1969) Automatic control and optimization of fermentation processes — glutamic acid. In *Fermentation Advances*, pp. 441–463 (Editor Perlman, D.). Academic Press, New York.

ZABRISKIE, D. W. (1985) Data analysis. In *Comprehensive Biochemistry*, Vol. 2, pp. 175–190 (Editors Cooney, C. L. and Humphrey, A. E.). Pergamon Press, Oxford.

# CHAPTER 9

# Aeration and Agitation

## INTRODUCTION

THE MAJORITY of fermentation processes are aerobic and, therefore, require the provision of oxygen. If the stoichiometry of respiration is considered, then the oxidation of glucose may be represented as:

$$C_6H_{12}O_6 + 6O_2 = 6H_2O + 6CO_2$$

Thus, 192 grams of oxygen are required for the complete oxidation of 180 grams of glucose. However, both components must be in solution before they are available to a micro-organism and oxygen is approximately 6000 times less soluble in water than is glucose (a fermentation medium saturated with oxygen contains approximately 7.6 mg $dm^{-3}$ of oxygen at 30°C). Thus, it is not possible to provide a microbial culture with all the oxygen it will need for the complete oxidation of the glucose (or any other carbon source) in one addition. Therefore, a microbial culture must be supplied with oxygen during growth at a rate sufficient to satisfy the organisms' demand.

The oxygen demand of an industrial fermentation process is normally satisfied by aerating and agitating the fermentation broth. However, the productivity of many fermentations is limited by oxygen availability and, therefore, it is important to consider the factors which affect a fermenter's efficiency in supplying microbial cells with oxygen. This chapter considers the requirement for oxygen in fermentation processes, the quantification of oxygen transfer and the factors which will influence the rate of oxygen transfer into solution.

## THE OXYGEN REQUIREMENTS OF INDUSTRIAL FERMENTATIONS

Although a consideration of the stoichiometry of respiration gives an appreciation of the problem of oxygen supply, it gives no indication of an organism's true oxygen demand as it does not take into account the carbon that is converted into biomass and products. A number of workers have considered the overall stoichiometry of the conversion of oxygen, a source of carbon and a source of nitrogen into biomass and have used such relationships to predict the oxygen demand of a fermentation. A selection of such equations is shown in Table 9.1. From these determinations it may be seen that a culture's demand for oxygen is very much dependent on the source of carbon in the medium. Thus, the more reduced the carbon source, the greater will be the oxygen demand. From Darlington's and Johnson's equations (Table 9.1) it may be seen that the production of 100 grams of biomass from hydrocarbon requires approximately three times the amount of oxygen to produce the same amount of biomass from carbohydrate. This point is also illustrated in Table 9.2. However, it must be remembered that the high carbon content of hydrocarbon substrates means that high yield factors (g biomass $g^{-1}$ substrate consumed) are obtained and the decision to use such substrates is based on the balance between the advantage of high biomass yield and the disadvantage of high oxygen demand and heat generation. These points are discussed in more detail in Chapter 4.

Darlington's, Johnson's and Mateles' equations only include biomass production and do not consider product formation, whereas Cooney's and Righelato's equations consider product formation. Ryu and Hospodka (1980) used Righelato's approach to calculate that the production of 1 g penicillin consumes 2.2 g of oxygen.

However, it is inadequate to base the provision of oxygen for a fermentation simply on an estimation of overall demand, because the metabolism of the culture is affected by the concentration of dissolved oxygen in the broth. The effect of dissolved oxygen concentration

TABLE 9.1. *Stoichiometric equations describing oxygen demand in a fermentation*

| Equation | Terms used | Reference |
|---|---|---|
| $6.67CH_2O + 2.1O_2 = C_{3.92}H_{6.5}O_{1.94}$ $+ 2.75CO_2 + 3.42H_2O$ | $C_{3.92}H_{6.5}O_{1.94}$ is 100 g (dry weight) of yeast cells; $CH_2O$ is carbohydrate | Darlington (1964) |
| $7.14CH_2 + 6.135O_2 = C_{3.92}H_{6.5}O_{1.94}$ $+ 3.22CO_2 + 3.89H_2O$ | $CH_2$ is hydrocarbon | Darlington (1964) |
| $(A/Y) - B = C$ | $A$ = Amount of oxygen for combustion of 1 g of substrate to $CO_2$, $H_2O$ and $NH_3$, if nitrogen is present in the substrate<br>$B$ = Amount of oxygen required for the combustion of 1 g cells to $CO_2$, $H_2O$ and $NH_3$<br>$Y$ = Cell yield (g cells $g^{-1}$ substrate)<br>$C$ = g oxygen consumed for the production of 1 g of cells | Johnson (1964) |
| $(dx/dt)/Y + mx + p = (dCO_2/dt) =$ $-(dO_2/dt)RQ$ | $x$ = biomass concentration<br>$t$ = time<br>$Y$ = g biomass $g^{-1}$ carbon substrate<br>$m$ = maintenance<br>$p$ = allowance for antibiotic production | Righelato *et al.* (1968) |
| $Y_{O/P} = (0.53/Y_{P/G}) - (0.6X/P) - 0.43$ | $Y_{O/P}$ = g oxygen consumed $g^{-1}$ glucose<br>$Y_{P/G}$ = g sodium penicillin $G$ produced $g^{-1}$ glucose<br>$X$ = g cells (dry weight) produced<br>$P$ = g sodium penicillin $G$ produced | Cooney (1979) |
| $Y_O = \{(32C + 8H - 16O)/YM\} - 1.58$ | $Y_O$ = g oxygen consumed $g^{-1}$ cells produced<br>$Y$ = Cell yield (g cells $g^{-1}$ substrate)<br>$M$ = Molecular weight of the carbon source<br>C, H and O = Number of atoms of carbon, hydrogen and oxygen per molecule of carbon source | Mateles (1971) |

on the specific oxygen uptake rate ($Q_{O_2}$, mmoles of oxygen consumed per gram dry weight of cells per hour) has been shown to be of the Michaelis–Menten type, as shown in Fig. 9.1.

From Fig. 9.1 it may be seen that the specific oxygen uptake rate increases with increase in the dissolved oxygen concentration up to a certain point (referred to as $C_{crit}$) above which no further increase in oxygen uptake rate occurs. Some examples of the critical oxygen levels for a range of micro-organisms are given in Table 9.3. Thus, maximum biomass production may be achieved by satisfying the organism's maximum specific oxygen demand by maintaining the dissolved oxygen concentration greater than the critical level. If the dissolved oxygen concentration were to fall below the critical level then the cells may be metabolically disturbed. However, it must be remembered that it is frequently the objective of the fermentation technologist to produce a product of the micro-organism rather than the organism itself and that metabolic

TABLE 9.2. *Oxygen requirements of a range of micro-organisms grown on a range of substrates (After Mateles, 1979)*

| Substrate | Organism | Oxygen requirement (g $O_2$ $g^{-1}$ dry wt.) | Reference |
|---|---|---|---|
| Glucose | *Escherichia coli* | 0.4 | Schulze and Lipe (1964) |
| Methanol | *Pseudomonas C* | 1.2 | Goldberg *et al.* (1976) |
| Octane | *Pseudomonas* sp. | 1.7 | Wodzinski and Johnson (1968) |

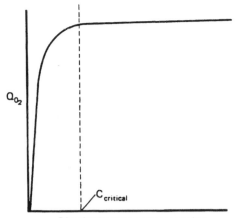

FIG. 9.1. The effect of dissolved oxygen concentration on the $Q_{O_2}$ of a micro-organism.

disturbance of the cell by oxygen starvation may be advantageous to the formation of certain products. Equally, provision of a dissolved oxygen concentration far greater than the critical level may have no influence on biomass production, but may stimulate product formation. Thus, the aeration conditions necessary for the optimum production of a product may be different from those favouring biomass production.

Hirose and Shibai's (1980) investigations of amino acid biosynthesis by *Brevibacterium flavum* provide an excellent example of the effects of the dissolved oxygen concentration on the production of a range of closely related metabolites. These workers demonstrated the critical dissolved oxygen concentration for *B. flavum* to be 0.01 mg dm$^{-3}$ and considered the extent of oxygen supply to the culture in terms of the degree of 'oxygen satisfaction', that is the respiratory rate of the culture expressed as a fraction of the maximum respiratory rate. Thus, a value of oxygen satisfaction below unity implied that the dissolved oxygen concentration was below the critical level. The effect of the degree of

oxygen satisfaction on the production of a range of amino acids is shown in Fig. 9.2. From Fig. 9.2 it may be seen that the production of members of the glutamate and aspartate families of amino acids was affected detrimentally by levels of oxygen satisfaction below 1.0, whereas optimum production of phenylalanine, valine and leucine occurred at oxygen satisfaction levels of 0.55, 0.60 and 0.85, respectively. The biosynthetic routes of the amino acids are shown in Fig. 9.3, from which it may be seen that the glutamate and aspartate families are all produced from tricarboxylic acid (TCA) cycle intermediates, whereas phenylalanine, valine and leucine are produced from the glycolysis intermediates, pyruvate and phosphoenol pyruvate. Oxygen excess should give rise to abundant TCA cycle intermediates, whereas oxygen limitation should result in less glucose being oxidized via the TCA cycle, allowing more intermediates to be available for phenylalanine, valine and leucine biosynthesis. Thus, some degree of metabolic disruption results in greater production of pyruvate derived amino acids.

An example of the effect of dissolved oxygen on secondary metabolism is provided by Zhou *et al.*'s (1992) work on cephalosporin C synthesis by *Cephalosporium acremonium*. These workers demonstrated that the critical oxygen concentration for cephalosporin C synthesis during the production phase was 20% saturation. At dissolved oxygen concentrations below 20% cephalosporin C concentration declined and penicillin N increased. The biosynthetic pathway to cephalosporin C is shown in Fig. 9.4, from which it may be seen that there are three oxygen-consuming steps in the pathway:

(i)   Cyclization of the tripeptide, α-amino-adipyl-cysteinyl-valine into isopenicillin N.
(ii)  The ring expansion of penicillin N into deacetoxycephalosporin C (DAOC).
(iii) The hydroxylation of DAOC to give deacetylcephalosporin C.

TABLE 9.3. *Critical dissolved oxygen concentrations for a range of micro-organisms* (Riviere, 1977)

| Organism | Temperature | Critical dissolved oxygen concentration (mmoles dm$^{-3}$) |
|---|---|---|
| *Azotobacter* sp. | 30 | 0.018 |
| *Escherichia coli* | 37 | 0.008 |
| *Saccharomyces* sp. | 30 | 0.004 |
| *Penicillium chrysogenum* | 24 | 0.022 |

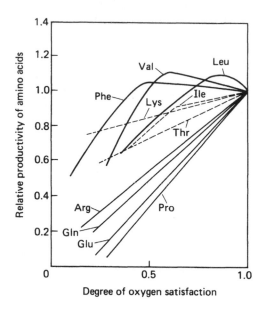

FIG. 9.2. The effect of dissolved oxygen on the production of amino acids by *Brevibacterium flavum* (Hirose and Shibai, 1980).

DAOC did not accumulate at low oxygen concentrations and, thus, it appears that the most oxygen sensitive step in the pathway is the ring expansion enzyme (expandase) resulting in the accumulation of penicillin N under oxygen limitation.

The requirement for a high dissolved oxygen concentration by many fermentations has resulted in the development of process techniques to ensure that the fermentation does not exceed the oxygen-supply capabilities of the fermentation vessel. The oxygen demand of a fermentation largely depends on the concentration of the biomass and its respiratory activity, which is related to the growth rate. By limiting the initial concentration of the medium, the biomass in the vessel may be kept at a reasonable level and by supplying some nutrient component as a feed, the rate of growth, and hence the respiratory rate, may be controlled. These techniques of medium design and nutrient feed are discussed in Chapters 2 and 4 and later in this chapter.

## OXYGEN SUPPLY

Oxygen is normally supplied to microbial cultures in the form of air, this being the cheapest available source

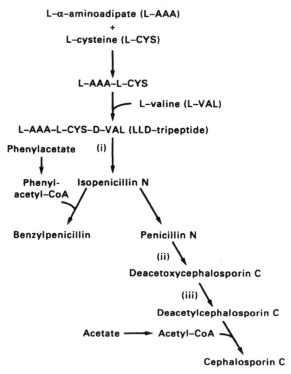

FIG. 9.4. The biosynthesis of cephalosporin C, indicating the oxygen consuming steps:
(i) isopenicillin-N-synthase,
(ii) deacetoxycephalosporin C synthase (commonly called expandase),
(iii) deacetyl cephalosporin C synthase (commonly called hydroxylase).

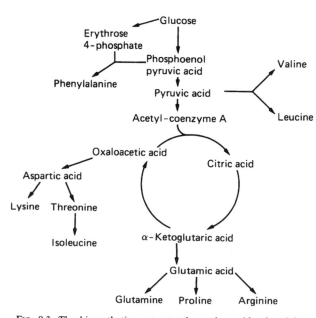

FIG. 9.3. The biosynthetic routes to the amino acids phenylalanine, valine, leucine, lysine, threonine, *l*-leucine, glutamic acid, proline, glutamine and arginine in *B. flavum*.

of the gas. The method for provision of a culture with a supply of air varies with the scale of the process:

(i) Laboratory-scale cultures may be aerated by means of the shake-flask technique where the culture (50 to 100 $cm^3$) is grown in a conical flask (250 to 500 $cm^3$) shaken on a platform contained in a controlled environment chamber.

(ii) Pilot- and industrial-scale fermentations are normally carried out in stirred, aerated vessels, termed fermenters, of the type described in Chapter 7. However, it is often advantageous to culture relatively small volumes (1 $dm^3$) in a stirred, aerated vessel as this enables the cultural conditions to be better monitored and controlled, and facilitates the addition of supplements and the removal of samples. Some fermenters are so designed that adequate oxygen transfer is obtained without agitation and the design of these systems (termed bubble columns and air-lift fermenters) is also discussed in Chapter 7.

Bartholomew *et al.* (1950) represented the transfer of oxygen from air to the cell, during a fermentation, as occurring in a number of steps:

(i) The transfer of oxygen from an air bubble into solution.
(ii) The transfer of the dissolved oxygen through the fermentation medium to the microbial cell.
(iii) The uptake of the dissolved oxygen by the cell.

These workers demonstrated that the limiting step in the transfer of oxygen from air to the cell in a *Streptomyces griseus* fermentation was the transfer of oxygen into solution. These findings have been shown to be correct for non-viscous fermentations but it has been demonstrated that transfer may be limited by either of the other two stages in certain highly viscous fermentations. The difficulties inherent in such fermentations are discussed later in this chapter.

The rate of oxygen transfer from air bubble to the liquid phase may be described by the equation:

$$dC_L/dt = K_L a(C^* - C_L) \qquad (9.1)$$

where $C_L$     is the concentration of dissolved oxygen in the fermentation broth (mmoles $dm^{-3}$),

$t$     is time (hours),

$dC_L/dt$     is the change in oxygen concentration over a time period, i.e. the oxygen-transfer rate (mmoles $O_2$ $dm^{-3}$ $h^{-1}$),

$K_L$     is the mass transfer coefficient (cm $h^{-1}$),

$a$     is the gas/liquid interface area per liquid volume ($cm^2$ $cm^{-3}$),

$C^*$     is the saturated dissolved oxygen concentration (mmoles $dm^{-3}$).

$K_L$ may be considered as the sum of the reciprocals of the resistances to the transfer of oxygen from gas to liquid and $(C^* - C_L)$ may be considered as the 'driving force' across the resistances. It is extremely difficult to measure both $K_L$ and '$a$' in a fermentation and, therefore, the two terms are generally combined in the term $K_L a$, the volumetric mass-transfer coefficient, the units of which are reciprocal time ($h^{-1}$). The volumetric mass-transfer coefficient is used as a measure of the aeration capacity of a fermenter. The larger the $K_L a$, the higher the aeration capacity of the system. The $K_L a$ value will depend upon the design and operating conditions of the fermenter and will be affected by such variables as aeration rate, agitation rate and impeller design. These variables affect '$K_L$' by reducing the resistances to transfer and affect '$a$' by changing the number, size and residence time of air bubbles. It is convenient to use $K_L a$ as a yardstick of fermenter performance because, unlike the oxygen-transfer rate, it is unaffected by dissolved oxygen concentration. However, the oxygen transfer rate is the critical criterion in a fermentation and, as may be seen from equation 9.1, it is affected by both $K_L a$ and dissolved oxygen concentration. The dissolved oxygen concentration reflects the balance between the supply of dissolved oxygen by the fermenter and the oxygen demand of the organism. If the $K_L a$ of the fermenter is such that the oxygen demand of the organism cannot be met, the dissolved oxygen concentration will decrease below the critical level ($C_{crit}$). If the $K_L a$ is such that the oxygen demand of the organism can be easily met the dissolved oxygen concentration will be greater than $C_{crit}$ and may be as high as 70 to 80% of the saturation level. Thus, the $K_L a$ of the fermenter must be such that the optimum oxygen concentration for product formation can be maintained in solution throughout the fermentation.

## DETERMINATION OF $K_L a$ VALUES

The determination of the $K_L a$ of a fermenter is essential in order to establish its aeration efficiency and to quantify the effects of operating variables on the provision of oxygen. This section considers the merits and limitations of the methods available for the determination of $K_L a$ values. It is important to remember at this stage that dissolved oxygen is usually monitored using a dissolved oxygen electrode (see Chapter 8) which records dissolved oxygen activity or dissolved oxygen tension (DOT) whilst the equations describing oxygen transfer are based on dissolved oxygen concentration. The solubility of oxygen is affected by dissolved solutes so that pure water and a fermentation medium saturated with oxygen would have different dissolved oxygen concentrations yet have the same DOT, i.e. an oxygen electrode would record 100% for both. Thus, to translate DOT into concentration the solubility of oxygen in the fermentation medium must be known and this can present difficulties.

### The sulphite oxidation technique

Cooper *et al.* (1944) were the first to describe the determination of oxygen-transfer rates in aerated vessels by the oxidation of sodium sulphite solution. This technique does not require the measurement of dissolved oxygen concentrations but relies on the rate of conversion of a 0.5 M solution of sodium sulphite to sodium sulphate in the presence of a copper or cobalt catalyst:

$$Na_2SO_3 + 0.5O_2 = Na_2SO_4$$

The rate of reaction is such that as oxygen enters solution it is immediately consumed in the oxidation of sulphite, so that the sulphite oxidation rate is equivalent to the oxygen-transfer rate. The dissolved oxygen concentration, for all practical purposes, will be zero and the $K_L a$ may then be calculated from the equation:

$$OTR = K_L a \cdot C^* \qquad (9.2)$$

(where OTR is the oxygen transfer rate).

The procedure is carried out as follows: the fermenter is batched with a 0.5 M solution of sodium sulphite containing $10^{-3}$ M $Cu^{2+}$ ions and aerated and agitated at fixed rates; samples are removed at set time intervals (depending on the aeration and agitation rates) and added to excess iodine solution which reacts with the unconsumed sulphite, the level of which may be determined by a back titration with standard sodium thiosulphate solution. The volumes of the thiosulphate titrations are plotted against sample time and the oxygen transfer rate may be calculated from the slope of the graph.

The sulphite oxidation method has the advantage of simplicity and, also, the technique involves sampling the bulk liquid in the fermenter and, therefore, removes some of the problems of conditions varying through the volume of the vessel. However, the method is time consuming (one determination taking up to 3 hours, depending on the aeration and agitation rates) and is notoriously inaccurate. Bell and Gallo (1971) demonstrated that minor amounts of surface-active contaminants (such as amino acids, proteins, fatty acids, esters, lipids, etc.) could have a major effect on the accuracy of the technique and apparent differences in aeration efficiency between vessels could be due to differences in the degree of contamination. Also, the rheology of a sodium sulphite solution is completely different from that of a fermentation broth, especially a mycelial one so that it is impossible to relate the results of sodium sulphite determinations to real fermentations. To quote Van't Riet and Tramper (1991) "It can safely be said that the application of this method should be strongly discouraged".

### Gassing-out techniques

The estimation of the $K_L a$ of a fermentation system by gassing-out techniques depends upon monitoring the increase in dissolved oxygen concentration of a solution during aeration and agitation. The oxygen transfer rate will decrease during the period of aeration as $C_L$ approaches $C^*$ due to the decline in the driving force $(C^* - C_L)$. The oxygen transfer rate, at any one time, will be equal to the slope of the tangent to the curve of values of dissolved oxygen concentration against time of aeration, as shown in Fig. 9.5.

To monitor the increase in dissolved oxygen over an adequate range it is necessary first to decrease the oxygen level to a low value. Two methods have been employed to achieve this lowering of the dissolved oxygen concentration — the static method and the dynamic method.

### THE STATIC METHOD OF GASSING OUT

In this technique, first described by Wise (1951), the oxygen concentration of the solution is lowered by gassing the liquid out with nitrogen gas, so that the solution is 'scrubbed' free of oxygen. The deoxygenated liquid is then aerated and agitated and the increase in dissolved oxygen monitored using some form of dissolved oxygen probe. The increase in dissolved oxygen

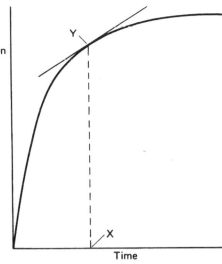

FIG. 9.5. The increase in dissolved oxygen concentration of a solution over a period of aeration. The oxygen transfer rate at time X is equal to the slope of the tangent at point Y.

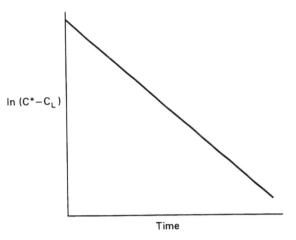

FIG. 9.6. A plot of the $\ln(C^* - C_L)$ against time of aeration, the slope of which equals $-K_L a$.

concentration has already been described by equation (9.1), i.e.:

$$dC_L/dt = K_L a(C^* - C_L)$$

and depicted in Fig. 9.5. Integration of equation (9.1) yields:

$$\ln(C^* - C_L) = -K_L at. \qquad (9.3)$$

Thus, a plot of $\ln (C^* - C_L)$ against time will yield a straight line of slope $-K_L a$, as shown in Fig 9.6. This technique has the advantage over the sulphite oxidation method in that it is very rapid (normally taking up to 15 minutes) and may utilize the fermentation medium, to which may be added dead cells or mycelium at a concentration equal to that produced during the fermentation. However, employing the fermentation medium with, or without, killed biomass necessitates the use of a membrane-type electrode, the response time of which may be inadequate to reflect the true change in the rate of oxygenation over a short period of time. The probe response time $(T_p)$ is defined as the time needed to record 63% of a stepwise change and this should be much smaller than the mass transfer response time of the system $(1/K_L a)$. According to Van't Riet (1979), the use of commercially available electrodes, with a response time of 2 to 3 seconds, should enable a $K_L a$ of up to 360 h$^{-1}$ to be measured with little loss of accuracy. However, for estimations of higher $K_L a$ values it would be necessary to incorporate a correction factor into the calculation, as discussed by Taguchi and Humphrey (1966), Heineken (1970, 1971)

and Wernau and Wilke (1973). It is not necessary to know the oxygen solubility in the medium because DOT values may be used directly in order to calculate the rates, i.e. $C^*$ is taken as 100%.

Whilst the method is acceptable for small scale vessels, there are severe limitations to its use on large scale fermenters which have high gas residence times. When the air supply to such a vessel is resumed after deoxygenation with nitrogen, the oxygen concentration in the gas phase may change with time as the nitrogen is replaced with air. Thus, $C^*$ will no longer be constant. Although correction factors have been derived to compensate for this phenomenon, Van't Riet and Tramper (1991) concluded that the method should not be used for vessels over 1-metre high.

THE DYNAMIC METHOD OF GASSING OUT

Taguchi and Humphrey (1966) utilized the respiratory activity of a growing culture in the fermenter to lower the oxygen level prior to aeration. Therefore, the estimation has the advantage of being carried out during a fermentation which should give a more realistic assessment of the fermenter's efficiency. Because of the complex nature of fermentation broths the probe used to monitor the change in dissolved oxygen concentration must be of the membrane-covered type which may necessitate the use of the response-correction factors referred to previously. The procedure involves stopping the supply of air to the fermentation which results in a linear decline in the dissolved oxygen concentration due to the respiration of the culture, as shown in Fig. 9.7. The slope of the line AB in Fig. 9.7 is

249

a measure of the respiration rate of the culture. At point B the aeration is resumed and the dissolved oxygen concentration increases until it reaches concentration X. Over the period, BC, the observed increase in dissolved oxygen concentration is the difference between the transfer of oxygen into solution and the uptake of oxygen by the respiring culture as expressed by the equation:

$$dC_L/dt = K_L a(C^* - C_L) - xQ_{O_2} \quad (9.4)$$

where $x$ is the concentration of biomass and
$Q_{O_2}$ is the specific respiration rate (mmoles of oxygen g$^{-1}$ biomass h$^{-1}$).

The term $xQ_{O_2}$ is given by the slope of the line AB in Fig. 9.7. Equation (9.4) may be rearranged as:

$$C_L = -1/K_L a \left\{ (dC_L/dt) + xQ_{O_2} \right\} + C^* \quad (9.5)$$

Thus, from equation (9.5), a plot of $C_L$ versus $dC_L/dt + xQ_{O_2}$ will yield a straight line, the slope of which will equal $-1/K_L a$, as shown in Fig. 9.8. This technique is convenient in that the equations may be applied using DOT rather than concentration because it is the rates of transfer and uptake that are being monitored so that the percentage saturation readings generated by the electrode may be used directly.

The dynamic gassing-out method has the advantage over the previous methods of determining the $K_L a$ during an actual fermentation and may be used to determine $K_L a$ values at different stages in the process. The technique is also rapid and only requires the use of a dissolved-oxygen probe, of the membrane type. A major limitation in the operation of the technique is the range over which the increase in dissolved oxygen

concentration may be measured. It is important not to allow the oxygen concentration to drop below $C_{crit}$ during the deoxygenation step as the specific oxygen uptake rate will then be limited and the term $xQ_{O_2}$ would not be constant on resumption of aeration. The occurrence of oxygen-limited conditions during deoxygenation may be detected by the deviation of the decline in oxygen concentration from a linear relationship with time, as shown in Fig. 9.9.

When the oxygen demand of a culture is very high it may be difficult to maintain the dissolved oxygen concentration significantly above $C_{crit}$ during the fermentation so that the range of measurements which could be used in the $K_L a$ determination would be very small. Thus, it may be difficult to apply the technique during a fermentation which has an oxygen demand close to the supply capacity of the fermenter.

Although the difficulty presented by nitrogen degassing does not arise with the dynamic method it, also, is not suitable for use with vessels in excess of one metre high. Van't Riet and Tramper (1991) pointed out that in such vessels the time taken to establish an equilibrium population of air bubbles would be significant and the gas–liquid interface area would change over the aeration period resulting in a considerable underestimate of the $K_L a$ value achievable under normal operating conditions. Both the dynamic and static methods are also unsuitable for measuring $K_L a$ values in viscous systems. This is due to the very small bubbles (< 1 mm diameter) formed in a viscous system which have an extended residence time compared with 'nor-

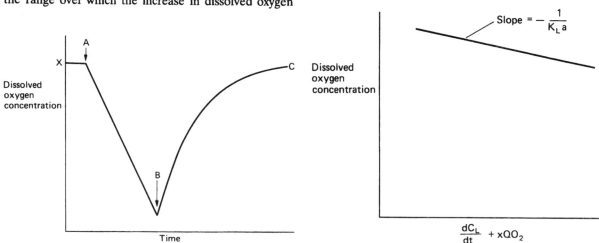

FIG. 9.7. Dynamic gassing out for the determination of $K_L a$ values. Aeration was terminated at point A and recommenced at point B.

FIG. 9.8. The dynamic method for determination of $K_L a$ values. The information is gleaned from Fig. 9.7. by taking tangents of the curve, BC, at various values of $C_L$.

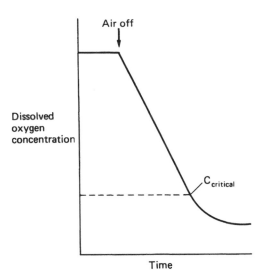

FIG. 9.9. The occurrence of oxygen limitation during the dynamic gassing out of a fermentation.

mal' sized bubbles. Thus, the gassing out techniques are only useful on a small scale with non-viscous systems.

### The oxygen-balance technique

The $K_L a$ of a fermenter may be measured during a fermentation by the oxygen balance technique which determines, directly, the amount of oxygen transferred into solution in a set time interval. The procedure involves measuring the following parameters:

(i) The volume of the broth contained in the vessel, $V_L$ (dm$^3$).

(ii) The volumetric air flow rates measured at the air inlet and outlet, $Q_i$ and $Q_o$, respectively (dm$^3$ min$^{-1}$).

(iii) The total pressure measured at the fermenter air inlet and outlet, $P_i$ and $P_o$, respectively (atm. absolute).

(iv) The temperature of the gases at the inlet and outlet, $T_i$ and $T_o$, respectively (K).

(v) The mole fraction of oxygen measured at the inlet and outlet, $y_i$ and $y_o$, respectively.

The oxygen transfer rate may then be determined from the following equation (Wang et al., 1979):

$$OTR = (7.32 \times 10^5 / V_L)(Q_i P_i y_i / T_i - Q_o P_o y_o / T_o)$$
$$(9.6)$$

where $7.32 \times 10^5$ is the conversion factor equalling (60 min h$^{-1}$) [mole/22.4 dm$^3$ (STP)] (273 K/1 atm).

These measurements require accurate flow meters, pressure gauges and temperature-sensing devices as well as gaseous oxygen analysers (see Chapter 8). The ideal gaseous oxygen analyser is a mass spectrometer analyser which is sufficiently accurate to detect changes of 1 to 2%.

The $K_L a$ may be determined, provided that $C_L$ and $C^*$ are known, from equation (9.1):

$$OTR = K_L a(C^* - C_L) \text{ or } K_L a = OTR/(C^* - C_L)$$

$C_L$ may be determined using a membrane-type dissolved-oxgen electrode and in this case the slow response time is not an important factor because a rate of change is not being measured, simply the steady-state oxygen concentration. However, it should be remembered that an electrode simply measures the oxygen tension at one point and it is, therefore, advisable to monitor the oxygen tension at a number of points in the vessel with a number of electrodes and to use an average value. Also, the DOT reading must be converted to concentration, which necessitates knowing the oxygen solubility in the fermentation medium. The value of $C^*$ is frequently taken as that value which is in equilibrium with the oxygen concentration of the gas outlet. Wang et al. (1979) claimed that this approach was adequate for small-scale fermenters but on a large scale there may be a considerable difference between the dissolved oxygen concentration in equilibrium with the inlet and outlet gases. Therefore, these workers suggested that the behaviour of the gas in transit in the fermenter would approximate to plug flow conditions and a logarithmic mean value for the dissolved oxygen concentration should be used.

The oxygen-balance technique appears to be the simplest method for the assessment of $K_L a$ and has the advantage of measuring aeration efficiency during a fermentation. The sulphite oxidation and static gassing-out techniques have the disadvantage of being carried out using either a salt solution or an uninoculated, sterile fermentation medium. Although, as Banks (1977) suggests, these techniques are adequate for the comparison of equipment or operating variables, it should not be assumed that the values obtained are those actually operating during a fermentation. This may be the case for bacterial or yeast fermentations where the rheology of the suspended cells in the broth is similar to that in a sterile medium or a salt solution, but it is certainly not true for fungal and streptomycete processes where the rheology is quite different.

Tuffile and Pinho (1970) compared a number of methods for the determination of $K_L a$ values in viscous streptomycete fermentations. The techniques used were static gassing-out, dynamic gassing-out and the oxygen-balance method. Tuffile and Pinho did not make it clear whether non-respiring mycelium was present during their static gassing-out procedure, but from their results it would appear that it was present in the vessel. Thus, the rheology of the fermenter contents would appear to have been similar for the different determinations. The $K_L a$ values, determined by the different techniques, for a 300-dm$^3$ fermenter containing a 90-hour culture of *Streptomyces aureofaciens* are shown in Table 9.4.

From Table 9.4 it may be seen that the $K_L a$ values for the two gassing-out techniques were very similar but there was a considerable difference between the oxygen-uptake rates and the $K_L a$s determined by the dynamic method and the balance method. Tuffile and Pinho (1970) claimed that the low oxygen-uptake rate determined by the dynamic method was due to air bubbles remaining in suspension in the mash during the dynamic gassing-out period. Thus, the decline in oxygen concentration after the cessation of aeration was not a measure of the oxygen-uptake rate but the difference between oxygen uptake and the transfer of oxygen from entrapped bubbles. It was demonstrated that a large number of bubbles remained suspended in the medium 15 minutes after aeration had been stopped. The use of the low oxygen-uptake rate in the calculation of the $K_L a$ would result in an artificially low $K_L a$ being determined. Heijnen *et al.* (1980) also observed anomalies in determining $K_L a$ values in viscous systems due to the presence of very small bubbles having a much longer residence time than the more abundant large bubbles in the vessel.

Overall, it would appear that the balance method is the most desirable technique to use and the extra cost of the monitoring equipment involved should be a worthwhile investment.

Before considering the factors which may affect the $K_L a$ of a fermenter it is necessary to consider the behaviour of fluids in agitated systems.

## FLUID RHEOLOGY

Fluids may be described as Newtonian or non-Newtonian depending on whether their rheology (flow) characteristics obey Newton's law of viscous flow. Consider a fluid contained between two parallel plates, area $A$ and distance $x$ apart. If the lower plate is moved in one direction at a constant velocity, the fluid adjacent to the moving plate will move in the same direction and impart some of its momentum to the 'layer' of liquid directly above it causing it, also, to move in the same direction at a slightly lower velocity. Newton's law of viscous flow states that the viscous force, $F$, opposing motion at the interface between the two liquid layers, flowing with a velocity gradient of $dv/dx$, is given by the equation:

$$F = \mu A (dv/dx) \tag{9.7}$$

where $\mu$ is the fluid viscosity, which may be considered as the resistance of the fluid to flow.

Equation (9.7) may be written as:

$$F/A = \mu (dv/dx)$$

$F/A$ is termed the shear stress ($\tau$) and is the applied force per unit area, $dv/dx$ is termed the shear rate ($\gamma$) and is the velocity gradient. Thus:

$$\tau = \mu \gamma \tag{9.8}$$

Equation (9.8) conforms to the general relationship:

$$\tau = K \gamma^n \tag{9.9}$$

where $K$ is the consistency coefficient and
$n$ is the flow behaviour index or power law index.

For a Newtonian fluid $n$ is 1 and the consistency coefficient is the viscosity which is the ratio of shear stress to shear rate. Thus, a plot of shear stress against

TABLE 9.4. $K_L a$ *values for a 300-dm$^3$ fermenter containing a 90-h culture of* S. *aureofaciens (Tuffile and Pinho, 1970)*

| Method of $K_L a$ determination | Measured oxygen uptake rate (mmoles dm$^{-3}$h$^{-1}$) | $K_L a$ (h$^{-1}$) |
|---|---|---|
| Static gassing out | — | 58.2 |
| Dynamic gassing out | 6.6 | 58.2 |
| Oxygen balance | 20.1 | 108.0 |

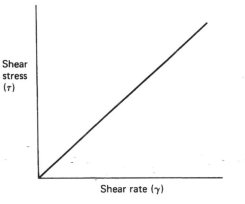

FIG. 9.10. A rheogram of a Newtonian fluid.

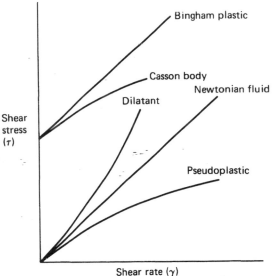

FIG. 9.11. Rheogram of fluids of different properties.

shear rate, for a Newtonian fluid, would produce a straight line, the slope of which would equal the viscosity. Such a plot is termed a rheogram (as shown in Fig. 9.10).

Thus, a Newtonian liquid has a constant viscosity regardless of shear, so that the viscosity of a Newtonian fermentation broth will not vary with agitation rate. However, a non-Newtonian liquid does not obey Newton's law of viscous flow and does not have a constant viscosity. The value for $n$ (equation (9.9)) of such a fluid deviates from 1 and its behaviour is said to follow a power law model. Thus, the viscosity of a non-Newtonian fermentation broth will vary with agitation rate and is described as an apparent viscosity ($\mu_a$). A plot of shear stress against shear rate for a non-Newtonian liquid will deviate from the relationship depicted in Fig. 9.10, depending on the nature of the liquid. Several types of non-Newtonian liquids are recognized and typical rheograms of types important in the study of culture fluids are given in Fig. 9.11, and their characteristics are discussed below.

### Bingham plastic rheology

Bingham plastics are similar to Newtonian liquids apart from the fact that shear rate will not increase until a threshold shear stress is exceeded. The threshold shear stress is termed the yield stress or yield value, $\tau_0$. A linear relationship of shear stress to shear rate is given once the yield stress is exceeded and the slope of this line is termed the coefficient of rigidity or the plastic viscosity. Thus, the flow of a Bingham plastic is described by the equation:

$$\tau = \tau_0 + n\gamma$$

where  $n$   is the coefficient of rigidity and
$\tau_0$   is the yield stress.

There have been some claims of mycelial fermentation broths displaying Bingham plastic characteristics (Table 9.5). Everyday examples of these fluids include toothpaste and clay.

### Pseudoplastic rheology

The apparent viscosity of a pseudoplastic liquid decreases with increasing shear rate. Most polymer solutions behave as pseudoplastics. The decrease in apparent viscosity is explained by the long chain molecules tending to align with each other at high shear rates resulting in easier flow. The flow of a pseudoplastic liquid may be described by the power law model, equation (9.9), i.e.:

$$\tau = K(\gamma)^n$$

$K$ has the same units as viscosity and may be taken as the apparent viscosity. The flow-behaviour index is less than unity for a pseudoplastic liquid, the smaller the value of n, the greater the flow characteristics of the liquid deviate from those of a Newtonian fluid. Equation (9.9) may be converted to the logarithmic form as:

$$\log\tau = \log K + n\log\gamma \qquad (9.10)$$

Thus, a plot of log shear stress against log shear rate will produce a straight line, the slope of which will

TABLE 9.5. *Some examples of the rheological nature of fermentation broths*

| Organism | Rheological type | Reference |
|----------|------------------|-----------|
| *Penicillium chrysogenum* | Bingham plastic | Deindoerfer and Gaden (1955) |
| *Streptomyces kanamyceticus* | Bingham plastic | Sato (1961) |
| *Penicillium chrysogenum* | Pseudoplastic | Deindoerfer and West (1960) |
| *Endomyces* sp. | Pseudoplastic | Taguchi *et al.* (1968) |
| *Penicillium chrysogenum* | Casson body | Roels *et al.* (1974) |

equal the flow-behaviour index and the intercept on the shear stress axis will be equal to the logarithm of the consistency coefficient.

Many workers have demonstrated that mycelial fermentation broths display pseudoplastic properties as shown in Table 9.5.

### Dilatant rheology

The apparent viscosity of a dilatant liquid increases with increasing shear rate. The flow of a dilatant liquid may also be described by equation (9.9) but in this case the value of the flow-behaviour index is greater than 1, the greater the value the greater the flow characteristics deviate from those of a Newtonian fluid. Thus, the values of $K$ and $n$ may be obtained from a plot of log shear stress against log shear rate. Fortunately this type of behaviour is not exhibited by fermentation broths — an everyday example is liquid cement slurry.

### Casson body rheology

Casson (1959) described a type of non-Newtonian fluid, termed a Casson body, which behaved as a pseudoplastic in that the apparent viscosity decreased with increasing shear rate but displayed a yield stress and, therefore, also resembled a Bingham plastic. The flow characteristics of a Casson body may be described by the following equation:

$$\sqrt{\tau} = \sqrt{\tau_0} + K_c \sqrt{\gamma} \qquad (9.11)$$

where $K_c$ is the Casson viscosity.

A plot of $\sqrt{\tau}$ against $\sqrt{\gamma}$ will give a straight line, the slope of which will equal the Casson viscosity and the intercept of the $\sqrt{\tau}$ axis will equal $\sqrt{\tau_0}$.

Roels *et al.* (1974) claimed that the rheology of a penicillin broth could be best described in terms of a Casson body.

Therefore, to determine the rheological nature of a fluid it is necessary to construct a rheogram which requires the use of a viscometer which is accurate over a wide range of shear rates. Furthermore, the testing of mycelial suspensions may present special difficulties. These problems have been considered in detail by Van't Riet and Tramper (1991), whose book should be consulted for methods of assessing the rheological properties of mycelial fluids.

### FACTORS AFFECTING $K_L a$ VALUES IN FERMENTATION VESSELS

A number of factors have been demonstrated to affect the $K_L a$ value achieved in a fermentation vessel. Such factors include the air-flow rate employed, the degree of agitation, the rheological properties of the culture broth and the presence of antifoam agents. If the scale of operation of a fermentation is increased (so-called 'scale-up') it is important that the optimum $K_L a$ found on the small scale is employed in the larger scale fermentation. The same $K_L a$ value may be achieved in different sized vessels by adjusting the operational conditions on the larger scale and measuring the $K_L a$ obtained. However, quantification of the relationship between operating variables and $K_L a$ should enable the prediction of conditions necessary to achieve a particular $K_L a$ value. Thus, such relationships should be of considerable value in scaling-up a fermentation and in fermenter design.

### The effect of air-flow rate on $K_L a$

MECHANICALLY AGITATED REACTORS

The effect of air flow rate on $K_L a$ values in conventional agitated systems is illustrated in Fig. 9.12. The quantitative relationships between aeration and $K_L a$ for agitated vessels are considered in the subsequent

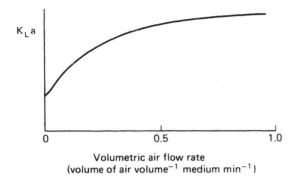

FIG. 9.12. The effect of air-flow rate on the $K_L a$ of an agitated, aerated vessel.

section on power consumption. The air-flow rate employed rarely falls outside the range of 0.5–1.5 volumes of air per volume of medium per minute and this tends to be maintained constant on scale-up. If the impeller is unable to disperse the incoming air then extremely low oxygen transfer rates may be achieved due to the impeller becoming 'flooded'. Flooding is the phenomenon where the air-flow dominates the flow pattern and is due to an inappropriate combination of air flow rate and speed of agitation (see also Chapter 7). Nienow et al. (1977) categorized the different flow patterns produced by a disc turbine that occur under a range of aeration and agitation conditions (Fig. 9.13) and these have been discussed further by Van't Riet and Tramper (1991). Figure 9.13 A shows the flow profile of a non-aerated vessel and Figs 9.13 B to F the profiles with increasing air flow rate. As air-flow rate increases the flow profile changes from one dominated by agitation (Fig. 9.13 B) to one dominated by air flow (Figs

9.13 D to F) until finally the air flow rate is such that the air escapes without being distributed by the agitator (Fig. 9.13 F). Different workers have used different criteria to define the onset of flooding with Nienow et al. (1977) claiming it to be represented by Fig. 9.13 D whereas Biesecker (1972) suggested Fig. 9.13 F. However, the desired pattern is represented by Fig. 9.13 C.

Several workers have produced empirical quantitative descriptions of flooding systems which may assist in avoiding the phenomenon:

(i)   Westerterp et al. (1963) calculated that the minimum impeller tip speed to avoid flooding should be between 1.5 and 2.5 m second$^{-1}$.

(ii)  Biesecker (1972) claimed that flooding occurs when the energy dissipated by the air flow is greater than that dissipated by the agitator. Van't Riet and Tramper (1991) modified this approach to consider the balance between the two energy dissipating systems in the lower compartment of the vessel because the energy dissipated by the agitator in the upper compartment of a large vessel is not related to gas dispersion.

(iii) Feijen et al. (1987) claimed that flooding could be avoided if:

$$F_s/ND^3 < 0.3\, N^2 D/g \qquad (9.12)$$

where $F_s$ is the volumetric air flow rate at the pressure conditions of the lower stirrer (m$^3$ second$^{-1}$),
   $N$ is the stirrer speed (second$^{-1}$),
   $D$ is the stirrer diameter (m),
   $g$ is the gravitational acceleration (m second$^{-2}$).

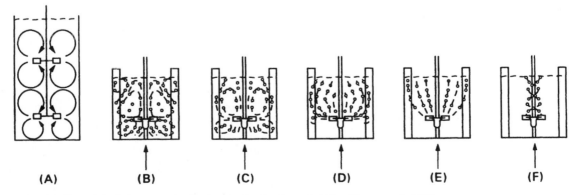

(A)        (B)        (C)        (D)        (E)        (F)

FIG. 9.13. The effect of air flow rate on the flow pattern in stirred vessels (After Nienow et al., 1977).
A. Non-aerated;
B to F, increasing air flow rates.

## NON-MECHANICALLY AGITATED REACTORS

Bubble columns and air-lift reactors are not mechanically agitated and, therefore, rely on the passage of air to both mix and aerate.

### (i) *Bubble columns*

The flow pattern of bubbles through a bubble column reactor is dependent on the gas superficial velocity (cm second$^{-1}$). At gas velocities of below 1–4 cm second$^{-1}$ the bubbles will rise uniformly through the medium (Van't Riet and Tramper, 1991) and the only mixing will be that created in the bubble wake. This type of flow is referrred to as homogeneous. At higher gas velocities bubbles are produced unevenly at the base of the vessel and bubbles coalesce resulting in local differences in fluid density. The differences in fluid density create circulatory currents and flow under these conditions is described as heterogeneous as shown in Fig. 9.14.

Flooding in a bubble column is the situation when the air flow is such that it blows the medium out of the vessel. This requires superficial gas velocities approaching 1 m second$^{-1}$ which are not attainable on commercial scales (Van't Riet and Tramper, 1991).

The volumetric mass transfer coefficient ($K_L a$) in a bubble column is essentially dependent on the superficial gas velocity. Heijnen and Van't Riet (1984) reviewed the subject and demonstrated that the precise mathematical relationship between $K_L a$ and superficial gas velocity is dependent on the coalescent properties of the medium, the type of flow and the bubble size. Unfortunately these characteristics are rarely known for a commercial process which makes the ap-

plication of these equations problematical. However, Van't Riet and Tramper (1991) claimed that the relationship derived for non-coalescing, non-viscous, large bubbles (6 mm diameter) will give a reasonably accurate estimation for most non-viscous situations:

$$K_L a = 0.32 \left( V_s^c \right)^{0.7} \tag{9.13}$$

where $V_s^c$ is the superficial air velocity corrected for local pressure.

However, viscosity has an overwhelming influence on $K_L a$ in a bubble column which Deckwer *et al.* (1982) expressed as:

$$K_L a = c \pi^{-0.84} \tag{9.14}$$

where $\pi$ is the liquid dynamic viscosity (N s m$^{-2}$).

The practical implication of this equation is that bubble columns cannot be used with highly viscous fluids. Van't Riet and Tramper (1991) suggested that the upper viscosity limit for a bubble column was 100 × 10$^{-3}$ N s m$^{-2}$ at which point the $K_L a$ would have decreased 50 fold compared with a reactor batched with water.

### (ii) *Air-lift reactors*

The structure of air-lift reactors is discussed in Chapter 7. The difference between a bubble column and an air-lift reactor is that liquid circulation is achieved in the air-lift in addition to that caused by the bubble flow. The reactor consists of a vertical loop of two connected compartments, the riser and downcomer. Air is introduced into the base of the riser and escapes at the top. The degassed liquid is more dense than the gassed liquid in the riser and flows down the downcomer. Thus, a circulatory pattern is established in the vessel — gassed liquid going up in the riser and degassed liquid coming down the downcomer.

For a given air-lift reactor and medium $K_L a$ varies linearly with superficial air velocity on a log–log scale over the normal range of velocities (Chen, 1990). However, it should be remembered that the circulation in an air-lift results in the bubbles being in contact with the liquid for a shorter time than in a corresponding bubble column. Thus, the $K_L a$ obtained in an air-lift will be less than that obtained in a bubble column at the same superficial air velocity, i.e. less than 0.32 $(V_s^c)^{0.7}$. The advantage of the air-lift lies in the circulation achieved, but this is at the cost of a lower $K_L a$ value.

As for a bubble column flooding will not occur within the normal operating superficial air velocities and should not be a problem on a large scale.

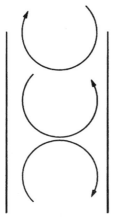

Fig. 9.14. Schematic representation of a heterogeneous flow regime in a bubble column.

## The effect of the degree of agitation on $K_L a$

The degree of agitation has been demonstrated to have a profound effect on the oxygen-transfer efficiency of an agitated fermenter. Banks (1977) claimed that agitation assisted oxygen transfer in the following ways:

(i) Agitation increases the area available for oxygen transfer by dispersing the air in the culture fluid in the form of small bubbles.

(ii) Agitation delays the escape of air bubbles from the liquid.

(iii) Agitation prevents coalescence of air bubbles.

(iv) Agitation decreases the thickness of the liquid film at the gas–liquid interface by creating turbulence in the culture fluid.

The degree of agitation may be measured by the amount of power consumed in stirring the vessel contents. The power consumption may be assessed by using a dynamometer, by using strain gauges attached to the agitator shaft and by measuring the electrical power consumption of the agitator motor (see Chapter 8). The assessment of electrical consumption is suitable only for use with large-scale vessels.

### THE RELATIONSHIP BETWEEN $K_L a$ AND POWER CONSUMPTION

A large number of empirical relationships have been developed between $K_L a$, power consumption and superficial air velocity which take the form of:

$$K_L a = k \left( P_g / V \right)^x V_s^y$$

where $P_g$    is the power absorption in an aerated system

$V$    is the liquid volume in the vessel

$V_s$    is the superficial air velocity

$k, x$

and   $y$    are empirical factors specific to the system under investigation.

Cooper et al. (1944) measured the $K_L a$s of a number of agitated and aerated vessels (up to a volume of 66 dm$^3$) containing one impeller, using the sulphite oxidation technique, and derived the following expression:

$$K_L a = k \left( P_g / V \right)^{0.95} V_s^{0.67}. \qquad (9.15)$$

Thus, it may be seen from equation (9.15) that the $K_L a$ value was claimed to be almost directly proportional to the gassed power consumption per unit volume. However, Bartholomew (1960) demonstrated that the relationship depended on the size of the vessel and the exponent on the term $P_g / V$ varied with scale as follows:

| Scale | Value of exponent on $P_g/V$ |
|---|---|
| Laboratory | 0.95 |
| Pilot plant | 0.67 |
| Production plant | 0.5 |

Bartholomew's vessels contained more than one impeller, whereas those of Cooper et al. contained only one. It is probable that the upper impellers would consume more power relative to their contribution to oxygen transfer than would the lowest impeller, thus affecting the value of the exponent term. Thus, it is important to appreciate that such relationships are scale-dependent when using them in scale-up calculations.

Many workers have produced similar correlations and these have been reviewed by Van't Riet (1983) and Winkler (1990). Van't Riet (1983) summarized the various correlations for coalescing air–water dispersion systems as falling within 20–40% of:

$$k = 0.026,$$
$$x = 0.4,$$
$$y = 0.5.$$

The common feature of these relationships is that the values of $x$ and $y$ are less than unity. Winkler (1990) pointed out that this means that increasing power input or air flow becomes progressively less efficient as the inputs rise. Thus, high oxygen-transfer rates are achieved at considerable expense.

From this discussion it is evident that the $K_L a$ of an aerated, agitated vessel is affected significantly by the consumption of power during stirring and, hence, the degree of agitation. Although it is not possible to derive a relationship between $K_L a$ and power consumption which is applicable to all situations it is possible to derive a relationship between the two which is operable within certain limits and should be a useful guide in practical design problems. If it is accepted that such relationships between power consumption and $K_L a$ are of some practical significance, it is of considerable importance to relate power consumption to operating variables which may affect it. Quantitative relationships between power consumption and operating variables may be useful in:

(i) Estimating the amount of power that an agitation system will consume under certain circumstances, which could assist in fermenter design.

(ii) In providing similar degrees of power consumption (and, hence, agitation and, therefore, $K_L a$s) in vessels of different size.

## THE RELATIONSHIP BETWEEN POWER CONSUMPTION AND OPERATING VARIABLES

Rushton *et al.* (1950) investigated the relationship between power consumption and operating variables in baffled, agitated vessels using the technique of dimensional analysis. They demonstrated that power absorption during agitation of non-gassed Newtonian liquids could be represented by a dimensionless group termed the power number, defined by the expression:

$$N_p = P/(\rho N^3 D^5) \qquad (9.16)$$

where $N_p$ is the power number,
$\quad P$ is the external power from the agitator,
$\quad \rho$ is the liquid density,
$\quad N$ is the impeller rotational speed,
$\quad D$ is the impeller diameter.

Thus, the power number is the ratio of external force exerted ($P$) to the inertial force imparted ($\rho N^3 D^5$) to the liquid. The motion of liquids in an agitated vessel may be described by another dimensionless number known as the Reynolds number which is a ratio of inertial to viscous forces:

$$N_{Re} = (\rho D^2 N)/\mu \qquad (9.17)$$

where $N_{Re}$ is the Reynolds number and
$\quad \mu$ is the liquid viscosity.

Yet another dimensionless number, termed the Froude number, relates inertial force to gravitational force and is given the term:

$$N_{Fr} = (\rho N D^2)/g \qquad (9.18)$$

where $N_{Fr}$ is the Froude number and
$\quad g$ is the gravitational force.

Rushton *et al.* (1950) demonstrated that the power number was related to the Reynolds and Froude numbers by the general expression:

$$N_p = c(N_{Re})^x (N_{Fr})^y \qquad (9.19)$$

where $c$ is a constant dependent on vessel geometry but independent of vessel size,
$\quad x$ and $y$ are exponents.

Examples of the values of $c$, $x$ and $y$ are considered later.

However, in a fully baffled agitated vessel the effect of gravity is minimal so that the relationship between the power number and the other dimensionless num-

bers becomes:

$$N_p = c(N_{Re})^x \qquad (9.20)$$

Therefore substituting from equations (9.16) and (9.17):

$$P/(\rho N^3 D^5) = c(\rho D^2 N/\mu)^x \qquad (9.21)$$

Values for $P$ at various values of $N$, $D$, $\mu$ and $\rho$ may be determined experimentally and the Reynolds and power numbers for each experimental situation may then be calculated. A plot of the logarithm of the power number against the logarithm of the Reynolds number yields a graph termed the power curve. A typical power curve for a baffled vessel agitated by a flat-blade turbine is illustrated in Fig. 9.15 and such a curve would apply to geometrically similar vessels regardless of size.

From Fig. 9.15 it may be seen that a power curve is divisible into three clearly defined zones depicting different types of fluid flow:

(i) The laminar or viscous flow zone where the logarithm of the power number decreases linearly with an increase in the logarithm of the Reynolds number. The slope of the graph is equal to $x$, the exponent in equation (9.21) and is obviously equal to $-1$. The power absorbed in this region is a function of the viscosity of the liquid and the Reynolds number is less than 10.

(ii) The transient or transition zone, where there is no consistent relationship between the power and Reynolds numbers. The value of $x$ (that is, the slope of the plot) is variable and the value of the Reynolds number is between 10 and $10^4$.

(iii) The turbulent flow zone, where the power number is a constant, independent of the

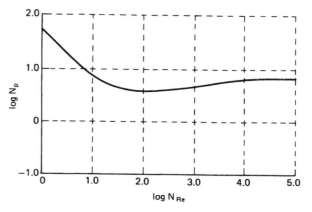

FIG. 9.15. A typical power curve for a baffled vessel agitated by a flat-blade turbine.

Reynolds number so that the value of $x$ is zero and the value of the Reynolds number is in excess of $10^4$.

If the values of the exponent, $x$ are substituted into equation (9.21) for the zones of viscous and turbulent flow, then the following terms are given:

$$\text{For viscous flow } P = c\mu N^2 D^3. \quad (9.22)$$

$$\text{For turbulent flow } P = c\rho N^3 D^5. \quad (9.23)$$

From these equations it may be seen that power consumption is dependent only on the viscosity of the liquid in the region of viscous flow and that increased speed of agitation, or an increase in the impeller diameter, results in a proportionally greater increase in power transmission to a liquid in turbulent flow than to one in viscous flow. Conditions of viscous flow are rare in fermentation processes, the majority of fermentations exhibiting flow characteristics in either the turbulent or transition zones. If turbulent flow is demonstrated to occur in a fermentation then equation (9.23) may be used to predict its power requirements and to predict the operating conditions of different sized vessels to achieve the same agitation conditions, as outlined by Banks (1979). Power consumption on the small scale may be represented as:

$$P_{sm} = c\rho N_{sm}^3 D_{sm}^5 \quad (9.24)$$

and on the large scale as:

$$P_L = c\rho N_L^3 D_L^5$$

where the subscripts $sm$ and $L$ refer to the small and large scales respectively. Maintaining the same power input per unit volume:

$$P_{sm}/P_L = V_{sm}/V_L = (c\rho N_{sm}^3 D_{sm}^5)/ (c\rho N_L^3 D_L^5) \quad (9.25)$$

where $V$ is the volume.

Assuming the vessels to be geometrically similar then $c$ will be the same regardless of scale and as the same broth would be employed $\rho$ would remain the same for both systems

$$V_{sm}/V_L = (N_{sm}^3 D_{sm}^5)/(N_L^3 D_L^5) \quad (9.26)$$

For geometrically similar vessels

$$D_{sm}/D_L = (V_{sm}/V_L)^{1/3}$$

Therefore, substituting for $D_{sm}/D_L$ in (9.26)

$$N_L = N_{sm}(V_{sm}/V_L)^{2/9}. \quad (9.27)$$

If transient flow conditions occur in a fermentation then it is necessary to construct a complete power curve for such predictions and this is discussed later in the chapter.

The work of Rushton et al. (1950) was carried out using ungassed liquids whereas the vast majority of fermentations are aerated. It is widely accepted that aeration of a liquid decreases the power consumption during agitation because an aerated liquid, containing suspended air bubbles, is less dense than an unaerated one and large gas-filled cavities generated behind the agitator blades decrease the hydrodynamic resistance of the blades. A number of workers have produced correlations of gassed power consumption, ungassed power consumption and operating variables, that of Michel and Miller (1962) being widely used:

$$P_g = k(P^2 ND^3/Q^{0.56})^{0.45}$$

where $Q$ is the volumetric air flow rate.

However, more recent correlations have been elucidated which are applicable over a wider range of operating conditions than that of Michel and Miller. Hughmark (1980) produced the following correlation from 248 sets of published data:

$$P_g/P = 0.1(Q/NV)^{-0.25}(N^2 D^4/gWV^{0.67})^{-0.2}$$

where $Q$ is the volumetric air flow rate,
$g$ is the acceleration due to gravity,
and $W$ is the impeller blade width.
Using dimensional analysis:

$$P_g/P = 0.0312 \cdot Fr^{-1.6} \cdot Re^{0.064} N_a^{-0.38} \cdot (T/D)^{0.8}$$

where $N_a$ is the aeration number and equals $Q/ND$
and $T$ is the vessel diameter.

Provided it is remembered that these expressions are not particularly accurate they may be used to predict power consumption in gassed systems where turbulent flow is known to be operating. However, it should be remembered that in non-mycelial fermentations the greatest power demands often occur during agitation when the system is not gassed, that is during the sterilization of the medium in situ or if the air supply were to fail. Thus, in designing the system care must be taken to ensure that the agitator motor is sufficiently powerful to agitate the ungassed system and for fixed speed motors the operating speed should be specified with respect to the ungassed power draw (Gbewonyo et al., 1986).

From the foregoing account it may be seen that reasonable techniques exist to relate operating variables to power consumption and, hence, to the degree of agitation which may be shown to have a proportional effect on $K_L a$. However, these techniques apply to Newtonian fluids and are not directly applicable to the study of non-Newtonian systems. Non-Newtonian fluids do not have constant viscosities, which creates difficulties in utilizing relationships which rely on being able to determine the fluid viscosity. These difficulties may be avoided if the agitation system is capable of maintaining turbulent-flow conditions during the fermentation, because under such conditions power consumption is independent of the Reynolds number and, hence, of viscosity. However the high viscosities of the majority of mycelial fermentation broths make fully turbulent flow conditions impossible, or extremely difficult, to achieve. Such fermentations tend to exhibit transient zone flow conditions which necessitate the construction of complete power curves to correlate power consumption with operating variables. The fact that the viscosity of a non-Newtonian liquid is affected by shear rate means that the viscosity of a non-Newtonian fermentation broth will not be uniform throughout the fermenter because the shear rate will be higher near the agitator than elsewhere in the vessel. Thus, the determination of the impeller Reynolds number is made difficult by not knowing the viscosity of the fermentation broth. Metzner and Otto (1957) proposed a solution to this paradox by introducing the concept of average shear rate $(y)$ related to the agitator shaft speed in the vessel, by the equation:

$$y = kN \qquad (9.28)$$

where $k$ is a proportionality constant.

Metzner and Otto determined the value of the proportionality constant to be 13 for pseudoplastic fluids in conventional, baffled reactors agitated by single, flat-blade turbines. Several groups of workers have determined values of $k$ under a wide range of operating variables; the values range from approximately 10 to 13. Metzner *et al.* (1961) suggested that a compromise value of 11 could be used for calculation purposes, with relatively little loss of accuracy, which would obviate the necessity to determine $k$ for each circumstance. Therefore, provided that the rheological properties of a fermentation broth are known, an apparent average viscosity of the fluid may be calculated using the average shear rate which would enable the calculation of the impeller Reynolds number for each value of the impeller rotational speed, thus enabling a power curve to be constructed. Such a power curve may be used to predict the power requirements of a fermentation and to scale up a fermentation on the basis of

power consumption per unit volume. Metzner and Otto's approach has not been widely applied but there are some recent examples of its adoption. Nienow and Elson (1988) suggested that a repetition of their work would be very valuable using the more sophisticated instrumentation now available. An example of the use of the technique is considered in a later section considering the operation of viscous polysaccharide fermentations.

## The effect of medium and culture rheology on $K_L a$

As can be seen from the previous section, the rheology of a fermentation broth has a marked influence on the relationship between $K_L a$ and the degree of agitation. The objective of this section is to discuss the effects of medium and culture rheology on oxygen transfer during a fermentation. A fermentation broth consists of the liquid medium in which the organism grows, the microbial biomass and any product which is secreted by the organism. Thus, the rheology of the broth is affected by the composition of the original medium and its modification by the growing culture, the concentration and morphology of the biomass and the concentration and rheological properties of the microbial products. Therefore, it should be apparent that fermentation broths vary widely in their rheological properties and significant changes in broth rheology may occur during a fermentation.

### MEDIUM RHEOLOGY

Fermentation media frequently contain starch as a carbon source which may render the medium non-Newtonian and relatively viscous. However, as the organism grows it will degrade the starch and thus modify the rheology of the medium and reduce its viscosity. Such a situation was described by Tuffile and Pinho (1970) in their study of the growth of *Streptomyces aureofaciens* on a starch-containing medium. Before inoculation, the medium displayed Bingham plastic characteristics with a well-defined yield stress and an apparent viscosity of approximately 18 pseudopoise; after 22 hours the organism's activity had decreased the medium viscosity to a few pseudopoise and modified its behaviour to that of a Newtonian liquid; from 22 hours onwards the apparent viscosity of the broth gradually increased, due to the development of the mycelium, up to a maximum of approximately 90 pseudopoise and the rheology of the broth became increasingly pseudoplastic in nature. Thus, this example suggests that the rheological problems presented by the medium are minor compared with those presented by a high mycelial biomass, especially when it is considered that the total oxygen demand is relatively low in the early stages of a fermentation. However, it is worth remembering that in non-viscous unicellular fermentations the highest power draw will occur when the medium is sterilized *in situ* when the vessel is not being aerated and this will correspond with the time when a starch-based medium is at its most viscous.

Berkman-Dik *et al.* (1992) observed that the medium composition could affect the rheological properties of mycelium suspensions by affecting the interactions between the hyphae.

## THE EFFECT OF MICROBIAL BIOMASS ON $K_L a$

### (i) *Agitator design for non-Newtonian fermentations*

The biomass concentration and its morphological form in a fermentation has been shown to have a profound effect on oxygen transfer. Most bacterial and yeast fermentations tend to give rise to relatively non-viscous Newtonian broths in which conditions of turbulent flow may be achieved. Such fermentations present relatively few oxygen-transfer problems. However, the highly viscous non-Newtonian broths of fungal and streptomycete fermentations present major difficulties

in oxygen provision, the productivities of many such fermentations being limited by oxygen availability. Banks (1977) stressed the difference in the pattern of oxygen uptake between unicellular and mycelial fermentations as illustrated in Fig. 9.16. In both unicellular and mycelial fermentations the pattern of total oxygen uptake is very similar during the exponential growth phase, up to the point of oxygen limitation. However, during oxygen limitation, when arithmetic growth occurs, the oxygen uptake rate remains constant in a unicellular system whereas it decreases in a mycelial one. Banks claimed that the only possible explanation for such a decrease is the increasing viscosity of the culture caused by the increasing mycelial concentration.

Several groups of workers have demonstrated the detrimental effect of the presence of mycelium on oxygen transfer. Figure 9.17 represents some of the data of Deindoerfer and Gaden (1955) illustrating the effect of *Penicillium chrysogenum* mycelium on $K_L a$. Buckland *et al.* (1988), using different agitator systems, reported that the $K_L a$ decreased approximately in proportion with the square root of the broth viscosity, i.e:

$$K_L a \ \alpha \ 1/\sqrt{\text{viscosity}}.$$

Steel and Maxon (1966) investigated the problem of oxygen provision to mycelial clumps in the *Streptomyces*

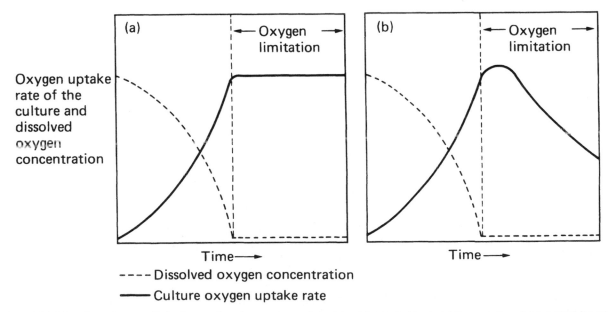

FIG. 9.16. The effect of oxygen limitation on the culture oxygen uptake rate: (a) A typical bacterial fermentation. (b) A typical fungal fermentation (Banks, 1977).

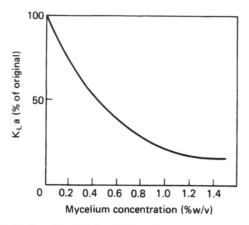

FIG. 9.17. The effect of *Penicillium chrysogenum* mycelium on $K_L a$ in a stirred fermenter (Deindoerfer and Gaden, 1955).

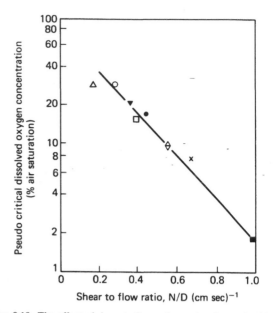

FIG. 9.18. The effect of shear to flow ratio on the observed critical oxygen concentration of *S. niveus* (Wang and Fewkes, 1977).

*niveus* novobiocin fermentation and demonstrated that high dissolved oxygen levels (60–80%) occurred in oxygen-limited cultures. It was concluded that, although oxygen was being transferred into solution, the dissolved gas was not reaching a large proportion of the biomass. Thus, as well as $K_L a$ being affected adversely by a high viscosity broth, efficient mixing also becomes extremely important in these systems. These workers also demonstrated that, at constant power input, small impellers were superior to large impellers in transferring oxygen from the gas phase to the microbial cells. Wang and Fewkes (1977) confirmed Steel and Maxon's work by demonstrating that the critical dissolved oxygen concentration ($C_{crit}$) for *S. niveus* in a fermentation varied depending on the degree of agitation and the size of the impeller. Remember that $C_{crit}$ is the dissolved oxygen concentration below which oxygen uptake is limited, i.e. it is a physiological characteristic of the organism. It was concluded that the limiting factor was the diffusion of oxygen to the cell surface through a dense mycelial mass. At higher agitation rates biomass within clumps would be receiving oxygen and would thus contribute to the measured respiration rate whereas at low agitation rates such mycelium would be oxygen limited, i.e the heterogeneity of the system increased at low agitation rates. Wang and Fewkes examined their results in terms of the impeller's ability to produce turbulent shear stress (oxygen transfer into solution) and pumping power (mixing). Turbulent shear stress is proportional to $N^2 D^2$ and impeller pumping power is proportional to $ND^3$ (where $N$ is the impeller rotational speed and $D$ is the impeller diame-

ter). Thus, the ratio of impeller turbulent shear stress to impeller pumping is proportional to:

$$N^2 D^2 / ND^3 \text{ or } N/D \text{ (cm sec)}^{-1}.$$

It was demonstrated that the observed critical dissolved-oxygen concentration decreased exponentially as the shear stress to pumping ratio increased, over the range 0.2 to 1.0 (cm sec)$^{-1}$, as shown in Fig. 9.18. Thus, an increase in the ratio of impeller shear stress to impeller pumping decreases the transport resistance of oxygen to the cell surface resulting in a lower dissolved oxygen concentration maintaining a higher respiration rate. Wang and Fewkes' analysis quantifies Steel and Maxon's observation that smaller impellers gave better oxygen transfer to the cells of *S. niveus*, in that the smaller impeller would have a larger shear stress to impeller pumping power ratio.

Wang and Fewkes' correlations are particularly relevant when it is considered that many mycelial broths are pseudoplastic. The viscosity of a pseudoplastic broth will decrease with increasing shear stress so that viscosity increases with increasing distance from the agitator. Air introduced into the fermenter tends to rise through the vessel by the route of least resistance, that is, through the well-stirred, less viscous central zone. Thus, stagnant zones, receiving little oxygen, may occur in the vessel. Therefore, it is essential that the agitation regime employed creates the correct balance of turbulence (and hence the transfer of oxygen into solution) and pumping power (mixing) to circulate the broth through the region of high shear.

The quantification of the problem of oxygen transfer and mixing is also considered by Van't Riet and Van Sonsberg (1992) in the context of the critical time for mass transfer. It is assumed that oxygen transfer into solution in a stirred, aerated reactor takes place only in the stirrer region. If one considers an aliquot of aerated broth leaving the agitator zone, it will be circulated through the vessel and eventually return to the agitator. The dissolved oxygen imparted to the broth should sustain the respiration of the organisms in that aliquot during the circulation. The time it takes for the oxygen in the aliquot to be exhausted will be:

$$t_{cro} = C_L(\text{ag})/OUR$$

where $t_{cro}$ is the time for oxygen to be exhausted,
$C_L(\text{ag})$ is the dissolved oxygen concentration in the zone of the agitator,
$OUR$ is the oxygen uptake rate.

If the circulation time for the vessel exceeds $t_{cro}$ then oxygen starvation will occur in the aliquot before it returns to the agitator. To prevent this occurring the dissolved oxygen concentration at the agitator should be high, but this would reduce the driving force of oxygen into solution and the oxygen transfer rate would decrease. The alternative approach is to achieve the balance of mass transfer and pumping power (broth circulation) already discussed.

As discussed in Chapter 7 the most widely used fermenter agitator is a disc turbine (Rushton turbine). Van't Riet (1979) and Chapman et al. (1983) demonstrated that *for non-viscous broths* the $K_La$ is dependent only on the power dissipated in the vessel and is independent of impeller type (at least those impellers included in the study). However, it is obvious from the foregoing discussion that impeller type is particularly relevant for viscous, non-Newtonian fermentations and this realization has resulted in the development of a range of agitators which address the dual problems of oxygen transfer and mixing in viscous fermentations.

Legrys and Solomons (1977) approached the problem of combining adequate pumping power and mass transfer in mycelial fermentations by using two impellers, a bottom-mounted disc turbine and a top-mounted curled-blade (hydrofoil) impeller. The bottom turbine produced a high degree of turbulence and radial mixing while the top-mounted impeller produced axial mixing with a high flow velocity, resulting in the circulation of one tank volume in 20–30 seconds. Thus, the mycelium was re-circulated through the oxygenation zone of the

vessel before it became oxygen limited. Cooke et al. (1988) extended Legrys and Solomon's approach using a combination of radial flow and axial flow agitators in a 60-dm³ fermenter intended for non-Newtonian fermentations. The radial flow agitator was an ICI Gasfoil which is similar to the Scaba SRGT illustrated in Fig. 9.19, being a disc turbine with concave blades. However, in this case the combination was not successful due to minimal fluid movement at the vessel walls which would have created significant cooling problems.

Gbewonyo et al. (1986) evaluated the performance of a hydrofoil impeller, the Prochem Maxflo (Fig. 9.19) in the avermectin fermentation employing *Streptomyces*

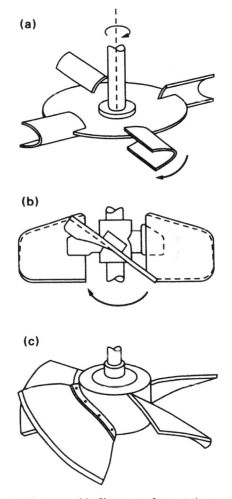

**(a)**

**(b)**

**(c)**

FIG. 9.19. Agitators used in filamentous fermentations:
(a) Scaba agitator;
(b) Lightnin' A315;
(c) Prochem Maxflo;
(Nienow, 1990).

*avermitilis* in a 600-dm$^3$ working volume vessel. The avermectin process is challenging because the broth is extremely viscous, the fermentation requires fairly high oxygen-transfer rates and the situation is complicated by the shear sensitivity of the mycelium. The results of this investigation may be summarized as follows:

(i) The impeller pumped the broth axially, that is from the top to the bottom of the fermenter, which is very different from the Rushton turbine which pumped radially, outwards from the agitator.

(ii) The Prochem agitator supported a significantly higher oxygen uptake rate than did the Rushton turbine.

(iii) The power number of the Prochem was 1.1 compared with 6.5 for the Rushton turbine. The equation (9.16) for power number was given previously as:

$$N_p = P/(\rho N^3 D^5)$$

where $N_p$    is the power number,
$P$     is the external power from the agitator,
$\rho$     is the liquid density,
$N$     is the impeller rotational speed,
$D$     is the impeller diameter.

Thus, a low power number indicates a low power draw and, hence, the Prochem agitator drew significantly less power than did the Rushton turbine, making the former far more economical to operate. This observation is strengthened by Nienow's work on a similar hydrofoil impeller, the Lightnin' A315, which gave a power number of 0.75 compared with 5.2 for a Rushton turbine.

(iv) The relationships between $K_L a$ and power consumption per unit volume at a viscosity of 700 cp were as follows:

Rushton $K_L a = 51(P/V)^{0.58}$,
Prochem $K_L a = 129(P/V)^{0.59}$.

These figures reinforce the previous point, demonstrating that the power requirement for the Prochem agitator is approximately 50% of that for the Rushton.

(v) Raising the power of the Prochem had a greater effect on oxygen transfer at high viscosity than it did at low viscosity. This points to the key role that bulk mixing plays in a viscous fermentation and suggests that it is at least as impor-

tant as bubble breakup (at which the Prochem is mediocre).

(vi) Unlike a Rushton turbine, the Prochem agitator did not generate high shear forces, which is advantageous for a shear sensitive organism.

(vii) The avermectin yields were slightly better in the Prochem fermenter, but these were achieved with approximately 40% less power consumption.

The same group (Buckland *et al.*, 1988, 1989) performed similar experiments on viscous fungal fermentations in 800-dm$^3$ and 19-m$^3$ vessels and came to the same basic conclusions that bulk mixing is extremely important in viscous fermentations and that an axial flow hydrofoil impeller results in lower power costs. Data generated from non-Newtonian polysaccharide fermentations using hydrofoil impellers are considered in a subsequent section of this chapter.

### (ii) *The manipulation of mycelial morphology*

The previous section considered engineering solutions to the problem of oxygen transfer in mycelial fermentations. However, this is not the only approach to improve oxygen transfer in such processes; it is possible to modify the morphology of the process organism. As discussed in Chapter 6, the biomass of mycelial organisms grown in submerged culture may vary from the filamentous type, in which the hyphae form a homogeneous suspension dispersed through the medium, to the 'pellet' type consisting of compact, discrete masses of hyphae. The filamentous form tends to give rise to a highly viscous, non-Newtonian broth whereas the pellet form tends to produce an essentially Newtonian system with a much lower viscosity making oxygen transfer much easier. Buckland (1993) reported that the $K_L a$ attained in the lovastatin *Aspergillus terreus* fermentation was 20 h$^{-1}$ with a filamentous culture and 80 h$^{-1}$ with a pelleted one at the same power input. Not all pelleted cultures are Newtonian: Metz *et al.* (1979) demonstrated that pellet suspensions could be non-Newtonian but confirmed that they did give rise to low viscosity broths. Also, it should be appreciated that the terms 'filamentous' and 'pelleted' each describe a range of morphology and the form of filamentous or pelleted growth may be affected by both the genetic makeup of the organism and the environment. Thus, the morphological form of a mycelial organism in submerged culture has a major effect on the broth rheology and may, therefore, be expected to influence aeration efficiency.

Carilli *et al.*'s work (1961) provides a good example

of both the effect of different filamentous form on process performance and the behaviour of filamentous and pelleted cultures in 3000-dm³ fermenters. Two strains of *P. chrysogenum* were employed, one which grew as short, highly branched hyphae and the other as long, relatively unbranched hyphae. The short, branched hyphae gave rise to a relatively low viscosity broth in which the oxygen transfer rate was approximately twice that achieved with the more viscous broth of the unbranched form. By manipulating the cultural conditions of *A. niger*, Carilli *et al.* were able to produce the fungus in either filamentous or pellet form and demonstrated that the pellet form gave rise to a broth exhibiting half the viscosity of the filamentous broth. Also, oxygen limitation occurred far earlier in the fermentation when the organism grew in the filamentous form.

Although the pellet type of growth tends to produce a low viscosity Newtonian broth in which turbulent flow conditions may be achieved, it may also give rise to problems of oxygen availability if the pellets become too large. A large pellet may be so compact that its centre may be unaffected by the turbulent forces occurring in the bulk of the fermentation broth so that the passage of oxygen within the pellet is dependent on simple diffusion; this may result in the centre of the pellet being oxygen limited. Thus, to maintain the intra-pellet oxygen concentration at an adequate level it would be necessary to maintain a high dissolved oxygen concentration to ensure an effective diffusion gradient. A similar situation was described by Steel and Maxon (1966) and Wang and Fewkes (1977). Kobayashi *et al.* (1973) demonstrated this phenomenon in pellets of *A. niger* where large pellets required a higher dissolved oxygen concentration to maintain the same specific oxygen-uptake rate as smaller pellets. If oxygen limitation does occur within a pellet then only its outer layer would contribute to its growth and the centre may autolyse. The diffusion of oxygen into the centre of a pellet will be influenced by the size of the pellet, and thus it is important to control pellet size.

Schugerl *et al.* (1988) monitored the dissolved oxygen concentration within pellets of *P. chrysogenum* and demonstrated that, provided they were smaller than 400 μm in diameter, the oxygen concentration in the centre of the pellet was not limiting. Similarly, Buckland (1993) reported that pellets of *Aspergillus terreus* in the lovastatin fermentation had to be smaller than 180 μm in diameter to avoid oxygen limitation in the centre of the pellet. It should be appreciated that the pellet sizes recommended by both groups are very small and it is possible to obtain fungal pellets which

are at least 1 cm in diameter. Pellet size may be influenced by the inoculum, the medium and the cultural conditions. As discussed in Chapter 6, pellet size is reduced at high spore inoculum concentrations, but it is unlikely that this alone would produce pellets of less than 400 μm in diameter. Schugerl *et al.* (1988) controlled the pellet size of the inoculum by physical means by either incorporating glass beads in inoculum shake flasks or using high agitator speeds in seed fermenters. Metz and Kossen (1977) also claimed that, once pellets are formed, strong agitation tends to give rise to smaller, more compact pellets. Buckland (1993) reported that the conditions of the lovastatin fermentation are carefully controlled to maintain the optimum pellet size but the cultural conditions used to achieve this end were not revealed.

Righelato (1979) discussed the effects of mycelium morphology on culture rheology and oxygen transfer and came to the conclusion that the most desirable way for a mycelium to grow in submerged culture is in the form of short, hyphal fragments which would produce a broth less susceptible to diffusion limitation than a pelleted one, and less viscous than one containing long filaments. However, attempts to encourage the formation of the desirable short hyphal fragment morphology (as compared with the long filaments) by increasing the shear stress on the mycelium has met with only limited success. Even if a less viscous broth is obtained the damage done to the mycelium may well be counterproductive. Dion *et al.* (1954) showed that the morphology of *P. chrysogenum* was influenced by the degree of agitation in that short, branched mycelium was produced at high agitation rates compared with long hyphae produced at low agitation rates. Lilly *et al.* (1992) extended this observation at 10 dm³ and 100 dm³ scales and related the mean main hyphal length and the penicillin specific production rate ($q_p$) to the term $P/D_i^3 t_c$, where $P$ is the agitator power, $D_i$ is the impeller diameter and $t_c$ is the calculated circulation time. This term is a measure of the maximum shear stress due to agitator power dissipation and the frequency with which mycelia pass through the high shear region. Both mean hyphal length and $q_p$ decreased with increasing $P/D_i^3 t_c$, implying that increased shear is disadvantageous at this scale. At 1000 dm³ it was not possible to introduce enough power into the fermenter to decrease $q_p$, which suggests that it is very difficult to disrupt the mycelium at this scale, thus confirming Van Suijdam and Metz's (1981) observation that an enormous amount of energy is required to reduce the hyphal length of *P. chrysogenum*. Righelato (1979) also claimed that it is unlikely that shear forces could account for the break up of mycelia and that autolysis and lysis of some hyphal compartments may be more important controlling factors, perhaps implying that the phenomenon may be more under genetic, rather than physical, control. This leads us on to strain improvement of morphologically favourable strains, as discussed in more detail in Chapter 3. However, Belmar-Beiny and Thomas (1990) demonstrated in 9-dm$^3$ fermenters that increased stirrer speed did result in the production of shorter, less branched hyphal fragments of *Streptomyces clavuligerus* and clavulanic acid synthesis was unaffected. This suggests that this approach may be used to influence rheological properties in clavulanic acid fermentations.

Other cultural conditions which have been claimed to influence mycelial morphology include medium composition (see Chapter 5), growth rate, dissolved oxygen concentration, polymer additives and temperature. Kuenzi (1978) reported that the viscosity of a *Cephalosporium* broth was considerably reduced by growing the organism at 27° rather than 25°C. Olsvik and Kristiansen (1992) investigated the influence of specific growth rate and dissolved oxygen concentration on the viscosity of *Aspergillus niger* in continuous culture. $K$, the consistency index (indicative of apparent viscosity, see equation (9.9)) was measured over a range of conditions. At dissolved oxygen (DO) concentrations above 10% saturation, $K$ increased with increasing dilution rate whereas at DO concentrations below 10% saturation, $K$ decreased with increasing dilution rate. The effect of DO on $K$ was particularly evident at low DO values and at low growth rates where a 2% change in the DO could give a 25% change in $K$. These observations may be particularly relevant in the late stages of batch or fed-batch processes where low growth rates, nutrient limitation, high biomass levels and low oxygen concentrations occur, all contributing to complex changes in morphology, viscosity and oxygen transfer rate.

Dispersed growth can be encouraged in certain organisms by incorporating polymeric compounds into the medium. Such anionic polymers include Junlon PW110 and Junlon 111 (cross-linked polyacrylic acids) and Carbopol-934 (carboxypolymethylene). It is claimed that these polymers modify the electrical charges on the spore surface and thus prevent the aggregation of spores into clumps, thus preventing the initiation of pellet formation. These agents have been used to increase the homogeneity of both fungal (Trinci, 1983) and streptomycete (Hobbs *et al.*, 1989) broths. Although these agents would not be practical to use on a large scale they may be useful in the early stages of an inoculum development programme if a dispersed morphology is desirable.

Several workers have discussed the possible advantages of reducing the viscosity of a mycelial fermentation, in its later stages, by diluting the broth with either water or fresh medium. Sato (1961) increased the yield of a kanamycin fermentation, displaying Bingham plastic rheology, by 20% by diluting the broth 5% by volume with sterile water. Taguchi (1971) achieved a 50% reduction in the viscosity of an *Endomyces* broth by diluting 10% with water or fresh medium. A scheme has been put forward for the control of viscosity and dissolved oxygen concentration in a hypothetical fermentation. These workers proposed that, as the critical dissolved oxygen concentration is approached, a set volume of broth could be removed from the fermenter and replaced with fresh medium. The process could be repeated in a step-wise manner as the system became oxygen limited, which could be determined by dissolved oxygen concentration or viscosity measurements. Thus, by using such techniques the viscosity may be controlled and maintained below the level which may cause oxygen limitation. Kuenzi (1978) reported an instance where the very slow feeding of medium to a *Cephalosporium* culture resulted in the organism growing in the form of long filaments which produced a highly viscous culture which could not be adequately aerated. The design of fed-batch processes such that efficient control may be achieved over the process is discussed in a subsequent section of this chapter and in Chapter 2.

The production of *Fusarium graminae* biomass for human food in the ICI-RHM mycoprotein (Quorn®) fermentation (see Chapter 1) presents a very different problem from those of most other fungal fermentations. It is essential that the organism grows as long hyphae so that the biomass can be processed into a textured food product. Long hyphae are susceptible to shear forces, so to maintain the morphological form of the organism an air-lift reactor is used, despite the fact that the viscous broth severely limits the attainable oxygen transfer rate. This limitation of the air-lift fermenter means that only a relatively low biomass concentration may be maintained in the vessel compared with that in a stirred system, but this is an acceptable penalty to pay for the correct morphological form.

## THE EFFECT OF MICROBIAL PRODUCTS ON AERATION

Generally speaking, the product of a fermentation contributes relatively little to the viscosity of the culture broth. However, the exception is the production of bacterial polysaccharides, where the broths tend to be highly viscous (30,000 cp, Sutherland and Ellwood, 1979) and non-Newtonian. Charles (1978) demonstrated that the bacterial cells in a polysaccharide fermentation made a minimal contribution to the high culture viscosity which was due primarily to the polysaccharide product. Normally, microbial polysaccharides tend to behave as pseudoplastic fluids, although some have also been shown to exhibit a yield stress. The yield stress of a polysaccharide can make the fermentation particularly difficult because, beyond a certain distance from the impeller, the broth will be stagnant and productivity in these regions will be practically zero (Gallindo and Nienow, 1992). Thus, bacterial polysaccharide fermentations present problems of oxygen transfer and bulk mixing similar to those presented by mycelial fermentations. Thus, similar stirrer configurations to those discussed in the previous section have been used in polysaccharide fermentations. Gallindo and Nienow (1992) investigated the behaviour of a hydrofoil impeller, the Lightnin' A315, in a simulated xanthan fermentation. These workers adopted Metzner and Otto's approach to construct power curves. Better agitator performance was achieved when its pumping direction was upwards rather than downwards resulting in lower power loss on aeration and less torque fluctuations. It was concluded that such agitators may give improved mixing in a xanthan fermentation provided that the polysaccharide concentration is below 25 kg m$^{-3}$.

A novel solution to the problem was proposed by Oosterhuis and Koerts (1987). These workers designed an air-lift loop reactor incorporating a pump to circulate the highly viscous broth. The system was operated on a 4-m$^3$ scale and proved to be much more efficient than a stirred tank reactor.

### The effect of foam and antifoams on oxygen transfer

The high degree of aeration and agitation required in a fermentation frequently gives rise to the undesirable phenomenon of foam formation. In extreme circumstances the foam may overflow from the fermenter via the air outlet or sample line resulting in the loss of medium and product, as well as increasing the risk of contamination. The presence of foam may also have an adverse effect on the oxygen-transfer rate. Hall *et al.* (1973) pointed out that Waldhof and vortex-type fermenters (see Chapter 7) were particularly affected due to the bubbles becoming entrapped in the continuously

recirculating foam, resulting in high bubble residence times and, therefore, oxygen-depleted bubbles. The presence of foam in a conventional agitated, baffled fermenter may also increase the residence time of bubbles and therefore result in their being depleted of oxygen. Furthermore, the presence of foam in the region of the impeller may prevent adequate mixing of the fermentation broth. Thus, it is desirable to break down a foam before it causes any process difficulties and, as discussed in Chapter 7, this may be achieved by the use of mechanical foam breakers or chemical antifoams. However, mechanical foam control consumes considerable energy and is not completely reliable so that chemical antifoams are preferred (Van't Riet and Van Sonsberg, 1992).

All antifoams are surfactants and may, themselves, be expected to have some effect on oxygen transfer. The predominant effect observed by most workers is that antifoams tend to decrease the oxygen-transfer rate, as discussed by Aiba *et al.* (1973) and Hall *et al.* (1973). Antifoams cause the collapse of bubbles in foam but they may favour the coalescence of bubbles within the liquid phase, resulting in larger bubbles with reduced surface area to volume ratios and hence a reduced rate of oxygen transfer (Van't Riet and Van Sonsberg, 1992). Thus, a balance must be struck between the necessity for foam control and the deleterious effects of the controlling agent. Foam formation has a particular influence on the liquid height in the fermenter at which it is practical to operate. If inadequate space is provided above the liquid level for foam control, then copious amounts of antifoam must be used to prevent loss of broth from the vessel. Van't Riet and Van Sonsberg (1992) observed that, above a critical liquid height, the $K_L a$ value decreases dramatically due to the excessive use of antifoams. Thus, it may be more productive to operate a vessel at a lower working volume.

Methods for foam control are considered in Chapter 8 and antifoams are discussed in Chapter 4.

## THE BALANCE BETWEEN OXYGEN SUPPLY AND DEMAND

Both the demand for oxygen by a micro-organism and the supply to the organism by the fermenter have been considered in this chapter. This section attempts to bring these two aspects together and considers how processes may be designed such that the oxygen uptake rate of the culture does not exceed the oxygen transfer rate of the fermenter.

The volumetric oxygen uptake rate of a culture is described by the term, $Q_{O_2}x$, where $Q_{O_2}$ is the specific oxygen uptake rate (mmoles $O_2$ $g^{-1}$ biomass $h^{-1}$) and $x$ is biomass concentration (g $dm^{-3}$). Thus, the units of $Q_{O_2}x$ are mmoles oxygen $dm^{-3}$ $h^{-1}$.

The volumetric oxygen transfer rate (also measured as mmoles $O_2$ $dm^{-3}$ $h^{-1}$) of a fermenter is given by equation (9.1), i.e.:

$$dC_L/dt = K_L a(C^* - C_L).$$

It will also be recalled that the dissolved oxygen concentration during the fermentation should not fall below the critical dissolved oxygen concentration ($C_{crit}$) or the dissolved oxygen concentration which gives optimum product formation. Thus, it is necessary that the oxygen-transfer rate of the fermenter matches the oxygen uptake rate of the culture whilst maintaining the dissolved oxygen above a particular concentration. A fermenter will have a maximum $K_L a$ dictated by the operating conditions of the fermentation and thus, to balance supply and demand it must be the demand that is adjusted to match the supply. This may be achieved by:

(i)  Controlling biomass concentration.
(ii)  Controlling the specific oxygen uptake rate.
(iii)  A combination of (i) and (ii).

### Controlling biomass concentration

Mavituna and Sinclair (1985a) developed a method to predict the highest biomass concentration (termed the critical biomass or $x_{crit}$) which can be maintained under fully aerobic conditions in a fermenter of known $K_L a$. Thus, $x_{crit}$ is the biomass concentration which gives a volumetric uptake rate ($Q_{O_2}x_{crit}$) equal to the maximum transfer rate of the fermenter, i.e. $K_L a (C^* - C_{crit})$. If $C_{crit}$ is defined as the dissolved oxygen concentration when:

$$Q_{O_2} = 0.99 Q_{O_2}max$$

then the volumetric oxygen uptake rate when the dissolved oxygen concentation is $C_{crit}$ will be:

$$0.99 Q_{O_2}max \cdot x_{crit}.$$

If the oxygen transfer rate were equal to the uptake rate when the dissolved oxygen concentration equals $C_{crit}$, then:

$$K_L a(C^* - C_{crit}) = 0.99 Q_{O_2}max \cdot x_{crit}. \quad (9.29)$$

Equation (9.29) may be used to calculate $x_{crit}$ for a fermenter with a particular $K_L a$ value:

$$x_{crit} = K_L a(C^* - C_{crit})/0.99 Q_{O_2}max \quad (9.30)$$

Equation (9.30) may also be modified to calculate the biomass concentration which may be maintained at any fixed dissolved oxygen concentration above $C_{crit}$:

$$x = K_L a(C^* - C_L)/Q_{O_2}max.$$

Mavituna and Sinclair presented this model graphically as shown in Fig. 9.20. The upper graph represents the relationship between the dissolved oxygen concentration and the volumetric oxygen transfer rate achievable in three fermenters (plots 1, 2 and 3 represent fermenters of increasing $K_L a$ values) whilst the lower graph represents the relationship between biomass and the volumetric oxygen uptake rate of the culture. The $x$ axes of both graphs are drawn to the same scale. A construction is drawn on the upper graph linking $C_{crit}$ to the oxygen-transfer rates attainable in each of the three fermenters. This construction is extended to the

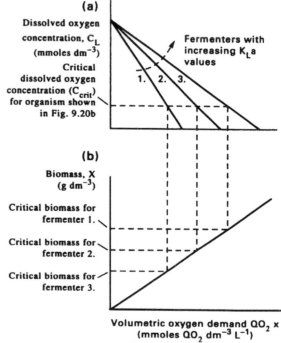

FIG. 9.20. (a) The relationship between dissolved oxygen concentration and the oxygen transfer rate attainable in 3 fermenters with increasing $K_L a$ values. (b) The relationship between biomass concentration and oxygen uptake rate of a process organism. The same scales are used for $dC_L/dt$ and $Q_{O_2}x$ allowing $x_{crit}$ to be determined (Mavituna and Sinclair, 1985).

lower graph indicating the oxygen uptake rates equal to the transfer rates attainable at $C_{crit}$. Finally, from the lower graph the biomass concentrations $(x_{crit})$ which would give rise to the uptake rates equal to the transfer rates may be determined. Again, this figure may be used to predict the maximum biomass concentration which may be maintained at any dissolved oxygen concentration above $C_{crit}$.

It should be appreciated that these authors intended this model to be used only as a method for preliminary design (Mavituna and Sinclair, 1985b). Thus, $x_{crit}$ is interpreted as a target which cannot be exceeded and, in practice, oxygen limitation will probably occur below this value. The mechanism for limiting the biomass concentration will be the concentration of the limiting substrate in the medium which, for batch culture, may be determined from the equation:

$$S_R = x_{crit}/Y$$

where $S_R$ is the initial limiting substrate concentration and $Y$ is the yield factor and it is assumed that the limiting substrate is exhausted on entry into the stationary phase.

The technique may also be applied to continuous and fed-batch culture but it must be appreciated that $Q_{O_2}$ is affected by specific growth rate and the relevant $Q_{O_2}$ value for the growth rate employed would have to be utilized in the calculations. The method should be very useful for the initial design of unicellular bacterial or yeast fermentations where biomass has no effect on $K_L a$. However, in viscous fermentations the biomass concentration influences the $K_L a$ considerably, as discussed in a previous section. Thus, the $K_L a$ will decline with increasing biomass concentration which makes the application of the technique more problematical.

### Controlling the specific oxygen uptake rate

Specific oxygen-uptake rate is directly proportional to specific growth rate so that, as $\mu$ increases, so does $Q_{O_2}$. Thus, $Q_{O_2}$ may be controlled by the dilution rate in continuous culture. Although very few commercial fermentations are operated in continuous culture, fed-batch culture is widely used in industrial fermentations and provides an excellent tool for the control of oxygen demand. The kinetics and applications of fed-batch culture are discussed in Chapter 2. The most common way in which the technique is applied to control oxygen demand is to link the nutrient addition system to a feed-back control loop using a dissolved oxygen electrode as the sensing element (see Chapter 8). If the dissolved oxygen concentration declines below the set point then the feed rate is reduced and when the dissolved oxygen concentration rises above the set point the feed rate may be increased. A pH electrode may also be used as a sensing unit in a fed-batch control loop for the control of oxygen demand — oxygen limitation being detected by the development of acidic conditions. These techniques are particularly important in the growth-stage of a secondary metabolite mycelial fermentation prior to product production when the highest growth rate commensurate with the oxygen transfer rate of the fermenter is required. A full discussion of the operation of fed-batch systems is given in Chapter 2.

### SCALE-UP AND SCALE-DOWN

Scale-up means increasing the scale of a fermentation, for example from the laboratory scale to the pilot plant scale or from the pilot plant scale to the production scale. Increase in scale means an increase in volume and the problems of process scale-up are due to the different ways in which process parameters are affected by the size of the unit. It is the task of the fermentation technologist to increase the scale of a fermentation without a decrease in yield or, if a yield reduction occurs, to identify the factor which gives rise to the decrease and to rectify it. The major factors involved in scale-up are:

(i) *Inoculum development*. An increase in scale may mean that extra stages have to be incorporated into the inoculum development programme. This aspect is considered in Chapter 6.

(ii) *Sterilization*. Sterilization is a scale dependent factor because the number of contaminating micro-organisms in a fermenter must be reduced to the same absolute number regardless of scale. Thus, when the scale of a process is increased the sterilization regime must be adjusted accordingly, which may result in a change in the quality of the medium after sterilization. This aspect is considered in detail in Chapter 5.

(iii) *Environmental parameters*. The increase in scale may result in a changed environment for the organism. These environmental parameters may be summarized as follows:

(a) nutrient availability,
(b) pH,

(c)  temperature,
(d)  dissolved oxygen concentration,
(e)  shear conditions,
(f)  dissolved carbon dioxide concentration,
(g)  foam production.

All the above parameters are affected by agitation and aeration, either in terms of bulk mixing or the provision of oxygen. Points a, b, c and e are related to bulk mixing whilst d, e, f and g are related to air flow and oxygen transfer. Thus, agitation and aeration tends to dominate the scale-up literature. However, it should always be remembered that inoculum development and sterilization difficulties may be the reason for a decrease in yield when a process is scaled up and that achieving the correct aeration/agitation regime is not the only problem to be addressed.

### Scale-up of aeration/agitation regimes in stirred tank reactors

From the list of environmental parameters affected by aeration and agitation it will be appreciated that it is extremely unlikely that the conditions of the small-scale fermentation will be replicated precisely on the large scale. Thus, the most important criteria for a particular fermentation must be established and the scale-up based on reproducing those characteristics. The problem of aeration/agitation scale-up has been extremely well illustrated by Fox (1978) in his description of the 'scale-up window'. The scale-up window represents the boundaries imposed by the environmental parameters and cost on the aeration/agitation regime and is shown in Fig. 9.21. Suitable conditions of mixing and oxygen transfer can be obtained with a range of aeration/agitation combinations. The two axes of Fig. 9.21 are agitation and aeration and the zone within the hexagon represents suitable aeration/agitation regimes. The boundary of the hexagon is defined by the limits of oxygen supply, carbon dioxide accumulation, shear damage to the cells, cost, foam formation and bulk mixing. For example, the agitation rate must fall between a minimum and maximum value — mixing is inadequate below the minimum level and shear damage to the cells is too great above the maximum value. The limits for aeration are determined at the minimum end by oxygen limitation and carbon dioxide accumulation and at the maximum end by foam formation. The shape of the window will depend on the fermentation — for exam-

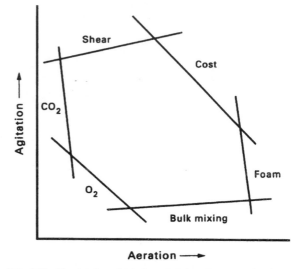

Fig. 9.21. The 'scale-up' window defining the operating boundaries for aeration and agitation in the scale-up of a fermentation. After Fox (1978) reproduced from Lilly (1983).

ple, the supply of oxygen would be irrelevant in an anaerobic fermentation, whereas the limitation due to shear would be of major importance in the scale-up of animal cell fermentations.

The solution of the scale-up problem is three-fold:

(i)   The identification of the principal environmental domain affected by aeration and agitation in the fermentation, e.g. oxygen concentration, shear, bulk mixing.
(ii)  The identification of a process variable (or variables) which affects the identified environmental domain.
(iii) The calculation of the value of the process variable to be used on the large scale which will result in the replication of the same environmental conditions on both scales.

The process variables which affect mixing and mass transfer are summarized in Table 9.6 (Oldshue, 1985; Scragg, 1991). Thus, if dissolved oxygen concentration is perceived as the over-riding environmental condition then power consumption per unit volume and volumetric air flow rate per unit volume should be maintained constant on scale-up. However, as a result, the other parameters will not be the same in the larger scale and, therefore, neither will the environmental factors which they influence. This phenomenon is well illustrated by Oldshue's example summarized in Table 9.7 where a 125 fold increase in scale is represented. If power

TABLE 9.6. *The effect of process variables on mass transfer or mixing characteristics*

| Process variable | Mass transfer or mixing characteristic affected |
|---|---|
| Power consumption per unit volume | Oxygen-transfer rate |
| Volumetric air flow rate | Oxygen-transfer rate |
| Impeller tip speed | Shear rate |
| Pumping rate | Mixing time |
| Reynolds number (see previous section) | Heat transfer |

consumption per unit volume is kept constant then impeller tip speed (i.e. shear) increases and flow min$^{-1}$ vol$^{-1}$ (i.e. mixing) decreases. If mixing is kept constant, an enormous (and totally uneconomic) increase in power is required and shear increases 5 fold. If impeller tip speed (shear) is kept constant then power consumption (hence, $K_L a$) and mixing decrease. This analysis indicates that it is economically impossible to maintain the same degree of mixing on scale-up and, therefore, a decrease in yield may be due to mixing anomalies.

The most important environmental domains affected by aeration and agitation for the majority of fermentations are oxygen concentration and shear. Thus, the most widely used scale-up criteria are the maintenance of a constant $K_L a$ or constant shear conditions. Constant shear may be achieved by scaling up on the basis of constant impeller tip speed. Constant $K_L a$ may be achieved on the basis of constant power consumption per unit volume and constant volumetric air-flow rate. The operating variable dictating constant power consumption in geometrically similar vessels is the agitator speed. The agitator speed on the large scale is then calculated from the correlations between $K_L a$ and power consumption and between power consumption and operating variables. An example of this approach is given in the previous section describing the effects of operating variables on power consumption.

Hubbard (1987) and Hubbard *et al.* (1988) summarized the procedure for scaling up both Newtonian and non-Newtonian fermentations and proposed two methods to determine the large scale conditions:

*Method 1*

(i) Determine the volumetric air flow rate ($Q$) on the large scale based on maintaining $Q/V$ constant ($V$ = working volume of the fermenter).

(ii) Calculate the agitator speed that will give the same $K_L a$ on the large scale; this is achieved using the correlations between power consumption and $N$ and between $K_L a$ and power consumption.

*Method 2*

(i) Calculate the agitator speed keeping the impeller tip speed constant, $\pi N D_i$.

(ii) Calculate $Q$ from power correlations and $K_L a$ correlations.

The accuracy of these scale-up techniques is only as good as the power and $K_L a$ correlations, so it is worth expending some considerable time to test the validity of potential correlations for the fermentation in question.

TABLE 9.7. *The effect of the choice of scale-up criteria on operating conditions in the scaled-up vessel. Based on scale-up from 80 dm$^3$ to 10$^4$ dm$^3$ (Based on Oldshue, 1985)*

| Criterion used in scale-up | Effect on the operating conditions on the large scale (Large scale value/Small scale value) | | | |
|---|---|---|---|---|
| | $P$ | $P/V$ | Flow min$^{-1}$ vol$^{-1}$ | $ND_i$ |
| $P/V$ | 125.0 | 1.0 | 0.34 | 1.7 |
| Flow min$^{-1}$ vol$^{-1}$ | 3125.0 | 25.0 | 1.0 | 5.0 |
| $ND_i$ (Impeller tip speed) | 25.0 | 0.2 | 0.2 | 1.0 |
| Reynolds number | 0.2 | 0.0016 | 0.04 | 0.2 |

### The scale-up of air-lift reactors

Bubble columns and air-lift vessels tend to be scaled-up on the basis of geometric similarity and constant gas velocity (Scragg, 1991). Under these conditions the $K_L a$ and shear rate in the two scales will be similar. The major difference will be the height of the vessels resulting in increased pressure at the base of the larger vessel. This would result in higher oxygen and carbon dioxide solubility which would give a higher $K_L a$ but might result in carbon dioxide inhibition. The other problem in the scale-up of air-lift systems is that the organism is exposed to extremes of oxygen levels in the riser and downcomer and the effects of these conditions should be investigated on the laboratory scale.

### Scale-down methods

Scale-down is the situation where laboratory- or pilot-scale experiments are conducted under conditions which mimic the industrial-scale conditions. This approach is important in both the development of a new product and the improvement of an existing full-scale fermentation. The procedure has been reviewed by Jem (1989). Frequently, conditions achievable on a laboratory scale are impractical on an industrial scale, which means that if inappropriate conditions have been used in the laboratory unrealistic yield objectives may be set for the scaled-up process. The aspects to consider in the design of laboratory- or pilot-plant experiments in the context of scale-down may be summarized as follows:

(i) *Medium design.* Media relevant to the industrial situation should be used in development experiments.

(ii) *Medium sterilization.* If the medium is to be batch sterilized on the large scale its exposure time at a high temperature will be much greater than that experienced in the laboratory or pilot plant. Thus, the sterilization times on the smaller scales should be increased to mimic the industrial situation. Alternatively, medium sterilized in the production fermenter may be used in the laboratory and pilot plant. This highlights the advantage of continuous sterilization where little loss of medium quality occurs. Furthermore, the same continuous sterilizer may be used for both full-scale and pilot scale vessels.

(iii) *Inoculation procedures.* Due to a range of circumstances, it may not always be possible to inoculate every production fermentation with inoculum in optimum condition. The scale-down approach can be used to predict the consequences of such events by mimicking these situations in the laboratory, for example by storing inoculum or using inocula of different ages.

(iv) *Number of generations.* An industrial scale fermentation requires a greater number of generations than does a laboratory one; this may place more severe stability criteria on the process strain than may have been appreciated on the small scale. The industrial situation may be modelled in the laboratory by using serial sub-culture to ensure that the strain is sufficiently stable. This approach is particularly pertinent in the development of recombinant fermentations.

(v) *Mixing.* As indicated in the previous section it is almost inevitable that the degree of mixing will decrease with an increase in scale. Thus, it is possible to model inadequate mixing in the laboratory by subjecting the organism to pulse medium feeds or fluctuating process conditions such as oxygen concentration, pH and temperature. Such scaled-down experiments then allow predictions to be made about the suitability of new strains for industrial exploitation.

(vi) *Oxygen transfer rate.* Far higher oxygen transfer rates can be achieved in laboratory fermenters than in industrial-scale ones. Thus, unrealistic demands may be made of a fermentation plant if the development work has been done at very high oxygen-transfer rates. Therefore, the laboratory and pilot fermenters should reflect the oxygen transfer rates achievable in the full-scale fermenters.

The adoption of these simple approaches to small scale experimentation can prevent many scale-up problems before they even occur!

### REFERENCES

AIBA, S., HUMPHREY, A. E. and MILLIS, N. (1973) *Biochemical Engineering.* Academic Press, London.

BANKS, G. T. (1977) Aeration of moulds and streptomycete culture fluids. *Topics in Enzyme and Fermentation Biotechnology,* Vol. 1, pp. 72–110 (Editor Wiseman, A.). Ellis Horwood, Chichester.

BANKS, G. T. (1979) Scale-up of fermentation processes. *Topics in Enzyme and Fermentation Biotechnology*, Vol. 3, pp. 170–267 (Editor Wiseman, A.). Ellis Horwood, Chichester.

BARTHOLOMEW, W. H. (1960) Scale-up of submerged fermentations. *Adv. App. Micro.* **2**, 289–300.

BARTHOLOMEW, W. H., KARROW, E. O., SFAT, M. R. and WILHELM, R. H. (1950) Oxygen transfer and agitation in submerged fermentations. Mass transfer of oxygen in submerged fermentations of *Streptomyces griseus*. *Ind. Eng. Chem.* **42** (9), 1801–1809.

BELL, G. H. and GALLO, M. (1971) Effect of impurities on oxygen transfer. *Process Biochem.* **6** (4), 33–35.

BELMAR-BEINY, M. T. and THOMAS, C. R. (1990) Morphology and clavulanic acid production of *Streptomyces clavuligerus*: effect of stirred speed in batch fermentation. *Biotech. Bioeng.* **37** (5), 456–462.

BERKMAN-DIK, T., OZILGEN, M. and BOZOGLU, T. F. (1992) Salt, EDTA, and pH effects on rheological behavior of mold suspensions. *Enzyme Microb. Technol.* **14**, 944–948.

BIESECKER, B. O. (1972) *Begasen von Flussigkeiten mit Ruhrern*. VDI, Forschungsheft.

BUCKLAND, B. C. (1993) Mevinolin production. Paper presented at the *Soc. Gen. Microbiol. 124th Meeting*, University of Kent, Canterbury, January 1993.

BUCKLAND, B. C., GBEWONYO, K., JAIN, D., GLAZOMITSKY, K., HUNT, G. and DREW, S. W. (1988) Oxygen transfer efficiency of hydrofoil impellers in both 800L and 1900L fermenters. In *Proceedings of the 2nd. International Conference on Bioreactor Fluid Dynamics*, pp. 1–16 (Editor, King, R.). Elsevier, London.

BUCKLAND, B. C., GBEWONYO, K., HALLADA, T., KAPLAN, L. and MASUREKAR, P. (1989) In *Novel Microbial Products in Medicine and Agriculture*, pp. 161–169 (Editor, Demain, A. L.). Elsevier, Amsterdam.

CARILLI, A., CHAIN, E. B., GAULANDI, G. and MORISI, G. (1961) Aeration studies III. Continuous measurements of dissolved oxygen in fermentation broths — Rheology and mass transfer. *Biotech. Bioeng.* **18**, 745–790.

CASSON, N. (1959) *Rheology of Disperse Systems*. Pergamon Press, Oxford.

CHAPMAN, C. M., NIENOW, A. W., COOKE, M. and MIDDLETON, J. C. (1983) Particle–gas–liquid mixing in stirred vessels. Part 3. *Chem. Eng. Res. Des.* **61** (3), 167–181.

CHARLES, M. (1978) Technical aspects of the rheological properties of microbial cultures. *Adv. Biochem. Eng.* **8**, 1–62.

CHEN, N. Y. (1990) The design of airlift fermenters for use in biotechnology. *Biotechnol. Gen. Eng. Rev.* **8**, 379–396.

COOKE, M., MIDDLETON, J. C. and BUSH, J. R. (1988) Mixing and mass transfer in filamentous fermentations. In *Proceedings of the 2nd International Conference on Bioreactor Fluid Dynamics*, pp. 37–64 (Editor, King, R.). Elsevier, London.

COONEY, C. L. (1979) Conversion yields in penicillin production: Theory versus practice. *Process Biochem.* **14** (5), 31–33.

COOPER, C. M., FERNSTROM, G. A. and MILLER, S. A. (1944) Performance of agitated gas–liquid contacters. *Ind. Eng. Chem.* **36**, 504–509.

DARLINGTON, W. A. (1964) Aerobic hydrocarbon fermentation — A practical evaluation. *Biotech. Bioeng.* **6** (2), 241–242.

DECKWER, W. D., NGUYEN-TIEN, K., SCHUMPE, A. and SERPEMEN, Y. (1982) Oxygen mass transfer into aerated CMC solutions in a bubble column. *Biotech. Bioeng.* **24**, 461–481.

DEINDOERFER, F. H. and GADEN, E. L. (1955) Effects of liquid physical properties on oxygen transfer in penicillin fermentation. *Appl. Micro.* **3**, 253–257.

DEINDOERFER, F. H. and WEST, J. M. (1960) Rheological examination of some fermentation broths. *J. Microbiol. Biochem. Technol. Eng.* **2**, 165–175.

DION, W. M., CASILLI, A., SERMONTI, G. and CHAIN, E. B. (1954) The effect of mechanical agitation on the morphology of *Penicillium chrysogenum* Thom in stirred fermenters. *Rend. 1st Super Sanita* **17**, 187–205.

FEIJEN, J., HEIJNEN, J. J. and VAN'T RIET, K. (1987) Gas hold-up and flooding in large-scale fermenters. In *Proceedings of Symposium on mixing and dispersion processes. Inst. Chem. Eng. and KIVI / NIRIA*. Delft Technical University.

FOX, R. I. (1978) The applicability of published scale-up criteria to commercial fermentation processes. *Proc. 1st Eur. Cong. Biotechnol.*, Part 1, pp. 80–83.

GALLINDO, E. and NIENOW, A. W. (1992) Mixing of highly viscous simulated fermentation broths with the Lightnin' A-315 impeller. *Biotechnol. Prog.* **8**, 233–239.

GBEWONYO, K., DIMASI, D. and BUCKLAND, B. C. (1986) The use of hydrofoil impellers to improve oxygen transfer efficiency in viscous mycelial fermentations. In *International Conference on Biorector Fluid Dynamics*, pp. 281–299. BHRA, Cranfield, U.K.

GOLDBERG, I., ROCK, J. S., BEN-BASSAT, A. and MATELES, R. I. (1976) Bacterial yields on methanol, methylamine, formaldehyde and formate. *Biotech. Bioeng.* **18**, 1657–1668.

HALL, M. J., DICKINSON, S. D., PRITCHARD, R. and EVANS, J. I. (1973) Foams and foam control in fermentation processes. *Prog. Ind. Micro.* **12**, 171–234.

HEIJNEN, J. J., RIET, K. W. and WOLTHUIS, A. J. (1980) Influence of very small bubbles on the dynamic $K_L a$ measurement in viscous gas–liquid systems. *Biotech. Bioeng.* **22**, 1945–1956.

HEIJNEN, J. J. and VAN'T RIET, K. (1984) Mass transfer, mixing and heat transfer phenomena in low viscosity bubble column reactors. *Chem. Eng. J.* **28**, B21.

HEINEKEN, F. G. (1970) Use of fast-response dissolved oxygen probes for oxygen transfer studies. *Biotech. Bioeng.* **12**, 145–154.

HEINEKEN, F. G. (1971) Oxygen mass transfer and oxygen respiration rate measurements utilising fast response oxygen electrodes. *Biotech. Bioeng.* **13**, 599–618.

HIROSE, Y. and SHIBAI, H. (1980) Effect of oxygen on amino acids fermentation. *Advances in Biotechnology*, Vol. 1, pp.

329–333 (Editors Moo-Young, M., Robinson, C. W. and Vezina, C.). Pergamon Press, Toronto.

HOBBS, G., FRAZER, C. M., GARDNER, D. C. J., CULLUM, J. A. and OLIVER, S. G. (1989) Dispersed growth of *Streptomyces* in liquid culture. *Appl. Microbiol. Biotechnol.* **31** (3), 272–277.

HUBBARD, D. W. (1987) Scale-up strategies for bioreactors containing non-Newtonian broths. *Ann. N.Y. Acad. Sci.*, **506**, 600–607.

HUBBARD, D. W., HARRIS, L.R. and WIERENGA, M. K. (1988) Scale-up for polysaccharide fermentation. *Chem. Eng. Prog.* **84** (8), 55–61.

HUGHMARK, G. A. (1980) Power requirements and interfacial area in gas–liquid turbine agitated systems. *Ind. Eng. Chem. Proc. Des. Dev.* **19**, 638–645.

JEM, K. J. (1989) Scale-down techniques in fermentation. *Biopharmacology* **2**, 30–39.

JOHNSON, M. J. (1964) Utilisation of hydrocarbons by microorganisms. *Chem. Ind.* **36**, 1532–1537.

KOBAYASHI, T., VAN DEDEM, G. and MOO-YOUNG, M. (1973) Oxygen transfer into mycelial pellets. *Biotech. Bioeng.* **15**, 27–45.

KUENZI, M. T. (1978) Process design and control in antibiotic fermentation. In *Antibiotics and Other Secondary Metabolites, Biosynthesis and Production, FEMS Symp.* Vol. **5**, pp. 39–56 (Editors Hutter, R., Leisinger, T., Neusch and Wehrli, W.). Academic Press, London.

LEGRYS, G. A. and SOLOMONS, G. L. (1977) US patent application 23128.

LILLY, M. D. (1983) Problems in process scale-up. In *Bioactive Microbial Products 2. Development and Production*, pp. 79–90 (Editors Nisbet, L. J. and Winstanley, D. J.). Academic Press, London.

LILLY, M. D, ISON, A. and SHAMLOU, P. A. (1992) The influence of the physical environment in fermenters on antibiotic production by micro-organisms. In *Harnessing Biotechnology in the 21st Century* (Editors Ladisch, M. R. and Bose, A.). American Chemical Society, Washington, DC.

MATELES, R. I. (1971) Calculation of the oxygen required for cell production. *Biotech. Bioeng.* **13** (4), 581–582.

MATELES, R. I. (1979) The physiology of single cell protein (SCP) production. Soc. Gen. Microbiology Symposium, **29**, *Microbial Technology: Current State, Future Prospects*, pp. 29–52 (Editors, Bull. A. T., Ellwood, D. C. and Ratledge, C.). Cambridge University Press, Cambridge.

MAVITUNA, F. and SINCLAIR, C. G. (1985a) A graphical method for the determination of critical biomass concentration for non-oxygen limited growth. *Biotechnol. Lett.* **7**, 69–74.

MAVITUNA, F. and SINCLAIR, C. G. (1985b) Reply to a comment on 'A graphical method for the determination of critical biomass concentration for non-oxygen limited growth'. *Biotechnol. Lett.* **7**, 813–814.

METZ, B. and KOSSEN, N. W. F. (1977) Pellet growth of moulds. *Biotech. Bioeng.* **19**, 781–799.

METZ, B., KOSSEN, N. W. F. and VAN SUIJDAM, J. C. (1979) Rheology of mould suspensions. *Adv. Biochem. Eng.* **11**, 103–156.

METZNER, A. B. and OTTO, R. E. (1957) Agitation of non-Newtonian fluids. *A.I.Ch.E.J.* **3** (1), 3–10.

METZNER, A. B., FEEHS, R. H., RAMOS, H. L., OTTO, R. E. and TOOTHILL, J. D. (1961) Agitation of viscous Newtonian and non-Newtonian fluids. *A.I.Ch.E.J.* **7**, 3–9.

MICHELL, B. J. and MILLER, S. A. (1962) Power requirements of gas–liquid agitated systems. *A.I.Ch.E.J.* **8**, 262–266.

NIENOW, A. W. and ELSON, T. P. (1988) Aspects of mixing in rheologically complex liquids. *Chem. Eng. Res. Des.*, **66**, 5–15.

NIENOW, A. W., WISDOM, D. J. and MIDDLETON, J. C. (1977) The effect of scale and geometry on flooding, recirculation and power in gassed stirred vessels. Paper F1, *2nd. Eur. Conf. on Mixing*, March 1977, Cambridge.

OLDSHUE, J. Y. (1985) Current trends in mixer scale-up techniques. In *Mixing of Liquids by Mechanical Agitation* pp. 309–341 (Editors Ulbrecht, J. and Patterson, G. K.). Gordon and Breach, New York.

OLSVIK, E. S. and KRISTIANSEN, B. (1992) Influence of oxygen tension, biomass concentration and specific growth rate on the rheological properties of a filamentous fermentation broth. *Biotech. Bioeng.* **40**, 1293–1299.

OOSTERHUIS, N. M. G and KOERTS, K. (1987) Method and reactor vessel for the fermentation production of polysaccharides in particular, xanthan. *Eur. Pat. Appl.* EP 249,288.

RIGHELATO, R. C. (1979) The kinetics of mycelial growth. In *Fungal Walls and Hyphal Growth*, pp. 1385–1402 (Editors Burnett, J. H. and Trinci. A. P. J.). Cambridge University Press, Cambridge.

RIGHELATO, R. C., TRINCI, A. P. J., PIRT, S. J. and PEAT, A. (1968) Influence of maintenance energy and growth rate on the metabolic activity, morphology and conidiation of *Penicillium chrysogenum*. *J. Gen. Micro.* **50** (1), 394–412.

RIVIERE, J. (1977) *Industrial Application of Microbiology* (translated and edited by Moss, M. O. and Smith, J. E.). Surrey University Press, Guildford.

ROELS, J. A., VAN DENBERG, J. and VONKEN, R. M. (1974) The rheology of mycelial broths. *Biotech. Bioeng.* **16**, 181–208.

RUSHTON, J. H., COSTICH, E. W. and EVERETT, H. J. (1950) Power characteristics of mixing impellers. *Chem. Eng. Prog.* **46**, 395–404.

SATO, K (1961) Rheological studies on some fermentation broths. (IV) Effect of dilution rate on rheological properties of fermentation broth. *J. Ferment. Technol.* **39**, 517–520.

SCHUGERL, K., WITTLER, R. and LORENZ, T. (1988) The use of moulds in pellet form. *Trends Biotechnol.* **1** (4), 120–122.

SCHULZE, K. L. and LIPE, R. S. (1964) Relationship between substrate concentration, growth rate and respiration rate of *Escherichia coli* in continuous culture. *Archiv. Mikrobiol.*, **48**, 1–20.

SCRAGG, A. H. 1991 *Bioreactors in Biotechnology. A Practical Approach.* Ellis Horwood, Chichester.

STEEL, R. and MAXON, W. D. (1966) Studies with a multiple-rod mixing impeller. *Biotech. Bioeng.* **8**, 109–116.

SUTHERLAND, I. W. and ELLWOOD, D. C. (1979) Microbial exopolysaccharides — industrial polymers of current and future potential. In Soc. Gen. Microbiology Symposium, **29**, *Microbial Technology: Current State, Future Prospects*, pp. 107–150 (Editors, Bull, A. T., Ellwood, D. C. and Ratledge, C.). Cambridge University Press, Cambridge.

TAGUCHI, H. (1971) The nature of fermentation fluids. *Adv. Biochem. Eng.* **1**, 1–30.

TAGUCHI, H and HUMPHREY, A. E. (1966) Dynamic measurement of the volumetric oxygen transfer coefficient in fermentation systems. *J. Ferm. Technol.* **44** (12), 881–889.

TAGUCHI, H, IMANAKA, T., TERAMOTO, S., TAKATSU, M. and SATO, M. (1968) Scale-up of glucamylase fermentation by *Endomyces* sp. *J. Ferm. Technol.* **46** (10), 823–828.

TRINCI, A. P. J. (1983) Effect of Junlon on the morphology of *Aspergillus niger* and its use in making turbidity measurements of microbial growth. *Trans. Br. Mycol. Soc.* **58**, 467–473.

TUFFILE, C. M. and PINHO, F. (1970) Determination of oxygen transfer coefficients in viscous streptomycete fermentations. *Biotech. Bioeng.* **12**, 849–871.

VAN SUIJDAM, J. C. and METZ, B. (1981) Influence of engineering variables upon morphology of filamentous moulds. *Biotech. Bioeng.* **23**, 111–148.

VAN'T RIET, K. (1979) Review of measuring methods and results in non-viscous gas–liquid mass transfer in stirred vessels. *Ind. Eng. Chem. Process Des. Dev.* **18** (3), 357–360.

VAN'T RIET, K. (1983) Mass transfer in fermentation. *Trends Biotechnol.*, **1** (4), 113–116.

VAN'T RIET, K. and TRAMPER, J. (1991) *Basic Bioreactor Design*. Marcel Dekker, New York.

VAN'T RIET, K. and VAN SONSBERG, (1992) Foaming, mass transfer and mixing: Interrelations in large scale fermentations. In *Harnessing Biotechnology for the 21st Century*, pp. 189–192 (Editors Ladisch, M.R. and Bose, A.). American Chemical Society, Washington, DC.

WANG, D. I. C. and FEWKES, R. C. J. (1977) Effect of operating and geometric parameters on the behaviour of non-Newtonian mycelial antibiotic fermentations. *Dev. Ind. Micro.*, **18**, 39–57.

WANG, D. I. C., COONEY, C. L., DEMAIN, A. L., DUNNILL, P. HUMPHREY, A. E. and LILLY, M. D. (1979) *Fermentation and Enzyme Technology*. Wiley, New York.

WERNAU, W. C. and WILKE, C. R. (1973) New method for evaluation of dissolved oxygen probe response for $K_L a$ determination. *Biotech. Bioeng.* **15**, 571–578.

WESTERTERP, K. R., VAN DIERENDONCK, L. L. and DE KRAA, J. A. (1963) Interfacial areas in agitated gas liquid contactors. *Chem. Eng. Sci.* **18**, 157–169.

WINKLER, M. A. (1990) Problems in fermenter design and operation. In *Chemical Engineering Problems in Biotechnology*, pp. 215–350 (Editor Winkler, M. A.). SCI/Elsevier, London.

WISE, W. S. (1951) The measurement of the aeration of culture media. *J. Gen. Micro.* **5**, 167–177.

WODZINSKI, R. S. and JOHNSON, M. J. (1968) Yield of bacterial cells form hydrocarbons. *Appl. Micro.* **16**, 1886–1891.

ZHOU, W., HOLZHAUER-RIEGER, K., DORS. M. and SCHUGERL, K. (1992) Influence of dissolved oxygen concentration on the biosynthesis of cephalosporin C. *Enzyme Microb. Technol.* **14**, 848–854.

# CHAPTER 10

# The Recovery and Purification of Fermentation Products

## INTRODUCTION

THE EXTRACTION and purification of fermentation products may be difficult and costly. Ideally, one is trying to obtain a high-quality product as quickly as possible at an efficient recovery rate using minimum plant investment operated at minimal costs. Unfortunately, recovery costs of microbial products may vary from as low as 15% to as high as 70% of the total manufacturing costs (Aiba *et al.*, 1973; Swartz, 1979; Pace and Smith, 1981; Atkinson and Sainter, 1982; Datar, 1986). Obviously, the chosen process, and therefore its relative cost, will depend on the specific product. Atkinson and Mavituna (1991) indicate percentage of total costs being 15% for industrial ethanol, 20–30% for bulk penicillin G and up to 70% for enzymes. The high (and sometimes dominant) cost of downstream processing will affect the overall objective in some fermentations.

If a fermentation broth is analysed at the time of harvesting it will be discovered that the specific product may be present at a low concentration in an aqueous solution that contains intact micro-organisms, cell fragments, soluble and insoluble medium components and other metabolic products. The product may also be intracellular, heat labile and easily broken down by contaminating micro-organisms. All these factors tend to increase the difficulties of product recovery. To ensure good recovery or purification, speed of operation may be the overriding factor because of the labile nature of a product. The processing equipment must therefore be of the correct type and also the correct size to ensure that the harvested broth can be processed within a satisfactory time limit.

The choice of recovery process is based on the following criteria:

1. The intracellular or extracellular location of the product.
2. The concentration of the product in the fermentation broth.
3. The physical and chemical properties of the desired product (as an aid to selecting separation procedures).
4. The intended use of the product.
5. The minimal acceptable standard of purity.
6. The magnitude of bio-hazard of the product or broth.
7. The impurities in the fermenter broth.
8. The marketable price for the product.

The main objective of the first stage for the recovery of an extracellular product is the removal of large solid particles and microbial cells usually by centrifugation or filtration (Fig. 10.1). In the next stage, the broth is fractionated or extracted into major fractions using ultrafiltration, reverse osmosis, adsorption/ion-exchange/gel filtration or affinity chromatography, liquid–liquid extraction, two phase aqueous extraction or precipitation. Afterwards, the product-containing fraction is purified by fractional precipitation, further more precise chromatographic techniques and crystallization to obtain a product which is highly concentrated and essentially free from impurities. Other products are isolated using modifications of this flow-stream.

Attempts to simplify this outline extraction procedure for antibiotic recovery using 'whole broth' processing have met with limited success. The technique of

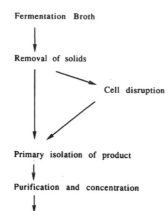

Fermentation Broth

Removal of solids

Cell disruption

Primary isolation of product

Purification and concentration

Final product isolation

FIG. 10.1. Stages in the recovery of product from a harvested fermentation broth.

1.  Harvest broth from fermenter

2.  Chill to 5-10°C

3.  Filter off *P. chrysogenum* mycelium using rotary vacuum filter

4.  Acidify filtrate to pH 2.0-2.5 with $H_2SO_4$

5.  Extract penicillin from aqueous filtrate into butyl acetate in a centrifugal counter-current extractor (treat/dispose aqueous phase)

6.  Extract penicillin from butyl acetate into aqueous buffer (pH 7.0) in a centrifugal counter-current extractor (recover and recycle butyl acetate)

7.  Acidify the aqueous fraction to pH 2.0-2.5 with $H_2SO_4$ and re-extract penicillin into butyl acetate as in stage 5

8.  Add potassium acetate to the organic extract in a crystallization tank to crystallize the penicillin as the potassium salt

9.  Recover crystals in a filter centrifuge (recover and recycle butyl acetate)

10. Further processing of penicillin salt

FIG. 10.2. Recovery and partial purification of penicillin G.

'whole broth' processing involves initial removal of large particles, which is then followed by passage of the broth (including cells) through, for example, well mixed ion-exchange columns or counter-current liquid–liquid extraction units to extract the product directly. This topic will be discussed in more detail in a later section of this chapter.

It may be possible to modify the handling characteristics of the broth so that it can be handled faster with simpler equipment making use of a number of techniques:

1.  Selection of a micro-organism which does not produce pigments or undesirable metabolites.
2.  Modification of the fermentation conditions to reduce the production of undesirable metabolites.
3.  Precise timing of harvesting.
4.  pH control after harvesting.
5.  Temperature treatment after harvesting.
6.  Addition of flocculating agents.
7.  Use of enzymes to attack cell walls.

It must be remembered that the fermentation and product recovery are integral parts of an overall process. Because of the interactions between the two, neither stage should be developed independently, as this might result in problems and unnecessary expense. Darbyshire (1981) has considered this problem with reference to enzyme recovery. The parameters to consider included time of harvest, pigment production, ionic strength and culture medium constituents. Large volumes of supernatants containing extracellular enzymes need immediate processing while harvesting times and enzyme yields might not be predictable. This can make recovery programmes difficult to plan. Pigment production might make some recovery procedures difficult, when the pigment binds to the same resin as the enzyme. Changes in fermentation conditions may reduce pigment formation. Certain antifoams remain in the supernatant and affect ultrafiltration or ion-exchange resins used in recovery stages. Trials may be needed to find the most suitable antifoam (see also Chapter 4). The ionic strength of the production medium may be too high, resulting in the harvested supernatant needing dilution with demineralized water before it can be processed. Such a negative procedure should be avoided if possible by unified research and development programmes. Media formulation is dominated by production requirements, but the protein content of complex media should be critically examined in view of subsequent enzyme recovery. This view is also shared by Topiwala and Khosrovi (1978), when considering water recycle in biomass production. They stated that the interaction between the different unit operations in a recycle process made it imperative that commercial plant design and operation should be viewed in an integrated fashion.

Flow sheets for recovery of penicillin, cephamycin C,

1. Fermenter broth containing cephamycin C
   ↓
2. Adjust pH to 2.5
   ↓
3. Conventional filtration
   ↓ Waste filter cake
4. Cation exchange (sulfonic acid resin)
   ↓
5. Pyridine elution
   ↓
6. Concentration by evaporation (remove pyridine)
   ↓
7. Adjust pH to 5 - 7
   ↓
8. Anion exchange (tertiary amine)
   ↓
9. Alkanoic acid wash
   ↓
10. Pyridine or phosphate buffer solution

FIG. 10.3. Purification of cephamycin C: sequential ion exchange process (Omstead et al., 1985).

1. Harvested broth
   ↓
2. Filter off A. niger mycelium using a rotary vacuum filter
   ↓
3. Add Ca(OH)$_2$ to filtrate until pH 5.8
   ↓
4. Calcium citrate
   ↓
5. Add H$_2$SO$_4$ while at 60°C
   ↓
6. Filter on rotary vacuum filter to recover CaSO$_4$
   ↓
7. Activated charcoal to decolourise
   ↓
8. Cation and anion exchange resins
   ↓
9. Evaporate to point of crystallization at 36°C
   ↓
10. Crystals of citric monohydrate separated in continuous centrifuges
   ↓
11. Driers at 50-60°C

FIG. 10.4. Recovery and purification of citric acid (Sodesk et al., 1981).

citric acid and micrococcal nuclease are given in Figs 10.2, 10.3, 10.4 and 10.5, to illustrate the range of techniques used in microbiological recovery processes. A series of comprehensive flow sheets for alcohols, organic acids, antibiotics, carotenoids, polysaccharides, intra- and extra-cellular enzymes, single-cell proteins and vitamins have been produced by Atkinson and Mavituna (1991). Other reviews on separation and purification are available for penicillin (Swartz, 1979), amino acids (Samejima, 1972), enzymes (Aunstrup, 1979; Darbyshire, 1981), single-cell protein (Hamer, 1979) and polysaccharides (Pace and Righelato, 1980; Smith and Pace, 1982). In the selection of processes for the recovery of biological products it should always be understood that recovery and production are interlinked, and that good recovery starts in the fermentation by the selection of, amongst other factors, the correct media and time of harvesting.

The recovery and purification of many compounds may be achieved by a number of alternative routes. The decision to follow a particular route involves comparing the following factors to determine the most appropriate under a given set of circumstances:

Capital costs.
Processing costs.
Throughput requirements.
Yield potential.
Product quality.
Technical expertise available.

1. Supernatant; 400 dm$^3$ pH 8.8
   ↓
2. Dilute; with 400 dm$^3$ demineralised water
   ↓
3. Acidify; with glacial acetic acid to pH 5.2
   ↓
4. Adsorption; add SP-sephadex C25 (~750g), stir 1hr, settling 2hrs
   ↓
5. Collect resin; pack into an adjustable column, remove unadsorbed protein by passing 0.3M ammonium acetate pH 6.0
   ↓
6. Elution; nuclease eluted with 2M ammonium sulphate
   ↓
7. Dialysis; overnight against demineralised water
   ↓
8. Concentration; CH$_3$ hollow fibre unit (Amicon) mol. wt. cut off 10,000
   ↓
9. Centrifugation; 10,000g for 30 minutes
   ↓
10. Gel filtration; Sephadex G75, 0.01% acetic acid plus 0.1N ammonium acetate
   ↓
11. Freeze drying

FIG. 10.5. Purification of micrococcal nuclease (Darbyshire, 1981).

Conformance to regulatory requirements.
Waste treatment needs.
Continuous or batch processing.
Automation.
Personnel health and safety (Wildfeuer, 1985).

The major problem currently faced in product recovery is the large-scale purification of biologically active molecules. For a process to be economically viable large-scale production is required, and therefore large-scale separation, recovery, and purification. This then requires the transfer of small-scale preparative/analytical technologies (e.g. chromatographic techniques) to the production scale whilst maintaining efficiency of the process, bio-activity of the product and purity of the product so that it conforms with safety legislation and regulatory requirements. Developments in this field, and remaining areas for development are documented by Pyle (1990).

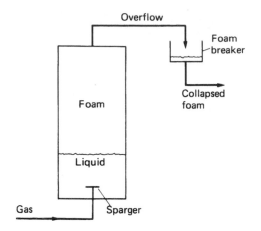

FIG. 10.6. Schematic flow diagram for foam fractionation (Wang and Sinskey, 1970).

## REMOVAL OF MICROBIAL CELLS AND OTHER SOLID MATTER

Microbial cells and other insoluble materials are normally separated from the harvested broth by filtration or centrifugation. Because of the small size of many microbial cells it will be necessary to consider the use of filter aids to improve filtration rates, while heat and flocculation treatments are employed as techniques for increasing sedimentation rates in centrifugation. The methods of cell and cell debris separation described in the following sections have been practised for many years. Bowden *et al.* (1987) review some potential developments in cell recovery. These include the use of electrophoresis and dielectrophoresis to exploit the charged properties of microbial cells, ultrasonic treatment to improve flocculation characteristics and magnetic separations. All these techniques suffer from high cost and scale-up difficulties and currently are not appropriate technologies. Of more current interest is the use of two-phase liquid extraction. Though still most appropriately used for separation of selected soluble components, it is easy to scale up and uses conditions which are gentle on the product.

## FOAM SEPARATION

Foam separation depends on using methods which exploit differences in surface activity of materials. The material may be whole cells or molecular such as a protein or colloidal, and is selectively adsorbed or attached to the surface of gas bubbles rising through a liquid, to be concentrated or separated and finally removed by skimming (Fig. 10.6). It may be possible to make some materials surface active by the application of surfactants such as long-chain fatty acids, amines and quaternary ammonium compounds. Materials made surface active and collected are termed colligends whereas the surfactants are termed collectors. When developing this method of separation, the important variables which may need experimental investigation are pH, air-flow rates, surfactants and colligend–collector ratios.

Rubin *et al.* (1966) investigated foam separation of *E. coli* starting with an initial cell concentration of $7.2 \times 10^8$ cells $cm^{-3}$. Using lauric acid, stearyl amine or *t*-octyl amine as surfactants, it was shown that up to 90% of the cells were removed in 1 minute and 99% in 10 minutes. The technique also proved successful with *Chlorella* sp. and *Chlamydomonas* sp. In other work with *E. coli*, Grieves and Wang (1966) were able to achieve cell enrichment ratios of between 10 and $1 \times 10^6$ using ethyl-hexadecyl-dimethyl ammonium bromide.

## PRECIPITATION

Precipitation may be conducted at various stages of the product recovery process. It is a particularly useful process in that it allows enrichment and concentration in one step, thereby reducing the volume of material for further processing.

It is possible to obtain some products (or to remove certain impurities) directly from the broth by precipitation, or to use the technique after a crude cell lysate has been obtained.

Typical agents used in precipitation render the compound of interest insoluble, and these include:

(a) Acids and bases to change the pH of a solution until the isoelectric point of the compound is reached and pH equals pI, when there is then no overall charge on the molecule and its solubility is decreased.

(b) Salts such as ammonium and sodium sulphate are used for the recovery and fractionation of proteins. The salt removes water from the surface of the protein revealing hydrophobic patches which come together causing the protein to precipitate. The most hydrophobic proteins will precipitate first, thus allowing fractionation to take place.

(c) Organic solvents. Dextrans can be precipitated out of a broth by the addition of methanol. Chilled ethanol and acetone can be used in the precipitation of proteins mainly due to changes in the dielectric properties of the solution.

(d) Non-ionic polymers such as polyethylene gylcol (PEG) can be used in the precipitation of proteins and are similar in behaviour to organic solvents.

(e) Polyelectrolytes can be used in the precipitation of a range of compounds, in addition to their use in cell aggregation.

(f) Protein binding dyes (triazine dyes) bind to and precipitate certain classes of protein (Lowe and Stead, 1985).

(g) Affinity precipitants are an area of much current interest in that they are able to bind to, and precipitate, compounds selectively (Niederauer and Glatz, 1992).

# FILTRATION

Filtration is one of the most common processes used at all scales of operation to separate suspended particles from a liquid or gas, using a porous medium which retains the particles but allows the liquid or gas to pass through. Gas filtration has been discussed in detail elsewhere (Chapters 5 and 7). It is possible to carry out filtration under a variety of conditions, but a number of factors will obviously influence the choice of the most suitable type of equipment to meet the specified requirements at minimum overall cost, including:

1. The properties of the filtrate, particularly its viscosity and density.

2. The nature of the solid particles, particularly their size and shape, the size distribution and packing characteristics.
3. The solids : liquid ratio.
4. The need for recovery of the solid or liquid fraction or both.
5. The scale of operation.
6. The need for batch or continuous operation.
7. The need for aseptic conditions.
8. The need for pressure or vacuum suction to ensure an adequate flow rate of the liquid.

## Theory of filtration

A simple filtration apparatus is illustrated in Fig. 10.7, which consists of a support covered with a porous filter cloth. A filter cake gradually builds up as filtrate passes through the filter cloth. As the filter cake increases in thickness the resistance to flow will gradually increase. Thus, if the pressure applied to the surface of the slurry is kept constant the rate of flow will gradually diminish. Alternatively, if the flow rate is to be kept constant the pressure will gradually have to be increased. The flow rate may also be reduced by blocking of holes in the filter cloth and closure of voids between particles, if the particles are soft and compressible. When particles are compressible it may not be feasible to apply increased pressure.

Flow through a uniform and constant depth porous bed can be represented by the Darcy equation:

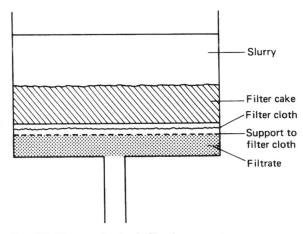

FIG. 10.7. Diagram of a simple filtration apparatus.

$$\text{Rate of flow} = \frac{dV}{dt} = \frac{KA\Delta P}{\mu L} \qquad (10.1)$$

where  $\mu$   = liquid viscosity,
$L$  = depth of the filter bed,
$\Delta P$  = pressure differential across the filter bed,
$A$  = area of the filter exposed to the liquid,
$K$  = constant for the system.

$K$ itself is a term which depends on the specific surface area $s$ (surface area/unit volume) of the particles making up the filter bed and the voidage $\Sigma$ when they are packed together. The voidage is the amount of filter-bed area which is free for the filtrate to pass through. It is normally 0.3 to 0.6 of the cross-sectional area of the filter bed. Thus $K$ (Kozeny's constant) can be expressed as

$$K = \frac{\Sigma^2}{5(1 - \Sigma)^2 s^2}.$$

Unfortunately, $s$ and $\Sigma$ are not easily determined.

In most practical cases $L$ is not readily measured but can be defined in terms of:
$V$ = volume of filtrate passed in time $t$ and
$v$ = volume of cake deposited per unit volume of filtrate.

Then
$$L = \frac{vV}{A}.$$

Substituting in equation (10.1):

$$\frac{dV}{dt} = \frac{KA^2 \Delta P}{\mu v V}. \qquad (10.2)$$

This is a general equation relating rate of filtration to pressure drop, cross-sectional area of the filter and filtrate retained. Equation 10.2 can be integrated for filtration at constant pressure.

$$VdV = \frac{KA^2 \Delta P \, dt}{\mu v} \qquad (10.3)$$

Integrating equation (10.3):

$$V^2 = \frac{2KA^2 \Delta P t}{\mu v}. \qquad (10.4)$$

Now in equation (10.4), $\Delta P$ is constant, $\mu$ is generally equal to 1, $v$ can be determined by laboratory investigation and $A^2$ remains approximately constant. Thus there is a linear relationship between $V^2$ and $t$. By carrying out small-scale filtration trials it is therefore possible to obtain a value for $K$. It is then possible to reapply the equation for large-scale filtration calculations.

Although it is also possible to derive the equation for the pressure necessary to maintain a constant filtration rate, it has little practical application. The pressure is made up of two components. Firstly the pressure needed to pass the constant volume through the filter resistance and, secondly, an increasing pressure component which is proportional to the resistance from the increasing cake depth. This filtration procedure would be complex to perform practically, and other methods of filtration are used to achieve constant flow rates, e.g. vacuum drum filters.

### The use of filter aids

It is common practice to use filter aids when filtering bacteria or other fine or gelatinous suspensions which prove slow to filter or partially block a filter. Kieselguhr (diatomaceous earth) is the most widely used material. It has a voidage of approximately 0.85, and, when it is mixed with the initial cell suspension, improves the porosity of a resulting filter cake leading to a faster flow rate. Alternatively, it may be used as an initial bridging agent in the wider pores of a filter to prevent or reduce blinding. The term 'blinding' means the wedging of particles which are not quite large enough to pass through the pores, so that an appreciable fraction of the filter surface becomes inactive.

The minimum quantity of filter aid to be used in filtration of a broth should be established experimentally. Kieselguhr is not cheap, and it will also absorb some of the filtrate, which will be lost when the filter cake is disposed. The main methods of using the filter aid are:

1. A thin layer of kieselguhr is applied to the filter to form a precoat prior to broth filtration.
2. The appropriate quantity of filter aid is mixed with the harvested broth. Filtration is started, to build up a satisfactory filter bed. The initial raffinate is returned to the remaining broth prior to starting the true filtration.
3. When vacuum drum filters are to be used which are fitted with advancing knife blades, a thick precoat filter is initially built up on the drum (later section in this chapter).

In some processes such as microbial biomass production, filter aids cannot be used and cell pretreatment by flocculation or heating must be considered (see later section in this chapter). In addition it is not normally practical to use filter aids when the product is intracellular and its removal would present a further stage of purification.

## Batch filters

## PLATE AND FRAME FILTERS

A plate and frame filter is a pressure filter in which the simplest form consists of plates and frames arranged alternately. The plates are covered with filter cloths (Fig. 10.8) or filter pads. The plates and frames are assembled on a horizontal framework and held together by means of a hand screw or hydraulic ram so that there is no leakage between the plates and frames which form a series of liquid-tight compartments. The slurry is fed to the filter frame through the continuous channel formed by the holes in the corners of the plates and frames. The filtrate passes through the filter cloth or pad, runs down grooves in the filter plates and is then discharged through outlet taps to a channel. Sometimes, if aseptic conditions are required, the outlets may lead directly into a pipe. The solids are retained within the frame and filtration is stopped when the frames are completely filled or when the flow of filtrate becomes uneconomically low.

On an industrial scale the plate and frame filter is one of the cheapest filters per unit of filtering space and requires the least floor space, but it is intermittent in operation (a batch process) and there may be considerable wear of filter cloths as a result of frequent dismantling. This type of filter is most suitable for fermentation broths with a low solids content and low resistance to filtration. It is widely used as a 'polishing' device in breweries to filter out residual yeast cells following initial clarification by centrifugation or rotary vacuum filtration. It may also be used for collecting high value solids that would not justify the use of a continuous filter. Because of high labour costs and the time involved in dismantling, cleaning and reassembly, these filters should not be used when removing large quantities of worthless solids from a broth.

## PRESSURE LEAF FILTERS

There are a number of intermittent batch filters usually called by their trade names. These filters incorporate a number of leaves, each consisting of a metal framework of grooved plates which is covered with a fine wire mesh, or occasionally a filter cloth and often precoated with a layer of cellulose fibres. The process slurry is fed into the filter which is operated under pressure or by suction with a vacuum pump. Because the filters are totally enclosed it is possible to sterilize them with steam. This type of filter is particularly suitable for 'polishing' large volumes of liquids with low solids content or small batch filtrations of valuable solids.

### (i) *Vertical metal-leaf filter*

This filter consists of a number of vertical porous metal leaves mounted on a hollow shaft in a cylindrical pressure vessel. The solids from the slurry gradually build up on the surface of the leaves and the filtrate is removed from the plates via the horizontal hollow shaft. In some designs the hollow shaft can be slowly rotated during filtration. Solids are normally removed at the end of a cycle by blowing air through the shaft and into the filter leaves.

### (ii) *Horizontal metal-leaf filter*

In this filter the metal leaves are mounted on a vertical hollow shaft within a pressure vessel. Often, only the upper surfaces of the leaves are porous. Filtration is continued until the cake fills the space between the disc-shaped leaves or when the operational pressure has become excessive. At the end of a process cycle, the solid cake can be discharged by releasing the pressure and spinning the shaft with a drive motor.

### (iii) *Stacked-disc filter*

One kind of filter of this type is the Metafilter. This is a very robust device and because there is no filter cloth and the bed is easily replaced, labour costs are low. It consists of a number of precision-made rings which are stacked on a fluted rod (Fig. 10.9). The rings

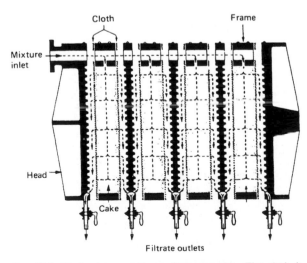

FIG. 10.8. Flush plate and frame filter assembly. The cloth is shown away from the plates to indicate flow of filtrate in the grooves between pyramids (Purchas, 1971).

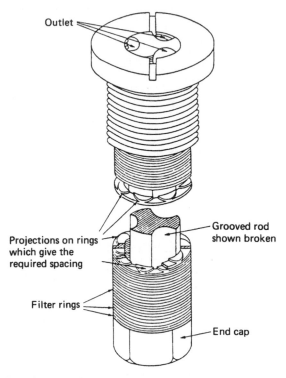

FIG. 10.9a. Metafilter pack (Coulson and Richardson, 1991).

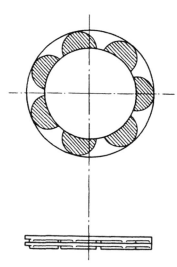

FIG. 10.9b. Rings for metafilter (Coulson and Richardson, 1991).

(22 mm external diameter, 16 mm internal diameter and 0.8 mm thick) are normally made from stainless steel and precision stamped so that there are a number of shoulders on one side. This ensures that there will be clearances of 0.025 mm to 0.25 mm when the rings

are assembled on the rods. The assembled stacks are placed in a pressure vessel which can be sterilized if necessary. The packs are normally coated with a thin layer of kieselguhr which is used as a filter aid. During use, the filtrate passes between the discs and is removed through the grooves of the fluted rods, while solids are deposited on the filter coating. Operation is continued until the resistance becomes too high and the solids are removed from the rings by applying back pressure via the fluted rods. Metafilters are primarily used for 'polishing' liquids such as beer.

**Continuous filters**

### ROTARY VACUUM FILTERS

Large rotary vacuum filters are commonly used by industries which produce large volumes of liquid which need continuous processing. The filter consists of a rotating, hollow, segmented drum covered with a fabric or metal filter which is partially immersed in a trough containing the broth to be filtered (Fig. 10.10). The slurry is fed on to the outside of the revolving drum and vacuum pressure is applied internally so that the filtrate is drawn through the filter, into the drum and finally to a collecting vessel. The interior of the drum is divided into a series of compartments, to which the vacuum pressure is normally applied for most of each revolution as the drum slowly revolves (~ 1 rpm). However, just before discharge of the filter cake, air pressure may be applied internally to help ease the filter cake off the drum. A number of spray jets may be carefully positioned so that water can be applied to rinse the cake. This washing is carefully controlled so that dilution of the filtrate is minimal.

It should be noted that the driving force for filtration (pressure differential across the filter) is limited to one atmosphere (100 kN m$^{-2}$) and in practice it is significantly less than this. In contrast, pressure filters can be operated at many atmospheres pressure. A number of rotary vacuum drum filters are manufactured, which differ in the mechanism of cake discharge from the drum:

(i)   String discharge.
(ii)  Scraper discharge.
(iii) Scraper discharge with precoating of the drum.

#### (i) String discharge

Fungal mycelia produce a fibrous filter cake which can easily be separated from the drum by string discharge (Fig. 10.11). Long lengths of string 1.5 cm apart

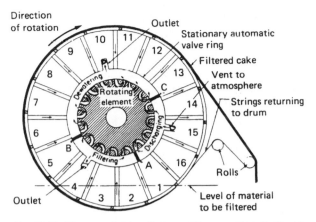

FIG. 10.10. Diagram of string-discharge filter operation. Sections 1 to 4 are filtering; sections 5 to 12 are dewatering; and section 13 is discharging the cake with the string discharge. Sections 14, 15 and 16 are ready to start a new cycle. A, B and C represent dividing members in the annular ring (Miller *et al.*, 1973).

are threaded over the drum and round two rollers. The cake is lifted free from the upper part of the drum when the vacuum pressure is released and carried to the small rollers where it falls free.

(ii) *Scraper discharge*

Yeast cells can be collected on a filter drum with a knife blade for scraper discharge (Fig. 10.12). The filter cake which builds up on the drum is removed by an accurately positioned knife blade. Because the knife is close to the drum, there may be gradual wearing of the filter cloth on the drum.

(iii) *Scraper discharge with precoating of the drum*

The filter cloth on the drum can be blocked by bacterial cells or mycelia of actinomycetes. This prob-

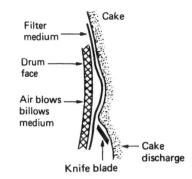

FIG. 10.12. Cake discharge on a drum using a scraper (Talcott *et al.*, 1980).

lem is overcome by precoating the drum with a layer of filter-aid 2 to 10 cm thick. The cake which builds up on the drum during operation is cut away by the knife blade (Fig. 10.13) which mechanically advances towards the drum at a controlled slow rate. Alternatively, the blade may be operated manually when there is an indication of 'blinding' which may be apparent from a reduction in the filtration rate. In either case the cake is removed together with a very thin layer of precoat. A study of precoat drum filtration has been made by Bell and Hutto (1958). The operating variables studied included drum speed, extent of drum submergence, knife advance speed and applied vacuum. The work indicated that optimization for a new process might require prolonged trials. Although primarily used for the separation of micro-organisms from broth, studies have indicated (Gray *et al.*, 1973) that rotary vacuum filters can be effective in the processing of disrupted cells.

## Cross-flow filtration (tangential filtration)

In the filtration processes previously described, the flow of broth was perpendicular to the filtration mem-

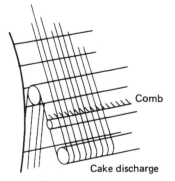

FIG. 10.11. Cake discharge on a drum filter using strings (Talcott *et al.*, 1980)

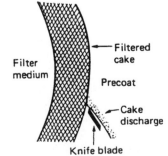

FIG. 10.13. Cake discharge on a precoated drum filter (Talcott *et al.*, 1980).

brane. Consequently, blockage of the membrane led to lower rates of productivity and/or the need for filter aids to be added, and these were serious disadvantages.

In contrast, an alternative which is rapidly gaining prominence both in the processing of whole fermentation broths (Tanny *et al.*, 1980; Brown and Kavanagh, 1987; Warren *et al.*, 1991) and cell lysates (Gabler and Ryan, 1985; Le and Atkinson, 1985) is cross-flow filtration. Here, the flow of medium to be filtered is tangential to the membrane (Fig. 10.14(a)), and no filter cake builds up on the membrane.

The benefits of cross-flow filtration are:

(a) Efficient separation, > 99.9% cell retention.
(b) Closed system; for the containment of organisms with no aerosol formation (see also Chapter 7).
(c) Separation is independent of cell and media densities, in contrast to centrifugation.
(d) No addition of filter aid (Zahka and Leahy, 1985).

The major components of a cross-flow filtration system are a media storage tank (or the fermenter), a pump and a membrane pack (Fig. 10.14(b)). The membrane is usually in a cassette pack of hollow fibres or flat sheets in a plate and frame type stack or a spiral cartridge (Strathmann, 1985). In this way, and by the introduction of a much convoluted surface, large filtration areas can be attained in compact devices. Two types of membrane may be used; microporous membranes with a specific pore size (0.45, 0.22 $\mu$m etc.) or an ultrafiltration membrane (see later section) with a specified molecular weight cut-off (MWCO). The type of membrane chosen is carefully matched to the product being harvested, with microporous and 100,000 MWCO membranes being used in cell separations.

The output from the pump is forced across the membrane surface; most of this flow sweeps the membrane, returning retained species back to the storage tank and generally less than 10% of the flow passes through the membrane (permeate). As this process is continued the cells, or other retained species are concentrated to between 5 and 10% of their initial volume. More complex variants of the process can allow *in-situ* washing of the retentate and enclosed systems for containment and sterilization (Mourot *et al.*, 1989).

Many factors influence filtration rate. Increased pressure drop will, up to a point increase flow across the membrane, but it should be remembered that the system is based on a swept clean membrane. Therefore, if the pressure drop is too great the membrane may become blocked. The filtration rate is therefore influenced by the rate of tangential flow across the membrane; by increasing the shear forces at the membrane's surface retained species are more effectively removed, thereby increasing filtration rate. Higher temperatures will increase filtration rate by lowering the viscosity of the media, though this is clearly of limited application in biological systems. Filtration rate is inversely proportional to concentration, and media constituents can influence filtration rate in three ways. Low molecular weight compounds increase media viscosity and high molecular weight compounds decrease shear at the membrane surface, both leading to a reduction in filtration rate. Finally, broth constituents can 'foul' the

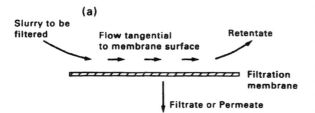

**(a)**

Slurry to be filtered

Flow tangential to membrane surface

Retentate

Filtration membrane

Filtrate or Permeate

FIG. 10.14a. Schematic diagram of cross-flow filtration.

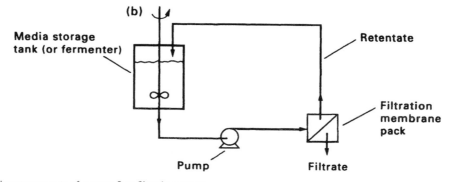

**(b)**

Media storage tank (or fermenter)

Retentate

Filtration membrane pack

Pump

Filtrate

FIG. 10.14b. Major components of a cross-flow filtration system.

membrane, primarily by adsorption onto the membrane's surface, causing a rapid loss in efficiency. This can be controlled by modification of the membrane or media formulation in particular by reducing the use of antifoaming agents. Lee *et al.* (1993) have shown that the pulses of air injected into the flow to a cross-flow filter increase the shear rate at the membrane surface reducing the effects of membrane fouling.

## CENTRIFUGATION

Micro-organisms and other similar sized particles can be removed from a broth by using a centrifuge when filtration is not a satisfactory separation method. Although a centrifuge may be expensive when compared with a filter it may be essential when:

1. Filtration is slow and difficult.
2. The cells or other suspended matter must be obtained free of filter aids.
3. Continuous separation to a high standard of hygiene is required.

Non-continuous centrifuges are of extremely limited capacity and therefore not suitable for large-scale separation. The centrifuges used in harvesting fermentation broths are all operated on a continuous or semi-continuous basis. Some centrifuges can be used for separating two immiscible liquids yielding a heavy phase and light phase liquid, as well as a solids fraction. They may also be used for the breaking of emulsions.

According to Stoke's law, the rate of sedimentation of spherical particles suspended in a fluid of Newtonian viscosity characteristics is proportional to the square of the diameter of the particles, thus the rate of sedimentation of a particle under gravitational force is:

$$V_g = \frac{d^2 g \, (\rho_P - \rho_L)}{18 \mu} \qquad (10.5)$$

where $V_g$ = rate of sedimentation (m s$^{-1}$)
  $d$ = particle diameter (m)
  $g$ = gravitational constant (m s$^{-2}$)
  $\rho_P$ = particle density (kg m$^{-3}$)
  $\rho_L$ = liquid density (kg m$^{-3}$)
  $\mu$ = viscosity (kg m$^{-1}$ s$^{-1}$)

This equation can then be modified for sedimentation in a centrifuge:

$$V_c = \frac{d \omega^2 r \, (\rho_P - \rho_L)}{18 \mu} \qquad (10.6)$$

where $V_c$ = rate of sedimentation in the centrifuge (m s$^{-1}$),
  $\omega$ = angular velocity of the rotor (s$^{-1}$),
  $r$ = radial position of the particle (m).

Dividing equation (10.6) by equation (10.5) yields

$$\frac{\omega^2 r}{g}.$$

This is a measure of the separating power of a centrifuge compared with gravity settling. It is often referred to as the relative centrifugal force and given the symbol 'Z'.

It is evident from this formula that factors influencing the rate of sedimentation over which one has little or no control are the difference in density between the cells and the liquid (increased temperature would lower media density but is of little practical use with fermentation broths), the diameter of the cells (could be increased by coagulation/flocculation) and the viscosity of the liquid. Ideally, the cells should have a large diameter, there should be a large density difference between cell and liquid and the liquid should have a low viscosity. In practice, the cells are usually very small, of low density and are often suspended in viscous media. Thus it can be seen that the angular velocity and diameter of the centrifuge are the major factors to be considered when attempting to maximize the rate of sedimentation (and therefore throughput) of fermentation broths.

### Cell aggregation and flocculation

Following an industrial fermentation it is quite common to add flocculating agents to the broth to aid de-watering (Wang, 1987). The use of flocculating agents is widely practised in the effluent-treatment industries for the removal of microbial cells and suspended colloidal matter (Delaine, 1983).

It is well known that aggregates of microbial cells, although they have the same density as the individual cells, will sediment faster because of the increased diameter of the particles (Stokes law). This sedimentation process may be achieved naturally with selected strains of brewing yeasts, particularly if the wort is chilled at the end of fermentation, and leads to a natural clearing of the beer.

Micro-organisms in solution are usually held as discrete units in three ways. Firstly, their surfaces are negatively charged and therefore repulse each other. Secondly, because of their generally hydrophilic cell walls a shell of bound water is associated with the cell which acts as a thermodynamic barrier to aggregation. Finally, due to the irregular shapes of cell walls (at the macromolecular level) steric hindrance will also play a part.

During flocculation one or more mechanisms besides temperature can induce cell flocculation:

(a) Neutralization of anionic charges, primarily carboxyl and phosphate groups, on the surfaces of the microbial cells, thus allowing the cells to aggregate. These include changes in the pH and the presence of a range of compounds which alter the ionic environment.

(b) Reduction in surface hydrophilicity.

(c) The use of high molecular weight polymer bridges. Anionic, non-ionic and cationic polymers can be used, though the former two also require the addition of a multivalent cation.

Flocculation usually involves the mixing of a process fluid with the flocculating agent under conditions of high shear in a stirred tank, although more compact and efficient devices have been proposed (Ashley, 1990). This stage is known as coagulation, and is usually followed by a period of gentle agitation when flocs developed initially are allowed to grow in size. The underlying theoretical principles of cell flocculation have been discussed by Atkinson and Daoud (1976).

Nakamura (1961) described the use of various compounds for flocculating bacteria, yeasts and algae, including alum, calcium salts and ferric salts. Other agents which are now used include tannic acid, titanium tetrachloride and cationic agents such as quaternary ammonium compounds, alkyl amines and alkyl pyridinium salts. Gasner and Wang (1970) reported a many hundred-fold increase in the sedimentation rate of *Candida intermedia* when recoveries of over 99% were readily obtained. They found that flocculation was very dependent on the choice of additive, dosage and conditions of floc formation, with the most effective agents being mineral colloids and polyelectrolytes. Nucleic acids, polysaccharides and proteins released from partly lysed cells may also bring about agglomeration. In SCP processes, phosphoric acid has been used as a flocculating agent since it can be used as a nutrient in medium recycle with considerable savings in water usage (Hamer, 1979).

The majority of flocculating agents currently in use are polyelectrolytes, which act by charge neutralization and hydrophobic interactions to link cells to each other. In processes where the addition of some toxic chemicals is to be avoided, alternative techniques have been adopted. One method is to coagulate microbial protein which has been released from the cells by heating for short periods. Kurane (1990) reports the use of bioflocculants obtained from *Rhodococcus erythropolis*. They are suggested as being safer alternatives to conventional flocculants. Warne and Bowden (1987) suggest the use of genetic manipulation to alter cell surface properties to aid aggregation. Flocculating agents such as cross-linked cationic polymers may also be used in the processing of cell lysates and extracts prior to further downstream processing (Fletcher *et al.*, 1990). Bentham *et al.* (1990) utilized borax as a flocculating agent for yeast cell debris prior to decanter centrifugation.

### The range of centrifuges

A number of centrifuges will be described which vary in their manner of liquid and solid discharge, their unloading speed and their relative maximum capacities. When choosing a centrifuge for a specific process it is important to ensure that the centrifuge will be able to perform the separation at the planned production rate, and operate reliably with minimum manpower. Large-scale tests may therefore be necessary with fermentation broths or other materials to check that the correct centrifuge is chosen.

#### THE BASKET CENTRIFUGE (PERFORATED-BOWL BASKET CENTRIFUGE)

Basket centrifuges are useful for separating mould mycelia or crystalline compounds. The centrifuge is most commonly used with a perforated bowl lined with a filter bag of nylon, cotton, etc. (Fig. 10.15). A continuous feed is used, and when the basket is filled with the filter cake it is possible to wash the cake before removing it. The bowl may suffer from blinding with soft biological materials so that high centrifugal forces cannot be used. These centrifuges are normally operated at speeds of up to 4000 rpm for feed rates of 50 to 300 $dm^3$ $min^{-1}$ and have a solids holding capacity of 30 to 500 $dm^3$. The basket centrifuge may be considered to be a centrifugal filter.

#### THE TUBULAR-BOWL CENTRIFUGE

This is a centrifuge to consider using for particle size ranges of 0.1 to 200 $\mu$m and up to 10% solids in the in-going slurry. Figure 10.16a shows an arrangement used in a Sharples Super-Centrifuge. The main component of the centrifuge is a cylindrical bowl (or rotor) (A in Fig. 10.16), which may be of a variable design de-

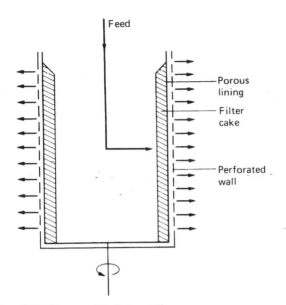

FIG. 10.15. Diagram of basket centrifuge.

pending on application, suspended by a flexible shaft (B), driven by an overhead motor or air turbine (C). The inlet to the bowl is via a nozzle attached to the bottom bearing (D). The feed which may consist of solids and light and heavy liquid phases is introduced by the nozzle (E). During operation solids sediment on the bowl wall while the liquids separate into the heavy phase in zone (G) and the light phase in the central zone (H). The two liquid phases are kept separate in their exit from the bowl by an adjustable ring, with the heavy phase flowing over the lip of the ring. Rings of various sizes may be fitted for the separation of liquids of various relative densities. Thus the centrifuge may be altered to use for:

(a) Light-phase/heavy-phase liquid separation.
(b) Solids/light-liquid phase/heavy-liquid phase separation.
(c) Solids/liquid separation (using a different rotor, Fig. 10.16b).

The Sharples laboratory centrifuge with a bowl radius of approximately 2.25 cm can be operated with an air turbine at 50,000 rpm to produce a centrifugal force of approximately 62,000 $g$, but has a bowl capacity of only 200 cm$^3$ with a throughput of 6 to 25 dm$^3$ h$^{-1}$. The largest size rotor is the Sharples AS 26, which has a bowl radius of 5.5 cm and a capacity of 9 dm$^3$, a solids capacity of 5 dm$^3$ and a throughput of 390 to 2400 dm$^3$ h$^{-1}$.

The advantages of this design of centrifuge are the

high centrifugal force, good dewatering and ease of cleaning. The disadvantages are limited solids capacity, difficulties in the recovery of collected solids, gradual loss in efficiency as the bowl fills, solids being dislodged from the walls as the bowl is slowing down and foaming. Plastic liners can be used in the bowls to help improve batch cycle time. Alternatively a spare bowl can be changed over in about 5 minutes.

### THE SOLID-BOWL SCROLL CENTRIFUGE (DECANTER CENTRIFUGE)

This type of centrifuge is used for continuous handling of fermentation broths, cell lysates and coarse materials such as sewage sludge (Fig. 10.17). The slurry is fed through the spindle of an archimedean screw within the horizontal rotating solids bowl. Typically the speed differential between the bowl and the screw is in the range 0.5 to 100 rpm (Coulson and Richardson, 1991). The solids settling on the walls of the bowl are scraped to the conical end of the bowl. The slope of the cone helps to remove excess liquid from the solids before discharge. The liquid phase is discharged from the opposite end of the bowl. The speed of this type of centrifuge is limited to around 5000 rpm in larger models because of the lack of balance within the bowl, with smaller models having bowl speeds of up to 10000 rpm. Bowl diameters are normally between 0.2 and 1.5 metres, with the length being up to five times the diameter. Feed rates range from around 200 dm$^3$ h$^{-1}$ to 200 m$^3$ h$^{-1}$ depending on scale of operation and material being processed. A number of variants on the basic design are available:

(a) Cake washing facilities (screen bowl decanters).
(b) Vertical bowl decanters.
(c) Facility for in-place cleaning.
(d) Bio-hazard containment features; steam sterilization *in-situ*, two or three stage mechanical seals, control of aerosols, containment casings and the use of high pressure sterile gas in seals to prevent the release of micro-organisms.

### THE MULTICHAMBER CENTRIFUGE

Ideally, this is a centrifuge for a slurry of up to 5% solids of particle size 0.1 to 200 $\mu$m diameter. In the multichamber centrifuge (Fig. 10.18), a series of concentric chambers are mounted within the rotor chamber. The broth enters via the central spindle and then takes a circuitous route through the chambers. Solids collect on the outer faces of each chamber. The smaller particles collect in the outer chambers where they are subjected to greater centrifugal forces (the greater the radial position of a particle, the greater the rate of sedimentation).

Although these vessels can have a greater solids capacity than tubular bowls and there is no loss of efficiency as the chamber fills with solids, their mechanical strength and design limits their speed to a maximum of 6500 rpm for a rotor 46-cm diameter with a holding capacity of up to 76 dm$^3$. Because of the time needed to dismantle and recover the solids fraction, the size and number of vessels must be of the correct volume for the solids of a batch run.

## THE DISC-BOWL CENTRIFUGE

This centrifuge relies for its efficiency on the pres-

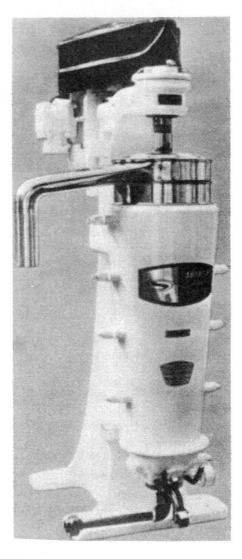

FIG. 10.16b. A Sharples Super-Centrifuge assembled for discharge of one liquid phase (Alfa Laval Sharples, Camberley, U.K.).

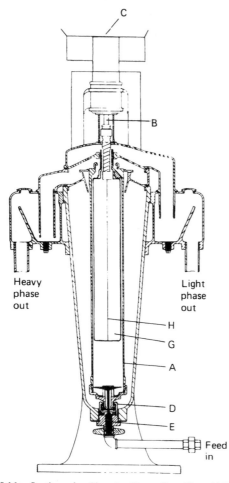

FIG. 10.16a. Section of a Sharples Super-Centrifuge (Alfa Laval Sharples, Camberley, U.K.).

ence of discs in the rotor or bowl (Fig. 10.19). A central inlet pipe is surrounded by a stack of stainless-steel conical discs. Each disc has spacers so that a stack can be built up. The broth to be separated flows outwards from the central feed pipe, then upwards and inwards beween the discs at an angle of 45° to the axis of rotation. The close packing of the discs assists rapid sedimentation and the solids then slide to the edge of the bowl, provided that there are no gums or fats in the slurry, and eventually accumulate on the inner wall of the bowl. Ideally, the sediment should form a sludge which flows, rather than a hard particulate or lumpy sediment. The main advantages of these centrifuges are

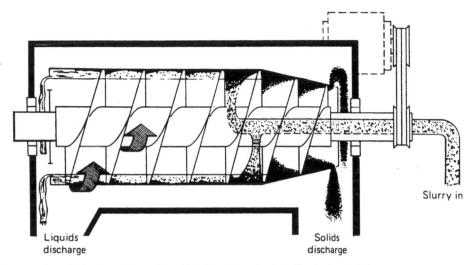

FIG. 10.17a. Diagram of a solid-bowl scroll centrifuge (Alfa Laval Sharples Ltd, Camberley, U.K.).

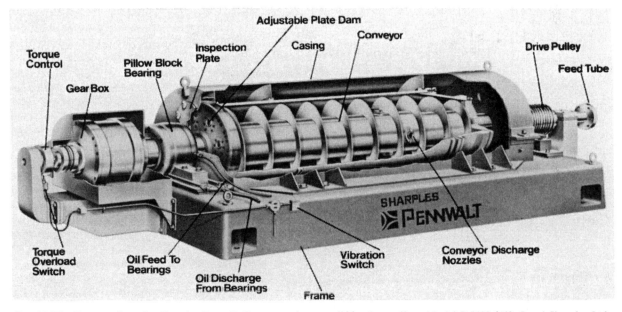

FIG. 10.17b. Cutaway view of a Sharples Super-D-Canter continuous solid-bowl centrifuge, Model P-5400 (Alfa Laval Sharples Ltd, Camberley, U.K.).

their small size compared with a bowl without discs for a given throughput. Some designs also have the facility for continuous solids removal through a series of nozzles in the circumference of the bowl or intermittent solids removal by automatic opening of the solids collection bowl. The arrangement of the discs makes this type of centrifuge laborious to clean. However, recent models such as the Alfa Laval BTUX 510 (Alfa Laval Sharples Ltd, Camberley, Surrey, U.K.) system (Fig. 10.20) are designed to allow for cleaning *in-situ*. In addition this and similar plant have the facility for *in-situ* steam sterilization and total containment, incorporating double seals to comply with containment regulations (see also Chapter 7). Feed rates range from 45 to 1800 dm$^3$ min$^{-1}$, with rotational speeds typically between 5000 and 10,000 rpm. The Westfalia CSA 19-47-476 is also steam sterilizable and has been used for the sterile collection of organisms (Walker *et al.*,

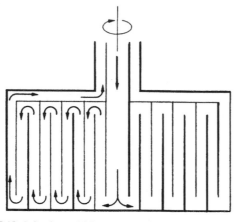

FIG. 10.18. L.S. of a multichamber centrifuge.

1987). Similarly, the Westfalia CSA 8 can be modified for contained operation and steam sterilization (Frude and Simpson, 1993).

## CELL DISRUPTION

Micro-organisms are protected by extremely tough cell walls. In order to release their cellular contents a number of methods for cell disintegration have been developed (Wimpenny, 1967; Hughes *et al.*, 1971). Any potential method of disruption must ensure that labile materials are not denatured by the process or hydrolysed by enzymes present in the cell. Huang *et al.* (1991) report the use of a combination of different techniques to release products from specific locations within yeast cells. In this way the desired product can be obtained with minimum contamination. Although many techniques are available which are satisfactory at laboratory scale, only a limited number have been proved to be suitable for large-scale applications, particularly for intracellular enzyme extraction (Wang *et al.*, 1979; Darbyshire, 1981). Containment of cells can be difficult or costly to achieve in many of the methods described below and thus containment requirements will strongly influence process choice. Methods available fall into two major categories:

*Physico-mechanical methods*

(a) Liquid shear.
(b) Solid shear.
(c) Agitation with abrasives.
(d) Freeze-thawing.
(e) Ultrasonication.

*Chemical methods*

(a) Detergents.
(b) Osmotic shock.
(c) Alkali treatment.
(d) Enzyme treatment.

### Physical–mechanical methods

LIQUID SHEAR

Liquid shear is the method which has been most widely used in large scale enzyme purification procedures (Scawen *et al.*, 1980). High-pressure homogenizers used in the processing of milk and other products in the food industry have proved to be very effective for microbial cell disruption. One machine, the APV-Manton Gaulin-homogenizer (The APV Co. Ltd, Crawley, Surrey, U.K.), which is a high-pressure positive displacement pump, incorporates an adjustable valve with a restricted orifice (Fig. 10.21). The smallest model has one plunger, while there are several in larger models. During use, the microbial slurry passes through a non-return valve and impinges against the operative valve set at the selected operating pressure. The cells then pass through a narrow channel between the valve and an impact ring followed by a sudden pressure drop at the exit to the narrow orifice. The large pressure drop across the valve is believed to cause cavitation in the slurry and the shock waves so produced disrupt the cells. Brookman (1974) considered the size of the pressure drop to be very important in achieving effective disruption, and as with all mechanical methods, cell size and shape influence ease of disruption (Wase and Patel, 1985). The working pressures are extremely high. Hetherington *et al.* (1971) used a pressure of 550 kg cm$^{-2}$ for a 60% yeast suspension. A throughput of 6.4 kg soluble protein h$^{-1}$ with 90% disruption could be achieved with a small industrial machine. In larger models, flow rates of up to 600 dm$^3$ h$^{-1}$ are now possible and operating pressures of 1200 bar are utilized in some processes (Asenjo, 1990). Darbyshire (1981) has stressed the need for cooling the slurry to between 0 and 4°C to minimize loss in enzyme activity because of heat generation during the process. The increase in slurry temperature is approximately proportional to the pressure drop across the valve. Because of problems caused by heat generation and because cell suspensions can be surprisingly abrasive, it is common practice to operate such homogenizers in a multi-pass

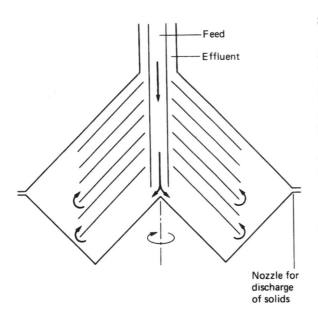

FIG. 10.19a. L.S. of disc-bowl centrifuge with nozzle discharge.

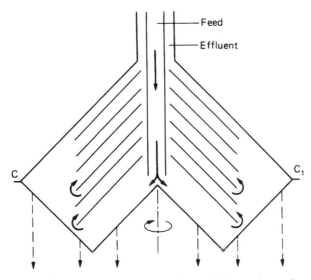

FIG. 10.19b. L.S. of disc-bowl centrifuge with intermittent discharge. (Solids discharged when rotor opens intermittently along the section C–C$_1$.)

mode but at a lower pressure. The degree of disruption and consequently the amount of protein released will influence the ease of subsequent separation of the product from the cell debris in high-pressure homogenizers and bead mills (Agerkvist and Enfors, 1990). A careful balance must therefore be made between percentage release of product and the difficulty and cost of further product purification.

## SOLID SHEAR

Pressure extrusion of frozen micro-organisms at around −25°C through a small orifice is a well established technique at a laboratory scale using a Hughes press or an X-press to obtain small samples of enzymes or microbial cell walls. Disruption is due to a combination of liquid shear through a narrow orifice and the presence of ice crystals. Magnusson and Edebo (1976) developed a semi-continuous X-press operating with a sample temperature of −35°C and an X-press temperature of −20°C. It was possible to obtain 90% disruption with a single passage of *S. cerevisiae* using a throughput of 10 kg yeast cell paste h$^{-1}$. This technique might be ideal for microbial products which are very temperature labile.

## AGITATION WITH ABRASIVES

Mechanical cell disruption can also be achieved in a disintegrator containing a series of rotating discs and a charge of small beads. The beads are made of mechanically resistant materials such as glass, alumina ceramics and some titanium compounds (Fig. 10.22). In a small disintegrator, the Dyno-Muhle KD5 (Wiley A. Bachofen, Basle, Switzerland), using a flow rate of 180 dm$^3$ h$^{-1}$, 85% disintegration of an 11% w/v suspension of *S. cerevisiae* was achieved with a single pass (Mogren *et al.*, 1974). Although temperatures of up to 35°C were recorded in the disintegrator, the specific enzyme activities were not considered to be very different from values obtained by other techniques. Dissipation of heat generated in the mill is one of the major problems in scale up, though this can generally be overcome with the provision of a cooling jacket. In another disintegrator, the Netsch LM20 mill (Netzsch GmbH, Selb, Germany), the agitator blades were alternately mounted vertically and obliquely on the horizontal shaft (Fig. 10.23). A flow rate of up to 400 dm$^3$ h$^{-1}$ was claimed for a vessel with a nominal capacity of 20 dm$^3$ (Rehacek and Schaefer, 1977).

## FREEZING–THAWING

Freezing and thawing of a microbial cell paste will inevitably cause ice crystals to form and their expansion followed by thawing will lead to some subsequent disruption of cells. It is slow, with limited release of cellular materials, and has not often been used as a technique on its own, although it is often used in combination with other techniques. β-Glucosidase has been obtained from *S. cerevisiae* by this method (Honig and Kula, 1976). A sample of 360 g of frozen yeast

FIG. 10.20. Alfa Laval BTUX 510 disc stack centrifuge (Alfa Laval Sharples Ltd, Camberley, U.K.).

paste was thawed at 5° for 10 hours. This cycle was repeated twice before further processing.

## ULTRASONICATION

High frequency vibration (~ 20 kHz) at the tip of an ultrasonication probe leads to cavitation, and shock waves thus produced cause cell disruption. The method can be very effective on a small scale, but a number of serious drawbacks make it unsuitable for large-scale operations. Power requirements are high, there is a large heating effect so cooling is needed, the probes

have a short working life and are only effective over a short range. Continuous laboratory sonicators with hold-up volumes of around 10 cm³ have been shown to be effective (James *et al.*, 1972).

## Chemical methods

### DETERGENTS

A number of detergents will damage the lipoproteins of the microbial cell membrane and lead to release of intracellular components. The compounds which can

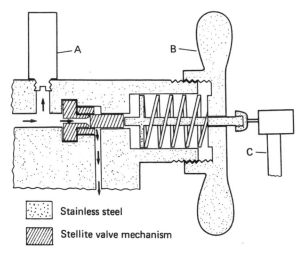

Stainless steel

Stellite valve mechanism

FIG. 10.21. Details of homogonizer valve assembly (Brookman, 1974). (A) 0–50,000 psi pressure transducer; (B) pressure-control handwheel; (C) linear variable displacement transformer; (→) direction of flow.

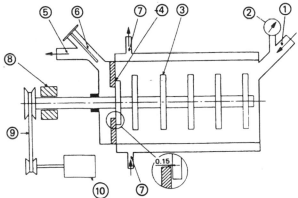

FIG. 10.22. Simplified drawing of the Dyno-Muhle KD5 (Mogren et al., 1974). (1) Inlet of suspension; (2) manometer; (3) rotating disc; (4) slit for separation of glass beads from the suspension; (5) outlet of suspension; (6) thermometer; (7) cooling water, inlet and outlet; (8) bearings; (9) variable V-belt drive; (10) drive motor. Cylinder dimensions: inside length 33 cm; inside diameter 14 cm.

be used for this purpose include quaternary ammonium compounds, sodium lauryl sulphate, sodium dodecyl sulphate (SDS) and Triton X-100. Unfortunately, the detergents may cause some protein denaturation and may need to be removed before further purification stages can be undertaken. Pullulanase is an enzyme which is bound to the outer membrane of *Klebsiella pneumoniae*. The cells were suspended in pH 7.8 buffer and 1% sodium cholate was added. The mixture was stirred for 1 hour to solubilize most of the enzyme (Kroner *et al.*, 1978). The use of Triton X-100 in combination with guanidine-HCl is widely and effectively used for the release of cellular protein (Naglak and Wang, 1992; Hettwer and Wang, 1989), Hettwer and Wang obtaining greater than 75% protein release in less than one hour from *Escherichia coli* under fermentation conditions.

OSMOTIC SHOCK

Osmotic shock caused by a sudden change in salt concentration will cause disruption of a number of cell types. However, the effect on microbial cells is normally minimal. It has proved to be a successful technique for the extraction of luciferase from *Photobacterium fischeri* (Hastings *et al.*, 1965). A batch of 120 dm³ of broth was harvested and the cells collected as a cell paste in a Sharples centrifuge. Enzyme extraction was achieved by osmotic lysis using a ratio of 1 g of cell paste to 4 cm³ of cold distilled water with stirring for

15 to 30 minutes. A second extraction gave a small additional yield of enzyme. Only low levels of soluble protein were released using this technique.

ALKALI TREATMENT

Alkali treatment might be used for hydrolysis of microbial cell wall material provided that the desired enzyme will tolerate a pH of 11.5 to 12.5 for 20 to 30 minutes. Darbyshire (1981) has reported the use of this technique in the extraction of L-asparaginase.

ENZYME TREATMENT

There are a number of enzymes which hydrolyse

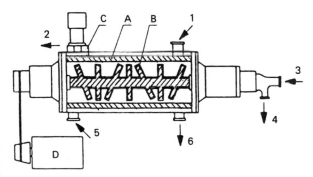

FIG. 10.23. Simplified drawing of the Netzsch model LM-20 mill (Rehacek and Schafer, 1977): A, cylidrical grinding vessel with cooling jacket; B, agitator with cooled shaft and discs; C, annular vibrating slot operator; D, variable-speed-drive motor; 1 and 2, product inlet and outlet; 3 and 4, agitator cooling inlet and outlet; 5 and 6, vessel-cooling inlet and outlet.

specific bonds in cell walls of a limited number of micro-organisms. Enzymes shown to have this activity include lysozyme and enzyme extracts from leucocytes, *Streptomyces* spp., *Micromonospora* spp. *Penicillium* spp., *Trichoderma* spp., and snails. Although this is probably one of the most gentle methods available, unfortunately it is relatively expensive and the presence of the enzyme(s) may complicate futher downstream purification processes. The use of immobilized lysozyme has been investigated by a number of workers and may provide the solution to such problems (Crapisi *et al.*, 1993). Chemical and enzymic methods for the release of intracellular products have not been used widely on a large scale, with the exception of lysozyme. However, their potential for the selective release of product and that they often yield a cleaner lysate mean that they are potentially invaluable tools in the recovery of fermentation products (Andrews and Asenjo, 1987; Andrews *et al.*, 1990). Enzymes may also be used as a pretreatment to partially hydrolyse cell walls prior to cell disruption by mechanical methods.

## LIQUID–LIQUID EXTRACTION

The separation of a component from a liquid mixture by treatment with a solvent in which the desired component is preferentially soluble is known as liquid–liquid extraction. The specific requirement is that a high percentage extraction of product must be obtained but concentrated in a smaller volume of solvent.

Prior to starting a large-scale extraction, it is important to find out on a small scale the solubility characteristics of the product using a wide range of solvents. A simple rule to remember is that 'like dissolves like'. The important 'likeness' as far as solubility relations are concerned is in the polarities of molecules. Polar liquids mix with each other and dissolve salts and other polar solids. The solvents for non-polar compounds are liquids of low or nil polarity.

The dielectric constant is a measure of the degree of molar polarization of a compound. If this value is known it is then possible to predict whether a compound will be polar or non-polar, with a high value indicating a highly polar compound. The dielectric constant $D$ of a substance can be measured by determining the electrostatic capacity $C$ of a condenser containing the substance between the plates. If $C_0$ is the value for the same condenser when completely evacuated then

$$D = \frac{C}{C_0}.$$

Experimentally, dielectric constants are obtained by comparing the capacity of the condenser when filled with a given liquid with the capacity of the same condenser containing a standard liquid whose dielectric constant is known very accurately. If $D_1$ and $D_2$ are the dielectric constants of the experimental and standard liquids and $C_1$ and $C_2$ are the electrostatic capacities of a condenser when filled with each of the liquids, then

$$\frac{D_1}{D_2} = \frac{C_1}{C_2}.$$

The value of $D_1$ can be calculated since $C_1$ and $C_2$ can be measured and $D_2$ is known. The dielectric constants for a number of solvents are given in Table 10.1.

The final choice of solvent will be influenced by the distribution or partition coefficient $K$ where

$$K = \frac{\text{Concentration of solute in extract}}{\text{Concentration of solute in raffinate}}.$$

The value of $K$ defines the ease of extraction. When there is a relatively high $K$ value, good stability of product and good separation of the aqueous and solvent phases, then it may be possible to use a single-stage extraction system (Fig. 10.24). A value of 50 indicates that the extraction should be straightforward whereas a value of 0.1 shows that the extraction will be difficult and that a multistage process will be necessary. Unfortunately, in a number of systems the value of $K$ is low and co-current or counter-current multistage systems

TABLE 10.1. *Dielectric constants of solvents at 25°C (arranged in order of increasing polarity)*

| Solvent | Dielectric constant |
|---|---|
| Hexane | 1.90 (least polar) |
| Cyclohexane | 2.02 |
| Carbon tetrachloride | 2.24 |
| Benzene | 2.28 |
| Di-ethyl ether | 4.34 |
| Chloroform | 4.87 |
| Ethyl acetate | 6.02 |
| Butan-2-ol | 15.8 |
| Butan-1-ol | 17.8 |
| Propan-1-ol | 20.1 |
| Acetone | 20.7 |
| Ethanol | 24.3 |
| Methanol | 32.6 |
| Water | 78.5 (most polar) |

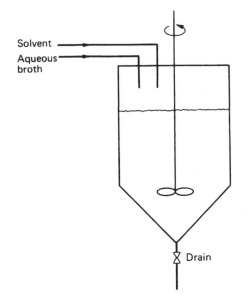

FIG. 10.24. Diagram of a single-stage extraction unit.

have to be utilized. The co-current system is illustrated in Fig. 10.25. There are $n$ mixer/separator vessels in line and the raffinate goes from vessel 1 to vessel $n$. Fresh solvent is added to each stage, the feed and extracting solvent pass through the cascade in the *same* direction. Extract is recovered from each stage. Although a relatively large amount of solvent is used, a high degree of extraction is achieved.

A counter-current system is illustrated in Fig. 10.26. There are a number of mixer/separators connected in series. The extracted raffinate passes from vessel 1 to vessel $n$ while the product-enriched solvent is flowing from vessel $n$ to vessel 1. The feed and extracting solvent pass through the cascade in *opposite* directions. The most efficient system for solvent utilization is counter-current operation, showing a considerable advantage over batch and co-current systems. Unless there

are special reasons the counter-current system should be used. In practice, the series of counter-current extractions are conducted in a single continuous extractor using centrifugal forces to separate the two liquid phases. The two liquid streams are forced to flow counter-current to each other through a long spiral of channels within the rotor.

The Podbielniak centrifugal extractor (Fig. 10.27) consists of a horizontal cylindrical drum revolving at up to 5000 rpm about a shaft passing through its axis. The liquids to be run counter-current are introduced into the shaft, with the heavy liquid entering the drum at the shaft while the light liquid is led by an internal route to the periphery of the drum. As the drum rotates, the heavy liquid is forced to the periphery of the drum by centrifugal action where it contacts the light liquid. The solute is transferred between the liquids and the light liquid is displaced back towards the axis of the drum. The heavy liquid is returned to the drum's axis via internal channels. The two liquid streams are then discharged via the shaft. Flow rates in excess of 100,000 $dm^3$ $h^{-1}$ are possible in the largest models. Probably the most useful property of this type of extractor is the low hold-up volume of liquid in the machine compared with the throughput.

Penicillin G is an antibiotic which is recovered from fermentation broths by centrifugal counter-current solvent extraction. At neutral pHs in water penicillin is ionized:

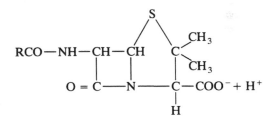

In acid conditions this ionization is suppressed and the

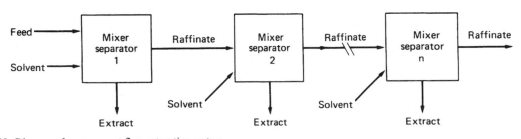

FIG. 10.25. Diagram of a co-current flow extraction system.

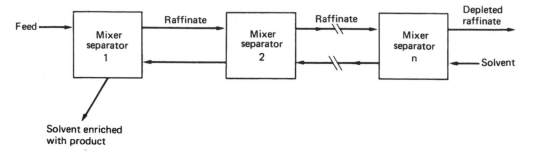

FIG. 10.26. Diagram of a counter-current extraction system.

penicillin is more soluble in organic solvents. At pH 2 to 3 the distribution ratio of total acid will be

$$K = \frac{(RCOOH)org}{(RCOOH)aq + (RCOO^-)aq}.$$

For penicillin this value may be as high as 40 in a suitable solvent (Podbielniak *et al.*, 1970). The penicillin extraction process may involve the four following stages:

1. Extraction of the penicillin G from the filtered broth into an organic solvent (amyl or butyl acetate or methyl iso-butyl ketone).

2. Extraction from the organic solvent into an aqueous buffer.
3. Extraction from aqueous buffer into organic solvent.
4. Extraction of the solvent to obtain the penicillin salt.

At each extraction stage progressively smaller volumes of extractant are used to achieve concentration of the penicillin (see also Fig. 10.2). Unfortunately, penicillin G has a half-life of 15 minutes at pH 2.0 at 20°. The harvested broth is therefore initially cooled to 0° to 3°. The cooled broth is then acidified to pH 2 to 3 with sulphuric or phosphoric acid immediately before extraction. This acidified broth is quickly passed through

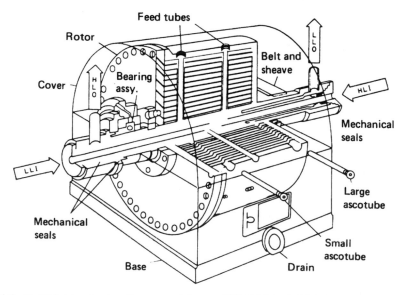

FIG. 10.27. Diagram of the Podbielniak extractor (Queener and Swartz, 1979). HLI, LLI, HLO and LLO indicate heavy and light liquid in and out.

a Podbielniak centrifugal counter-current extractor using about 20% by volume of the solvent in the counter flow. Ideally, the hold-up time should be about 60 to 90 seconds. The penicillin-rich solvent then passes through a second Podbielniak extractor counter-current to an aqueous NaOH or KOH solution (again about 20% by volume) so that the penicillin is removed to the aqueous phase (pH 7.0 to 8.0) as the salt.

$$RCOOH(org) + NaOH(aq) \rightarrow RCOO^-Na^+ + H_2O.$$

These two stages may be sufficient to concentrate the penicillin adequately from a broth with a high titre. Penicillin will crystallize out of aqueous solution at a concentration of approximately $1.5 \times 10^6$ units $cm^{-3}$. If the broth harvested initially contains 60,000 units $cm^{-3}$, and two five-fold concentrations are achieved in the two extraction stages, then the penicillin liquor should crystallize. If the initial broth titre is lower than 60,000 units $cm^{-3}$ or the extractions are not so effective, the solvent and buffer extractions will have to be repeated. At each stage the spent liquids should be checked for residual penicillin and solvent usage carefully monitored. Since the solvents are expensive and their disposal is environmentally sensitive they are recovered for recirculation through the extraction process. The success of a process may depend on efficient solvent recovery and reuse.

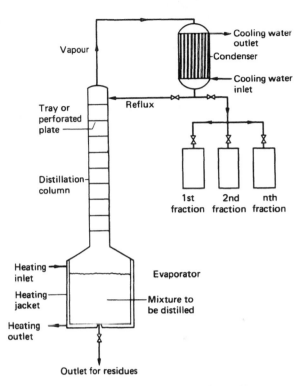

FIG. 10.28. Diagram of a batch distillation plant with a tray or perforated-plate column.

## SOLVENT RECOVERY

A major item of equipment in an extraction process is the solvent-recovery plant which is usually a distillation unit. It is not normally essential to remove all the raffinate from the solvent as this will be recycled through the system. In some processes the more difficult problem will be to remove all the solvent from the raffinate because of the value of the solvent and problems which might arise from contamination of the product.

Distillation may be achieved in three stages:

1. Evaporation, the removal of solvent as a vapour from a solution.
2. Vapour–liquid separation in a column, to separate the lower boiling more volatile component from other less volatile components.
3. Condensation of the vapour, to recover the more volatile solvent fraction.

Evaporation is the removal of solvent from a solution by the application of heat to the solution. A wide range of evaporators is available. Some are operated on a batch basis and others continuously. Most industrial evaporators employ tubular heating surfaces. Circulation of the liquid past the heating surfaces may be induced by boiling or by mechanical agitation. In batch distillation (Fig. 10.28) the vapour from the boiler passes up the column and is condensed. Part of the condensate will be returned as the reflux for counter-current contact with the rising vapour in the column. The distillation is continued until a satisfactory recovery of the lower-boiling (more volatile) component(s) has been accomplished. The ratio of condensate returned to the column as reflux to that withdrawn as product is, along with the number of plates or stages in the column, the major method of controlling the product purity. A continuous distillation (Fig. 10.29) is initially begun in a similar way as with a batch distillation, but no condensate is withdrawn initially. There is total reflux of the condensate until ideal operating conditions have been established throughout the column. At this stage the liquid feed is fed into the column at an intermediate level. The more volatile components move upwards as vapour and are condensed, followed by partial reflux of

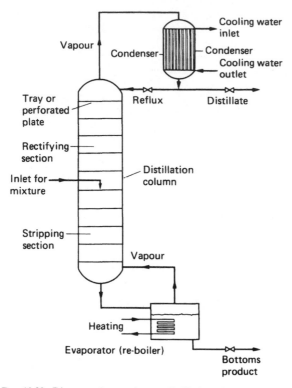

Fig. 10.29. Diagram of a continuous distillation plant with a tray or perforated-plate column.

the condensate. Meanwhile, the less volatile fractions move down the column to the evaporator (reboiler). At this stage part of the bottoms fraction is continuously withdrawn and part is reboiled and returned to the column.

Counter-current contacting of the vapour and liquid streams is achieved by causing:

(a) vapour to be dispersed in the liquid phase (plate or tray column),
(b) liquid to be dispersed in a continuous vapour phase (packed column).

The plate or tray column consists of a number of distinct chambers separated by perforated plates or trays. The rising vapour bubbles through the liquid which is flowing across each plate, and is dispersed into the liquid from perforations (sieve plates) or bubble caps. The liquid flows across the plates and reaches the reboiler by a series of overflow wiers and down pipes.

A packed tower is filled with a randomly packed material such as rings, saddles, helices, spheres or beads. Their dimensions are approximately one-tenth

to one-fiftieth of the diameter of the column and are designed to provide a large surface area for liquid–vapour contacting and high voidage to allow high throughput of liquid and vapour.

The heat input to a distillation column can be considerable. The simplest ways of conserving heat are to preheat the initial feed by a heat exchanger using heat from:

(a) the hot vapours at the top of the column,
(b) heat from the bottoms fraction when it is being removed in a continuous process,
(c) a combination of both.

Since it is beyond the scope of this text to consider the distillation process more fully the reader is therefore directed to Coulson and Richardson (1991).

## TWO-PHASE AQUEOUS EXTRACTION

Liquid–liquid extraction is a well established technology in chemical processing and in certain sectors of biochemical processing. However, the use of organic solvents has limited application in the processing of sensitive biologicals. Aqueous two-phase systems, on the other hand, have a high water content and low interfacial surface tension and are regarded as being biocompatible (Mattiasson and Ling, 1987).

Two-phase aqueous systems have been known since the late nineteenth century, and a large variety of natural and synthetic hydrophilic polymers are used today to create two (or more) aqueous phases. Phase separation occurs when hydrophilic polymers are added to an aqueous solution, and when the concentrations exceed a certain value two immiscible aqueous phases are formed. Settling time for the two phases can be prolonged, depending on the components used and vessel geometry. Phase separation can be improved by using centrifugal separators (Huddlestone et al., 1991), or novel techniques such as magnetic separators (Wikstrom et al., 1987).

Many systems are available:

(i) Non-ionic polymer/non-ionic polymer/water, e.g. polyethylene glycol/dextran.
(ii) Polyelectrolyte/non-ionic polymer/water, e.g. sodium carboxymethyl cellulose/polyethylene glycol.
(iii) Polyelectrolyte/polyelectrolyte/water, e.g. sodium dextran sulphate/sodium carboxy-methyl cellulose

(iv) Polymer/low molecular weight compo-
nent/water, e.g. dextran/propyl alcohol.

The distribution of a solute species between the
phases is characterized by the partition coefficient, and
is influenced by a number of factors such as temperature, polymer (type and molecular weight), salt concentration, ionic strength, pH and properties (e.g. molecular weight) of the solute. As the goal of any extraction
process is to selectively recover and concentrate a
solute, affinity techniques such as those applied in
chromatographic processes can be used to improve
selectivity. Examples include the use of PEG–NADH
derivatives in the extraction of dehydrogenases, *p*-
aminobenzamidine in the extraction of trypsin and
cibacron blue in the extraction of phosphofructokinase.
It is possible to use different ligands in the two phases
leading to an increase in selectivity or the simultaneous
recovery and separation of several species (Cabral and
Aires-Barros, 1993).

Two phase aqueous systems have found application
in the purification of many solutes; proteins, enzymes
(Gonzalez *et al.*, 1990; Guan *et al.*, 1992), cells and
subcellular particles, and in extractive bioconversions.
Several aqueous two-phase systems for handling large-
scale protein separation have emerged, the majority of
which use PEG as the upper phase forming polymer
with either dextran, concentrated salt solution or hy-
droxypropyl starch as the lower phase forming material
(Mattiasson and Kaul, 1986). Hustedt *et al.* (1988)
demonstrated the application of continuous cross-current extraction of enzymes (fumarase and penicillin
acylase) by aqueous two-phase systems at production
scale.

## SUPERCRITICAL FLUID EXTRACTION

The technique of supercritical fluid extraction utilizes the dissolution power of supercritical fluids, i.e.
fluids above their critical temperature and pressure. Its
advantages include the use of moderate temperatures,
and that several cheap and non-toxic fluids are available.

Supercritical fluids are used in the extraction of hop
oils, caffeine, vanilla, vegetable oils and $\beta$-carotene. It
has also been shown experimentally that the extraction
of certain steroids and chemotherapeutic drugs can be
achieved using supercritical fluids. Other current and
potential uses include the removal of undesirable substances such as pesticide residues, removal of bacte-
riostatic agents from fermentation broths, the recovery

of organic solvents from aqueous solutions, cell disintegration, destruction and treatment of industrial wastes
and liposome preparation. There are, however, a number of significant disadvantages in the utilization of this
technology:

(i) Phase equilibria of the solvent/solute system is
complex, making design of extraction conditions difficult.
(ii) The most popular solvent (carbon dioxide) is
non-polar and is therefore most useful in the
extraction of non-polar solutes. Though co-
solvents can be added for the extraction of
polar compounds, they will complicate further
downstream processing.
(iii) The use of high pressures leads to capital costs
for plant, and operating costs may also be high.

Thus, the number of commercial processes utilizing
supercritical fluid extraction is relatively small, due
mainly to the existence of more economical processes.
However, its use is likely to increase in some sectors,
for example the recovery of high value biologicals,
when conventional extractions are inappropriate, and
in the treatment of toxic wastes (Bruno *et al.*, 1993).

## CHROMATOGRAPHY

In many fermentation processes, chromatographic
techniques are used to isolate and purify relatively low
concentrations of metabolic products. In this context,
chromatography will be concerned with the passage
and separation of different solutes as liquid is passed
through a column, i.e. *liquid chromatography*. Depending on the mechanism by which the solutes may be
differentially held in a column, the techniques can be
grouped as follows:

(a) Adsorption chromatography.
(b) Ion-exchange chromatography.
(c) Gel permeation chromatography.
(d) Affinity chromatography.
(e) Reverse phase chromatography.
(f) High performance liquid chromatography.

Chromatographic techniques are also used in the final
stages of purification of a number of products. The
scale-up of chromatographic processes can prove difficult, and there is much current interest in the use of
mathematical models and computer programmes to

translate data obtained from small-scale processes into operating conditions for larger scale applications (Cowan *et al.*, 1986, 1987).

### Adsorption chromatography

Adsorption chromatography involves binding of the solute to the solid phase primarily by weak Van de Waals forces. The materials used for this purpose to pack columns include inorganic adsorbants (active carbon, aluminium oxide, aluminium hydroxide, magnesium oxide, silica gel) and organic macro-porous resins. Adsorption and affinity chromatography are mechanistically identical, but are strategically different. In affinity systems selectivity is designed rationally whilst in adsorption selectivity must be determined empirically.

Di-hydro-streptomycin can be extracted from filtrates using activated charcoal columns. It is then eluted with methanolic hydrochloric acid and purified in further stages (Nakazawa *et al.*, 1960). Some other applications for small-scale antibiotic purification are quoted by Weinstein and Wagman (1978). Active carbon may be used to remove pigments to clarify broths. Penicillin-containing solvents may be treated with 0.25 to 0.5% active carbon to remove pigments and other impurities (Sylvester and Coghill, 1954).

Macro-porous adsorbants have also been tested. The first synthetic organic macro-porous adsorbants, the Amberlite XAD resins, were produced by Rohm and Haas in 1965. These resins have surface polarities which vary from non-polar to highly polar and do not possess any ionic functional groups. Voser (1982) considers their most interesting application to be in the isolation of hydrophilic fermentation products. He stated that these resins would be used at Ciba-Geigy in recovery of cephalosporin C (acidic amino acid), cefotiam (basic amino acid), desferrioxamine B (basic hydroxamic acid) and paramethasone (neutral steroid).

### Ion exchange

Ion exchange can be defined as the reversible exchange of ions between a liquid phase and a solid phase (ion-exchange resin) which is not accompanied by any radical change in the solid structure. Cationic ion-exchange resins normally contain a sulphonic acid, carboxylic acid or phosphonic acid active group. Carboxy-methyl cellulose is a common cation exchange resin. Positively charged solutes (e.g. certain proteins) will bind to the resin, the strength of attachment de-

pending on the net charge of the solute at the pH of the column feed. After deposition solutes are sequentially washed off by the passage of buffers of increasing ionic strength or pH. Anionic ion-exchange resins normally contain a secondary amine, quaternary amine or quaternary ammonium active group. A common anion exchange resin, DEAE (diethylaminoethyl) cellulose is used in a similar manner to that described above for the separation of negatively charged solutes. Other functional groups may also be attached to the resin skeleton to provide more selective behaviour similar to that of affinity chromatography. The appropriate resin for a particular purpose will depend on various factors such as bead size, pore size, diffusion rate, resin capacity, range of reactive groups and the life of the resin before replacement is necessary. Weak-acid cation ion-exchange resins can be used in the isolation and purification of streptomycin, neomycin and similar antibiotics.

In the recovery of streptomycin, the harvested filtrate is fed on to a column of a weak-acid cationic resin such as Amberlite IRC 50 which is in the sodium form. The streptomycin is adsorbed on to the column and the sodium ions are displaced.

$$RCOO^-Na^+_{(resin)} + streptomycin$$

$$\rightarrow RCOO^- streptomycin^+_{(resin)} + NaOH$$

Flow rates of between 10 and 30 bed volumes per hour have been used. The resin bed is now rinsed with water and eluted with dilute hydrochloric acid to release the bound streptomycin.

$$RCOO^- streptomycin^+_{(resin)} + HCl$$

$$\rightarrow RCOOH_{(resin)} + streptomycin^+ Cl^-$$

A slow flow is used to ensure the highest recovery of streptomycin using the smallest volume of eluant. In one step the antibiotic has been both purified and concentrated, maybe more than 100-fold. The resin column is regenerated to the sodium form by passing an adequate volume of NaOH slowly through the column and rinsing with distilled water to remove excess sodium ions.

$$RCOOH_{(resin)} + NaOH \rightarrow RCOO^-Na^+_{(resin)} + H_2O$$

The resin can have a capacity of 1 g of streptomycin $g^{-1}$ resin. Commercially, it is not economic to regenerate the resin completely, therefore the capacity will be reduced. In practice, the filtered broth is taken through two columns in series while a third is being eluted and regenerated. When the first column is saturated, it is isolated for elution and regeneration while the third column is brought into operation.

Details for isolation of some other antibiotics are given in Weinstein and Wagman (1978). Ion-exchange chromatography may be combined with HPLC in, for example, the purification of somatotropin using DEAE cellulose columns and β-urogastrone in multi-gram quantities using a cation exchange column (Brewer and Larsen, 1987).

## Gel permeation

This technique is also known as gel exclusion and gel filtration. Gel permeation separates molecules on the basis of their size. The smaller molecules diffuse into the gel more rapidly than the larger ones, and penetrate the pores of the gel to a greater degree. This means that once elution is started, the larger molecules which are still in the voids in the gel will be eluted first. A wide range of gels are available, including cross-linked dextrans (Sephadex and Sephacryl) and cross-linked agarose (Sepharose) with various pore sizes depending on the fractionation range required.

One early industrial application, although on a relatively small scale, was the purification of vaccines (Latham et al., 1967). Tetanus and diphtheria broths for batches of up to 100,000 human doses are passed through a 13 dm$^3$ column of G 100 followed by a 13 dm$^3$ column of G 200. This technique yields a fairly pure fraction which is then concentrated ten-fold by pressure dialysis to remove the eluant buffer ($Na_2HPO_4$).

## Affinity chromatography

Affinity chromatography is a separation technique with many applications since it is possible to use it for separation and purification of most biological molecules on the basis of their function or chemical structure. This technique depends on the highly specific interactions between pairs of biological materials such as enzyme–substrate, enzyme–inhibitor, antigen–antibody, etc. The molecule to be purified is specifically adsorbed from, for example, a cell lysate applied to the affinity column by a binding substance (ligand) which is immobilized on an insoluble support (matrix). Eluent is then passed through the column to release the highly purified and concentrated molecule. The ligand is at-

tached to the matrix by physical absorption or chemically by a covalent bond. The pore size and ligand location must be carefully matched to the size of the product for effective separation. The latter method is preferred whenever possible. Porath (1974) and Yang and Tsao (1982) have reviewed methods and coupling procedures.

Coupling procedures have been developed using cyanogen bromide, bisoxiranes, disaziridines and periodates, for matrixes of gels and beads. Four polymers which are often used for matrix materials are agarose, cellulose, dextrose and polyacrylamide. Agarose activated with cyanogen bromide is one of the most commonly used supports for the coupling of amino ligands. Silica based solid phases have been shown to be an effective alternative to gel supports in affinity chromatography (Mohan and Lyddiatt, 1992).

Purification may be several thousand-fold with good recovery of active material. The method can however be quite costly and time consuming, and alternative affinity methods such as affinity cross-flow filtration, affinity precipitation and affinity partitioning may offer some advantages (Janson, 1984; Luong et al., 1987). Affinity chromatography was used initially in protein isolation and purification, particularly enzymes. Since then many other large-scale applications have been developed for enzyme inhibitors, antibodies, interferon and recombinant proteins (Janson and Hedman, 1982; Ostlund, 1986; Folena-Wasserman et al., 1987; Nachman et al., 1992), and on a smaller scale for nucleic acids, cell organelles and whole cells (Yang and Tsao, 1982). In the scale-up of affinity chromatographic processes (Katoh, 1987) bed height limits the superficial velocity of the liquid, thus scale-up requires an increase in bed diameter or adsorption capacity.

## Reverse phase chromatography (RPC)

This chromatographic method utilizes a solid phase (e.g. silica) which is modified so as to replace hydrophilic groups with hydrophobic alkyl chains. This allows the separation of proteins according to their hydrophobicity. More-hydrophobic proteins bind most strongly to the stationary phase and are therefore eluted later than less-hydrophobic proteins. The alkyl groupings are normally eight or eighteen carbons in length ($C_8$ and $C_{18}$). RPC can also be combined with affinity techniques in the separation of, for example, proteins and peptides (Davankov et al., 1990).

## High performance liquid chromatography (HPLC)

HPLC is a high resolution column chromatographic technique. Improvements in the nature of column packing materials for a range of chromatographic techniques (e.g. gel permeation and ion-exchange) yield smaller, more rigid and more uniform beads. This allows packing in columns with minimum spaces between the beads, thus minimizing peak broadening of eluted species. It was originally known as high *pressure* liquid chromatography because of the high pressures required to drive solvents through silica based packed beds. Improvements in performance led to the name change and its widespread use in the separation and purification of a wide range of solute species, including bio-molecules. HPLC is distinguished from liquid chromatography by the use of improved media (in terms of their selectivity and physical properties) for the solid (stationary) phase through which the mobile (fluid) phase passes.

The stationary phase must have high surface area/unit volume, even size and shape and be resistant to mechanical and chemical damage. However, it is factors such as these which lead to high pressure requirements and cost. This may be acceptable for analytical work, but not for preparative separations. Thus, in preparative HPLC some resolution is often sacrificed (by the use of larger stationary-phase particles) to reduce operating and capital costs. For very high value products large-scale HPLC columns containing analytical media have been used.

Affinity techniques can be merged with HPLC to combine the selectivity of the former with the speed and resolving power of the latter (Forstecher *et al.*, 1986; Shojaosadaty and Lyddiatt, 1987).

## Continuous chromatography

Although the concept of continuous enzyme isolation is well established (Dunnill and Lilly, 1972), the stage of least development is continuous chromatography. Fox *et al.* (1969) developed a continuous-fed column for this purpose (Fig. 10.30). It consisted of two concentric cylindrical sections clamped to a base plate. The space (1 cm wide) between the two sections was packed with the appropriate resin or gel giving a total column capacity of 2.58 dm$^3$. A series of orifices in the circumference of the base plate below the column space led to collecting vessels. The column assembly was rotated in a slow-moving turntable (0.4–2.0 rpm). The mixture for separation was fed to the apparatus by

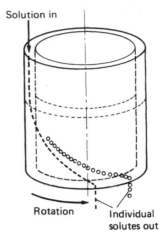

FIG. 10.30. The principle of continuous-partition chromatography. –––, faster-moving component; O O, slower-moving component (Fox, 1969).

an applicator rotating at the same speed as the column, thus allowing application at a fixed point, while the eluent was fed evenly to the whole circumference of the column. The components of a mixture separated as a series of helical pathways, which varied with the retention properties of the constituent components. This method gave a satisfactory separation and recovery but the consumption of eluent and the unreliable throughput rate were not considered to be satisfactory for a large-scale method (Nicholas and Fox, 1969; Dunnill and Lilly, 1972). However, the development of such continuous separation equipment suitable for large-scale extraction would considerably simplify the use of chromatographic separation.

## MEMBRANE PROCESSES

### Ultrafiltration and reverse osmosis

Both processes utilize semi-permeable membranes to separate molecules of different sizes and therefore act in a similar manner to conventional filters.

### Ultrafiltration

Ultrafiltration can be described as a process in which solutes of high molecular weight are retained when the solvent and low molecular weight solutes are forced under hydraulic pressure (around 7 atmospheres) through a membrane of a very fine pore size. It is therefore used for product concentration and purification. A range of membranes made from a variety of polymeric materials, with different molecular weight cut-offs (500 to 500,000), are available which makes possible the separation of macro-molecules such as proteins, enzymes, hormones and viruses. It is practical only to separate molecules whose molecular weights are a factor of ten different due to variability in pore size (Heath and Belfort, 1992). Because the flux through such a membrane is inversely proportional to its thickness, asymmetric membranes are used where the membrane ($\sim 0.3$ $\mu$m thick) is supported by a mesh around 0.3 mm thick.

When considering the feasibility of ultrafiltration it is important to remember that factors other than the molecular weight of the solute affect the passage of molecules through the membranes (Melling and Westmacott, 1972). There may be concentration polarization caused by accumulation of solute at the membrane surface which can be reduced by increasing the shear forces at the membrane surface either by conventional agitation or by the use of a cross-flow system (see previous section). Secondly a slurry of protein may accumulate on the membrane surface forming a gel layer which is not easily removed by agitation. Formation of the gel layer may be partially controlled by careful choice of conditions such as pH (Bailey and Ollis, 1986). Finally, equipment and energy costs may be considerable because of the high pressures necessary; this also limits the life of ultrafiltration membranes.

There are numerous examples of the use of ultrafiltration for the recovery of bio-molecules: viruses (Weiss, 1980), enzymes (Atkinson and Mavituna, 1991), antibiotics (Pandey et al., 1985). Details of large scale applications are given by Lacey and Loeb (1972) and by Ricketts et al. (1985). Affinity ultrafiltration (Luong et al., 1987; Luong and Nguyen, 1992) is a novel separation process developed to circumvent difficulties in affinity chromatography. It offers high selectivity, yield and concentration, but it is an expensive batch process and scale up is difficult.

### Reverse osmosis

Reverse osmosis is a separation process where the solvent molecules are forced by an applied pressure to flow through a semi-permeable membrane in the opposite direction to that dictated by osmotic forces, and hence is termed reverse osmosis. It is used for the concentration of smaller molecules than is possible by ultrafiltration. Concentration polarization is again a problem and must be controlled by increased turbulence at the membrane surface.

### Liquid membranes

Liquid membranes are insoluble liquids (e.g. an organic solvent) which are selective for a given solute and separate two other liquid phases. Extraction takes place by the transport of solute from one liquid to the other. They are of great interest in the extraction and purification of biologicals for the following reasons:

(a) Large area for extraction.
(b) Separation and concentration are achieved in one step.
(c) Scale-up is relatively easy.

Their use has been reported in the extraction of lactic acid (Chaudhuri and Pyle, 1990) and citric acid using a supported liquid membrane (Sirman et al., 1990). The utilization of selective carriers to transport specific components across the liquid membrane at relatively high rates has increased interest in recent years (Strathmann, 1991). Liquid membranes may also be used in cell and enzyme immobilization, and thus provide the opportunity for combined production and isolation/extraction in a single unit (Mohan and Li, 1974, 1975). The potential use of liquid membranes has also been described for the production of alcohol reduced beer as having little effect on flavour or the physico-chemical properties of the product (Etuk and Murray, 1990).

### DRYING

The drying of any product (including biological products) is often the last stage of a manufacturing process (McCabe et al., 1984; Coulson and Richardson, 1991). It involves the final removal of water from a heat-sensitive material ensuring that there is minimum loss in viability, activity or nutritional value. Drying is undertaken because:

(a) The cost of transport can be reduced.
(b) The material is easier to handle and package.
(c) The material can be stored more conveniently in the dry state.

A detailed review of the theory and practice of drying can be found in Perry and Green (1984). It is important that as much water as possible is removed initially by centrifugation or in a filter press to minimize heating costs in the drying process. Driers can be classified by the method of heat transfer to the product and the degree of agitation of the product. In contact driers the product is contacted with a heated surface. An example of this type is the drum drier (Fig. 10.31), which may be used for more temperature stable bio-products. A slurry is run onto a slowly rotating steam heated drum, evaporation takes place and the dry product is removed by a scraper blade in a similar manner as for rotary vacuum filtration. The solid is in contact with the heating surface for 6–15 seconds and heat transfer coefficients are generally between 1 and 2 kW m$^{-2}$ K$^{-1}$. Vacuum drum driers can be used to lower the temperature of drying.

A spray drier (Fig. 10.32) is most widely used for drying of biological materials when the starting material is in the form of a liquid or paste. The material to be dried does not come into contact with the heating surfaces, instead, it is atomized into small droplets through for example a nozzle or by contact with a rotating disc. The wide range of atomizers available is described in Coulson and Richardson (1991). The droplets then fall into a spiral stream of hot gas at 150° to 250°. The high surface area:volume ratio of the droplets results in a rapid rate of evaporation and complete drying in a few seconds, with drying rate and product size being directly related to droplet size produced by the atomizer. The evaporative cooling effect prevents the material from becoming overheated and damaged. The gas-flow rate must be carefully regulated

so that the gas has the capacity to contain the required moisture content at the cool-air exhaust temperature (75° to 100°). In most processes the recovery of very small particles from the exit gas must be conducted using cyclones or filters. This is especially important for containment of biologically active compounds. The jet spray drier is particularly suited to handling heat sensitive materials. Operating at a temperature of around 350°, residence times are approximately 0.01 seconds because of the very fine droplets produced in the atomizing nozzle.

Spray driers are the most economical available for handling large volumes, and it is only at feed rates below 6 kg min$^{-1}$ that drum driers become more economic.

Freeze drying is an important operation in the production of many biologicals and pharmaceuticals. The material is first frozen and then dried by sublimation in a high vacuum. The great benefit of this technique is that it does not harm heat sensitive materials. The process is often termed lyophilization when the solvent being evaporated is water.

Fluidized bed driers are used increasingly in the

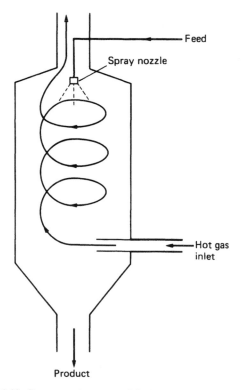

FIG. 10.32. Counter-current spray drier.

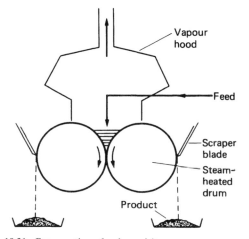

FIG. 10.31. Cross-section of a drum drier.

pharmaceutical industry. Heated air is fed into a chamber of fluidized solids, to which wet material is continuously added and dry material continuously removed. Very high mass-transfer rates are achieved, giving rapid evaporation and allowing the whole bed to be maintained in a dry condition.

## CRYSTALLIZATION

Crystallization is an established method used in the initial recovery of organic acids and amino acids, and more widely used for final purification of a diverse range of compounds.

In citric acid production, the filtered broth is treated with $Ca(OH)_2$ so that the relatively insoluble calcium citrate crystals will be precipitated from solution. Checks are made to ensure that the $Ca(OH)_2$ has a low magnesium content, since magnesium citrate is more soluble and would remain in solution. The calcium citrate is filtered off and treated with sulphuric acid to precipitate the calcium as the insoluble sulphate and release the citric acid. After clarification with active carbon, the aqueous citric acid is evaporated to the point of crystallization (Lockwood and Irwin, 1964; Sodeck et al., 1981; Atkinson and Mavituna, 1991). Crystallization is also used in the recovery of amino acids; Samejima (1972) has reviewed methods for glutamic acid, lysine and other amino acids. The recovery of cephalosporin C as its sodium or potassium salt by crystallization has been described by Wildfeuer (1985).

## WHOLE BROTH PROCESSING

The concept of recovering a metabolite directly from an unfiltered fermentation broth is of considerable interest because of its simplicity, the reduction in process stages and the potential cost savings. It may also be possible to remove the desired fermentation product continuously from a broth during fermentation so that inhibitory effects due to product formation and product degradation can be minimized throughout the production phase (Roffler et al., 1984; Diaz, 1988).

Bartels et al. (1958) developed a process for adsorption of streptomycin on to a series of cationic ion-exchange resin columns directly from the fermentation broth, which had only been screened to remove large particles so that the columns would not become blocked. This procedure could only be used as a batch process. Belter et al. (1973) developed a similar process

for the recovery of novobiocin. The harvested broth was first filtered through a vibrating screen to remove large particles. The broth was then fed into a continuous series of well-mixed resin columns fitted with screens to retain the resin particles, plus the absorbed novobiocin, but allow the streptomycete filaments plus other small particulate matter to pass through. The first resin column was removed from the extraction line after a predetermined time and eluted with methanolic ammonium chloride to recover the novobiocin.

Karr et al. (1980) developed a reciprocating plate extraction column (Fig. 10.33) to use for whole broth processing of a broth containing 1.4 g $dm^{-3}$ of a slightly soluble organic compound and 4% undissolved solids provided that chloroform or methylene chloride were used for extraction. Methyl-iso-butyl ketone, diethyl ketone and iso-propyl acetate were shown to be more efficient solvents than chloroform for extracting the active compound, but they presented problems since they also extracted impurities from the mycelia, making it necessary to filter the broth before beginning the solvent extraction. Considerable economies were claimed in a comparison with a process using a Podbielniak extractor, in investment, maintenance costs, solvent usage and power costs but there was no significant difference in operating labour costs.

An alternative approach is to remove the metabolite continuously from the broth during the fermentation. Cycloheximide production by Streptomyces griseus has been shown to be affected by its own feed-back regulation (Kominek, 1975). Wang et al. (1981) have tested two techniques at laboratory scale for improving production of cycloheximide. In a dialysis method (Fig. 10.34), methylene chloride was circulated in a dialysis tubing loop which passed through a 10 $dm^{-3}$ fermenter. Cycloheximide in the fermentation broth was extracted into the methylene chloride. It was shown that the product yield could be almost doubled by this dialysis-solvent extraction method to over 1200 $\mu$g $cm^{-3}$ as compared with a control yield of approximately 700 $\mu$g $cm^{-3}$. In a resin method, sterile beads of XAD-7, an acrylic resin, as dispersed beads or beads wrapped in an ultrafiltration membrane, were put in fermenters 48 hours after inoculation. Some of the cycloheximide formed in the broth is absorbed by the resin. Recovery of the antibiotic from the resin is achieved by solvents or by changing the temperature or pH. When assayed after harvesting, the control (without resin) had a bioactivity of 750 $\mu$g $cm^{-3}$. Readings of total bioactivity (from beads and broth) for the bead treatment and the membrane-wrapped bead treatments

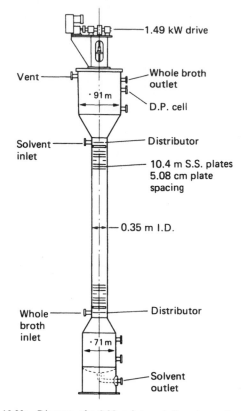

FIG. 10.33a. Diagram of a 0.35-m internal diameter reciprocating plate column (Karr *et al.*, 1980).

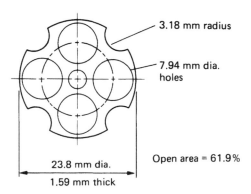

FIG. 10.33b. Plan of a 23.8-m stainless-steel plate for a 25-mm diameter reciprocating plate test column (Karr *et al.*, 1980).

were 1420 $\mu$g cm$^{-3}$ and 1790 $\mu$g cm$^{-3}$ respectively.

Roffler *et al.* (1984) reviewed the use of a number of techniques for the *in-situ* recovery of fermentation products:

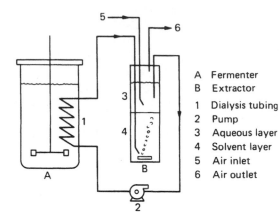

FIG. 10.34. Dialysis-extraction fermentation system (Wang *et al.*, 1981).

(a) Vacuum and flash fermentations for the direct recovery of ethanol from fermentation broths.
(b) Extractive fermentation (liquid–liquid and two-phase aqueous) for the recovery of ethanol, organic acids and toxin produced by *Clostridium tetani*.
(c) Adsorption for the recovery of ethanol and cycloheximide.
(d) Ion-exchange in the extraction of salicylic acid and antibiotics.
(e) Dialysis fermentation in the selective recovery of lactic acid, salicylic acid and cycloheximide.

Hansson *et al.* (1994) have used an expanded adsorption bed for the recovery of a recombinant protein produced by *E. coli* directly from the fermentation broth. The protein was produced in high yields (550 mg dm$^{-3}$) and > 90% recovery together with concentration (volume reduction) and removal of cells was achieved on the expanded bed. Affinity chromatography was used for further purification, and again an overall yield of > 90% obtained.

## REFERENCES

AIBA, S., HUMPHREY, A. E. and MILLIS, N. F. (1973) Recovery of fermentation products. In *Biochemical Engineering* 2nd edition, pp. 346–392. Academic Press, New York.

AGERKVIST, I. and ENFORS, S.-O. (1990) Characterisation of *E. coli* cell disintegrates from a bead mill and high pressure homogenizers. *Biotech. Bioeng.* **36**(11), 1083–1089.

ANDREWS, B. A. and ASENJO, J. A. (1987) Enzymatic lysis and

disruption of microbial cells. *Trends Biotechnol.* 5, 273–277.

ANDREWS, B. A., HUANG, R. B. and ASENJO, J. A. (1990) Differential product release from yeast cells by selective enzymatic lysis. In *Separations for Biotechnology 2*, pp. 21–28 (Editor Pyle, D. L.). Elsevier, London.

ASENJO, J. A. (1990) Cell disruption and removal of insolubles. In *Separations for Biotechnology 2*, pp. 11–20. (Editor Pyle, D. L.). Elsevier, London.

ASHLEY, M. H. J. (1990) Conceptual design of a novel flocculation device. In *Separations for Biotechnology 2*, pp. 29–37 (Editor Pyle, D. L.). Elsevier, London.

ATKINSON, B. and DAOUD, I. S. (1976) Microbial flocs and flocculation in fermentation process engineering. *Adv. Biochem. Eng.* 4, 41–124.

ATKINSON, B. and MAVITUNA, F. (1991) *Biochemical Engineering and Biotechnology Handbook* (2nd edition), Chapters 16 and 20. Macmillan, London.

ATKINSON, B. and SAINTER, P. (1982) Development of downstream processing. *J. Chem. Tech. Biotechnol.* 32, 100–108.

AUNSTRUP, K. (1979) Production, isolation and economics of extracellular enzymes. In *Applied Biochemistry and Bioengineering*, Vol. 2, pp. 27–69. (Editors Wingard, L. B. and Katzir-Katchalski, E.). Academic Press, New York.

BAILEY, J. E. and OLLIS, D. F. (1986) *Biochemical Engineering Fundamentals* (2nd edition), pp. 768–769. McGraw-Hill, New York.

BARTELS, C. R., KLEIMAN, G., KORZUN, J. N. and IRISH, D. B. (1958) A novel ion-exchange method for the isolation of streptomycin. *Chem. Eng. Prog.* 54, 49–51.

BELL, G. B. and HUTTO, F. B. (1958) Analysis of rotary pre-coat filter operation. 1. New concepts. *Chem. Eng. Prog.* 54, 69–76.

BELTER, P. A., CUNNINGHAM, F. L. and CHEN, J. W. (1973) Development of a recovery process for novobiocin. *Biotech. Bioeng.* 15, 533–549.

BENTHAM, C. C., BONNERJEA, J., ORSBORN, C. B., WARD, P. N. and HOARE, M. (1990) The separation of affinity flocculated yeast cell debris using a pilot plant scroll decanter centrifuge. *Biotech. Bioeng.* 36(4), 397–401.

BOWDEN, C. P., LEAVER, G., MELLING, J., NORTON, M. G. and WHITTINGTON, P. N. (1987) Recent and novel developments in the recovery of cells from fermentation broths. In *Separations for Biotechnology*, pp. 49–61 (Editors Verrall, M. S. and Hudson, M. J.). Ellis Horwood, Chichester.

BREWER, S. J. and LARSEN, B. R. (1987) Isolation and purification of proteins using preparative HPLC. In *Separations for Biotechnology*, pp. 113–126 (Editors Verrall, M. S. and Hudson, M. J.). Ellis Horwood, Chichester.

BROOKMAN, J. S. G. (1974) Mechanism of cell disintegration in a high pressure homogenizer. *Biotech. Bioeng.* 16, 371–383.

BROWN, D. E. and KAVANAGH, P. R. (1987) Cross-flow separation of cells. *Process Biochem.* 22(4), 96–101.

BRUNO, T. J., NIETO DE CASTRO, C. A., HAMEL, J.-F. P. and PALAVRA, A. M. F. (1993) Supercritical fluid extraction of biological products. In *Recovery Processes for Biological Materials*, pp. 303–354 (Editors Kennedy, J. F. and Cabral, J. M. S.). Wiley, Chichester.

CABRAL, J. M. S. and AIRES-BARROWS, M. R. (1993) Liquid–liquid extraction of biomolecules using aqueous two-phase systems. In *Recovery Processes for Biological Materials*, pp. 273–302 (Editors Kennedy, J.F. and Cabral, J.M.S.). Wiley, Chichester.

CHAUDHUR, J.B. and PYLE, D.L. (1990) A model for emulsion liquid membrane extraction of organic acids. In *Separations for Biotechnology 2*, pp. 112–121 (Editor Pyle, D.L.). Elsevier Applied Science, London.

COULSON, J. M. and RICHARDSON, J. F. (1991) *Chemical Engineering* (4th edition), Vol. 2. Pergamon Press, Oxford.

COWAN, G. H., GOSLING, I. S. and LAWS, J. F. (1986) Physical and mathematical modelling to aid scale-up of liquid chromatography. *J. Chromatog.* 363, 37–56.

COWAN, G. H., GOSLING, I. S. and SWEETENHAM, W. P. (1987) Modelling for scale-up and optimisation of packed-bed columns in adsorption and chromatograpy. In *Separations for Biotechnology*, pp. 152–175 (Editors Verrall, M. S. and Hudson, M. J.). Ellis Horwood, Chichester.

CRAPISI, A., LANTE, A., PASINI, G. and SPETTOLI, P. (1993) Enhanced microbial cell lysis by the use of lysozyme immobilised on different carriers. *Process Biochem.* 28(1), 17–21.

DARBYSHIRE, J. (1981) Large scale enzyme extraction and recovery. In *Topics in Enzyme and Fermentation Biotechnology*, Vol. 5, pp. 147–186 (Editor Wiseman, A.). Ellis Horwood, Chichester.

DATAR, R. (1986) Economics of primary separation steps in relation to fermentation and genetic engineering. *Process Biochem.* 21(1), 19–26.

DAVANKOV, V. A., KURGANOV, A. A. and UNGER, K. K. (1990) Reversed phase high performance liquid chromatography of proteins and peptides on polystyrene coated silica supports. *J. Chromatog.* 500, 519–530.

DELAINE, J. (1983) Physico-chemical pretreatment of process effluents. I.Chem.E. Symp. Ser. No. 77, 183–203.

DIAZ, M. (1988) Three-phase extractive fermentation. *TIBTECH.* 6(6), 126–130.

DUNNILL, P. and LILLY, M. D. (1972) Continuous enzyme isolation. *Biotech. Bioeng. Symp.* 3, 97–113.

FILIK, B. R. and MURRAY, K. R. (1990) Potential use of liquid membranes for alcohol reduced beer production. *Process Biochem.* 25(1), 24–32.

FLETCHER, K., DELEY, S., FLEISCHAKER, R. J., FORRESTER, I. T., GRABSKI, A. C. and STRICKLAND, W. N. (1990) Clarification of tissue culture fluid and cell lysates using Biocryl bioprocessing aids. In *Separations for Biotechnology 2*, pp. 142–151 (Editor Pyle, D. L.). Elsevier, London.

FOLENA-WASSERMAN, G., INACKER, R. and ROSENBLOOM, J. (1987) Assay, purification and characterisation of a recombinant malaria circumsporozoite fusion protein by high performance liquid chromatography. *J. Chromatog.* 411, 345–354.

FORSTECHER, P., HAMMADI, H., BOUZERNA, N. and DAUTRE-VAUX, M. (1986) Rapid purification of antisteriod antibodies by high performance liquid affinity chromatography. *J. Chromatog.* **369**, 379–390.

FOX, J. B., CALHOUN, R. C. and EGLINTON, W. J. (1969) Continuous chromatography apparatus. 1. Construction. *J. Chromatog.* **43**, 48–54.

FRUDE, M. J. and SIMPSON, M. T. (1993) Steam sterilisation of a Westfalia CSA 8 centrifuge. *Process Biochem.* **28**, 297–303.

GABLER, R. and RYAN, M. (1985) Processing cell lysate with tangential flow filtration. In *Purification of Fermentation Products*, pp. 1–20. (Editors LeRoith, D., Shiloach, J. and Leahy, T. J.). ACS Symposium Series, 271. ACS, Washington, DC.

GASNER, C. C. and WANG, D. I. C. (1970) Microbial recovery enhancement through flocculation. *Biotech. Bioeng.* **12**, 873–887.

GONZALEZ, M., PENCS, C. and CASAS, L. T. (1990) Partial purification of β-galactosidase from yeast by an aqueous two phase system method. *Proc. Biochem.*, **25**(5), 157–161.

GRAY, P. P., DUNNILL, P. and LILLY, M. D. (1973) The clarification of mechanically disrupted yeast suspensions by rotary vacuum precoat filtration. *Biotech. Bioeng.* **15**, 309–320.

GRIEVES, R. B. and WANG, S. L. (1966) Foam separation of *Escherichia coli* with a cationic surfactant. *Biotech. Bioeng.* **8**, 323–336.

GUAN, Y., WU, X.-Y., TREFFRY, T. E. and LILLEY, T. H. (1992) Studies on the isolation of penicillin acylase from *Escherichia coli* by aqueous two-phase partitioning. *Biotech. Bioeng.* **40**(4), 517–524.

HAMER, G. (1979) Biomass from natural gas. In *Economic Microbiology*, Vol. 4, pp. 315–360 (Editor Rose, A. H.). Academic Press, London.

HANSSON, M., STAHL, S., HJORTH, R., UHLEN, M. and MOKS, T. (1994) Single step recovery of a secreted recombinant protein by expanded bed adsorption. *Bio / Technol*, **12**(3), 285–288.

HASTINGS, J. W., RILEY, W. H. and MASSA, J. (1965) The purification, properties and chemiluminescent quantum yield of bacterial luciferase *J. Biol. Chem.* **240**, 1473–1481.

HEATH, C. A. and BELFORT, G. (1992) Synthetic membranes in biotechnology. *Adv. Biochem. Eng. Biotechnol.* **47**, 45–88.

HETHERINGTON, P.J., FOLLOWS, M., DUNHILL, P. and LILLY, M. D. (1971) Release of protein from baker's yeast (*Saccharomyces cerevisiae*) by disruption in an industrial homogenizer. *Trans. Ind. Chem. Eng.* **49**, 142–148.

HETTEWER, D. J. and WANG, H. Y. (1989) Protein release from *Escherichia coli* cells permeabilized with guanidine-HCl and Triton X-100. *Biotech. Bioeng.* **33**, 886–895.

HONIG, W. and KULA, M. R. (1976) Selectivity of protein precipitation with polyethylene glycol fractions of various molecular weights. *Anal. Biochem.* **72**, 502–512.

HUANG, R.-B., ANDREWS, B. A. and ASENJO, J. A. (1991) Differential product release (DPR) of proteins from yeast: a new technique for selective product recovery from microbial cells. *Biotech. Bioeng.* **38**(9), 977–985.

HUDDLESTONE, J., VEIDE, A., KOHLER, K., FLANAGAN, J., EN-

FORS, S.-O. and LYDDIATT, A. (1991) The molecular basis of partitioning in aqueous two-phase systems. *TIB / TECH.* **9**, 381–388.

HUGHES, D. E., WIMPENNY, J. W. T. and LLOYD, D. (1971) The disintegration of micro-organisms. In *Methods in Microbiology*, Vol. 5B, pp. 1–54 (Editors Norris, J. R. and Ribbons, D. W.). Academic Press, London.

HUSTEDT, H., KRONER, K.-H. and PAPAMICHAEL, N. (1988) Continuous cross-current aqueous two-phase extraction of enzymes from biomass. Automated recovery in production scale. *Process Biochem.* **23**(5), 129–137.

JAMES, C. J., COAKLEY, W. T. and HUGHES, D. E. (1972) Kinetics of protein release from yeast sonicated in batch and flow systems at 20 kHz. *Biotech. Bioeng.* **14**(1), 33–42.

JANSON, J.-C. (1984) Large scale purification — state of the art and future prospects. *Trends Biotechnol.* **2**(2), 31–38.

JANSON, J.-C. and HEDMAN, P. (1982) Large-scale chromatography of proteins. *Adv. Biochem. Eng.* **25**, 43–99.

KARR, A. E., GEBERT, W. and WANG, M. (1980) Extraction of whole fermentation broth with a Karr reciprocating plate extraction column. *Can. J. Chem. Eng.* **58**, 249–252.

KATOH, S. (1987) Scaling-up affinity chromatography. *Trends Biotechnol.* **5**, 328–331.

KOMINEK, L. A. (1975) Cycloheximide production by *Streptomyces griseus*; alleviation of end-product inhibition by dialysis extraction fermentation. *Antimicrobiol. Chemother.* **7**, 856–860.

KRONER, K. H., HUSTEDT, S., GRANDA, S. and KULA, M. R. (1978) Technical aspects of separation using aqueous two-phase systems in enzyme isolation processes. *Biotech. Bioeng.* **20**, 1967–1988.

KURANE, R. (1990) Separation of biopolymers: Separation of suspended solids by microbial flocculant. In *Separations for Biotechnology 2*, pp. 48–54 (Editor Pyle, D. L.). Elsevier, London.

LACEY, R. E. and LOEB, S. (1972) *Industrial Processing with Membranes*. Wiley-Interscience, New York.

LATHAM, W. C., MICHELSEN, C. B. and EDSALL, G. (1967) Preparative procedure for the purification of toxoids by gel filtration. *Appl. Microbiol.* **15**, 616–621.

LE, M. S. and ATKINSON, T. (1985) Crossflow microfiltration for recovery of intracellular products. *Process Biochem.* **20**(1), 26–31.

LEE, C.-K., CHANG, W.-G. and JU, Y.-H. (1993). Air slugs entrapped cross-flow filtration of bacterial suspensions. *Biotech. Bioeng.* **41**(5), 525–530.

LOCKWOOD, L. B. and IRWIN, W. E. (1964) Citric acid. In *Kirk-Othmer Encyclopedia of Chemical Technology* (2nd edition), Vol. 5, pp. 524–541. Wiley, New York.

LOWE, C. R. and STEAD, C. V. (1985) The use of reactive dyestuffs in the isolation of proteins. In *Discovery and Isolation of Microbial Products*, pp. 148–158 (Editor Verrall, M. S.). Ellis Horwood, Chichester.

LUONG, J. H. T. and NGUYEN, A.-L. (1992) Novel separations based on affinity interactions. *Adv. Biochem. Eng. Biotechnol.* **47**, 137–158.

LUONG, J. H. T., NGUYEN, A. L. and MALE, K. B. (1987)

Recent developments in downstream processing based on affinity interactions. *Trends Biotechnol.* **5**, 281–286.

MCCABE, W. L., SMITH, J. C. and HARRIOT, P. (1984) *Unit Operations in Chemical Engineering* (4th edition). McGraw-Hill, New York.

MAGNUSSON, K. E. and EDEBO, L. (1976) Large-scale disintegration of micro-organisms by freeze-pressing. *Biotech. Bioeng.* **18**, 975–986.

MATTIASSON, B. and KAUL, R. (1986) Use of aqueous two-phase systems for recovery and purification in biotechnology. In *Separation, Recovery and Purification in Biotechnology*, pp. 78–92 (Editors Asenjo, J.A. and Hong, J.). ACS Symposium Series 314, ACS, Washington, DC.

MATTIASSON, B. and LING, T. G. I. (1987) Extraction in aqueous two-phase systems for biotechnology. In *Separations for Biotechnology*, pp. 270–292 (Editors Verrall, M.S. and Hudson, M.J.). Ellis Horwood, Chichester.

MELLING, J. and WESTMACOTT, D. (1972) The influence of pH value and ionic strength on the ultrafiltration characteristics of a penicillinase produced by *Escherichia coli* strain W3310. *J. Appl. Chem.* **22**, 951–958.

MOGREN, H., LINDBLOM, M. and HEDENSKOG, G. (1974) Mechanical disintegration of micro-organisms in an industrial homogenizer. *Biotech. Bioeng.* **16**, 261–274.

MOHAN, S. B. and LYDDIATT, A. (1992) Silica based solid phases for affinity chromatography. Effect of pore size and ligand location upon biochemical productivity. *Biotech. Bioeng.* **40**(5), 549–563.

MOHAN, R. R. and LI, N. N. (1974) Reduction and separation of nitrate and nitrite by liquid membrane encapsulated enzymes. *Biotech. Bioeng.* **16**(4), 513–523.

MOHAN, R. R. and LI, N. N. (1975) Nitrate and nitrite reduction by liquid membrane-encapsulated whole cells. *Biotech. Bioeng.* **17**(8), 1137–1156.

MOUROT, P., LaFRANCE, M. and OLIVER, M. (1989) Aseptic concentration of microbial cells by cross-flow filtration. *Process Biochem.* **24**(1), 3–8.

NACHMAN, M., AZAD, A. R. M. and BAILON, P. (1992) Efficient recovery of recombinant proteins using membrane based immunoaffinity chromatography (MIC). *Biotech. Bioeng.* **40**(5), 564–571.

NAGLAK, T. J. and WANG, H. Y. (1992) Rapid protein release from *Escherichia coli* by chemical permeabilization under fermentation conditions. *Biotech Bioeng.* **39**(7), 733–740.

NAKAMURA, H. (1961) Chemical separation methods for common microbes. *J. Biochem. Microbiol. Technol. Eng.* **3**, 395–403.

NAKAZAWA, K., SHIBATA, M., TANABE, K. and YAMAMOTO, H. (1960) Di-hydrostreptomycin. U.S. patent 2,931,756.

NICHOLAS, R. A. and FOX, J. B. (1969) Continuous chromatography apparatus. II Application. *J. Chromatog.* **43**, 61–65.

NIEDERAUER, M. Q. and GLATZ, C.E. (1992) Selective precipitation. *Adv. Biochem. Eng. Biotechnol.* **47**, 159–188.

OMSTEAD, D. R., HUNT, G. R. and BUCKLAND, B. C. (1985) Commercial production of Cephamycin antibiotics. In *Comprehensive Biotechnology*, Vol. 3, pp. 187–210. (Edi-

tors Blanch, H. W., Drew, S. and Wang, D. I. C.). Pergamon, New York.

OSTLUND, C. (1986) Large-scale purification of monoclonal antibodies. *TIBTECH.* **4**(11), 288–293.

PACE, G. W. and RIGHELATO, R. H. (1980) Production of extracellular polysaccharides. *Adv. Biochem. Eng.* **15**, 41–70.

PACE, G. W. and SMITH, I. H. (1981) Recovery of microbial polysaccharides. In *Abstracts of Communications. Second European Congress of Biotechnology, Eastbourne*, p. 36. Society of Chemical Industry, London.

PANDEY, R. C., KALITA, C. C., GUSTAFSON, M. E., KLINE, M. C., LEIDHECKER, M. E. and ROSS, J. T. (1985) Process developments in the isolation of Largomycin F-II, a chemoprotein antitumor antibiotic. In *Purification of Fermentation Products*, pp 133–153 (Editors LeRoith, D., Shiloach, J. and Leahy, T. J.). ACS Symposium Series 271. ACS, Washington, DC.

PERRY, R. H. and CHILTON, C. H. (1974) *Chemical Engineers' Handbook* (5th edition), Section 19, p. 78. McGraw-Hill, New York.

PERRY, R. H. and GREEN, D. W. (1984) *Perry's Chemical Engineers' Handbook* (6th edition), Section 20. McGraw-Hill, New York.

PODBIELNIAK, W. J., KAISER, H. R. and ZIEGENHORN, G. J. (1970) Centrifugal solvent extraction. In *The History of Penicillin Production. Chem. Eng. Prog. Symp.* **66**(100), 44–50.

PORATH, J. (1974) Affinity techniques — General methods and coupling procedures. In *Methods in Enzymology*, Vol. 34, pp. 13–30 (Editors Jakoby, W. B. and Wilchek, M.). Academic Press, New York.

PRITCHARD, M., SCOTT, J. A. and HOWELL, J. A. (1990) The concentration of yeast suspensions by crossflow filtration. In *Separations for Biotechnology 2*, pp. 65–73 (Editor Pyle, D. L.) Elsevier, London.

PURCHAS, D. B. (1971) *Industrial Filtration of Liquids* (2nd edition), pp. 209. Leonard Hill, London.

PYLE, D. L. (1990) (Editor) *Separations for Biotechnology 2*, SCI/Elsevier Applied Science, London.

QUEENER, S. and SWARTZ, R. W. (1979) Penicillins: biosynthetic and semisynthetic. In *Secondary Products of Metabolism, Economic Microbiology*, Vol. 3, pp. 35–122 (Editor Rose, A. H.). Academic Press, London.

REHACEK, J. and SCHAEFER, J. (1977) Disintegration of microorganisms in an industrial horizontal mill of novel design. *Biotech. Bioeng.* **19**, 1523–1534.

RICKETTS, R. T., LEBHERZ, W. B., KLEIN, F., GUSTAFSON, M. E. and FLICKINGER, M. C. (1985) Application, sterilisation, and decontamination of ultrafiltration systems for large scale production of biologicals. In *Purification of Fermentation Products.* pp. 21–49. (Editors LeRoith, D., Shiloach, J. and Leahy, T. J.). ACS Symposium Series 271. ACS, Washington, DC.

ROFFLER, S. R., BLANCH, H. W. and WILKE, C. R. (1984) *In situ* recovery of fermentation products. *Trends Biotechnol.* **2**(5), 129–136.

RUBIN, A. J., CASSEL, E. A., HENDERSON, O., JOHNSON, J. D. and LAMB, J. C. (1966) Microflotation: new low gas-flow rate foam separation technique for bacteria and algae. *Biotech. Bioeng.* **8**, 135–151.

SAMEJIMA, H. (1972) Methods for extraction and purification. In *The Microbial Production of Amino Acids*, pp. 227–259 (Editors Yamada, K., Kinoshita, S., Tsunoda, T. and Aida, K.). Kodanska, Tokyo.

SCAWEN, M. D., ATKINSON, A. and DARBYSHIRE, J. (1980) Large scale enzyme purification. In *Applied Protein Chemistry*, pp. 281–324 (Editor Grant, R. A.). Applied Science Publishers, London.

SHOJAOSADATY, S. A. and LYDDIATT, A. (1987) Application of affinity HPLC to the recovery and monitoring operations in biotechnology. In *Separations for Biotechnology*, pp. 436–444. (Editors Verrall, M. S. and Hudson, M. J.). Ellis Horwood, Chichester.

SIRMAN, T., PYLE, D.L. and GRANDISON, A.S. (1990) Extraction of citric acid using a supported liquid membrane. In *Separations for Biotechnology 2*, pp. 245–254 (Editor Pyle, D.L.). Elsevier Applied Science, London.

SMITH, I. H. and PACE, G. W. (1982) Recovery of microbial polysaccharides. *J. Chem. Tech. Biotechnol.* **32**, 119–129.

SODECK, G., MODL, J., KOMINEK, J. and SALZBRUNN, G. (1981) Production of citric acid according to the submerged fermentation. *Process Biochem.* **16**(6), 9–11.

STRATHMANN, H. (1985) Membranes and membrane processes in biotechnology. *Trends Biotechnol.* **3**(5), 112–118.

STRATHMANN, H. (1991) Fundamentals of membrane separation processes. In *Chromatographic and Membrane Processes in Biotechnology*, pp. 153–175 (Editors Costa, C. A. and Cabral, J. S.). NATO ASI Series, Kluwer Academic Publishers, Dordrecht, The Netherlands.

SWARTZ, R. R. (1979) The use of economic analyses of penicillin G manufacturing costs in establishing priorities for fermentation process improvement. *Ann. Rep. Ferm. Processes* 3, 75–110.

SYLVESTER, J. C. and COGHILL, R. D. (1954) The penicillin fermentation. In *Industrial Fermentations*, Vol. 2, pp. 219–263 (Editors Underkofler, L. A. and Hickey, R. J.). Chemical Publishing Co., New York.

TALCOTT, R. M., WILLUS, C. and FREEMAN, M. P. (1980) Filtration. In *Kirk-Othmer Encyclopedia of Chemical Technology* 3rd edition, Vol. 10, pp. 284–337. Wiley, New York.

TANNY, G. B., MIRELMAN, D. and PISTOLE, T. (1980) Improved filtration technique for concentrating and harvesting bacteria. *Appl. Env. Microbiol.* **40**(2), 269–273.

TOPIWALA, H. H. and KHOSROVI, B. (1978) Water recycle in biomass production processes. *Biotech. Bioeng.* 20, 73–85.

VOSER, W. (1982) Isolation of hydrophylic fermentation products by adsorption chromatography. *J. Chem. Tech. Biotechnol.* **32**, 109–118.

WALKER, P. D., NARENDRANATHAN, T. J., BROWN, D. C., WOOLHOUSE, F. and VRANCH, S. P. (1987) Containment of micro-organisms during fermentation and downstream processing. In *Separations for Biotechnology*, pp. 467–482 (Editors Verall, M. S. and Hudson, M. J.). Ellis Horwood, Chichester.

WANG, D. I. C. (1987) Separation for biotechnology. In *Separations for Biotechnology*, pp. 30–48 (Editors Verall, M. S. and Hudson, M. J.). Ellis Horwood, Chichester.

WANG, D. I. C. and SINSKEY, A.J. (1970) Collection of microbial cells. *Adv. Appl. Microbiol.* **12**, 121–152.

WANG, D. I. C., COONEY, C. L., DEMAIN, A. L., DUNNILL, P., HUMPHREY, A. E. and LILLY, M. D. (1979) Enzyme isolation. In *Fermentation and Enzyme Technology*, pp. 238–310. Wiley, New York.

WANG, H. Y., KOMINEK, L. A. and JOST, J. L. (1981) On-line extraction fermentation processes. In *Advances in Biotechnology*, Vol. 1, pp. 601–607 (Editors Moo-Young, M., Robinson, C. W. and Vezina, C.). Pergamon Press, Toronto.

WARNE, S. R. and BOWDEN, C. P. (1987) Biosurface properties and their significance to primary separation: a genetic engineering approach to the utilisation of bacterial auto-aggregation. In *Separations for Biotechnology*, pp. 90–104 (Editors Verrall, M. S. and Hudson, M. J.). Ellis Horwood, Chichester.

WARREN, R. K., MACDONALD, D. G. and HILL, G. A. (1991). Cross-flow filtration of *Saccharomyces cerevisae*. *Process Biochem.* **26**(6), 337–342.

WASE, D. A. J. and PATEL, Y. R. (1985) Effect of cell volume on disintegration by ultrasonics. *J. Chem. Tech. Biotechnol.* **35**, 165–173.

WEINSTEIN, M. J. and WAGMAN, G. H. (1978) *Antibiotics, Isolation, Separation and Purification*. Elsevier, Amsterdam.

WEISS, S. A. (1980) Concentration of baboon endogenous virus in large scale production by use of hollow fibre ultrafiltration technology. *Biotech. Bioeng.* **22**(1), 19–31.

WIKSTROM, P., FLYGARE, S., GRONDALEN, A. and LARSSON, P. O. (1987) Magnetic aqueous two phase separation: A new technique to increase rate of phase separation using dextran-ferrofluid or larger iron particles. *Anal. Biochem.* **167**, 331–339.

WILDFEUER, M. E. (1985) Approaches to cephalosporin C purification from fermentation broth. In *Purification of Fermentation Products*, pp 155–174 (Editors LeRoith, D., Shiloach, J. and Leahy, T. J.). ACS Symposium Series 271. ACS, Washington, DC.

WIMPENNY, J. W. T. (1967) Breakage of micro-organisms. *Process. Biochem.* **2**(7), 41–44.

YANG, C. M. and TSAO, G. T. (1982) Affinity chromatography. *Adv. Biochem. Eng.* **15**, 19–42.

ZAHKA, J. and LEAHY, T. J. (1985) Practical aspects of tangential flow filtration in cell separations. In *Purification of Fermentation Products*, pp. 51–69 (Editors LeRoith, D., Shiloach, J. and Leahy, T. J.). ACS Symposium Series 271. ACS, Washington, DC.

# CHAPTER 11

# Effluent Treatment

## INTRODUCTION

EVERY fermentation plant utilizes raw materials which are converted to a variety of products. Depending on the individual process, varying amounts of a range of waste materials are produced. Typical wastes might include unconsumed inorganic and organic media components, microbial cells and other suspended solids, filter aids, waste wash water from cleansing operations, cooling water, water containing traces of solvents, acids, alkalis, human sewage, etc. Historically, it was possible to dispose of wastes directly to a convenient area of land or into a nearby watercourse. This cheap and simple method of disposal is now very rarely possible, nor is it environmentally desirable. With increasing density of population and industrial expansion, together with greater awareness of the damage caused by pollution, the need for treatment and controlled disposal of waste has, and will, continue to grow. Water authorities and similar bodies have become more active in combating pollution caused by domestic and industrial wastes. Legislation in all developed countries now regulates the discharge of wastes, be they gas, liquid or solid (Fisher, 1977; Hill, 1980; Masters, 1991; Brown, 1992). In the U.K., much of the legislation pertaining to waste disposal and pollution is embraced by the Environmental Pollution Act 1990 (HMSO, 1990, 1991). Futher information on legislation and environmental policy can be found in the following texts; Haigh (1990), Tromans (1991) and Hughes (1992).

With liquid wastes, it may be possible to dispose of untreated effluents to a municipal sewage treatment works (STW). Obviously, much will depend on the composition, strength and volumetric flow rate of the effluent. STWs are planned to operate with an effluent of a reasonably constant composition at a steady flow rate. Thus, if the discharge from an industrial process is large in volume and intermittently produced it may be necessary to install storage tanks on site to regulate the effluent flow. In some locations, municipal sewers are not available or the effluent may be of such a composition that the wastewater treatment company or regulatory authority requires some form of pretreatment before discharge to its sewers. In these cases an effluent-treatment plant will have to be installed at the factory. Whatever the pollutant load of the liquid effluent, its discharge to a sewer will be a cost centred activity, and will incur charges from the treatment company.

Normally, fermentation effluents do not contain toxic materials which directly affect the aquatic flora or fauna. Unfortunately, most of the effluents do contain high levels of organic matter which are readily oxidized by microbial attack and so drastically deplete the dissolved oxygen concentration in the receiving water unless there is a large dilution factor. This can be shown by the oxygen sag curve in Fig. 11.1. Different aquatic species have varying tolerances to depleted oxygen levels, and as a consequence some species will die off in specific stretches of the receiving water, and in other regions a different population capable of growth at lower oxygen levels will develop.

Effluents may be treated in a variety of ways, as will be outlined later in this chapter. In a number of processes it may be possible to recover waste organic material as a solid and sell it as a by-product which may be an animal feed supplement or a nutrient to use in fermentation media (Chapter 4). The marketable by-product helps to offset the cost of the treatment process. It is now recognized that water is no longer a cheap raw material (Chapters 4 and 12), hence there are considerable advantages in reducing the quantities

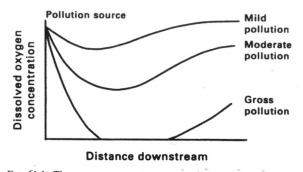

FIG. 11.1. The oxygen sag curve.

used and in recycling whenever it is feasible (Ashley, 1982). Obviously the introduction of good 'housekeeping' will lead to reductions in the volume of water used and the volume of effluent for treatment and final discharge. Recycling and reuse of materials, waste minimization, waste reduction at source and integrated pollution control are now very important factors to consider in the design and operation of any manufacturing facility, and may be the subject of new legislation in this field (Laing, 1992; Donaldson, 1993; McLeod and O'Hara, 1993).

## DISSOLVED OXYGEN CONCENTRATION AS AN INDICATOR OF WATER QUALITY

Since oxygen is essential for the survival of most macro-organisms, it is important to ensure that there are adequate levels of dissolved oxygen in rivers, lakes, reservoirs, etc., if they are to be managed satisfactorily. Ideally, the oxygen concentration should be at least 90% of the saturation concentration at the ambient temperature and salinity of the water. It is therefore important to know how effluents containing soluble and particulate organic matter can influence the dissolved oxygen concentration. One widely used method of assessment is the 'biochemical oxygen demand' (BOD), which is a measure of the quantity of oxygen required for the oxidation of organic matter in water, by micro-organisms present, in a given time interval at a given temperature. The oxygen concentration of the effluent, or a dilution of it, is determined before and after incubation in the dark at 20° for 5 days. The oxygen decrease can then be determined titrimetrically and the results presented as mg of oxygen consumed per $dm^3$ of sample. Mineral nutrients and a suitable bacterial inoculum are usually added to the initial sample to ensure optimal growth conditions. This test

is only an estimate of biodegradable material, hence recalcitrant or inhibitory compounds might be overlooked (SCA, 1989).

Because the BOD test takes 5 days it may be necessary to resort to the 'chemical oxygen demand' (COD), a chemical test which only takes a few hours to complete. The test is based on treating the sample with a known amount of boiling acidic potassium dichromate solution for 2.5 to 4 hours and then titrating the excess dichromate with ferrous sulphate or ferrous ammonium sulphate (HMSO, 1972). The oxidized organic matter is taken as being proportional to the potassium dichromate utilized. Most compounds are oxidized virtually to completion in this test, including those which are not biodegradable. In circumstances where substances are toxic to micro-organisms, the COD test may be the only suitable method available for assessing the degree of treatment required. The BOD:COD ratios for sewage are normally between 0.2:1 and 0.5:1. The ratio values for domestic sewage may be fairly steady. When industrial effluents of variable composition and loading are discharged, the ratio may fluctuate considerably. Very low BOD:COD ratios will indicate high concentrations of non biodegradable organic matter and consequently biological effluent treatment processes may be ineffective (Ballinger and Lishka, 1962; Davis, 1971). A number of alternative tests are available to indicate the 'oxygen demand' of a wastewater, including total organic carbon (TOC) and permanganate value (HMSO, 1972; American Public Health Association, 1992).

## SITE SURVEYS

A complete survey of industrial operations is essential for any individual site before an economical waste-treatment programme can be planned. It is desirable to divide the facility into as many units as possible, as knowledge of the various material streams may show unexpected losses of finished product, solvent wastage, excessive use of water or unnecessary contamination of water which might be recycled, recovered or reused within the site. The factors, and concentrations where appropriate, listed in Table 11.1 ought to be known at all production rates under which an individual unit may operate in a representative time period.

The survey may indicate a need for better control of water usage and should identify sources of uncontaminated and contaminated water that might be reused in the factory. Concentrated waste streams should be kept separate if they contain materials that can be profitably recovered. It is also often more economical to treat a concentrate rather than a large volume of a dilute effluent because of the saving on pumps and settling tank capacities, provided that concentrations do not reach toxic or inhibitory levels in biological treatment processes.

The various wastes may be tested in a laboratory and on a pilot scale to assess the best potential methods of chemical and biological treatment. Once the pHs of the effluents are known, samples may be mixed to see if a neutral pH is reached. A variety of tests may be used to establish methods for reducing salt concentrations, co-agulating suspended particles and colloidal materials, and for breaking emulsions.

The commonly used biological tests include respirometry, aeration-flask tests (Otto *et al.*, 1962) and continuous-culture experiments. Small flask respirometers (Warburg or Gilson) and oxygen electrodes are used initially to establish the conditions to use in bio-oxidation of the effluent, and to test for the presence of toxic materials. Large respirometers (Simpson and Anderson, 1967) are useful for predicting effluent treatment rates and oxygen requirements. The residues in the flasks can be analysed to see if there are any recalcitrant materials. The use of laboratory continuous-culture vessels fitted with sludge-return pumps and settling tanks can provide detailed information (Ramathan and Gaudy, 1969). Proposed large-scale operating conditions for feed and aeration rates can be tested and their effectiveness assessed. The results from all these experiments may help in the design of a full-scale plant.

If the survey is comprehensive it should be possible to plan an overall treatment programme for a site and to establish:

1. Water sources which can be combined or reused.

2. Concentrated waste streams which contain valuable wastes to be recovered as food, animal feed, fertilizer or fuel.
3. Toxic effluents needing special treatment, or acids or alkalis needing neutralization.
4. The effluent loading expected under maximum production conditions.
5. The effluent(s) which might be discharged directly, without treatment, on to land or to a watercourse and not cause any pollution.
6. The effluent(s) which might be discharged into municipal sewers.

When all the relevant information has been obtained one can predict the size and type of effluent treatment plant required, and thus its capital and operational costs. This can then be compared with water company charges to treat the waste at an STW with and without on-site treatment. It should be remembered that the water company may insist on on-site treatment before a waste is discharged to the sewer, and will in most cases set consent limits for maximum flow rates and concentrations of specific analytes.

## THE STRENGTHS OF FERMENTATION EFFLUENTS

It is already evident from earlier sections of this chapter that the presence of high levels of particulate or soluble organic matter in water will result in potential high BODs. This is precisely what is being achieved in all large-scale fermentation processes. An initial medium rich in organic matter is converted to biomass and primary and secondary metabolites. Unfortunately, the product often represents a small proportion of the initial raw material, even in an efficiently operated

TABLE 11.1. *Factors to investigate in a site survey*

Daily flow rate
Fluctuations in daily, weekly and seasonal flow
BOD/COD
Suspended solids
Turbidity
pH range
Temperature range
Odours and tastes
Colour
Hardness
Detergents
Radioactivity
Presence of specific toxins or inhibitors (e.g. heavy metals, phenolics etc.)

fermentation. The spent wastes remaining after the distillation of whisky may account for 90% of the initial raw organic materials, while in an antibiotic fermentation the effluent may represent in excess of 95%.

Data for a variety of fermentation effluents are summarized in Table 11.2. The BODs of many of these samples are much higher than that of domestic sewage and some may be comparable with strong effluents such as sulphite paper mill liquor. It is evident from these data that fermentation effluents may present serious potential pollution problems and may be expensive to dispose of unless well planned processes are used. A number of steps may be taken to reduce BODs in a process. Some of these will be discussed in this chapter. Careful selection of raw materials may have a significant effect on the type and quantities of effluent being produced. The cheapest raw material which meets the nutrient requirements of the micro-organisms may not be ideal if product yield, recovery cost, effluent disposal cost and possible by-product value are considered together. The high BOD value of fungal mycelium (40,000 to 70,000 mg dm$^{-3}$) would indicate that any biomass should normally be kept separate from the remainder of an effluent and some of it may be sold as a by-product. It may also be worthwhile to

concentrate liquid fractions, for example, industrial alcohol and distillery stillages (10,000 to 25,000 mg dm$^{-3}$) will both produce dried solubles fractions which can be sold.

Metabolites or components of some fermentation effluents may be extremely toxic and polluting and will require complete destruction, for example by chemical or thermal methods, before disposal. The need for such a treatment strategy will therefore make a significant contribution to the overall cost of the process. One such metabolite is avermectin produced by *Streptomyces avermitilis* fermentations. Here all effluent streams from the process are captured and any avermectin present chemically degraded (Omstead *et al.*, 1989).

## TREATMENT AND DISPOSAL OF EFFLUENTS

The effluent disposal procedure which is finally adopted by a particular manufacturer is obviously determined by a number of factors, of which the most important is the control exercised by the relevant authorities in many countries on the quantity and quality of the waste discharge and the way in which it might be

TABLE 11.2. *BOD strengths of effluents (mg dm$^{-3}$)*

| Effluent | BOD | Reference |
|---|---|---|
| Domestic sewage | 350 | Boruff (1953) |
| Sulphite liquor from paper mill | 20,000–40,000 | |
| Beer: | | |
| (a) spent grain press | 15,000 | Abson and |
| (b) hop-press liquor | 7430 | Todhunter (1967) |
| (c) yeast wash water | 7400 | |
| (d) spoil beer | up to 100,000 | |
| (e) bottle washings | 550 | |
| Maltings: | | |
| (a) suspended solids | 1240 | Koziorowski and |
| (b) wastes | 20–204 | Kucharski (1972) |
| (c) grain washings | 1500 | |
| Brewery effluent | 1,400–1,800 | Fang *et al.* (1990) |
| Industrial alcohol stillage | 10,000–25,000 | Blaine (1965) |
| Distillery stillage | 10,000–25,000 | Jackson (1960) |
| Yeast production | 3,000–14,000 | Boruff (1953) |
| Antibiotic waste | 5,000–30,000 | Jackson (1960) |
| Penicillin: | | |
| (a) wet mycelium from filter | 40,000–70,000 | Boruff (1953) |
| (b) filtrate | 2,150–10,000 | |
| (c) wash water | 210–13,800 | |
| Streptomycin spent liquor | 2,450–5,900 | Koziorowski and |
| Aureomycin spent liquor | 4,000–7,000 | Kucharski (1972) |
| Solvents | up to 2,000,000 | |

done (Fisher, 1977). The range of effluent-disposal methods which can be considered is:

1. The effluent is discharged to land, river or sea in an untreated state.
2. The effluent is removed and disposed of in a landfill site or is incinerated.
3. The effluent is partially treated on site (e.g. by lagooning) prior to further treatment or disposal by one of the other routes indicated.
4. Part of the effluent is untreated and discharged as in 1 or 2, the remainder is treated at a sewage works or at the site before discharge.
5. All of the effluent is sent to the sewage works for treatment, although there might be reluctance by the sewage works to accept it, possibly resulting in some preliminary on-site treatment being required, and discharge rates and effluent composition defined.
6. All the effluent is treated at the factory before discharge.

## DISPOSAL

### Seas and rivers

The simplest way of disposal will be on a sea coast or in a large estuary where the effluent is discharged through a pipeline (installed by the factory or local authorities) extending below the low-water mark. In such a case there may be little preliminary treatment and one relies solely on the degree of dilution in the sea water.

If effluents are to be discharged into a river they must meet the requirements of the local river or drainage authorities. In Britain there is a Royal Commission standard requiring a maximum BOD (5 days) of 20 mg dm$^{-3}$ and 30 mg dm$^{-3}$ of suspended solid matter (the 20:30 standard). Stricter standards are often applied, depending on the use of the receiving water, such as a 10:10 standard; in addition, levels of ammoniacal nitrogen may be stipulated. There are, as well, often stringent upper limits for toxic metals and chemicals which might kill the fauna (particularly fish) and flora, e.g. sulphites, cyanides, phenols, copper, zinc, cadmium, arsenic, etc. It is highly unlikely that one would be able to discharge an industrial waste today without some form of pretreatment.

### Lagoons (oxidation ponds)

Lagoons, holding ponds, oxidation ponds, etc., may be used by a number of industries if land is available at a reasonable cost. It is a method often used in seasonal industries where capital investment in effluent plant is difficult to justify. The lagoon normally consists of a volume of shallow water enclosed by watertight embankments. Oxidation ponds are typically 1–2 m deep. They can be designed to maintain aerobic conditions throughout, but more commonly decomposition at the surface is aerobic and that nearer the bottom is anaerobic and they are then known as facultative ponds. Oxygen for aerobic degradation is provided both from the surface of the pond, and from algal photosynthesis. Deeper ponds (known as lagoons) are mechanically agitated to provide aeration. Lagoons are simple to build and operate, but are expensive in terms of land requirements. They may be used as the sole method of treatment, incorporating both physical (sedimentation) and biological processes, but the effluent produced may not reach locally acceptable standards. Alternatively they can provide an initial pretreatment or can be used to 'polish' effluent from secondary treatment processes.

### Spray irrigation

Liquid wastes can be applied directly to land as irrigation water and fertilizer when they are claimed to have a number of beneficial effects on the soil and plants. If this method of disposal is to be used, then it is necessary to have a large area of land near to a manufacturing plant in an area of low to medium rainfall. Pipeline costs will often restrict use of this technique. Colovos and Tinklenberg (1962) described the disposal of antibiotic and steroid wastes with BODs of 5000 to 20,000 mg dm$^{-3}$. These wastes were initially chlorinated to lower the BOD and reduce unpleasant odours and then sprayed on to land until the equivalent of 38 mm of rainfall was reached. This process was repeated at monthly intervals and improved plant growth.

When appropriate, solid wastes may be spread onto land as a fertilizer and soil conditioner. This practice is common with sewage sludges, and Mbagwu and Ekwealor (1990) report the use of spent brewers grains to improve the productivity of fragile soils. Irrespective of whether the waste is liquid or solid, the concentration of heavy metals and certain organic components will require careful monitoring and control to safeguard the environment and public health.

## Well disposal

Disused wells, boreholes or mine shafts may provide an ideal, cheap method for disposal when the volume of waste is limited, the underground strata are suitable and the chances of contamination of water supplies utilized by water authorities are negligible (Zajic, 1971). Melcher (1962) has described the use of wells 500-m deep for the daily disposal of:

| | |
|---|---|
| Acetic acid | 900 kg |
| Ammonium acetate | 900 kg |
| Sodium acetate | 760 kg |
| Sodium chloride | 450 kg |
| Sodium and ammonium bromide | 225 kg |

Methanol, xylene, tars and organic compounds

This mixture had a pH of 4 to 5 and a COD of 40,000 to 60,000 mg $dm^{-3}$ rising to 100,000 mg $dm^{-3}$ and was pumped into the wells at 50 to 100 $dm^3$ $min^{-1}$.

Careful hydrogeological surveys will be needed to prove that waste disposal in wells will not cause pollution of aquifers and threaten groundwater supplies.

## Landfilling

Landfilling is a disposal method for municipal solid waste (MSW) and industrial waste. It utilizes natural or man made voids (e.g. disused clay pits) into which the waste is deposited. Both solid and liquid wastes can be deposited depending on restrictions imposed by the site licence. Strict controls exist on the amount of liquid and toxic materials which can be accepted because of the threat of groundwater pollution if leachate (a liquid having BOD levels up to 30,000 mg $dm^{-3}$) escapes from the site. Leachate is generated from liquid deposited in the site, water entering the site naturally via precipitation or surface run-off and by anaerobic microbial action as organic matter in the landfill is degraded. Microbial action similar to that in anaerobic digesters leads to the production of landfill gas (LFG) which, being 50–60% methane can, if collected efficiently, provide a useful source of energy (Freestone et al., 1994).

## Incineration

A number of designs exist for the incineration of solid and/or liquid wastes either on site or at a commercial incinerator, including rotary kilns, fluidized beds and multiple hearth furnaces. Combustion temperatures need to be carefully controlled to destroy and prevent the formation of dioxins and furans, formation of which occurs at between 300° and 800°, and total destruction is effected at temperatures above 1000° with a retention time of 1 second. Flue gases from the incinerator require cleaning to remove particulates, acid vapours, etc. using electrostatic precipitators, cyclones and wet scrubbers to comply with local environmental protection standards. Waste disposal by incineration is currently significantly more expensive than landfilling, with costs for the disposal of MSW being \$37 $tonne^{-1}$ for incineration compared with \$10 $tonne^{-1}$ for landfilling (Smith, 1993).

## Disposal of effluents to sewers

Municipal authorities and water treatment companies which accept trade effluents into their sewage systems will want to be sure that:

1. The sewage works has the capacity to cope with the estimated volume of effluent.
2. The effluent will not interfere with the treatment processes used at the sewage works.
3. There are no compounds present in the effluent which will pass through the sewage works unchanged and then cause problems when discharged into a watercourse.

It is common practice for local authorities to demand preliminary on-site pretreatment before discharge into sewers to minimize the effects of industrial wastes. The actual pretreatment required will depend on the precise nature of the waste and may range from simple sedimentation to complex chemical and biological processes.

## TREATMENT PROCESSES

Fermentation wastes may be treated on-site or at an STW by any or all of the three following methods:

1. Physical treatment.
2. Chemical treatment.
3. Biological treatment.

The final choice of treatment and disposal processes

used in each individual factory will depend on local circumstances.

Treatment processes may also be described in the following manner:

1. Primary treatment; physical and chemical methods, e.g. sedimentation, coagulation etc.
2. Secondary treatment; biological methods (e.g. activated sludge) conducted after primary treatment.
3. Tertiary treatment; physical, chemical or biological methods (e.g. microstrainers, sand filters and grass plot irrigation) used to improve the quality of liquor from previous stages (Forster, 1985).
4. Sludge conditioning and disposal; physical, chemical and biological methods. Anaerobic digestion is often used to condition (make it more amenable to dewatering) the sludge produced in previous stages. Following dewatering (e.g. by centrifugation using a decanter centrifuge) the sludge can then be disposed of by incineration, landfilling, etc.

### Physical treatment

The removal of suspended solids by physical methods before subsequent biological treatment will considerably reduce the BOD of the resulting effluent. In nearly all fermentation processes the cells are separated from the liquid fraction in recovery processes (Chapter 10). Obviously, biomass processes need not be considered. Yeast cells from other processes may be a marketable product, but microbial cells may not always be marketable, particularly when contaminated with filter aid. In these instances, when the cells and filter aid are a waste, the recovered material may be dealt with in two basic ways:

1. The waste is disposed of without any further treatment.
2. The waste bulk is reduced by mechanical dewatering with a filter press, centrifuge, rotary vacuum filter or belt press. The compressed waste is then incinerated (Grieve, 1978) or disposed of in a landfill site.

Solid wastes are produced in some processes before inoculation. In breweries, where malted grain is still used, coarse screens or 'whirlpool' centrifuges may be used to remove spent grain from the wort after it is mashed. About 5 kg (wet weight) of grain are produced per barrel (180 $dm^3$) of beer. If hops are used, rather than hop extracts, they will also be recovered on screens in a 'hop back'. This residue may amount to 250 g per barrel. Both the spent grain and hop waste may then be mechanically dewatered before being sold or dumped.

The stillage (after distillation) in whisky distilleries may be passed through screens (1 mm openings). These screenings are then removed, mechanically dewatered, and dried in rotary driers to yield a potentially marketable residue known as Distillers' grains. According to a survey in Scotland, about half the whisky distilleries were evaporating the spent waste to a syrup containing 45% solids, mixing with spent grain, drying and selling the final product, 'Distillers' Dark Grains', as a low-grade cattle food (Mackel, 1976).

Physical processes installed for primary effluent treatment may include the following stages:

1. Screens, to remove larger suspended and floating matter.
2. Comminutors, to reduce particle size.
3. Constant velocity channels ($\sim 0.3$ m s$^{-1}$) for grit removal to prevent damage to plant in later processes.
4. Sedimentation tanks for the removal of finer suspended matter. These are generally circular or rectangular continuous flow tanks operating at retention times of 6–15 hours (and designed to have a minimum retention time of 2 hours), with facility for the continuous removal of settled sludge. Sedimentation tanks can remove 70% of the incoming suspended solids and, depending on the nature of the waste, up to 40% of its BOD load (Forster, 1985). They can be operated with or without prior chemical coagulation/flocculation. Similar settlement processes are also conducted after secondary (biological) treatment.

Physical processes used in tertiary treatment to produce an effluent of better quality than the 30:20 standard include microstrainers, slow sand filters, upflow sand filters and rapid gravity sand filters. Throughputs vary between around 3 m$^3$ m$^{-2}$ day$^{-1}$ for slow sand filters and 700 m$^3$ m$^{-2}$ day$^{-1}$ for microstrainers. Suspended solids removal is generally 50–70% and BOD removal around 30–50%, depending on the technique used. A detailed description of tertiary treatment is given by Truesdale (1979) and Viessman and Hammer (1993).

## Chemical treatment

Fine suspended particles in an effluent may be removed by coagulation and/or flocculation (Cooper, 1975; see also Chapter 10). Coagulation is essentially instantaneous whereas flocculation requires some more time and gentle agitation to achieve 'aggregation' of the particles. Ferrous or ferric sulphate, aluminium sulphate (alum), calcium hydroxide (lime) and polyelectrolytes are often used as chemical coagulants. A solution of coagulant of the appropriate strength for effective treatment is added to the effluent in a vigorously mixed tank, a precipitate or floc forms almost immediately and carries down the suspended solids to form a sludge. This sludge may be drawn off, mechanically dewatered and subjected to further treatment. The flocs formed on coagulation may be small, and will therefore require an extended period to settle, and as a consequence, for a given throughput of effluent a large sedimentation tank will be needed. Increasing the particle diameter by encouraging small flocs to coalesce (flocculation) increases the rate of sedimentation and thus, for a given throughput, a smaller vessel can be operated. Polyelectrolytes are commonly used as flocculants, and following addition the effluent is gently mixed (turbulent mixing would break up the flocs) by passage through sinuous flocculation channels, hydrodynamic flocculators or mechanically mixed flocculators (Smethurst, 1988).

## Biological treatment

Most organic-waste materials may be degraded biologically. This process may be achieved aerobically or anaerobically in a number of ways. The most widely used aerobic processes are trickling filters, rotating disc contactors, activated sludge processes and their modifications. The anaerobic processes (digestion, filtration and sludge blankets) are used both in the treatment of specific wastewaters and in sludge conditioning.

### Aerobic processes

TRICKLING FILTERS

The term filter in this unit operation is a misnomer, as the action of a trickling filter is not one of filtration, but rather it is a fixed film bio-reactor. Settled effluent to be treated is passed down through a packed bed counter-current to a flow of air. Micro-organisms ad-

hering to the packing matrix adsorb oxygen from the upflowing air and organic matter from the downflowing effluent; the latter is then metabolized and the effluent stream's BOD reduced.

A conventional trickling filter (Fig. 11.2) usually consists of a cylindrical concrete tank 2 to 3 m in depth and 8 to 16 m in diameter. Some filters are rectangular in shape, but a rotary system allows more uniform hydraulic loading (Bruce and Hawkes, 1983; Viessman and Hammer, 1993). The tank is packed with a bed of stone (usually granite) or special plastic packings, the bed being underlaid with drains. The packing material diameter should be 50 to 100 mm to give a specific surface of around 100 $m^2$ $m^{-3}$, and the material should be packed to give a voidage (% air space of total bed volume) of 45 to 55%, which should minimize the risk of the spaces between the packing material becoming blocked by the microbial film. Synthetic packing material, although more expensive, has a higher surface area and voidage, allowing higher treatment rates per unit volume of bed and reducing the likelihood of blockages. The trickling filter is always followed by a secondary sedimentation tank or humus tank to remove suspended matter (e.g. biofilm sloughed off the packing) from the treated effluent. In conventionally loaded or low-rate filters, the effluent from which the suspended solids have been removed, is fed on to the upper surface of the bed by spray nozzles or mechanical distributor arms (McKinney, 1962; Higgins, 1968). The effluent trickles gradually through the bed and a slime layer of biologically active material (bacteria, fungi, algae, protozoa and nematodes) forms on the surface of the support material. The large surface area created in the bed permits close contact between air flowing upwards through the bed, the descending effluent and the biologically active growth. The bacteria in the biological film remove the majority of the organic loading. Complex organic materials are broken down and utilized, nitrogenous matter and ammonia are oxidized to nitrates and sulphides and other compounds are similarly oxidized. The higher organisms (protozoa etc.) control the accumulation of the biological film (prevents the filter from blocking) and improve the settling characteristics of the solids (humus) discharged with the filter effluent. In low-rate filters the scouring action of the hydraulic load normally has a minor role in removal of any loose microbial film. The active slime takes time to develop and can be poisoned by the addition of toxic chemicals. The simple filter is inefficient when operated at abnormally high organic loading rates. Initially, there is a very rapid build up of

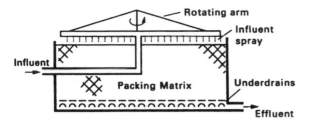

FIG. 11.2. Schematic diagram of a trickling filter.

bacteria, fungi and algae at the top of the filter which cannot be controlled by the resident population of worms and larvae. The voids, therefore, block up, resulting in ponding (untreated effluent accumulating on the surface of the filter bed). Film growth can be limited by reducing the dosing frequency, which gives better liquid distribution deeper into the bed.

A trickling filter bed should remove 75 to 95% of the BOD and 90 to 95% of the suspended solids at organic loading rates of 0.06–0.12 kg BOD $m^{-3}$ $day^{-1}$ for conventional trickling filtration. When part of the treated effluent is being recirculated to dilute the feed and increase the hydraulic load placed on the unit the organic loading can be increased to 0.9–0.15 kg BOD $m^{-3}$ $day^{-1}$. The increase in hydraulic load thus applied causes greater hydraulic scouring of the bed (preventing blockage), but does not reduce treatment efficiency due to improved wetting of the packing surface and, thus, more efficient use of the biofilm (Forster, 1985). To achieve the Royal Commission (20:30) Standard together with a high degree of nitrification, filters being supplied with domestic sewage should receive organic loading rates of 0.07–0.1 kg BOD $m^{-3}$ $day^{-1}$ and hydraulic loading rates of 0.12–0.6 $m^3$ $m^{-3}$ $day^{-1}$ (Gray, 1989).

It is possible to modify the trickling filter to increase the capacity for organic loading by the use of two sets of filters and settling tanks in series; this is known as alternating double filtration (ADF). Effluent is applied to the first filter at a high hydraulic and organic loading rate, it passes from this filter through the first settling tank and then on to a second filter and settler. After a period of one to two weeks the sequence of the filters is reversed and the second filter receives the higher loading. In this way heavy film growth is promoted in the first filter to receive the effluent, but when the filter sequence is reversed it becomes nutrient limited, encouraging excess film removal. Loading rates of 0.32–0.47 kg BOD $m^{-3}$ $day^{-1}$ have been claimed (Forster, 1977), but recommended rates for design purposes are 0.15–0.26 kg BOD $m^{-3}$ $day^{-1}$ (Forster, 1985).

Alternatively, enclosed deep beds of 3.5 to 5.5-m depth may be used in which air is blown through the beds by fans. Loading rates up to 12 times that of the ordinary filter have been claimed (Abson and Todhunter, 1967).

Cook (1978) has stressed the need to consider the possible intermittency of a factory wastewater treatment process. It was shown that starving of a laboratory-scale trickling filter beyond 48 hours resulted in near failure of the filter. This indicated the need to supplement or artificially load the filter to maintain a viable biomass in the system.

## TOWERS

Because trickling filters do not have both a high specific area and a high voidage, they are less suitable for the treatment of large volumes of strong industrial effluents (Table 11.2). Large areas of land would be required for the expensive and extensive traditional filter beds. Towers, 6–9 m in height, packed with lightweight (40–80 kg $m^{-3}$) plastic multi-faced modules or small random packing units have provided to be a space saving and relatively inexpensive solution to the problem. These packings have a relatively open structure for oxygen transfer (specific surface of 100–300 $m^2$ $m^{-3}$) and high voidage (90–98%), but are expensive compared with the conventional filter packings. They are capable of coping with high BOD loadings. At a loading of 3.2 kg BOD $m^{-3}$ $day^{-1}$ a 50% BOD removal may be achieved, and at 1.5 kg $m^{-3}$ $day^{-1}$ 70% removal is possible (Ripley, 1979). The biological film is similar to that formed on the conventional packing and scouring is due to the hydraulic load applied rather than the predation of higher organisms.

## BIOLOGICALLY AERATED FILTERS (BAFS)

Biologically aerated filters are a relatively recent development based on the trickling filter. They consist of a packed bed which provides sites for microbial growth through which air is passed but, unlike trickling filters, the reactor volume is flooded with the effluent to be treated which is passed upwards or downwards through the reactor (i.e. co- or counter-currently to the air supply) depending on the design. The packing matrix may be natural (e.g. pumice) or synthetic (e.g. polyethylene), and may be either a fixed structure or randomly packed.

The combination of aeration and filtration allows high rates of BOD and ammonia removal together with solids capture, so that sedimentation tanks may not be required. However, regular backwashing is essential to remove filtered solids and excess biomass. Organic loading rates for 90% BOD removal are significantly greater than those obtained for trickling filters, being in the range 0.7–2.8 kg BOD m$^{-3}$ day$^{-1}$ (Stephenson *et al.*, 1993). They are versatile treatment systems, and of those currently in operation, design capacities vary between 600 and 70,000 m$^3$ day$^{-1}$. In addition to their use as a secondary treatment process they can also be utilized for tertiary treatment or modified to allow denitrification in a manner similar to that of activated sludge systems.

## ROTATING BIOLOGICAL CONTACTORS (ROTATING DISC CONTACTORS)

In this treatment method (Fig. 11.3) a unit composed of closely spaced discs (2 to 3 m diameter with 1 to 2 cm spacing between discs), on a central drive shaft are rotated slowly (0.5 to 15 rpm) through the effluent so that 40 to 50% of the disc surfaces are submerged (Borchardt, 1970; Pretorious, 1973). The discs, usually made from synthetic material (e.g. polystyrene, PVC), are arranged in stages or groups separated by baffles to minimize short circuiting or surging (Forster, 1985) and to enhance specific treatment requirements such as nitrification. The discs may be flat or corrugated to increase surface area. A microbial film forms on the discs; this is aerated during the exposed part of the cycle and absorbs nutrients during the submerged part. Shear forces produced as the discs rotate through the liquid control the thickness of the biofilm, with excess biofilm being sloughed from the discs. A sedimentation tank following the biological stage is therefore required to remove biological solids. Loading rates of 13 g BOD m$^{-2}$ day$^{-1}$ for domestic sewage and partial treatment of loads of 400 g BOD m$^{-2}$ day$^{-1}$ have been used. To achieve the 20:30 standard the loading rate should not exceed 6 g BOD m$^{-2}$ day$^{-1}$. Rotating biological contactors are compact, easily covered for health and aesthetic reasons, available as packaged units, simpler to operate under varying loads than trickling filters (the biofilm being wetted at all times) and are easily added onto existing treatment processes. As such they can provide a cost effective method of on-site treatment.

Ware and Pescod (1989) describe the use of full scale anaerobic/aerobic rotating biological contactors for treating brewery wastewaters. Greater than 85% COD removal was obtained in the aerobic stage, but difficulties were experienced in maintaining anaerobic populations.

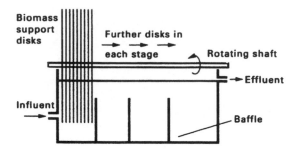

FIG. 11.3. Schematic diagram of a rotating biological contactor.

## ROTATING DRUMS

Large rotating drums packed with random plastic packing materials or spheres have been manufactured as a development of rotating discs. Loading rates for the random packing were similar to those of rotating discs, while plastic spheres used in partial treatment could cope with loads of 6 kg BOD m$^{-2}$ day$^{-1}$ (Water Pollution Research Centre, 1972).

## FLUIDIZED-BED SYSTEMS

Fluidized-bed reactors in wastewater treatment are relatively recent innovations. The support matrix (sand, anthracite, reticulated foam) has a large surface area on which the biofilm adheres and thus they are able to operate at high biomass concentrations with high rates of treatment. This allows strong wastewaters to be treated in small reactors. They are also useful for the treatment of industrial wastewaters when variable loadings are encountered (Cooper and Wheeldon, 1980, 1982). The support matrix is fluidized by the upflow of effluent through the reactor, and the degree of bed expansion is controlled by the flow rate of wastewater. The treated effluent can thus be decanted off without loss of the support matrix and with careful operation a secondary sedimentation tank may not be needed. The support matrix is regularly withdrawn to remove excess biomass. Fluidized-bed systems can be operated aerobically, anaerobically (see later section) or anoxically for denitrification.

## ACTIVATED SLUDGE PROCESSES

The basic activated-sludge process (Fig. 11.4) consists of aerating and agitating the effluent in the presence of a flocculated suspension of micro-organisms on particulate organic matter — the activated sludge. This process was first reported by Arden and Lockett (1914) and is now the most widely used biological treatment process for both domestic and industrial wastewaters. The raw effluent enters a primary sedimentation tank where coarse solids are removed. The partially clarified effluent passes to a second vessel, which can be of a variety of designs, into which air or oxygen is injected by bubble diffusers, paddles, stirrers, surface aerators etc. Vigorous agitation is used to ensure that the effluent and oxygen are in contact with the activated sludge. After a predetermined residence time of several hours, the effluent passes to a second sedimentation tank to remove the flocculated solids. Part of the sludge from the settlement tank is recycled to the aeration tank to maintain the biological activity. The overflow obtained from the settlement tank should be of a 20:30 standard or better and be suitable for discharge to inland waters. The excess sludge is dewatered and dried, to be sold as a fertilizer, incinerated or landfilled. In conventional activated sludge processes, organic loading rates are 0.5–1.5 kg BOD $m^{-3}$ $day^{-1}$ with hydraulic retention times of 5–14 hours depending on the nature of the wastewater, giving BOD reductions of 90–95%. High-rate activated-sludge processes can be used as a partial treatment for strong wastes prior to further treatment or discharge to a sewer and are widely used in the food processing and dairy industries. The organic loading rate is 1.5–3.5 kg BOD $m^{-3}$ $day^{-1}$, and with hydraulic retention times of only 1–2 hours, BOD reductions of 60–70% are possible (Gray, 1989).

A number of modifications of the basic process can be used to improve treatment efficiency, or for a more specific purpose such as denitrification (Winkler and Thomas, 1978; Gray, 1989). Tapered aeration and stepped feed aeration are used to balance oxygen demand (which is greatest at the point of wastewater entry to the aeration basin) with the amount of oxygen supplied. Contact stabilization exploits biosorption processes and thereby allows considerable reduction in basin capacity ($\sim 50\%$) for a given wastewater throughput. Denitrification (the biological reduction of nitrate to nitrite and on to nitrogen gas under anoxic conditions) can be accomplished in an activated-sludge plant when the first part of the basin is not aerated.

In advanced activated-sludge systems the amount of dissolved oxygen available for biological activity is increased to improve treatment rate. One vessel of this type is the 'Deep Shaft' (Hemming et al., 1977), which is quite distinctive from the other aeration tanks and has been developed from the ICI plc SCP process (Taylor and Senior, 1978; Chapter 7). The 'Deep Shaft' (Fig. 11.5) consists of a shaft 50 to 150 m deep, separated into a down-flow section (down-comer) and an up-flow section (riser). The shaft may be 0.5 to 10 m diameter, depending on capacity. Fresh effluent is fed in at the top of the 'Deep Shaft' and air is injected into the down-flow section at a suffficient depth to make the liquid circulate at 1 to 2 m $s^{-1}$. The driving force for circulation is created by the difference in density (due to air bubble volume) between the riser and down-flow sections. For starting up, circulation of liquid is stimulated by injecting air at the same depth in the riser. Air injection is then gradually all transferred to the air injection point in the down-comer. Because of the pressure created in the down-comer, oxygen-transfer rates of 10 kg $O_2$ $m^{-3}$ $h^{-1}$ can be achieved and bubble contact times of 3 to 5 minutes are possible instead of 15 seconds in diffused air systems. BOD removal rates of 90% are achievable at organic loadings of 3.7–6.6 kg BOD $m^{-3}$ $day^{-1}$ at hydraulic retention times of 1.17–1.75 hours (Gray, 1989). Sludge production was found to be much less than that for conventional sewage-treatment processes.

Two types of pure oxygen systems have also been developed to increase the rate of oxygen transfer:

(i) closed systems which operate in oxygen-rich atmospheres and,

(ii) open systems employing fine bubble diffusers.

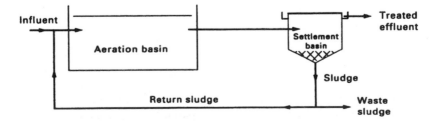

Fig. 11.4. Simplified cross-section of an activated sludge process.

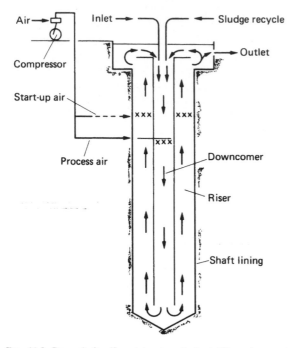

FIG. 11.5. Deep-shaft effluent treatment plant (Hemming *et al.*, 1977).

An example of the closed system is the UNOX Process developed by the Union Carbide Corporation in the U.S.A., and marketed in the U.K. by Wimpey Unox (Fig. 11.6). The enclosed oxygenation tank is compartmentalized by baffles. Settled wastewater and returned sludge are fed into the first stage and oxygen pumped into the headspace. The oxygen and wastewater move sequentially through the compartments, and the oxygen concentration in the gas phase decreases in each stage as it is consumed by the micro-organisms. As the nutrient concentration also falls stage by stage, oxygen supply and demand is balanced. Organic loading rates (when treating municiple wastewater) are 3–4 times higher than those of aerated systems at 2.5–4.0 kg BOD m$^{-3}$ day$^{-1}$. Pure oxygen systems also have shorter residence times and produce less sludge with better settling qualities than conventional systems. The UNOX process has been successfully used for the treatment of brewery wastewaters containing 2000 mg BOD dm$^{-3}$ at flow rates of 2269 m$^3$ day$^{-1}$ with a 6.6 hour retention time (Brooking *et al.*, 1990).

The VITOX aeration system (Fig. 11.7) developed by the British Oxygen Company is an example of an open tank oxygenation system. Its main advantage is that it can be used in existing aeration tanks either to replace or upgrade conventional aeration without the need to replace or modify existing units. Oxygenation is achieved by pumping settled wastewater at high pressure through *a venturi* where oxygen is injected. Turbulence and high pressures thus created ensure high levels of oxygen dissolution. The flexibility of this system means that it is particularly useful for the treatment of high strength intermittently produced wastewaters such as those generated by food processing industries (Gostick *et al.*, 1990).

### Anaerobic treatment

Anaerobic treatment of waste organic materials originated with the use of septic tanks and Imhoff tanks, which have now been replaced by a variety of high-rate digesters (Pohland and Ghosh, 1971). Loehr (1968) has listed the following reasons for using anaerobic processes for waste treatment:

1. Higher loading rates can be achieved than are possible for aerobic treatment techniques.
2. Lower power requirements may be needed per unit of BOD treated.
3. Useful end-products such as digested sludge and/or combustible gases may be produced.
4. Organic matter is metabolized to a stable form.
5. There is an alteration of water-binding characteristics to permit rapid sludge dewatering.
6. The reduced amount of microbial biomass leads to easier handling of sludge.
7. Low levels of microbial growth will decrease the possible need for supplementary nutrients with nutritionally unbalanced wastes.

### ANAEROBIC DIGESTION

Large volumes of wet sludge which are produced in primary and secondary sedimentation tanks may have to be reduced in volume before disposal. This volume of sludge can be reduced by anaerobic digestion. In sludges containing 20,000–60,000 mg dm$^{-3}$ solid matter, 80% of the degradable matter may be digested, which will reduce the solids content by 50%. During anaerobic digestion acid fermenting bacteria degrade the waste to free volatile fatty acids, mainly acetic and propionic acid, which are then converted to methane ($\sim 60\%$) and carbon dioxide ($\sim 40\%$). The gas produced (biogas) is a very useful by-product, and can be burnt as a heating fuel, fed to gas engines to generate

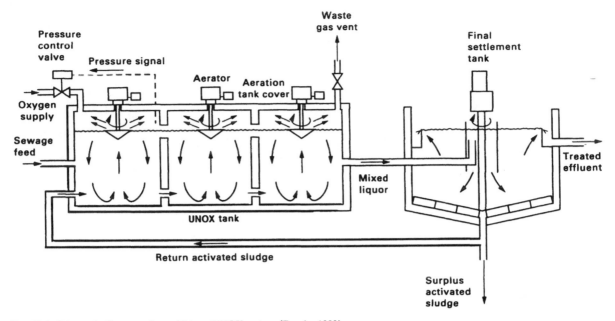

FIG. 11.6. Schematic diagram of a multistage UNOX system (Fuggle, 1983).

electricity or used as a vehicle fuel. As well as being used in sludge digestion and conditioning, anaerobic digesters are also used directly in the treatment of many high strength wastewaters, for example from the food and agricultural industries (Gray, 1989).

A number of anaerobic processes have been developed; completely mixed reactors (often these are simply described as anaerobic digesters) with or without sludge recycle, anaerobic filters, up-flow anaerobic sludge blankets (UASB) and anaerobic fluidized beds are the most common. Two stage anaerobic treatment systems utilizing a short retention time completely mixed acidification reactor followed by a UASB methanogenic reactor have been reported (Burgess and Morris, 1984; Ghosh et al., 1985). Benefits claimed

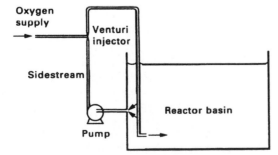

FIG. 11.7. Simplified VITOX sidestream aeration system.

include improved reliability, higher BOD/COD reductions and greater methane productivity.

### ANAEROBIC DIGESTERS

The digester tanks used for this process may be up to 12,000-$m^3$ volume and equipped with heating coils for accurate temperature control normally in the mesophilic range (25°–38°, usually 28°–32° in anaerobic digestion) to increase the rate of digestion, which is also improved by mechanical agitation (Loll, 1977). Some digesters are operated thermophilically (50°–70°, usually 55°–63° in anaerobic digestion) to increase the rate of degradation and biogas production, though much of the extra gas produced will be consumed in maintaining the digester temperature. The use of thermophilic anaerobic digestion has been investigated for the treatment of high strength (16–25 g COD $dm^{-3}$), low pH (~ 3.8) and high temperature winery effluents (Romero et al., 1988). Wiegant et al. (1985) reported the successful application of thermophilic anaerobic digestion of vinasse in completely mixed and UASB reactors, although methane production was found to decrease at high loading rates due to the presence of inhibitory compounds. Retention times are generally between 10 and 30 days, although with solids recycle retention times may be as low as 0.5 day. Most digesters operate on a semi-continuous basis with the appropriate volume of digested material being removed and replaced with 'fresh' material on a daily basis.

## ANAEROBIC FILTERS

In anaerobic filters, as with the aerobic filters, there is a microbial film growing on an inert support (Chian and DeWalle, 1977). Anaerobic filters may be operated in an upflow or downflow mode, and a wide variety of packings are used; both natural and synthetic (Young, 1983). The first full-scale anaerobic filters were constructed in the 1970s and in 1982 the Bacardi Corporation brought into operation a 100,000-$m^3$ plant to treat distillery effluents (Szendrey, 1983). Many other effluents are found to be amenable to treatment using anaerobic filter systems; antibiotic fermentation wastes, citric acid fermentation wastes, yeast production wastewater and brewery and winery wastewaters (Szendrey, 1983), molasses distillery slops (Silvero et al., 1986) and fermentation and pharmaceutical wastes (Bonastre and Paris, 1989).

## UP-FLOW ANAEROBIC SLUDGE BLANKETS (UASB)

In this system (originally developed in the Netherlands) high levels of active biomass are retained in the reactor by flocculation (Lettinga et al., 1980, 1983). No support media is added to the reactor, instead the flocculated sludge develops in the reactor and acts as a fluidized bed. Feed is pumped through the bed (the sludge blanket), above which fine particles flocculate and settle back to the blanket as sludge, thus preventing washout of organisms. Anaerobic sludge blankets have been found to be effective in the treatment of many wastewaters, including sugar-beet wastes, domestic sewage, slaughterhouse wastes, agricultural wastes (Lettinga et al., 1983), brewery wastes (Fang et al., 1989, 1990), winery wastes (Cheng et al., 1990) and distillery wastes (Burgess and Morris, 1984; Ghosh et al., 1985).

## BY-PRODUCTS

The marketing of wastes from fermentation processes has been established for at least 200 years. By the 1700s, brewers' grains, spent hops and surplus yeast from larger breweries were accumulating in sufficient quantities for specialized trades to develop (Mathias, 1959). Around London, cattle and pigs were fattened on 'wash' and brewers' grains. Excess yeast was supplied to bakers and gin distillers.

In any fermentation recovery process there is a need to recognize whether there is a potential marketable waste and, if necessary, to develop a market. Obviously the marketability and cost of reclamation of by-products will be very important in deciding upon a policy of waste recovery. Under favourable market conditions it has been claimed that the profit on animal feed by-products from a completely integrated distillery may almost pay the cost of the grain (Blaine, 1965). This claim would now be more difficult to substantiate, due to fuel costs and capital outlay on plant (Quinn and Marchant, 1980; Sheenan and Greenfield, 1980).

Some microbial processes yield residues which are difficult or impossible to sell (Blaine, 1965). In these processes it may be possible, and worth while, to change to low residue raw materials such as refined sugars or other pure compounds as sources of soluble nutrients.

### Distilleries

In grain-based distilleries it has been common practice to recover spent grain and stillage, the waste liquor after the alcohol has been distilled off (Boruff, 1953; Blaine, 1965). The stillage is first passed through screens with 1-mm openings. The screenings are then dewatered with mechanical filter presses and dried in rotary driers (Chapter 10). This product is termed Distillers' Dried Grains (light grains). The screened stillage is concentrated in evaporators to give 25 to 35% solids in a thick syrup. Stillages which have been prefiltered may be concentrated to 35 to 50% solids. This syrup can then be mixed with the pressed screenings and dried to give Distillers' Dried Grains with Solubles (dark grains). Alternatively the evaporated stillage may be dried completely in drum driers to produce Distillers' Dried Solubles. Dark and light grains and dried solubles have all been used as animal-feed supplements. Flachowsky et al. (1990) report their use following chemical treatment as a replacement feed for sheep. Dried solubles have also been used as a medium adjunct in the preparation of antibiotics (Chapter 4).

In distilleries using cane molasses as a feedstock, evaporated spent wash has been used as a fuel for boilers (Sheenan and Greenfield, 1980). It has proved worth while to recover potassium salts from sugar-beet stillage. The market for the evaporated product must be considered within a range of 50 km of the evaporation plant (Lewicki, 1978).

Boruff (1953) cited an unpublished 1949 survey of American distilleries which showed that 85% of the stillage solids were being recovered as dried feeds, 14% solids as wet grains and only 1% was waste. In recent years the traditional distillers' by-products from stillage have become increasingly uneconomic. A number of processes to produce SCP from stillage using *Geotrichum candidum, Candida utilis* or *C. tropicalis* have been evaluated (Quinn and Marchant, 1980; Sheenan and Greenfield, 1980).

## Breweries

The three marketable wastes from breweries are spent grain, spent hops and yeast. The spent grain is recovered from the mash tun and is then sold as animal feed either after pressing in a wet state or after drying in rotary driers. Alternatively, the wet grain may be used in the preparation of silage for cattle. The possible markets for hops are restricted. Some are used as a fertilizer or as a low-grade fuel.

Yeast can be separated from beer by filtration or centrifuging. The yeast slurry is then dried in drum driers. Some of the yeast may be mixed with brewers' spent grains to produce a feed material with a slightly higher protein content than normal brewers' grains. The yeast may also be used directly as a source of vitamins. If it is to be used as a human food it must be debittered to remove the hop bitter substances absorbed on to the yeast cells. The cells are then washed in an alkaline solution, washed with water and drum dried. Although bakers' yeast was originally obtained as a brewery by-product, this market has diminished considerably. Most bakers' yeast is now produced directly by a distinct production process. Dewatered sludge from brewery wastewater treatment operations has been reported to increase agricultural yields when used as fertilizer (Naylor and Severson, 1984). Lyons (1983) reports on the potential use of brewery effluents as a feedstock for fuel and industrial ethanol production.

## Amino acid wastes

The main wastes from glutamic acid or lysine fermentations are cells, a liquor with a high amino-acid content which can be used as an animal-feed supplement, and the salts removed from the liquor by crystallization, which is a good fertilizer (Renaud, 1980).

## Fuel alcohol wastes

The stillage from ethanol production and wastes from starchy fermentations are, following concentration, saleable as an animal-feed supplement. Wastes from sugar fermentations and distillation can be digested anaerobically and the methane generated used as an energy source (Essien and Pyle, 1983; Faust *et al.*, 1983; Singh *et al.*, 1983). Faust *et al.* (1983) also suggest the use of $CO_2$ rich off gases in the food and beverage industries.

## REFERENCES

ABSON, J. W. and TODHUNTER, K. H. (1967) Effluent treatment. In *Biochemical and Biological Engineering Science*, Vol. 1, pp. 309–343 (Editor Blakebrough, N.). Academic Press, London.

AMERICAN PUBLIC HEALTH ASSOCIATION (1992) *Standard Methods for the Examination of Water and Wastewater* (18th edition). (Editors Greenberg, A. E., Clesceri, L. S. and Eaton, A. D.). American Public Health Association, Washington, DC.

ARDEN, E. and LOCKETT, W. T. (1914) Experiments on the oxidation of sewage without the aid of filters. *Surveyor* 45, 610–620.

ASHLEY, M. H. J. (1982) The efficient use of water in fermentation processes. *I. Chem. E. Symp. Ser.* 78, 355–368. I. Chem. E., Rugby.

BALLINGER, D. G. and LISHKA, R. J. (1962) Reliability and precision of BOD and COD determinations. *JWPCF* 34, 470–474.

BLAINE, R. K. (1965) Fermentation products. In *Industrial Waste Water Control*, pp. 147–166 (Editor Gurnham, C. F.). Academic Press, New York.

BONASTRE, N. and PARIS, J. M. (1989) Survey of laboratory, pilot and industrial anaerobic filter operations. *Process Biochem.* 24(1), 15–20.

BORCHARDT, J. A. (1970) Biological waste treatment using rotating discs. In *Biological Waste Treatment. Biotechnol. Bioeng. Symp.* 2, 131–140.

BORUFF, C. S. (1953) The fermentation industries. In *Industrial Wastes. Their Disposal and Treatment*, pp. 99–131 (Editor Rudolfs, W.). Reinhold, New York.

BROOKING, J., BUCKINGHAM, C. and FUGGLE, R. (1990) Constraints on effluent plant design for a brewery. In *Effluent Treatment and Waste Disposal*, pp. 109–126. I. Chem. E. Symp. Ser. 116, I. Chem. E., Rugby.

BROWN, A. (1992) *The UK Environment*, pp. 75–104 (Editor Brown, A.). HMSO (Department of the Environment), London.

BRUCE, A. M. and HAWKES, H. A. (1983) Biological filters. In *Ecological Aspects of Used Water Treatment*, Vol. 3, *The Processes and Their Ecology*, pp. 1–111 (Editors Curds, C. R. and Hawkes, H. A.). Academic Press, London.

BURGESS, S. and MORRIS, G. G. (1984) Two-phase anaerobic digestion of distillery effluents. *Process Biochem. (supplement)*, 19(5), iv–v.

CHENG, S. S., LAY, J. J., WEI, Y. T., WIU, M. H., ROAM, G. D.

and CHANG, T. C. (1990) A modified UASB process treating winery wastewater. *Water Sci. Technol.* **22**(9), 167–174.

CHIAN, E. S. K. and DEWALLE, F. B. (1977) Treatment of high strength acidic waste water with a completely mixed anaerobic filter. *Water Res.* **11**, 295–304.

COLOVOS, G. C. and TINKLENBERG, N. (1962) Land disposal of pharmaceutical manufacturing wastes. *Biotech. Bioeng.* **4**, 153–160.

COOK, E. E. (1978) Effects of long term endogenous respiration (fasting) on the organic removal capacity of a trickling filter. *Biotech. Bioeng.* **20**, 293–296.

COOPER, P. F. (1975) Physical and chemical methods of sewage treatment. Review of present state of technology. *Water Pollut. Control* **74**, 303–311.

COOPER, P. F. and WHEELDON, D. H. V. (1980) Fluidised and expanded bed reactors for wastewater treatment. *Water Pollut. Control.* **79**, 286–306.

COOPER, P. F. and WHEELDON, D. H. V. (1982) Complete treatment of sewage in a two-stage fluidised bed system. *Water Pollut. Control.* **81**, 447–464.

DAVIS, E. M. (1971) BOD vs COD vs TOC vs TOD. *Water Wastes Eng.* **8**(2), 32–34, 38.

DONALDSON, J. (1993) Identifying opportunities for chemical waste recycling. *Wastes Management*, June 1993, 16–18.

ESSIEMN, D. and PYLE, D. L. (1983) Energy conservation in ethanol production by fermentation. *Process Biochem.* **18**(4), 31–37.

FANG, H. H. P., JINFU, Z. and GUOHUA, L. (1989) Anaerobic treatment of brewery effluent. *Biotechnol. Lett.* **11**(9), 673–678.

FANG, H. H. P., GUOHUA, L., JINFU, Z., BUTE, C. and GUOWEI, G. (1990) Treatment of brewery effluent by UASB process. *J. Environ. Eng.* **116**(3), 454–460.

FAUST, U., PRAVE, P. and SCHLINGMANN, M. (1983) An integral approach to power alcohol. *Process Biochem.* **18**(3), 31–37.

FISHER, N. S. (1977) Legal aspects of pollution. In *Treatment of Industrial Effluents*, pp. 18–29 (Editors Callely, A. G., Forster, C. F. and Stafford, D. A.). Hodder and Stoughton, London.

FLACHOWSKY, G., BALDEWEG, P., TIROKE, K., KONIG, H. and SCHNEIDER, A. (1990) Feed value and feeding of wastelage made from distillers solubles, pig slurry solids and ground straw treated with urea and NaOH. *Biol. Wastes* **34**(4), 271–280.

FORSTER, C. F. (1977) Bio-oxidation. In *Treatment of Industrial Effluents*, pp. 65–87 (Editors Callely, A. G., Forster, C. F. and Stafford, D. A.). Hodder and Stoughton, London.

FORSTER, C. F. (1985) *Biotechnology and Wastewater Treatment*. Cambridge University Press, Cambridge.

FREESTONE, N. P., PHILLIPS, P. S. and HALL, R. (1994) Having the last gas. *Chem. Britain*, Jan. 1994, 48–50.

FUGGLE, R. W. (1983) The application of the UNOX activated sludge processto the treatment of coal processing wastewaters. In *Effluent Treatment in the Process Industries*. pp. 105–124. *I. Chem. E. Symp. Ser.* **77**, I. Chem. E., Rugby.

GHOSH, S., OMBREGT, J. P. and PIPYN, P. (1985) Methane production from industrial wastes by two-phase anaerobic digestion. *Water Res.* **19**(9), 1083–1088.

GOSTICK, N. A., WHEATLEY, A. D., BRUCE, B. M. and NEWTON, P. E. (1990) Pure oxygen activated sludge treatment of a vegetable processing wastewater. In *Effluent Treatment and Waste Disposal*, pp. 69–84. I. Chem. E. Symp. Ser. No. 116, I. Chem. E., Rugby.

GRAY, N. F. (1989) *Biology of Wastewater Treatment*. Oxford Science Publications, Oxford.

GRIEVE, A. (1978) Sludge incineration with particular reference to the Coleshill plant. *Water Pollut. Control*, **77**, 314–321.

HMSO (1972) *Analysis of Raw, Potable and Waste Waters*, pp. 121–122. HMSO, London.

HMSO (1990) *Environmental Protection Act 1990*. HMSO, London.

HMSO (1991) *Waste Management; the Duty of Care, a Code of Practice: Environmental Protection Act 1990*. HMSO, London.

HAIGH, N. (1990) *EEC Environmental Policy and Britain* (2nd revised edition). Longman, Harlow.

HEMMING, M. L., OUSBY, J. C., PLOWRIGHT, D. R. and WALKER, J. (1977) 'Deep Shaft' — Latest position. *Water Pollut. Control* **76**, 441–451.

HIGGINS, P. M. (1968) Waste treatment by aerobic techniques. *Dev. Ind. Microbiol.* **9**, 146–159.

HILL, F. (1980) Effluent treatment in the Bakers' yeast industry. In *Efffluent Treatment in the Biochemical Industries*, Process Biochemistry's 3rd International Conference.

HUGHES, D. (1992) *Environmental Law (2nd edition)*. Butterworths, London.

JACKSON, J. (1960) The treatment of distillery and antibiotics wastes. In *Waste Treatment*, pp. 226–239 (Editor Isaac, P. C. G.). Pergamon Press, Oxford.

KOZIOROWSKI, B. and KUCHARSKI, J. (1972) *Industrial Waste Disposal*. Pergamon Press, Oxford.

LAING, I. G. (1992) Waste minimisation: The role of process development. *Chem. Ind.*, Sept. 1992, 682–686.

LETTINGA, G., VAN VELSEN, A. F. M., HOBMA, S. W., DE ZEEUW, W. and KLAPWIJK, A. (1980) Use of the upflow sludge blanket (USB) reactor concept for biological waste treatment, especially for anaerobic treatment. *Biotech. Bioeng.* **22**, 699–734.

LETTINGA, G., HULSHOFF POL, L. W., WIEGANT, W., DE ZEEUW, W., HOBMA, S. W., GRIN, P., ROERSMA, R., SAYAD, S. and VAN VELSEN, A. M. F. (1983) Upflow sludge blanket processes. In *Proceedings of the 3rd. Intl. Symp. on Anaerobic Digestion*. pp. 139–158. Aug. 14–19, 1983, Boston, MA.

LEWICKI, W. (1978) Production, application and marketing of concentrated molasses-fermentation-effluent (vinasses). *Proc. Biochem.* **14**(6), 12–13.

LOEHR, R. C. (1968) Anaerobic treatment of wastes. *Dev. Ind. Microbiol.* **9**, 160–174.

LOLL, U. (1977) Engineering, operation and economics of biodigestion. In *Microbial Energy Conversion*, pp. 361–378

(Editors Schlegel, H. G. and Barnes, J.). Pergamon Press, Oxford.

LYONS, T. P. (1983) Ethanol production in developing countries. *Process Biochem.* **18**(2), 18–25.

MACKEL, C. J. (1976) A study of the availability of distillery by-products to the agricultural industry, 46 pp. Report of the School of Agriculture, University of Aberdeen, Scotland.

McKINNEY, R. L. (1962) Complete mixing activated sludge treatment of antibiotic wastes. *Biotech. Bioeng.* **4**, 181–195.

McLEOD, G. and O'HARA, J. (1993) EC proposals for integrated pollution prevention and control. *Chem. Ind.,* Nov. 1993, 849–851.

MBAGWU, J. S. C. and EKWEALOR, G. C. (1990) Agronomic potential of brewers spent grains. *Biol. Wastes* **34**(4), 335–347.

MASTERS, G. M. (1991) *Introduction to Environmental Engineering and Science,* pp. 101–179. Prentice Hall, New Jersey.

MATHIAS, P. (1959) *The Brewing Industry in England, 1700–1830.* University Press, Cambridge.

MELCHER, R. R. (1962) Pharmaceutical waste disposal by soil injection. *Biotech. Bioeng.* **4**, 147–151.

NAYLOR, L. M. and SEVERSON, K. Y. (1984) Brewery sludge as a fertilizer. *Biocycle,* **25**(3), 48–51.

OMSTEAD, M. N., KAPLAN, L. and BUCKLAND, B. C. (1989) Fermentation development and process improvement. In *Ivermectin and Abamectin,* pp. 33–54 (Editor Campbell, W. C.). Springer, New York.

OTTO, R., BARKER, W., SCHWARZ, D. and TJARKSEN, B. (1962) Laboratory testing of pharmaceutical wastes for biological control. *Biotech. Bioeng.* **4**, 139–145.

POHLAND, F. G. and GHOSH, S. (1971) Developments in anaerobic treatment processes. In *Biological Waste Treatment. Biotech. Bioeng. Symp.* **2**, 85–106.

PRETORIOUS, W. A. (1973) The Rotating Disc unit. A waste treatment system for small communities. *Water Pollut. Control,* **72**, 721 – 724.

QUINN, J. P. and MARCHANT, R. (1980) The treatment of malt whisky distillery waste using the fungus *Geotrichum candidum. Water Res.* **14**, 545–551.

RAMATHAN, M. and GAUDY, A. F. (1969) Effect of high substrate concentration and cell feed-back on kinetic behaviour of heterogeneous populations in completely mixed systems. *Biotech. Bioeng.* **11**, 207–237.

RENAUD, C. (1980) Treatment of effluent from glutamic acid and lysine fermentation. In *Effluent Treatment in the Biochemical Industries,* Process Biochemistry's 3rd International Conference.

RIPLEY, P. (1979) Process engineering aspects of the treatment and disposal of distillery effluent. *Proc. Biochem.* **14**(1), 8–10.

ROMERO, L. I., SALES, D., CANTERO, D. and GALAN, M. A. (1988) Thermophilic anaerobic digestion of winery waste (vinasses): Kinetics and process optimisation. *Process Biochem.* **23**(4), 119–125.

SCA (Standing Committee of Analysts) (1989) 5 Day Biochemical Oxygen Demand (BOD$_5$). MEWAM, HMSO, London.

SHEENAN, G. J. and GREENFIELD, P. F. (1980) Utilisation, treatment and disposal of distillery waste water. *Water Res.* **14**, 257–277.

SILVERO, C. M., ANGLO, P. G., MONTERO, G. V., PACHECO, M. V., ALAMIS, M. L. and LUIS, V. S. Jr. (1986) Anaerobic treatment of distillery slops using an upflow anaerobic filter reactor. *Process Biochem.* **21**(6), 192–195.

SIMPSON, J. R. and ANDERSON, G. K. (1967) Large-volume respirometers with particular reference to waste-treatment. *Prog. Ind. Microbiol.* **6**, 141–167.

SINGH, V., HSU, C. C., CHEN, C. and TZENG, C. H. (1983) Fermentation processes for dilute food and dairy wastes. *Process Biochem.* **18**(2), 13–25.

SMETHURST, G. (1988) *Basic Water Treatment for Application World-Wide* (2nd Edition). Thomas Telford, London.

SMITH, C. (1993) Waste, incineration and the environment. *Waste Planning* **9**, 7–17.

STEPHENSON, T., ALLAN, M. and UPTON, J. (1993) The small footprint wastewater treatment process. *Chem. Ind.* **15**, 533–536.

SZENDREY, L. M. (1983) Startup and operation of the Bacardi Corporation anaerobic filter. In *Proceedings of the 3rd Intl. Symp. on Anaerobic Digestion.* pp. 365–377. 14–19 Aug., 1983, Boston, MA.

TAYLOR, I. J. and SENIOR, P. J. (1978) Single cell proteins: a new source of animal feeds. *Endeavour* (N.S.) **2**, 31–34.

TROMANS, S. (1991) *The Environmental Protection Act 1990, Text and Commentary.* Sweet & Maxwell, London.

TRUESDALE, G. A. (1979) Tertiary treatment. In *Water Pollution Control Technology,* pp. 84–91. HMSO, London.

VIESSMAN, W. and HAMMER, M. J. (1993) *Water Supply and Pollution Control* (5th edition), pp. 741–748. Harper-Collins, New York.

WARE, A. J. and PESCOD, M. B. (1989) Full scale studies with an anaerobic/aerobic RBC unit treating brewery waste-water. *Water Sci. Technol.* **21**(4–5), 197–208.

WATER POLLUTION RESEARCH CENTRE (1972) *Water Pollution Research: Report of the Director of Water Pollution Research, 1971.* HMSO, London.

WIEGANT, W. M., CLAASSEN, J. A. and LETTINGA, G. (1985) Thermophilic anaerobic digestion of high strength wastewaters. *Biotech. Bioeng.* **27**(9), 1374–1381.

WINKLER, W. A. and THOMAS, A. (1978) Biological treatment of aqueous wastes. In *Topics in Enzyme and Fermentation Biotechnology,* Vol. 2, pp. 200–279 (Editor Wiseman, A.). Ellis Horwood, Chichester.

YOUNG, J. C. (1983) The anaerobic filter — past, present and future. In *Proceedings of the 3rd Intl. Symp. on Anaerobic Digestion,* pp. 91–106. 14–19 Aug. 1983, Boston, MA.

ZAJIC, J . E. ( 1971 ) Deep well disposal. In *Water Pollution, Disposal and Re-use,* Vol. 2, pp. 574–589. Marcel Dekker, New York.

# CHAPTER 12

# Fermentation Economics

## INTRODUCTION

IF A FERMENTATION process is to yield a product at a competitive price, the chosen micro-organism or animal cell culture should give the desired end-product in predictable, and economically adequate, quantities. A number of basic objectives are commonly used in developing a successful process which will be economically viable.

1. The capital investment in the fermenter and ancillary equipment should be confined to a minimum, provided that the equipment is reliable and may be used in a range of fermentation processes.
2. Raw materials should be as cheap as possible and utilized efficiently. A search for possible alternative materials might be made, even when a process is operational.
3. The highest-yielding strain of micro-organism or animal cell culture should be used.
4. There should be a saving in labour whenever possible and automation should be used where it is feasible.
5. When a batch process is operated, the growth cycle should be as short as possible to obtain the highest yield of product and allow for maximum utilization of equipment. To achieve this objective it may be possible to use fed-batch culture (Chapters 2 and 4).
6. Recovery and purification procedures should be as simple and rapid as possible.
7. The effluent discharge should be kept to a minimum.
8. Heat and power should be used efficiently.
9. Space requirements should be kept to a minimum, but there should be some allowance for potential expansion in production capacity.
10. All the above must comply with safety guidelines and regulations.

The consideration of so many criteria means that there may have to be a compromise for the particular set of circumstances relating to an individual process. Winkler (1991) identified three key economic objectives in the fermentation stage: maximum product yield, process productivity and substrate utilization. However, these criteria may be overridden by the demands of subsequent purification stages for high product concentration and high product purity which may in turn be overridden by safety considerations.

In any process it is important to know the cost breakdown (Table 12.1), so that it may be seen where the biggest potential savings may be achieved. In a review of a number of processes, Nyiri and Charles (1977) concluded that four basic components contributed to the process cost in the following decreasing order: raw materials, fixed costs, utilities, labour. When raw materials are a major part of the total cost it is obvious that media and microbial strain-improvement research should form a major part of a development programme. Development work on components contributing little to the cost of the product could not be justified.

Costs quoted in this chapter have been taken from the source of information and are applicable only at the time of that publication. Atkinson and Mavituna (1991) have discussed the use of cost indices to update historical data.

Caution is necessary when examining some examples of potential fermentation processes, since the capital and operating costs normally have been estimated us-

331

TABLE 12.1. *Production costs breakdown expressed as percentages of total cost*

| Item | Product | | | | | | | |
|------|------|------|------|------|------|------|------|------|
| | Beer (Pratten, 1971) | Alcohol (batch) (Maiorella, 1984) | Acetic acid (Pape, 1977) | Citric acid (Schierholt, 1977) | Norprotein (Mogren, 1979) | SCP (Anon, 1974) | Penicillin (Swartz 1979) | tPA (Datar *et al.* 1993) |
| Raw materials | 38.4 | 76.7 | 42.2 | 39.7 | 70.0 | 62 | 58.0 | 39.8 |
| Utilities | * | 11.7 | 23.1 | 35.3 | 16.0 | 10 | 20.3 | 20.5 |
| Labour and supervision | 24.5 | 2.9 | 19.5 | 25.0 | 9.0 | 9 | 5.4 | 10.9 |
| Fixed charges | 7.2 | 4.8 | 10.5 | — | — | 19 | — | — |
| Maintenance | 29.9 | — | — | — | 5.0 | † | 14.9 | — |
| Operating supplies | — | — | — | — | — | — | 1.4‡ | — |
| Waste | — | — | — | — | — | — | — | 12.0 |
| Materials recovery | — | — | — | — | — | — | — | 13.3 |
| Other | — | — | — | — | — | — | — | 3.6 |

*Note* *Included in 29.9% for maintenance and operating supplies.
†Included in 9% for labour and supervision. ‡0.2% for laboratory costs included.

ing chemical engineering costing principles. Accurate detailed costing of industrial fermentations are rarely published.

Government aid or taxation can determine the viability of many fermentation processes. Support in construction of plant, development and production programmes of acetone–butanol and penicillin during wartime led to production of compounds at an earlier stage than might normally be regarded as economically feasible (Hastings, 1971, 1978). Agricultural-aid programmes in the United States of America made available low-cost supplies of grain and potatoes and enabled fermentations to be operated when they would not have been economically viable in a free market (Perlman, 1970). In 1980 carbohydrates as molasses or cane juice could be obtained in sugar-producing countries at half the price of molasses in the European Economic Community (Meers, 1980) and differences in starch prices between the EEC and the U.S.A. were about 40% (Gray, 1987). This price differential of carbon substrates discouraged investment and further research in the EEC. Policy changes were made to allow fermentation companies within the EEC to buy sugar and starch based substrates at a lower price to enable them to compete on a world-wide scale.

Some very useful reviews of process economics have been produced for penicillin G (Swartz, 1979), gibberellic acid (Vass and Jefferys, 1979), biomass from natural gas (Hamer, 1979), biomass from whey (Meyrath and Beyer, 1979), biomass from waste carbohydrates (Mateles, 1975), biomass from cane and coffee process-

ing by-products (Rolz, 1975), 6-aminopenicillanic acid (Harrison and Gibson, 1984), evaluation of 11 alternative ethanol fermentation processes (Maiorella *et al.*, 1984), tissue plasminogen activator (Datar *et al.*, 1993), primary separation steps (Datar, 1986) and chromatography (Sofer and Nystrom, 1989). A number of general reviews of fermentation economics are also available (Whitaker, 1973; Nyiri and Charles, 1977; Bartholomew and Reisman, 1979; Stowell and Bateson, 1984; Bailey and Ollis, 1986; Hacking, 1986; Kalk and Langlykke, 1986; Reisman, 1988; Atkinson and Mavituna, 1991).

## ISOLATION OF MICRO-ORGANISMS OF POTENTIAL INDUSTRIAL INTEREST

The most appropriate micro-organism for a potential process is usually found by isolation from a variety of sources, most commonly soil. The classical method of screening to obtain a suitable organism tended to be very time consuming, expensive and often without a very clear objective. Eli Lilly and Company Ltd. discovered three new antibiotics in 10 years while screening 400,000 micro-organisms (Nelson, 1961). More recently Berdy (1989) has speculated that the screening of 100,000 soil micro-organisms may lead to the isolation of 5 to 50 new compounds, but there is no guarantee after evaluation that a useful new drug or other product will be found.

If a desired characteristic, which gives the organism a selective advantage, has already been recognized, a screen might be designed incorporating this characteristic as a selective factor (Chapter 3). The isolation may begin with pretreatment of samples which favour the survival of the preferred organism. This is followed by growth on selective or non-selective media and often associated with batch or continuous enrichment. Important factors which will be of economic significance which might be selected in a well planned screen could include:

1. Growth on a simple cheap medium.
2. Growth at a higher temperature (to reduce cooling costs).
3. Better resistance to contamination.

However, the synthesis of other microbial products (e.g. antibiotics) does not give the producing organism any selective advantage which might be used in an isolation procedure (Chapter 3). Therefore a collection of these organisms must be made before testing for the desired characteristic. Because many isolation procedures will lead to the rediscovery of known organisms with known activities it is important to use well planned, efficient isolation procedures which can prove very productive.

A number of approaches are currently being used to improve isolation procedures (see also Chapter 3). Numerical taxonomic data bases are being exploited to design selective media for certain microbial taxa. Using these data bases it has been possible to design media to encourage the growth of uncommon streptomycetes or discourage the growth of common species (Vickers et al., 1984; Williams and Vickers, 1988). Knowledge of antibiotic sensitivity gained from taxonomic data bases has led to the design of other selective media which will select for resistant groups of organisms (Goodfellow and O'Donnell, 1989; Bull, 1992; Bull et al., 1992). The selection of antibiotic producing soil isolates has also been achieved using media designed by a stepwise discrimination analysis technique (Huck et al., 1991).

These 'designed' isolation media are now being used extensively for the isolation of novel and rare microorganisms. For example, within the actinomycetes, Streptomyces spp. have been extensively screened since the 1940s for antibiotic production with subsequent notable commercial success. More interest has now been shown in isolation of strains of the less common genera such as Actinomadura, Actinoplanes, Kitosatosporia, Streptoalloteichus, etc. which are producing other novel bioactive compounds (Goodfellow and O'Donnell, 1989; Bull, 1992).

It has become a common practice to obtain isolates from unusual habitats, which may include extreme environments, to ensure that the greatest microbial diversity is being examined (Bull, 1992; Bull et al., 1992; Chapter 3).

The screening tests which have been developed to detect new useful compounds of potential industrial interest have become much more selective and sensitive. Better knowledge of cell biochemistry has enabled the design of screens which are much more precisely targeted to detect the desired activity using specific detector strains (Chapter 3). Many large companies which undertake screening programmes have introduced some automation to enable high throughput rates which are cost effective and less labour intensive. In 1989, Nisbet and Porter considered that $10^5$ tests per year with up to 20 assay tests should be undertaken in a worthwhile screening programme.

## STRAIN IMPROVEMENT

Strain improvement using a mutation/selection programme (Chapter 3) for improving an organism being used in an established process or a potential process can be very cost effective. Historically, mutation/selection programmes to improve strains of *Penicillium chrysogenum* were time consuming, labour intensive and very random because of the lack of knowledge about penicillin biosynthesis. These mutation programmes did, however, contribute significantly to increases in penicillin yields from less than 100 units $cm^{-3}$ in the 1940s to over 51,000 units $cm^{-3}$ by 1976 (Queener and Swartz, 1979) and a four-fold increase in yields between 1970 and 1985 at Gist Brocades (Royce, 1993). Improvements for streptomycin, chlortetracycline and erythromycin are reported in Table 3.8.

It is always very important to decide if a strain improvement programme can be justified on financial grounds to improve the overall economy of a process. Lockwood and Streets (1966) quoted the example of a fermentation process making 453,600 kg year$^{-1}$ of product at 23.5 p kg$^{-1}$ in which a 1% increase in the product would produce an increased return of only £1070, which they considered insufficient to support a worthwhile mutation programme. If the output had been 10 times greater, a 1% increase would have produced £10,700 which was thought to be just sufficient to meet the costs of research. If this output could have been increased to 10%, the increased return would have been £107,000, which would have been much larger than the cost of a mutation programme at that time. Calam (1969) suggested that a graduate worker with two assistants are capable of operating one or two mutation research programmes and of making effective progress.

A better understanding of cell metabolism and its regulation has enabled the development of more logical targeted methods to be introduced to select for mutants with desirable 'blue prints' where there may be a need to block undesirable enzyme activities or eliminate negatively acting control mechanisms (Chapter 3). This approach is much more efficient and economic in terms of resources and time. It was first employed extensively in the preparation of mutant strains used in amino acid fermentations.

Although the main targets in strain improvement are normally to increase the product yield or specific production rates, it is also important to consider strain stability, resistance to phage infection, response to dissolved oxygen, tolerance to medium components, production of foam and the morphological form of the organism. Methods to achieve many of these changes are discussed in Chapter 3. These are very important in helping to achieve targets in a research and development programme as they can have a significant impact on the process and/or product (Table 12.2). A study of the range of targets given in this table, which might be aided by strain improvement, has economic effects on all aspects of a fermentation process.

There are a number of companies with special expertise, such as Cetus Ltd and Panlabs Inc., who will perform strain-development programmes or make cultures available for commercial clients. An example of work by Panlabs Inc.'s improvements in yields of penicillin from cultures of *P. chrysogenum* during 1973 to 1976 is shown in Table 12.3 (Queener and Swartz, 1979). These data make it possible to compare yields from the same culture cultured at various scales.

## MARKET POTENTIAL

The fermentation technologist should be aware of the problem of assessing market potential, although he/she may not be primarily involved in collating or assessing the necessary data. Some aspects have been considered by MacLennan (1976), Hepner (1977), Lawson and Sutherland (1978) and Keim and Venkatasubramanian (1989).

Four categories of microbial product can be recognized economically and it is important to consider to which category a compound belongs:

1. Low price bulk chemicals, e.g. solvents, biomass, high fructose syrups (US$$10^2$–$10^3$ tonne$^{-1}$).
2. Mid price chemicals, e.g. organic acids, amino acids, bipolymers ($$10^3$–$10^5$ tonne$^{-1}$).
3. High price microbial and animal-cell products, e.g. enzymes, vitamins, antibiotics, corticosteroids, vaccines, etc. ($$10^5$–$10^7$ tonne$^{-1}$).
4. Very high-price animal-cell products, e.g. monoclonals, tissue plasminogen activator, etc. ($$10^7$–$10^9$ tonne$^{-1}$).

The third and fourth groups can normally be produced only by a microbial or animal-cell based process and therefore do not have to compete with an alternative chemical process which is usually much cheaper. This includes compounds which have complicated structures, are chemically or thermally unstable or for which a multi-stage chemical synthesis would be expensive. Many microbial products are not exploited because cheaper synthetic processes are available.

Hepner (1977, 1978) has examined the factors that determine the feasibility of large-scale ethanol production by fermentation. He considered that ethanol produced by fermentation would only be competitive with synthetic ethanol from crude oil if the fermentation plant was in an area where cheap supplies of carbohydrate were available. In an example based on 1977 costs, if crude oil cost $100 tonne$^{-1}$, fermentation produced ethanol would be financially viable only if raw-sugar feedstock cost less than $109 tonne$^{-1}$ or molasses cost less than $75 tonne$^{-1}$(Hepner, 1978). In Brazil in 1975, an ethanol production programme using sugar-cane as a substrate was started so that the petroleum imports could be reduced. During 1977, $7 \times 10^8$ dm$^3$ of ethanol were produced and it was hoped to have increased production to $1.5 \times 10^9$ dm$^3$ during 1979. The ethanol, although subsidized by the Brazilian government, was selling at $1 per 4.5 dm$^3$, which was more than the cost of petrol refined from imported oil (Hammond, 1978). Other aspects of potential ethanol production have been discussed by Bu'Lock (1979).

It is necessary to estimate the size of the present and potential market and the increase in demand for a compound. This type of exercise was undertaken for single-cell protein (MacLennan, 1976; Taylor and Senior, 1978). Taylor and Senior (1978) gave summaries of major single-cell protein plants which were operational or planned, estimates of single-cell protein production in 1980 and 1985 and predicted world supply and demand for high-quality protein meal. It was estimated that by 1985 there would be a market of $5 \times 10^6$

TABLE 12.2. *Criteria for strain improvement* (Schwab, 1988)

| Target | Impact on process or product |
|---|---|
| Improvement of titre and/or specific production rate | General decrease of production costs, improved exploitation of reactor capacity, lower investment costs, increased efficiency in downstream processing steps |
| Improvement of yield | Lower costs for substrates, decreased production of heat and $CO_2$, lower cooling costs, less waste and pollution |
| Change in catabolic capabilities | Use of more favourable substrates (less expensive, better availability, etc.), omission of pretreatment steps (e.g. enzymatic hydrolysis of polysaccharides). |
| Improvement of technological features of micro-organisms (e.g. flocculation behaviour, structure of mycelium, sporulation, foaming, strain stability, etc.) | Less energy costs for mixing and oxygen transfer, improved separation characteristics, fewer problems in inoculum preparation or scale-up of the process |
| Improvement of product quality | Decreased production of specific by-products (fewer impurities), prevention of product degradation (e.g. pectinases) |
| Modification of products | Improvement of solubility in extraction solvents (e.g. addition of specific side chains), increased thermic stability of altered enzymatic properties of proteins |
| Changing of the locus of product accumulation (e.g. intracellular to extra-cellular) | Improved product recovery (e.g. omission of cell disruption), correct products (e.g. fully processed proteins). |

tonnes annum$^{-1}$ for single-cell protein world wide, whereas production would only be $2.9 \times 10^6$ tonnes annum$^{-1}$. Such predictions supported the establishment of a process by ICI plc. Unfortunately the product which was marketed as an animal protein feed during the 1980s could not compete for price with soya beans and the manufacturing plant was closed down (Sharp, 1989).

Ratafia (1987) estimated the world markets in 1991 for products that can be made by mammalian cell culture to be as follows ($ $\times 10^6$): diagnostics 6500, vaccines 5200, hormones 4900, lymphokines 1750, monoclonal antibodies 1700, other products 3040. The largest market for a cell culture product was indentified for tissue plasminogen activator (tPA) which was being used in clinical trials and proving effective in dissolving several forms of blood clots. Datar *et al.* (1993) have estimated the potential number of patients in the

TABLE 12.3 *Panlab's penicillin strain-development programme, P-line, showing yields in units cm$^{-3}$ from data supplied by Panlabs Inc.* (Queener and Swartz, 1979)

| Year | Culture | Penicillin | Shaker yield | | | Time (days) | Pilot-plant yield | | | Time (h) | Production yield | | | Time (h) |
|---|---|---|---|---|---|---|---|---|---|---|---|---|---|---|
| | | | Min. | Max. | Av. | | Min. | Max. | Av. | | Min. | Max. | Av. | |
| 1973 | P-1 | V | 6,100 | 7,500 | 7,100 | 5 | 14,500 | 17,600 | 15,900 | 140 | 11,700 | 14,400 | 13,560 | 130 |
| | P-3 | V | 11,000 | 13,100 | 12,000 | 8 | 13,900 | 20,700 | 18,000 | 180 | 14,200 | 19,100 | 16,500 | 180 |
| | | V | 17,400 | 21,300 | 19,800 | 9 | 15,400 | 21,800 | 18,200 | 182 | | | | |
| | | G | | | | | | | | | 17,600 | 17,900 | 17,700 | 132 |
| | P-4 | V | 23,300 | 25,400 | 24,400 | 10 | | | 22,100 | 182 | 29,700 | 20,700 | 20,120 | 182 |
| | | G | | | | | | | | | 15,800 | 20,600 | 17,900⁻ | 182 |
| 1974 | P-5 | V | 27,500 | 31,200 | 29,800 | 10 | 22,600 | 23,700 | 23,150 | 182 | 23,600 | 25,900 | 24,700 | 182 |
| | | G | | | | | | | | | 21,300 | 24,600 | 23,150 | 182 |
| | P-7 | V | 29,200 | 33,900 | 32,100 | 10 | 25,100 | 33,400 | 30,200 | 210 | 23,400 | 29,600 | 27,300 | 190 |
| | | G | | | | | 25,200 | 29,000 | 27,700 | 185 | 23,700 | 28,200 | 25,600 | 190 |
| 1975 | P-8 | V | 34,700 | 36,900 | 36,000 | 10 | 31,500 | 33,000 | 32,250 | 185 | 29,200 | 34,800 | 32,400 | 203 |
| | | G | | | | | | | | | 27,900 | 30,800 | 29,600 | 203 |
| | P-10 | V | 36,800 | 42,600 | 38,700 | | | | | | | | | |
| 1976 | P-10 | V | | | | | 32,500 | 36,500 | 33,700 | 185 | | | | |
| | P-12 | V | 40,911 | 42,800 | 41,900 | 9 | 32,300 | 38,000 | 35,000 | 185 | | | | |
| | P-13 | V | 33,700 | 39,000 | 37,200 | 7 | | | 41,500 | 185 | | | | |
| | | G | | | | | | | 42,250 | 185 | | | | |

U.S.A., Europe and Japan who would benefit from treatment with tPA to be $2.8 \times 10^6$ (about 0.58% of the human population). The price of the FDA licensed tPA is currently $2,200 per dose or $22,000 g$^{-1}$.

At this stage it is worthwhile comparing actual sales of recombinant products with those of some other fermentation products. Since their introduction in the 1980s, the total sales of four new secondary metabolites, cyclosporin (an immunoregulator), imipen (a broad spectrum antibiotic), lovastatin (controller of cholesterol levels) and ivermectin (antiparasitic compound), have been higher than those of all the recombinant products (Buckland, 1992).

The life expectancy of a compound will have to be predicted even when covered by a patent. This is sometimes difficult, as has been demonstrated with the industrial enzyme market (Aunstrup, 1977). In about 1965 detergent enzymes became widely used, which led to a general increase in sales of microbial enzymes. In 1971 allergic symptoms were discovered in workers handling enzymes in a detergent factory, resulting in the removal of enzymes from most detergent powders and a sudden drop in enzyme sales. Enzyme sales recovered after the introduction of improved process techniques.

## SOME EFFECTS OF LEGISLATION ON PRODUCTION OF ANTIBIOTICS AND RECOMBINANT PROTEINS IN THE U.S.A.

There are a number of differences in regulatory requirements for production facilities for antibiotics and recombinant proteins (Bader, 1992). In both cases, the products must be licensed by the FDA before they can be marketed.

The recombinant protein is produced by a precisely specified process using high quality substrates and processed and purified in an aseptic pharmaceutical facility that has been inspected and licensed by the CBER (Center of Biological Evaluation and Research) of the FDA. Steps must also be taken to ensure that there is no cross contamination if more than one product is produced in the same facility. Any changes in a process or facilities must be approved by the CBER before implementation. It may take 7 years to obtain approval by the CBER for a production plant. In contrast, the requirements for an antibiotic production plant are less stringent and it may be operational in 4 years. The delay in start up to produce the recombinant protein will result in much higher costs.

It is important to remember that the detailed clinical evaluation of a microbial compound as a drug, plus FDA approval, may take 8–10 years from initial discovery and cost up to $150 \times 10^6$.

## PLANT AND EQUIPMENT

It is most logical to build equipment as large as possible because of the economy of scale. There is an empirical relationship between cost and size of an item of equipment. According to this relationship, as facility size increases, its cost increases thus:

$$\frac{\text{cost}_1}{\text{cost}_2} = \left(\frac{\text{size}_1}{\text{size}_2}\right)^n$$

where $n$ is an exponent or scale factor. Scale factors have been estimated to be 0.6 for brewing (Pratten, 1971), 0.7 to 0.8 for a single cell protein plant (Humphrey, 1975; MacLennan, 1976), 0.6 for antibody production (Birch *et al.*, 1987), 0.75 for fermentation processes (Bartholomew and Reisman, 1979) and 0.6 for waste water treatment (Eckenfelder, 1989).

In brewing, when cylindrico-conical vessels are used, there would appear to be no economic advantage in scaling up above 108,000-dm$^3$ capacity (Hoggan, 1977), although vessels of 360,000 dm$^3$ have been installed. In some breweries a number of smaller vessels have been installed to allow for fluctuating demand of different beers.

The operational vessel volume is a critical factor when considering high volume–low cost products. In processes carried out at volumes greater than 100 m$^3$, the use of an air-lift fermenter is more economical as the relative investment costs per unit of output decreased more rapidly than for a stirred type of fermenter (Table 12.4).

However, a number of restraints have to be considered before deciding on the scale of operation. Such restrictions include cooling and aeration requirements and the method of fermentation vessel construction. The need for cooling provisions in many fermentations has already been described (Chapter 7). At this stage it is important to remember that the volume of a fermenter is proportional to $r^3$ (where $r$ is the fermenter radius), whereas the increase in surface area is proportional to $r^2$. Therefore the scaling up of a vessel will lead to a decrease in the surface area to volume ratio and therefore a decrease in the effectiveness of a cooling jacket. There may be a fermenter volume above

TABLE 12.4. *Estimate of relative costs of a fermentation process as a function of scale (100 m$^3$ = 1 unit of volume / unit of time)* (Van Suydam, 1987).

| Scale (m$^3$) | Cost scale factor |
| --- | --- |
| 1000 | 0.3 |
| 200 | 0.6 |
| 100 | 1.0 |
| 10 | 10.0 |
| 1 | 30.0 |

which it will not be possible to remove enough heat to maintain constant temperature, unless the cooling capacity is increased by incorporating internal cooling coils or by using external heat exchangers. These modifications may prove costly or else interfere with mixing in the vessels. There is possibly the alternative of using micro-organisms with higher optimum temperatures. The oxygen requirements of a process may limit the size of vessel which can be operated successfully. In acetic acid (vinegar) production, where very efficient aeration is critical, because *Acetobacter* spp. are very sensitive to oxygen depletion in the broth, the maximum acetator capacity which could be used was about 50,000 dm$^3$ (Pape, 1977).

There is an upper size limit for a custom-built fermenter that can be transported to a site, while much larger vessels may be built on site. Most countries limit the maximum size of unit that can be transported on the road. In 1979 the first ICI plc production vessel for single-cell protein was installed; this had a volume of $1.5 \times 10^6$ dm$^3$, was 80-m tall and cost £6 $\times 10^6$ (Sharp, 1989). It was constructed in France, floated on a barge to the River Tees, and transported a very short distance on land. At the time of erection it was the largest fabricated fermenter ever to be transported.

Because of the capital investment and operational costs there is now a trend to consider unconventional fermenter designs of simple construction with very efficient oxygen-transfer to be used for specific purposes, particularly single-cell protein. In this context, Schugerl *et al.* (1978) have considered the case of a single-cell protein plant for 100,000 tonnes annum$^{-1}$ using methanol as the main substrate. The plant investment was only 20% of the production cost. In this investment cost only 20 to 25% was due to the fermenter. Of this, the vessel accounted for 5 to 10%, the stirrer and aerator 5 to 10% and cooling 10 to 15%. Energy, water, aeration and auxiliaries were estimated to be a further 10% of the production costs. Since the fermenter investment costs are approximately 4 to 5% and the operating costs are 10%, it was concluded that the type of vessel could influence only 15% of the total production costs. If a tower fermenter were used, the main advantage would appear to be lower operating costs, particularly reduced energy and cooling-water costs, as the removal of mechanical stirring would diminish the cooling water requirements.

A useful guide has been prepared by the Institution of Chemical Engineers (1977) which outlined, in reasonably quantitative terms, many factors making up a check list which must be incorporated into the final capital costing of a chemical-plant project. It is possible from a knowledge of the proposed plant location, a sketch of the chemical plant process flow sheet, the size of the major items of equipment and the service requirements, to estimate capital and operating costs to ±15% (Backhurst and Harker, 1973). Unfortunately, comparable literature for fermentation plants is not available.

Some details of the cost of equipment used in fermentation processes has been discussed by Whitaker (1973), Humphrey (1975), Mateles (1975), Rolz (1975), Nyiri and Charles (1977), Maiorella et al. (1984), Hacking (1986), Kalk and Langlykke (1986), Reisman (1988) and Atkinson and Mavituna (1991). It is essential to remember that most of the data are estimates for proposed processes. Some equipment-cost breakdowns are quoted in Table 12.5.

Kalk and Landlykke (1986) quote 1985 costs of fermenters. In the unit size range of 1.0 to 45 m³, costs for modular units with control and recording ranged from $90,000 to $350,000. The cost of 70 to 250 m³ vessels, which needed site erection, ranged from $800,000 to $2,000,000.

If a vessel is to be used to produce genetically engineered products, the costs will increase significantly because of the extra containment provisions which must be incorporated during construction (see also Chapter 7). It has been estimated that the basic fermenter costs will increase by 10–30% for each increase in Containment level (Hambleton et al., 1991). At level 2 or B3 (see Chapter 7 for definition) the provision of containment of an instrumented 20-dm³ fermenter may increase the cost from £45,000 to £100,000. The cheaper alternative might be to enclose an ordinary fermenter in a suitable containment cabinet. Hambleton et al. (1991) have reported the use of a

TABLE 12.5. *Capital cost breakdown for fermentation plant*

| Item | % of total |
|---|---|
| (a) *Penicillin plant*, estimated for five 225,000 dm³ fermenters with ancillary equipment (Swartz, 1979) | |
| Process equipment | 23.6 |
| Installation | 5.2 |
| Insulation | 1.9 |
| Instruments | 2.7 |
| Piping | 11.8 |
| Electrical | 15.8 |
| Building | 11.3 |
| Utilities | 21.3 |
| Site | 2.4 |
| Laboratory equipment | 3.8 |
| Spare parts | 0.5 |
| (b) *Norprotein plant* (Mogren, 1979) | |
| Raw materials storage | 10 |
| Media preparation and utilities | 17 |
| Fermentation | 41 |
| Cell recovery and drying | 22 |
| Product storage | 10 |
| (c) *ICI plc. Single-cell protein plant* (Smith, 1980) | |
| Raw materials | 3 |
| Storage and packing | 12 |
| Off-site services | 16 |
| On-site services | 11 |
| Fermentation | 14 |
| Compression | 9 |
| Dewatering | 19 |
| Drying | 12 |
| Effluent treatment | 4 |

Class 3 microbiological cabinet for fermenters of up to 50 dm$^3$. Buckland (1992) has stated that it now costs as much to build a 3 m$^3$ facility for a recombinant protein which will need extensive containment provisions as for a 2000 m$^3$ scale facility for an antibiotic. In a containment facility only 30% may be usable space. The remaining 70% of the building is needed to accommodate and service the facility.

Reisman (1993) has stressed the need for the correct level of containment. More stringent requirements may exist in a laboratory than are strictly needed. The same high level of containment in a production facility may be onerous, costly and unnecessary. Containment needs should be determined jointly by a group of appropriately trained staff.

The designing of vessels which can be converted for multi-use (Reisman, 1993) may have economic advantages, but descriptions of such vessels except at pilot scale are not common (Hambleton *et al.*, 1991).

The life of a proposed plant has to be predicted. MacLennan (1976) has estimated the life of a single-cell protein plant to be 10 years. This is probably the lower limit. Hamer (1979) claimed that a planned SCP plant life of 15 years is typical. Acetone and butanol have been produced in 25-year-old vessels (Spivey, 1978), while in long-established breweries in Great Britain, equipment of 50 to 100 years old is still used.

Allowance must be made for the service provisions for a fermentation process. In a moderate sized antibiotic plant, there may be the need for the provision of 5 tonnes h$^{-1}$ of steam, 5000 kW h$^{-1}$ of electricity, 57,000 m$^3$ h$^{-1}$ of compressed air and 200,000 dm$^3$ h$^{-1}$ of water (Hastings and Jackson, 1965). When producing 1 tonne of acetic acid, 480 m$^3$ of cooling water, 10 m$^3$ of process water, 12 tonnes of steam and 570 kW of electricity would be utilized (Pape, 1977).

## MEDIA

The cost of the various components of a production medium can have a profound effect on the overall cost of a fermentation process, since these account for 38 to 73% of the total production cost (Table 12.1). The organic-carbon source in microbial processes is usually the most expensive component contributing to the cost of the process. Ratledge (1977) has made a detailed analysis of annual price and availability of major carbon substrates. The price of a natural material may fluctuate due to other competing demands and the annual variation in the quantity harvested. Big capital investment may be tied up in natural materials if they are seasonal and require storage. Hastings and Jackson (1965) stated that up to 23 × 10$^6$ dm$^3$ of molasses may have to be stored for an industrial alcohol-production process. A particular material may be selected because it is cheap locally, rather than the best substrate (Calam, 1967).

In a cost study analysis for tissue plasminogen activator by a mammalian-cell process, fermentation materials accounted for 75% of the total raw materials cost (Datar *et al.*, 1993). Of this cost, calf serum was estimated to contribute 73% or 55% of the total cost. However, it has been possible to develop a serum-free medium for economic production of antibodies from hybridoma cells (Maiorella, 1992).

In the search for suitable nutrient components for media, six criteria were considered in Chapter 4 which ought to be satisfied whenever possible. Ratledge (1977) has stressed the need to note the amount of carbon in a carbon substrate when costing. Higher yields might be achieved by changing from a carbohydrate to an alternative carbon source, but increased aeration and/or agitation rates may be necessary with the change in substrate. The cost of this extra provision must therefore be less than the savings from the change of substrate if the process is to be feasible. When cheapness is added as a further restraint the number of potential major nutrients which could be used on an industrial scale is limited. Carbohydrates from beet, sugar cane or grain are the major carbon and energy sources in most media and comply with the requirements of economy and those stated in Chapter 4. In 1993, sugar and starch based fermentation substrates were available within the EEC for approximately £200 tonne$^{-1}$. However, these are commodities whose cost fluctuates according to supply and demand. It is therefore worthwhile to develop a series of cost optimized media formulations so that the most cost effective growth medium can always be used when necessary (Winkler, 1991).

During the 1970s there was considerable interest in using petrochemicals as substrates for SCP production as protein animal feed by a number of major chemical and petroleum companies (Sharp, 1989). A number of processes were developed using methane or methanol as the main carbon substrate. None are now being operated. Major factors which contributed to making these processes uncompetitive included the increased cost of the substrate (see Chapter 4) and the availability of cheaper alternative animal feeds.

A variety of waste materials would seem to be potential cheap carbon sources. Unfortunately, it has been shown that their use is very restricted because they cannot compete economically with conventional substrates. This may be due to a number of possible reasons including variability of the material, impurities which make downstream processing more difficult, high water content making transport costly, geographical location, quantities produced and limited seasonal availability. The economics of biomass production from whey have been reviewed by Meyrath and Bayer (1979). The possible use of waste substrates may depend on the cost of alternative methods of disposal or on the availability of government grants to diminish pollution (Perlman, 1970).

Mineral components normally constitute a smaller part of the cost of media, e.g. they account for 4 to 14% of the manufacturing cost of single-cell protein (Cooney and Makiguchi, 1977). Although feed grades of phosphates are more expensive than fertilizer grades, they do not contain impurities such as iron, arsenic and fluoride. This is an important consideration in the production of foods and drugs (Litchfield, 1977). The hydroxides and sulphates of potassium, magnesium, manganese, zinc and iron are preferred to the chlorides to minimize corrosion of stainless steel. The source of basic materials can cause considerable variation in product yield. Corbett (1980) compared six samples of calcium carbonate, and found that five of them reduced the titre of penicillin G in a production medium.

Problems concerned with the storage, handling and mixing of media should not be neglected. Powders must be kept in dry conditions because of the possibility of substances becoming rock-like or glutinous. Some bulk liquids with a high solids content need to be kept warm to prevent them solidifying, e.g. glucose and corn-steep liquor. If storage temperatures are too high there could be degradation. It is also vital for workers to follow instructions for media preparation very carefully to prevent 'lumpy' media, etc. (Corbett, 1980).

## AIR STERILIZATION

The problems associated with producing large volumes of sterile air for aerobic fermentations are unique (see Chapter 5). Although sterilization by heating is technically possible, it has generally been regarded as too costly for full-scale operation (Cherry et al., 1963), although it might be used in the treatment of exhaust gases (Walker et al., 1987).

Absolute fixed-pore membrane systems using pleated membranes of PTFE are now widely used in the fermentation industry (Chapter 5) and have proved to be very reliable. This is very important when considering the costs associated with loss of fermentation batches due to contamination and production downtime due to filter failure. Banks (1979) has reported that a contamination probability of 1 in 1000 is economically acceptable for microbial batch fermentations, while in large scale animal-cell culture processes contamination rates as low as 2% are now achievable (Spier, 1988).

Operating costs will be based on the estimated life of the filters. Factors to consider include the cost of replacement filters or filter materials, servicing and labour. Even if the filters could be cleaned there must be an allowance for depreciation due to normal wear and tear. Savings may also be made by introducing series filtration whereby the major part of the foreign matter from the air stream is taken out by varying degrees of coarse filtration, thus reducing renewals of the more expensive high-efficiency filter media such as membrane filters.

The treatment of fermenter exhaust gases to satisfy containment requirements is also important (see also Chapters 5 and 7). Treatment is normally by filtration with 0.2-$\mu$m hydrofilters, but in-line incinerators may be an alternative approach. Filtration is usually cheaper but it may be necessary to supplement filtration with incineration depending on the process and scale (Walker et al., 1987).

## HEATING AND COOLING

Ideally there should be no heating or cooling at any stage in a fermentation process, but because this is virtually impossible, heat should be conserved and cooling minimized by careful process design. A fermentation may include the following heating or cooling stages:

1. Sterilization or boiling of the medium to 100° or above followed by cooling to 35° or below.
2. Heating the fermenter and ancillary equipment to sterilize it, followed by cooling.
3. Heat may be generated during the fermentation. This heat output has to be removed by cooling to maintain the growth temperature of the microorganism within prescribed limits.
4. After harvesting, heat may be required to remove water from the product.

Cooling requirements will be influenced by the size and type of an individual process (Chapter 7). British

Petroleum Ltd. estimated that the cooling requirements for a 100,000 tonnes annum$^{-1}$ single-cell protein plant to be $110 \times 10^6$ kcal h$^{-1}$ using n-alkanes or gas-oil as the primary substrate (Litchfield, 1977). To reduce cooling requirements, the specific energy input may be minimized through the use of air-lift fermenters (Schugerl et al., 1978). Cooling equipment has been estimated at 10 to 15% of the investment cost for single-cell protein (Moo-Young, 1977; Schugerl et al., 1978). Another way to minimize cooling costs is to use micro-organisms with higher optimum growth temperatures, if it is feasible. The selection and use of thermophiles and thermotolerant organisms would have obvious advantages to reduce cooling demands (Chapter 3).

## AERATION AND AGITATION

Nearly all fermentations require some form of mixing to maintain a constant environment, and many also need aerating (see also Chapter 9). Fermentations may be broadly classified into:

1. Fermentations which are anaerobic where oxygen is undesirable, e.g. acetone–butanol.
2. Fermentations which have a minimal oxygen demand, e.g. ethanol.
3. Fermentations which have a high oxygen demand, e.g. antibiotics, acetic acid, single-cell protein.

In categories 1 and 2, aeration is not generally regarded as a major economic consideration. During an acetone–butanol fermentation carbon dioxide and hydrogen are evolved. Once this gas production starts it will help to maintain anaerobic conditions and stir the mass of broth without the need for mechanical agitation. Anaerobic conditions are achieved initially in a production fermenter by maintaining a positive pressure of filtered carbon dioxide and hydrogen obtained from another established fermentation (Beesch, 1952).

For ethanol production, the yeast inoculum in the vessel is initially dispersed in the medium by compressed air or by mechanical stirring. Aeration or agitation is stopped once the biomass concentration reaches a predetermined level. A vigorous anaerobic fermentation commences, and the evolution of carbon dioxide bubbles stirs the contents of the vessel and disperses the cells in the medium so that mechanical agitation is unnecessary. In this process aeration and agitation are considered to be a minor component of the total production costs.

Fermentations having a high oxygen demand must be agitated with sufficient power to maintain a uniform environment and to disperse the stream of air introduced by aeration. In an early reference it was stated that the cost of energy necessary to compress air for yeast production proved that a considerable amount (10 to 20%) of the total production expenses was due to aeration (de Becze and Liebmann, 1944). Swartz (1979) has reported that the mixing costs in a penicillin fermentation are > 15% of the total production costs. Energy consumption for a stirred aerobic fermentation to provide agitation, air compression and chilled water is approximately 8.2 kW m$^{-3}$ (Curran and Smith, 1989). Assuming an electricity cost of $0.07 kW$^{-1}$, a 6-day antibiotic fermentation in a 100-m$^3$ fermenter with a 1-day turnaround would use $8,000 of power (Royce, 1993).

In single-cell protein processes, the carbon substrate yield coefficient is the most critical physiological factor (Hamer, 1979). It is also well documented that much higher carbon-substrate yield coefficients are obtained with methane or n-alkanes instead of carbohydrates (see Table 12.6). Unfortunately, cells grown on hydrocarbons have greater oxygen requirements. The oxygen requirements of a hydrocarbon yeast fermentation is

TABLE 12.6. *Effect of substrate and yield coefficients on SCP operating costs* (Abbott and Clamen, 1973)

| Substrate | Substrate costs | | O$_2$ transfer costs | Heat removal costs | Combined costs |
|---|---|---|---|---|---|
| | (¢ lb$^{-1}$ substrate) | (¢ lb$^{-1}$ cells) | (¢ lb$^{-1}$ cells) | (¢ lb$^{-1}$ cells) | (¢ lb$^{-1}$ cells) |
| Maleate (waste) | 0 | 0 | 0.46 | 0.75 | 1.2 |
| Glucose (molasses) | 2.0 | 3.9 | 0.23 | 0.54 | 4.7 |
| n-Paraffins | 4.0 | 4.0 | 0.97 | 1.4 | 6.4 |
| Methanol | 2.0 | 5.0 | 1.2 | 1.9 | 8.1 |
| Ethanol | 6.0 | 8.8 | 0.75 | 1.3 | 10.9 |
| Acetate | 6.0 | 16.7 | 0.62 | 1.1 | 18.4 |

almost triple that of a yeast fermentation grown on a carbohydrate substrate and producing an equal quantity of cells (Darlington, 1964; Chapter 4). Therefore, if there is to be effective utilization of a hydrocarbon substrate, which can account for over 50% of total production costs (Litchfield, 1977), the production fermenter must have a high oxygen-transfer capacity. The demands on fermenter design are further complicated by the hydrocarbon fermentation being highly exothermic, which necessitates the provision of good cooling facilities if a constant temperature is to be maintained in the fermenter.

A few companies developed SCP processes using mechanically stirred fermenters with sparged air. BP Ltd. constructed vessels of up to 1000 $m^3$ capacity for their n-alkane process in their Sardinian Ital protein project (Levi et al., 1979). In the Swedish Norprotein process it has been estimated that the total utilities costs, which included aeration and agitation (1978 prices) for 100,000 tons year$^{-1}$ of SCP would only be 16% of total production costs (Mogren, 1979).

A number of companies, including ICI plc (Taylor and Senior, 1978) and Hoechst (Knecht et al., 1977), decided to develop fermenters based on the air-lift principle (Chapter 7; Hamer, 1979; Levi et al., 1979; Sharp, 1989). The main advantages of these fermenters are simpler design and reduced energy and cooling water costs. Since the energy supplied to an air-lift fermenter is only supplied with the air, it is crucial to obtain a fermenter design which minimizes the energy requirement for biomass production yet creates high oxygen-transfer facilities to ensure efficient substrate utilization. In the ICI plc process, the estimated manufacturing costs for all utilities were 14%, with aeration accounting for 70% of fermentation utilities costs (Moo-Young, 1977).

## BATCH-PROCESS CYCLE TIMES

In a batch process productivity must be determined for the complete process cycle. Here productivity is defined as grams of product $dm^{-3}$ $h^{-1}$. This productivity is based on a combination of the time for the actual fermentation and the time to prepare the fermenter ready for the next run. Heijnen et al. (1992) estimated this to take 15–25 h. Thus the total time for a fermentation may be calculated (Wang et al., 1979) as:

$$t = \frac{1}{\mu_m} \cdot \ln \frac{X_f}{X_0} + t_T + t_L + t_D$$

where $\mu_m$ = maximum specific growth rate,
$X_0$ = initial cell concentration,

$X_f$ = final cell concentration,
$t_T$ = turn-around time (washing, sterilizing, filling with media),
$t_D$ = delay time until inoculation,
$t_L$ = lag time after inoculation.

The overall productivity $P$ is given by:

$$P = \frac{X_f}{\frac{1}{\mu_m} \ln \frac{X_f}{X_0} + t_T + t_D + t_L}.$$

It will be possible from this equation to determine the effects of process changes on the overall productivity. A larger initial inoculum would increase $X_0$ and shorten the process time. Actively growing inocula would reduce the lag time $t_L$. Aspects of this problem have been discussed in Chapter 6. It is also worthwhile to isolate faster-growing organisms and/or higher yielding strains (Chapter 3). In many processes the growth phase and/or the production phase has been extended by the use of fed-batch or continuous feed (Chapter 2) with improved productivity.

Richards (1968a,b), Geysen and Gray (1972) and Stowell and Bateson (1984) have given details for determining maximum production at minimum cost and the optimum time for harvesting (Fig. 12.1).

In fermentations with short growth cycles such as bakers' yeast (14 to 24 hours), the turnaround time will be as important as the time between inoculation and

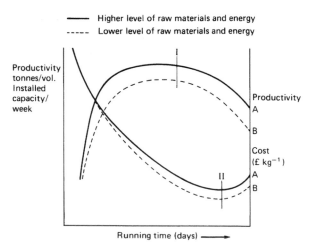

FIG. 12.1. Fed-batch process—effect of running time on productivity and product cost (Stowell and Bateson, 1984). Fermentation should be terminated at I for maximum productivity, II for minimum cost. Productivity is defined as the units of product generated per unit of fermenter volume in a given time.

harvesting. When the production cycle is long, as with penicillin (6 to 7 days), a few extra hours for turnaround will have little influence on overall productivity.

## CONTINUOUS CULTURE

At the present time, very few large-scale continuous-culture processes are being operated. These are primarily for the production of microbial biomass, glucose isomerase, buttermilk souring and yoghurt (Heijnen *et al.*, 1992).

It is appropriate at this stage to compare batch-culture productivity and continuous culture productivity. Wang *et al.* (1979) derived an equation to quantify this relationship:

$$\frac{\text{Continuous-culture productivity}}{\text{Batch-culture productivity}}$$
$$= \frac{\ln\ (X_m/X_0) + \mu_m t_L}{(X_m - X_0)/X_m} D_c Y$$

where $\mu_m$ = maximum specific-growth rate,
$X_0$ = initial cell density,
$X_m$ = maximum cell density,
$t_L$ = turn-around time,
$D_c$ = critical dilution rate,
$Y$ = cellular yield coefficient for the limiting nutrient.

In an example they used an inoculum size of 5% ($X_0/X_m$ = 0.05), a process turn-around time of 10 hours, a cellular yield of 0.5 g cells per g of substrate and a final cell concentration of 30 g dm$^{-3}$. Productivity was then calculated for a series of maximum specific growth rates (Table 12.7). It is clear from these data that the faster the growth of the organism, the more favourable is a continuous process over a batch process.

When assessing feasibility of a continuous process for product formation it is necessary to know: volumetric productivity, conversion yield of product from the

most expensive substrate in the medium and the product concentration (Wang *et al.*, 1979). In processes for cell biomass, some alcohols and organic acids where the major production cost is at the fermentation stage, volumetric productivity and conversion yield will be most important. A continuous process may be more economic than a batch process if higher productivities at higher efficiencies can be achieved. Unfortunately, in some processes, the final product concentration in the effluent broth from a continuous culture will be less than that obtained in a batch process, which will create a need for greater concentration at the recovery stage.

Heijnen *et al.* (1992) argue that low dilution rates are favoured if high product concentrations are to be obtained as recovery cost may be 50–80% of total costs. However, the concentration preference is not valid in processes which can start with a simple concentration step (SCP or intracellular glucose isomerase) or with products that need no concentrating (beer, yoghurt and buttermilk).

In a production plant, continuous culture offers the advantages of constant flow, product quality and simple automation and control. Disadvantages are due to specific production facilities that cannot be used for other purposes, lack of continuous recovery techniques and lack of constant market demand (Heijnen *et al.*, 1992).

The continued increase in efficiency of fed-batch culture processes for antibiotics and other non-growth-associated products makes manufacturers reluctant to make radical alterations to established processes such as the introduction of continuous culture (see also Chapter 2).

## RECOVERY COSTS

The costs of product recovery and purification are rarely quoted, though in some processes they are obviously considerable. It is now accepted that cost analy-

TABLE 12.7. *Comparison of productivity in batch and continuous culture* (Wang *et al.*, 1979)

| Maximum specific growth rate in batch culture (h$^{-1}$) | Continuous productivity |
|---|---|
| | Batch productivity |
| 0.05 | 0.09 |
| 0.10 | 0.21 |
| 0.20 | 0.53 |
| 0.40 | 1.5 |
| 0.80 | 4.6 |
| 1.0 | 6.8 |
| 1.2 | 9.5 |

ses of 'older' fermentation processes split approximately 1:1 between fermentation costs and isolation/purification costs (Reisman, 1993). However, the change in the last 10 years to use recombinant organisms to produce exotic compounds at extremely low titres has meant that the fermentation/ isolation–purification split may be 1:8 or 1:10. This means that the fermentation may only be 10% of the costs, while the recovery accounts for 90%. In this case the correct choice of the recovery–purification procedure can be crucial to the success of a process and early evaluation of alternative techniques may be very important.

Stowell and Bateson (1984) identified a number of factors contributing to these costs:

1. Yield losses, even if only modest, are certain to occur at each stage of the recovery process.
2. High energy and maintenance costs associated with running filtration and centrifugation equipment.
3. High costs of solvents and other raw materials used in recovery and refining of products.

Atkinson and Mavituna (1991) reported losses of 8% for citric acid and 4% for penicillin G in the recovery and purification stages before conversion to the potassium salt in production processes. They also stressed the importance of trying to reduce the number of downstream stages as much as possible to reduce capital and operating costs.

It is thought that depreciation, return on capital and maintenance can account for over 80% of the overall cost for a large-scale rotary filtration or centrifugation plant (Atkinson and Mavituna, 1983). However, it is considered that removing cells by filtration is less energy consuming than by centrifuging. If filter aids are to be used, in the most economical way, this will still add £9 tonne$^{-1}$ for a product at a 10% concentration in the broth.

One of the main factors affecting centrifuge economics is the size of the particle to be separated (Asenjo, 1990). Filtration costs are less dependent on particle size. At a particle size greater than 1 to 2 $\mu$m, centrifugation is more economical. Below this size ultrafiltration becomes more economical.

Taksen (1986) has given a case study for a moderate value product (£1.50 to 3.00 per kg) requiring the processing of 200,000 dm$^3$ of broth per day by rotary vacuum filtration or centrifugation. Capital investment for the rotary filter would be £500,000 with an operating cost of 1.15 p per dm$^3$ of broth. A three-stage

centrifuge would cost £667,000 and broth could be processed at 0.62p per dm$^3$. Because rotary filters were already available, it was decided that the new capital investment could not be justified, although the broth would cost more to process. In another case study on ultrafiltration vs. evaporation to pre-concentrate a moderately high-molecular-weight product, the cost of water removal by membrane techniques was estimated to be 40% lower than by evaporation.

Maiorella *et al.* (1984) have compared extraction processes in 11 alternative ethanol fermentation processes. Selective ethanol removal by flash distillation was thought to be the most economic technique.

When a product may be made by a microbial process or obtained from an alternative source there are cost limits on product recovery to be considered. Atkinson and Mavituna (1991) estimated:

1. A limit of about £40 tonne$^{-1}$ for ethanol selling at £220 tonne$^{-1}$ produced at 7% w/v in broth from a local source of carbohydrate.
2. A limit of about £100 tonne$^{-1}$ for SCP selling at £300 tonne$^{-1}$ produced from a petroleum-based substrate to give a 3% w/v yield.
3. A limit of about £300 tonne$^{-1}$ for organic acids or glycol selling at £700 tonne$^{-1}$, produced at 10% concentration in a carbohydrate medium.

When high-value end-products have been produced it has been acceptable to use relatively large weights of filter aid to achieve initial clarification to remove small amounts of solids. The solvents used in subsequent extraction have then been recovered in a high energy-consuming distillation plant (Atkinson and Sainter, 1982). In this case, the manufacturer has only to make his product as economically as those of other fermentation companies.

Before certain products may be marketed the extraction/purification procedure will have to be validated (approved) by the FDA (U.S.A.) or similar regulatory bodies. Any changes in extraction procedures will have to be checked for revalidation, which will incur further costs (Reisman, 1993). Therefore, if validation is to become an issue in processing it may be worthwhile having alternative procedures validated at an early stage in development.

## WATER USAGE AND RECYCLING

Many fermentations have a high daily water usage (Table 12.8). As charges for water increase, many of

TABLE 12.8. *Daily water usage in fermentation processes*

| Industry | m$^3$ of water used day$^{-1}$ | Reference |
|---|---|---|
| Maltings | 230 | Askew (1975) |
| Brewing | 10,000 | |
| Distilling | 320 | |
| Antibiotics | 245 | |
| Antibiotics | 5,200 | Hastings and Jackson (1965) |
| Acetic acid | 700 | Pape (1977) |
| Single cell protein (methanol substrate) | 4,000 to 12,000 | Taylor and Senior (1978) |
| Yeast (alkane substrate) | 18,200 | Ratledge (1975) |
| Bacteria (methanol substrate) | 45,500 | |
| Bacteria (methane substrate) | 18,200 | |

these processes will become vulnerable to cost escalation because of the relatively large volumes of water required per unit volume of product.

There is now a widespread interest in reducing overall consumption. In a bacitracin plant described by Inskeep *et al.* (1951), the water from the mash cooler was collected and reused to charge the mashing vessels and wash the fermenters. Water from the cooler coils was used to wash down the discharge cake from the filter presses. Bernstein *et al.* (1977) designed a 'closed-loop' system for fermenting cheese whey in which effluents were completely recycled.

Recycling of water was an integral stage of large scale SCP processes developed during the 1970s to minimize water consumption, reduce effluent treatment costs and reduce media costs by recycling of spent media (Sharp, 1989). When ICI plc's SCP Pruteen plant was operating, it was designed to recycle most of the fermenter medium water (Ashley and Rodgers, 1986). Under optimum conditions they claimed that the water loss could be reduced to 3% of the flow through the fermenter using water recovery systems.

## EFFLUENT TREATMENT

Before deciding on the most economic form of treatment for wastes it is important to make a factory survey. The information should include the water volume, the organic and solids loading, range of pH variation, nutrient level, temperature fluctuation, and the presence of any toxic compounds. It will also be necessary to consider company finance policies, the site location and government legislation for waste disposal (see also Chapter 11).

In the majority of fermentation processes it is impossible to dispose of effluents at zero cost. Whether the

waste is incinerated, dumped on waste land, or discharged to sewers, rivers or tidal waters, some expenditure will be necessary for treatment that ensures that minimal harm is done to the environment.

Since the 1980s the European Community has adopted a number of Directives to reinforce earlier legislation for the protection and improvement of inland and marine water quality. The standards of these Directives and consent for discharge are implemented in the United Kingdom through the monitoring of the National Rivers Authority and the water companies in England and Wales, the River Purification Boards in Scotland and the Department of the Environment in Northern Ireland (Brown, 1992). Costs to meet the requirements of these Directives will need to be included in process costings.

The various alternative disposal procedures may be compared using economic considerations. Pape (1977) claimed that the cheapest treatment method was controlled dumping, followed by waste incineration or dumping in salt mines. The most expensive method was biological degradation in a waste-water-treatment plant. The last method has often to be used because the effluents usually contain only a few percent of organic matter which would be costly to separate, concentrate and incinerate.

The possibility of direct disposal of pharmaceutical waste into the sea is now very restricted, especially if the waste is untreated, even though many of the large fermentation plants in the United Kingdom are in coastal locations. In 1972, Jackson and Lines stated that a pipeline of over 2.8 km overland and 2.8 km on the sea bed at a cost of £350,000 would be needed to dispose of 3000 dm$^3$ day$^{-1}$ of untreated antibiotic waste. The other options are to discharge the effluent direct to the sewers and pay a charge, to treat the waste in the plant, or to operate a combination of the two. Sewage works' charges for treating effluents have increased 1000% in less than 5 years in some instances (Forage, 1978). Those plants which treat all or part of their effluents have discovered that energy costs have risen and sludge disposal is more costly and difficult. Ripley (1979) estimated the costs for a treatment plant for a whisky distillery producing 4,500,000 $dm^3$ $year^{-1}$ of proof whisky to have a capital cost of £75,000 and operating costs of £9000 $year^{-1}$. Power was calculated at 0.9 kW $kg^{-1}$ of BOD removed while dosage of nutrients was £0.03 $kg^{-1}$ of effluent BOD.

Avermectin is an antihelmintic compound produced by *Streptomyces avermitilis* (Omstead *et al.*, 1989). During development of this compound it was recognized that it was very potent and could have a potential impact on aquatic fauna. All possible exit streams from the process at the factory, both fermentation and downstream purification, are therefore captured and chemically degraded. In this case, environmentally safe waste treatment is a major component in production costs. An alternative method for disposing of wastes of this type would be to absorb all waste streams in a suitable material and incinerate this and solid wastes at an appropriate temperature (see also Chapter 11).

## REFERENCES

ABBOTT, B. J. and CLAMEN, A. (1973) Relation of substrate, growth rate and maintenance coefficient to single cell protein production. *Biotech. Bioeng.* **15**, 117–127.

ANON (1974) *Eur. Chem. News*, **5**(3), 30.

ASENJO, J. A. (1990) Cell disruption and removal of insolubles. In *Separations for Biotechnology*, Vol. 2, pp.11–20 (Editor Pyle, D. L.). Elsevier, London.

ASHLEY, M. H. J. and RODGERS, B. L. F. (1986) The efficient use of water in single cell protein production. In *Perspectives in Biotechnology and Applied Microbiology*, pp. 71–79 (Editors Alani, D. I. and Moo-Young, M.). Elsevier, London.

ASKEW, M. W. (1975) Fermentation: water, waste and money. *Proc. Biochem.* **10**(1), 5–7, 13.

ATKINSON, B. and MAVITUNA, F. (1983) *Biochemical and Bioengineering Handbook* (1st edition), Chapter 12, pp. 890–931. Macmillan, London.

ATKINSON, B. and MAVITUNA, F. (1991) Principles of costing and economic evaluation for bioprocesses. In *Biochemical Engineering and Biotechnology Handbook* (2nd edition), pp. 1059–1109. Macmillan, London.

ATKINSON, B. and SAINTER, P. (1982) The development of down stream processing. *J. Chem. Tech. Biotechnol.* **32**, 100–108.

AUNSTRUP, K. (1977) Production of industrial enzymes. In *Biotechnology and Fungal Differentiation, FEMS Symp*, **4**, pp. 157–171 (Editors Meyrath, J. and Bu'Lock, J. D.). Academic Press, London.

BACKHURST, J. R. and HARKER, J. H. (1973) *Process Plant Design*. Heinemann, London.

BADER, F. G. (1992) Evolution in fermentation facility design from antibiotics to recombinant proteins. In *Harnessing Biotechnology for the 21st Century*, pp. 228–231 (Editor Ladisch, M. R. and Bose, A.). American Chemical Society, Washington, DC.

BAILEY, J. E. and OLLIS, D. F. (1986) Bioprocess economics. In *Biochemical Engineering Fundamentals* (2nd edition), pp. 798–853. McGraw-Hill, New York.

BANKS, G. T. (1979) Scale-up of fermentation processes. In *Topics in Enzyme and Fermentation Biotechnology*, pp. 170–266 (Editor Wiseman, A.). Ellis Horwood, Chichester.

BARTHOLOMEW, W. H. and REISMAN, H. B. (1979) Economics of fermentation processes. In *Microbial Technology* (2nd edition), Vol. 2, pp. 463–496 (Editors Peppler, H. J. and Perlman, D.). Academic Press, London.

DE BECZE, G. and LIEBMANN, A. J. (1944) Aeration in the production of compressed yeast. *Ind. Eng. Chem.* **36**, 882–890.

BEESCH, S. C. (1952) Acetone–butanol fermentation of sugars. *Ind. Eng. Chem.* **44**, 1677–1682.

BERDY, J. (1989) The discovery of new bioactive microbial metabolites; screening and identification. *Prog. Ind. Microbiol.* **27**, 3–25.

BERNSTEIN, S., TZENG, T. H. and SISSON, D. (1977) The commercial fermentation of cheese whey for the production of protein and/or alcohol. In *Single Cell Protein from Renewable and Non-Renewable Resources, Biotech. Bioeng. Symp.* **7**, 1–10.

BIRCH, J. R., LAMBERT, K., THOMPSON, P. W., KENNY, A. C. and WOOD, L. A. (1987) Antibody production with air lift fermenters. In *Large Scale Cell Culture Technology*, pp. 1–20 (Editor Lyderson, B. K.). Hanser, Munich.

BROWN, A. (1992) *The UK Environment*. HMSO, London.

BUCKLAND, B. C. (1992) Reduction to practice. In *Harnessing Biotechnology for the 21st Century*, pp. 215–218 (Editors Ladisch, M. R. and Bose, A.). American Chemical Society, Washington, DC.

BULL, A. T. (1992) Isolation and screening of industrially important microorganisms. In *Recent Advances in Biotechnology*, pp. 1–17 (Editors Vardar-Sukan, F. and Suha-Sukan, S.). Kluwer, Dordrecht.

BULL, A. T., GOODFELLOW, M. and SLATER, J. H. (1992) Biodiversity as a source of innovation in biotechnology. *Ann. Rev. Microbiol.* **46**, 219–252.

BU'LOCK, J. D. (1979) Industrial alcohol. In *Microbial Technology: Current State, Future Prospects, Soc. Gen. Microbiol. Symp.* **29**, pp. 309–325 (Editors Bull, A. T., Ellwood, D. C. and Ratledge, C.). University Press, Cambridge.

CALAM, C. T. (1967) Media for industrial fermentations. *Proc. Biochem.* **2**(6), 19–22, 46.

CALAM, C. T. (1969) Automation in screening. In *Fermentation Advances*, pp. 34–41 (Editor Perlman, D.). Academic Press, New York.

CHERRY, G. B., KEMP, S. D. and PARKER, A. (1963) The sterilization of air. *Prog. Ind. Microbiol.* **4**, 37–60.

COONEY, C. L. and MAKIGUCHI, N. (1977) An assessment of single cell protein from methanol-grown yeast. In *Single Cell Protein from Renewable and Non-Renewable Resources, Biotech. Bioeng. Symp.* **7**, 65–76.

CORBETT, K. (1980) Preparation, sterilization and design of media. In *Fungal Biotechnology, Brit. Myc. Soc. Symp.* **3**, pp. 25–41 (Editors Smith, J. E., Berry, D. R. and Kristiansen, B.). Academic Press, London.

CURRAN, J. S. and SMITH, J. (1989) Heat and power in industrial fermentation. *Appl. Energy*, **34**, 9–16.

DARLINGTON, W. A. (1964) Aerobic hydrocarbon fermentation-a practical evaluation. *Biotech. Bioeng.* **6**, 241–242.

DATAR, R. (1993) Economics of primary separation steps in relation to fermentation and genetic engineering. *Process Biochem.* **21**(1), 19–26.

DATAR, R., CARTWRIGHT, T. and ROSEN, C. (1993) Process economics of animal cell and bacterial fermentations; A case study analysis of tissue plasminogen activator. *Biotechnology* **11**(3), 349–357.

ECKENFELDER, W. W. (1989) *Industrial Water Pollution Control* (2nd edition). McGraw-Hill, New York.

FORAGE, A. J. (1978) Recovery of yeast from confectionary effluent. *Proc. Biochem.* **13**(1), 8, 11, 30.

FOSTER, J. W. (1961) Microbiological process discussion. A view of microbiological science in Japan. *Appl. Microbiol.* **9**, 434–451.

GEYSEN, H. M. and GRAY, P. P. (1972) A graphical method for optimizing fermentations. *Biotech. Bioeng.* **14**, 857–860.

GOODFELLOW, M. and O'DONNELL, A. G. (1989) Search and discovery of industrially significant actinomycetes. In *Microbial Products — New Approaches*, pp. 343–383 (Editors Baumberg, S., Hunter, I. S. and Rhodes, P. M.). University Press, Cambridge.

GRAY, P. S. (1987) Impact of EEC regulations on the economics of fermentation substrates. In *Carbon Substrates in Biotechnology*, pp. 1–12 (Editors Stowell, J. D., Beardsmore, A. J., Keevil, C. J. and Woodward, J. R.). IRL Press, Oxford.

HACKING, A. J. (1986) *Economic Aspects of Biotechnology.* University Press, Cambridge.

HAMBLETON, P., GRIFFITHS, B., CAMERON, D. R. and MELLING, J. (1991) A high containment polymodal pilot-plant fermenter-design concepts. *J. Chem. Tech. Biotechnol.* **50**, 167–180.

HAMER, G. (1979) Biomass from natural gas. In *Microbial Biomass, Economic Microbiology*, Vol. 4, pp. 315–360 (Editor Rose, A. H.). Academic Pres, London.

HAMMOND, A. L. (1978) Energy: elements of a Latin American policy. *Science* **200**, 733–754.

HARRISON, F. G. and GIBSON, E. D. (1984) Approaches for reducing the manufacturing costs of 6-amino penicillanic acid. *Process Biochem.* **19** (1), 33–36.

HASTINGS, J. J. H. (1971) Development of the fermentation industries in Great Britian. *Adv. Appl. Microbiol.* **16**, 1–45.

HASTINGS, J. J. H. (1978) Acetone–butanol fermentation. In *Primary Products of Metabolism, Economic Microbiology*, Vol. 2, pp. 31–45 (Editor Rose, A. H.). Academic Press, London.

HASTINGS, J. J. H. and JACKSON, T. (1965) The technical use of biochemical processes. In *Chemical Engineering*, Vol. 8, pp. 406–409 (Editors Cremer, W. H. and Watkins, S. B.). Butterworths, London.

HEIJNEN, J. J., TERWISSCHA VAN SCHELTINGA, A. H. and STRADTHOF, A. F. (1992) Fundemental bottlenecks in the application of continuous bioprocesses. *J. Biotechnol.* **22**, 3–20.

HEPNER, L. (1977) Feasibility of producing basic chemicals by fermentation. In *Microbial Energy Conversion*, pp. 531–554 (Editors Schlegel, H. G. and Barnea, J.). Pergamon Press, Oxford.

HEPNER, L. (1978) The feasibility of basic chemicals by fermentation processes. *Eng. Process Econ.* **3**, 17–23.

HOGGAN, J. (1977) Aspects of fermentation in conical vessels. *J. Inst. Brewing* **83**, 133–138.

HUCK, T. A., PORTER, N. and BUSHELL, M. E. (1991) Positive selection of antibiotic-producing soil isolates. *J. Gen. Microbiol.* **137**, 2321–2329.

HUMPHREY, A. E. (1975) Product outlook and technical feasibility of SCP. In *Single Cell Protein II*, pp. 1–23 (Editors Tannenbaum, S. R. and Wang, D. I. C.). MIT Press, MA.

HUMPHREY, A. E., MORIERA, A., ARMIGER, W. and ZABROMINSKIE, D. (1977) In *Single Cell Protein from Renewable and Non-Renewable Resources, Biotech. Bioeng. Symp.* **7**, 45–64.

INSKEEP, G. G., BENNETT, R. E., DUDLEY, J. F. and SHEPARD, M. W. (1951) Bacitracin, product of biochemical engineering. *Ind. Eng. Chem.* **43**, 1488–1498.

INSTITUTION OF CHEMICAL ENGINEERS (1977) *A New Guide to Capital Cost Estimation.* Institute of Chemical Engineers, London.

JACKSON, C. J. and LINES, G. T. (1972) Measures against water pollution in the fermentation industries. In *Industrial Waste Water*, pp. 381–393 (Editor Goransson, B.). Butterworths, London.

KALK, J. P. and LANGLYKKE, A. F. (1986) Cost estimation for biotechnology projects. In *Manual of Industrial Microbiology and Biotechnology*, pp. 363–385 (Editors Demain, A. L. and Solomon, N. A.). American Society for Microbiology, Washington, DC.

KEIM, C. R. and VENKATASUBRAMANIAN, K. (1989) Economics of current biotechnological methods of producing ethanol. *Trends Biotechnol.* **7**, 22–29.

KNECHT, R., PRAVE, P., SEIPENBUSCH, R. and SUKATSCH, D. A. (1977) Microbiology and biotechnology of SCP produced from n-parrafin. *Proc. Biochem.* **12**,(4), 11–14.

LAWSON, C. J. and SUTHERLAND, I. W. (1978) Polysaccharides. In *Primary Products of Metabolism, Economic Microbiology*, Vol. 2, pp. 337–392 (Editor Rose, A. H.). Academic Press, London.

LEVI, J. D. SHENNAN, J. L. and EBBON, G. P. (1979) Biomass from liquid n-alkanes. In *Microbial Biomass, Economic*

*Micobiology*, Vol. 4, pp. 361–419 (Editor Rose, A. H.). Academic Press, London.

LITCHFIELD, J. H. (1977) Comparative technical and economic aspects of single cell protein processes. *Adv. Appl. Microbiol.* 22, 267–305.

LOCKWOOD, L. B. and STREETS, B. W. (1966) Potentials in strain development. *Dev. Ind. Microbiol.* 7, 74–78.

MacLENNAN, D. G. (1976) Single cell protein from starch. In *Continuous Culture 6: Applications and New Fields*, pp. 69–84 (Editors Dean, A. C. R., Ellwood, D. C., Evans, C. G. T. and Melling, J.). Ellis Horwood, Chichester.

MAIORELLA, B. (1992) Hydridoma culture — Optimization, characterization and cost. In *Harnessing Biotechnology for the 21st Centuary*, pp. 26–29 (Editors Ladisch, M. R. and Bose, A.). American/Chemical Society, Washington, DC.

MAIORELLA, B. L., BLANCH, H. W. and WILKE, C. R. (1984) Economic evaluation of alternative fermentation processes. *Biotech. Bioeng.* 26, 1003–1025.

MATELES, R. I. (1975) Production of SCP in Isreal. In *Single Cell Protein II*, pp. 202–222 (Editors Tannenbaum, S. R. and Wang, D. I. C.). MIT Press, Massachusetts.

MEERS, J. L. (1980) Comment included in discussion of Smith (1980).

MEYRATH, J. and BAYER, K. (1979) Biomass from whey. In *Microbial Biomass, Economic Microbiology*, Vol. 4, pp. 207–269 (Editor Rose, A. H.). Academic Press, London.

MOGREN, H. (1979) SCP from methanol — the Norprotein process. *Proc. Biochem.* 12(3), 2–4, 7.

MOO-YOUNG, M. (1977) Economics of SCP production. *Proc. Biochem.* 12(4), 6–7, 9–10.

NELSON, T. C. (1961) *Mutation and Plant Breeding*, p. 331, Publ. No. 891, Nat. Acad. Sci. Nat. Res. Council, Washington, DC.

NISBET, L. J. and PORTER, N. (1989) The impact of pharmacology and molecular biology on the exploitation of microbial products. In *Microbial Products: New Approaches*, pp. 309–342 (Editors Baumberg, S., Hunter, I. S. and Rhodes, P. M.). University Press, Cambridge.

NYIRI, L. K. and CHARLES, M. (1977) Economic status of fermentation processes. *Ann. Rep. Ferm. Processes*, 1, 365–381.

OMSTEAD, M. N. , KAPLAN, L. and BUCKLAND, B. C. (1989) Fermentation development and process improvement. In *Ivermectin and Abamectin*, pp. 33–54 (Editor Campbell, W. C.). Springer-Verlag, New York.

PAPE, M. (1977) The competition between microbial and chemical processes for the manufacture of basic chemicals and intermediates. In *Microbial Energy Conversion*, pp. 510–530 (Editors Schlegel, H. G. and Barnea, J.). Pergamon Press, Oxford.

PERLMAN, D. (1970) Some prospects for the fermentation industries. *Wallerstein Comm.* 33, 165–173.

PRATTEN, C. F. (1971) *Economics of Scale in Manufacturing Industry*, Department of Applied Economics Occasional Papers 28. University Press, Cambridge.

QUEENER, S. and SWARTZ, R. W. (1979) Penicillins: biosynthetic and synthetic. In *Secondary Products of Metabolism,*

*Economic Microbiology*, Vol. 3, pp. 35–122 (Editor Rose, A. H.). Academic Press, London.

RATAFIA, M. (1987) Mammalian cell culture: Worldwide activities and markets. *Bio / Technology* 5 (7), 692–694.

RATLEDGE, C. (1975) The economics of single cell protein production; substrates and processes. *Chem. Ind.* 21, 918–920.

RATLEDGE, C. (1977) Fermentation substrates. *Ann. Rep. Ferm. Processes* 1, 49–71.

REISMAN, H. B. (1988) *Economic Analysis of Fermentation Processes*. CRC Press, Boca Raton.

REISMAN, H. B. (1993) Problems in scale-up of biotechnology production processes. *Crit. Rev. Biotechnol.* 13, 195–253.

RICHARDS, J. W. (1968a) Economics of fermenter operation. Part I. *Proc. Biochem.* 3(5), 28–31.

RICHARDS, J. W. (1968b) Economics of fermenter operation. Part 2. *Proc. Biochem.* 3(6), 56–58.

RIPLEY, P. (1979) Process engineering aspects of the treatment and disposal of distillery effluent. *Proc. Biochem.* 14(1), 8–10.

ROLZ, C. (1975) Utilization of cane and coffee processing by-products as microbial protein substrates. In *Single Cell Protein II*, pp. 273–313 (Editors Tannenbaum, S. R. and Wang, D. I. C.). MIT Press, Massachusetts.

ROYAL COMMISSION ON SEWAGE DISPOSAL (1912) 8th Report HMSO, London.

ROYCE, P. N. (1993) A discussion of recent developments in fermentation monitoring and control from a practical perspective. *Crit. Rev. Biotechnol.* 13, 117–149.

SCHIERHOLT, J. (1977) Fermentation processes for production of citric acid. *Proc. Biochem.* 12(9), 20–21.

SCHUGERL, K., LUCKE, J., LEHMANN, J. and WAGNER, F. (1978) Application of tower bioreactors in cell mass production. *Adv. Biochem. Eng.* 8, 63–131.

SCHWAB, H. (1988) Strain improvement in industrial microorganisms by recombinant DNA techniques. *Adv. Biochemical Eng. / Biotechnol.* 37, 129–168.

SHARP, D. H. (1989) *Bioprotein Manufacture: A Critical Assessment*. Ellis Horwood, Chichester.

SMITH, S. R. L. (1980) Single cell protein. *Phil. Trans. R. Soc. (London)*, B, 290, 341–354.

SOFER, G. K. and NYSTROM, L. E.(1989) Economics. *Process Chromatography: A Practical Guide*, pp. 107–116. Academic Press, London.

SPIER, R. (1988) Animal cells in culture; moving into the exponential phase. *Trends Biotechnol.* 6, 2–6.

SPIVEY, M. J. (1978) The acetone/butanol/ethanol fermentation. *Proc. Biochem.* 13(11), 2–4, 25.

STOWELL, J. D. and BATESON, J. B. (1984) Economic aspects of industrial fermentations. In *Bioactive Microbial Products*, Vol. 2, *Development and Production*, pp. 117–139 (Editors Nisbet, L. J. and Winstanley, D. J.). Academic Press, London.

SWARTZ, R. W. (1979) The use of economic analysis of penicillin G manufacturing costs in establishing priorities for fermentation process improvement. *Ann. Rep. Ferm. Processes* 3, 75–110.

TAKSEN, K. G. (1986) Industrial approaches to fermentation recovery R and D. In *Bioactive Microbial Products 3: Downstream Processing*, pp. 11–26 (Editors Stowell, J. D., Bailey, P. J. and Winstanley, D. J.). Society for General Microbiology /Academic Press, London.

TAYLOR, I. J. and SENIOR, P. J. (1978) Single cell proteins: a new source. *Endeavour (N.S.)* **2**, 31–34.

VAN SUYDAM, J. C. (1987) Trends in the development of bioreactors. *Chemisch Magazine*, (6). Cited by Oosterhuis, N. M G. (1992) The airlift bioreactor; a more and more common tool in bioprocessing. *Biotecteknowledge* (Applikon, Holland), **12**, 9–12.

VASS, R. C. and JEFFERYS, E. G. (1979) Gibberellic acid. In *Secondary Products of Metabolism, Economic Microbiology*, Vol. 3 (Editor Rose, A. H.). Academic Press, London.

VICKERS, J. C., WILLIAMS, S. T. and ROSS, G. W. (1984) A taxonomic approach to selective isolation of streptomycetes from soil. In *Biological, Biochemical and Biomedical Aspects of Actinomycetes*, pp. 553–561 (Editors Ortiz-Ortiz, L., Bojalil, L. F. and Yakeloff, V.). Academic Press, Orlando.

WALKER, P. D., NARENDRATHEN, T. J., BROWN, D. C., WOODHOUSE, F. and VRANCH, S. P. (1987) Containment of micro-organisms during fermentation and downstream processing. In *Separations for Biotechnology*, pp. 467–482 (Verrall, M. S. and Hudson, M. J.). Ellis Horwood, Chichester.

WANG, D. I. C., COONEY, C. L., DEMAIN, A. L., DUNNILL, P. and LILLY, M. D. (1979) *Fermentation and Enzyme Technology*. Wiley, New York.

WHITAKER, A. (1973) Fermentation economics. *Proc. Biochem.* **8**(9), 23–26.

WILLIAMS, S. T. and VICKERS, J. C. (1988) Detection of actinomycetes in natural habitats — problems and perspectives. In *Biology of Actinomycetes '88*, pp. 265–270 (Editors Okani, Y., Beppu, T. and Ogawara, H.). Japanese Science Society, Tokyo.

WINKLER, M.(1991) Time-profiling and environmental design in computer controlled fermentation and enzyme production. In *Genetically-Engineered Proteins and Enzymes from Yeast: Production Control*, pp. 96–146 (Editor Wiseman, A.). Ellis Horwood, London.

# Index